Bioinorganic Chemistry

A symposium co-sponsored by the Division of Inorganic Chemistry of the American Chemical Society and the Chemical Institute of Canada at Virginia Polytechnic Institute and State University, Blacksburg, Va., June 22–25, 1970.

Raymond Dessy
John Dillard
Larry Taylor
Symposium Co-Chairmen

ADVANCES IN CHEMISTRY SERIES **100**

AMERICAN CHEMICAL SOCIETY
WASHINGTON, D. C.   1971

Coden: ADCSHA

Copyright © 1971

American Chemical Society

All Rights Reserved

Library of Congress Catalog Card 70–153666

ISBN 8412–0138–2

PRINTED IN THE UNITED STATES OF AMERICA

Advances in Chemistry Series
Robert F. Gould, *Editor*

Advisory Board

Paul N. Craig

Thomas H. Donnelly

Gunther Eichhorn

Frederick M. Fowkes

Fred W. McLafferty

William E. Parham

Aaron A. Rosen

Charles N. Satterfield

Jack Weiner

AMERICAN CHEMICAL SOCIETY PUBLICATIONS

FOREWORD

ADVANCES IN CHEMISTRY SERIES was founded in 1949 by the American Chemical Society as an outlet for symposia and collections of data in special areas of topical interest that could not be accommodated in the Society's journals. It provides a medium for symposia that would otherwise be fragmented, their papers distributed among several journals or not published at all. Papers are refereed critically according to ACS editorial standards and receive the careful attention and processing characteristic of ACS publications. Papers published in ADVANCES IN CHEMISTRY SERIES are original contributions not published elsewhere in whole or major part and include reports of research as well as reviews since symposia may embrace both types of presentation.

CONTENTS

Preface ix

1. Model Studies in Nitrogen Fixation and Cobalamin Chemistry ... 1
   G. N. Schrauzer

2. Studies on Enzyme Models 21
   Ronald Breslow

3. Chemical Foundations for the Understanding of Natural Macrocyclic Complexes 44
   D. H. Busch, K. Farmery, V. Goedken, V. Katovic, A. C. Melnyk, C. R. Sperati, and N. Tokel

4. Developments in Inorganic Models of N_2 Fixation 79
   A. D. Allen

5. Fixation of Molecular Nitrogen Under Mild Conditions 95
   E. E. Van Tamelen

6. Uptake of Oxygen by Cobalt(II) Complexes in Solution 111
   Ralph G. Wilkins

7. The Effect of Metal Ions on the Structure of Nucleic Acids 135
   Gunther L. Eichhorn, Nathan A. Berger, James J. Butzow, Patricia Clark, Joseph M. Rifkind, Yong A. Shin, and Edward Tarien

8. Biochemistry of Group IA and IIA Cations 155
   R. J. P. Williams

9. Structure of the Second Coordination Sphere of Metal Complexes and its Role in Catalysis 174
   D. R. Eaton

10. Structure and Function of Metalloenzymes 187
    D. D. Ulmer and B. L. Vallee

11. The Biochemistry of N_2 Fixation 219
    R. W. F. Hardy, R. C. Burns, and G. W. Parshall

12. Structure–Function Relationships in Cytochrome *c* Oxidase and Other Hemeproteins 248
    Winslow S. Caughey

13. Low-Spin Compounds of Heme Proteins 271
    W. E. Blumberg and J. Peisach

14. Ceruloplasmin, A Link Between Copper and Iron Metabolism 292
    Earl Frieden

15. Clostridial Ferredoxin: An Iron–Sulfur Protein 322
    Jesse C. Rabinowitz

16. Studies on Mechanism of Action of B_{12} Coenzymes 346
    R. H. Abeles

17. Structural Models for Iron and Copper Proteins Based on Spectroscopic and Magnetic Properties 365
    Harry B. Gray

18. Nuclear Relaxation Studies of the Role of Metals in Enzyme-Catalyzed Enolization and Elimination Reactions 390
    A. S. Mildvan

19. Rapid Reaction Kinetics Involving the Iron–Porphyrin Site of Horseradish Peroxidase 413
    H. B. Dunford and B. B. Hasinoff

Index 427

This volume is dedicated to
Dr. Graeme Everett Cheney
(1931–1970)
Chairman, Inorganic Division
Chemical Institute of Canada

PREFACE

A symposium is made up of many things—ideas, people, organization, sponsorship, and finance. Since the immediate purpose of a symposium is to satisfy the urgent needs of the participants to meet commonly and discuss topics of current scientific interest, many of these building blocks become forgotten.

It seems most appropriate then, as the Bioinorganic Chemistry Symposium held at Virginia Polytechnic Institute and State University, June 22–25, 1970, is made more lasting by its publication in the ADVANCES IN CHEMISTRY SERIES, that appreciation be shown to those people and institutions that made it all possible.

The Bioinorganic Symposium was held under the joint sponsorship of the Inorganic Divisions of the Chemical Institute of Canada and the American Chemical Society. Such joint sponsorship of summer symposia dates back to 1968, when the first of the series was held at Banff, Alberta on the subject of Stereochemistry of Inorganic Compounds. By mutual agreement, the topic is chosen by one group, the site by the other. At the Minneapolis meeting of the American Chemical Society in 1969, A. D. Allan of Toronto University announced that the Canadians had chosen the subject—Bioinorganic Chemistry—as the topic for the 1970 meeting. The Executive Committee of the ACS then confirmed Virginia Polytechnic Institute and State University as the site. Graeme Cheney, of Acadia University, then current chairman of the CIC Inorganic Division, was chosen as co-chairman, along with Raymond Dessy of VPI, W. C. Cooper of Noranda Research Center, Montreal, and John Dillard and Larry Taylor of VPI assumed the task of co-organizers. It became readily apparent that there was much interest in a symposium at which inorganic and biochemists could meet on common ground. To aid in choosing a capable and affable set of discussion leaders and speakers, the organizing committee sought the help of outside consultants—in the form of Gunther Eichhorn of the National Institutes of Health, Ralph Hardy of the Dupont Co., and the faculty of the Department of Biochemistry at VPI. The list of active participants formed easily and, as a mark of the potential value of the conference and the need it so obviously filled, only two invitations were declined.

Some financing was already in hand from the CIC, but for a meeting of the size envisaged, external aid was sought successfully from the National Research Council of Canada and the National Science Foundation In addition, support for social functions was provided by Alpha Inorganics, Inc., a subsidiary of Ventron.

Response by chemists, biochemists, biophysicists, and others to attend the symposium was overwhelming. Indeed, it turned out to be necessary to limit attendance. In all, some 248 scientists attended the meeting.

Because of the fact that attendance was to be limited and a large number of people were interested in obtaining abstracts of the program, the organizing committee members at VPI decided to have the entire formal proceedings transcribed by Marcus Bieler, a court reporter in Roanoke, Va. He and his assistant, Miss Hopkins, provided daily copy of the talks given, while the slides used were duplicated in the VPI microphoto laboratory. Each manuscript, as it was received by the editing group set up by Phil Hall of VPI, was checked against the magnetic tape recording of the lecture and a corrected manuscript submitted to the review panel, selected by Robert F. Gould of the ADVANCES IN CHEMISTRY SERIES. The author then received an edited, reviewed copy of his manuscript before he left Blacksburg, along with a set of proof photos of his slides. The corrected final manuscript appears in this volume.

The volume is dedicated to Graeme Cheney, to whom much is owed for the organizational duties in the formative stages. Graeme was lost to us in April 1970, depriving us of a good organizer and an excellent companion. The Canadian thread was then picked up by W. C. Cooper and J. C. Thompson of the CIC. We at VPI owe them a debt we cannot repay.

We are also beholden to the members of the staff of the Biochemistry Department of VPI, under the headship of Bruce Anderson, who helped us edit the manuscripts at all stages. These were L. B. Barnett, R. D. Brown, J. L. Hess, K. M. Plowman, and J. Vercellotti.

We are also grateful to Jerry Hargis and the staff of the Continuing Education Center for their excellent cooperation and guidance in planning and helping make the conference a success.

Raymond Dessy
John Dillard
Larry Taylor

Virginia Polytechnic Institute and State University
Blacksburg, Va. 24061
September 1970

Model Studies in Nitrogen Fixation and Cobalamin Chemistry

G. N. SCHRAUZER

The University of California at San Diego, La Jolla, Calif. 92037

> *Complexes of inorganic molybdenum salts with certain thiols catalyze the reduction of most substrates of nitrogenase in striking analogy with the enzymatic reactions. Among thiol complexes of all metals, those of molybdenum exhibit the highest catalytic activity, while iron complexes are completely inactive. Essential reactions of the substrates of nitrogenase must therefore occur on a molybdenum-containing active site. The conversion of 1,2-diols to aldehydes by the coenzyme B_{12}-dependent diol dehydratase is postulated to occur via organocobalt intermediates. The base-catalyzed β-elimination reactions of β-substituted alkyl-cobalamins and -cobaloximes are d-orbital assisted. Initial products of elimination are π-complexes of the Co(I) nucleophile with the olefinic elimination products; 1,2-di-cyanoethyl(pyridine)cobaloxime undergoes reversible, base-assisted valence tautomerization in solution at room temperature.*

The American philosopher Mary B. Hesse distinguishes two kinds of scientists, to whom she refers as "Duhemists" (after the French physicist and philosopher P. Duhem, 1861–1916), and "Campbellians" (after the British physicist N. R. Campbell, 1880–) (1). Duhemists differ from Campbellians essentially by their attitudes toward the use of models in science. Since much of this conference deals with the development of model systems in bioinorganic chemistry, let me therefore briefly discuss the philosophical basis of the model approach in science.

P. Duhem was fundamentally disinclined to accept any kind of model in science and held the view that a scientific theory should be deductive, abstract, logical, and systematizing. Campbell opposed this approach and suggested that all theories should have an intelligible interpretation in terms of a model. He stated at one time (2) that "analogies and models are an utterly essential part of theories without which theories would be completely valueless and unworthy of the name." The Campbellian approach is supported by Mary Hesse, who concludes that the strength of a theory rests in the strength of its inductive support and on probability values assigned on the basis of evidence. A hypothesis based on a model is preferable to one which is not, on grounds of inductive support, falsifiability plus corroboration and simplicity.

The approach toward the solution of biochemical problems in the past has been largely, and of necessity, deductive. However, model building and the application of material analogues are becoming increasingly important for the elucidation of fundamental problems of biochemical structure and reactivity. While "models help us to derive theories supported by choice criteria which appeal to the models as empirical data" (1), they also provide independent information which often leads to new areas of research and to significant discoveries outside the initially set research objective.

This paper demonstrates the development and use of models in two areas of bioinorganic chemistry, the mechanism of biological nitrogen fixation and the enzymatic function of vitamin B_{12} coenzymes.

Nitrogenase Models

Nitrogenase, the molybdenum-, iron-, and sulfide-containing nitrogen-fixing enzyme present in organisms such as *Clostridium pasteurianum* or *Azotobacter vinelandii*, was recently isolated and studied mainly by Hardy and his coworkers (3). One of its most striking properties is its non-specificity with regard to the reduction of substrates other than nitrogen (Figure 1); e.g., acetylene, N_2O, N_3^-, various nitriles, and isonitriles are reduced. Acetylene is converted to ethylene presumably at the same binding site as is nitrogen and is for this reason generally employed for the determination of nitrogenase activity. The reaction with isonitriles is remarkable in that these give rise to methane, ethylene, ethane, C_3, and C_4 hydrocarbons. Acetylene behaves very much like molecular nitrogen on incubation with nitrogenase. This is exemplified in Table I, taken from a recent paper by Yoch and Arnon (4), who established the nitrogen-fixing ability of the photosynthetic organism Chromatium.

To develop a model that could be helpful for the development of an empirically supported mechanism of nitrogen fixation, we have first attempted to reduce acetylene catalytically with transition metal ions and various reducing agents such as sodium dithionite, sodium borohydride, etc. These experiments were not very successful since many metal ions are reduced to the metal under these conditions, making it impossible

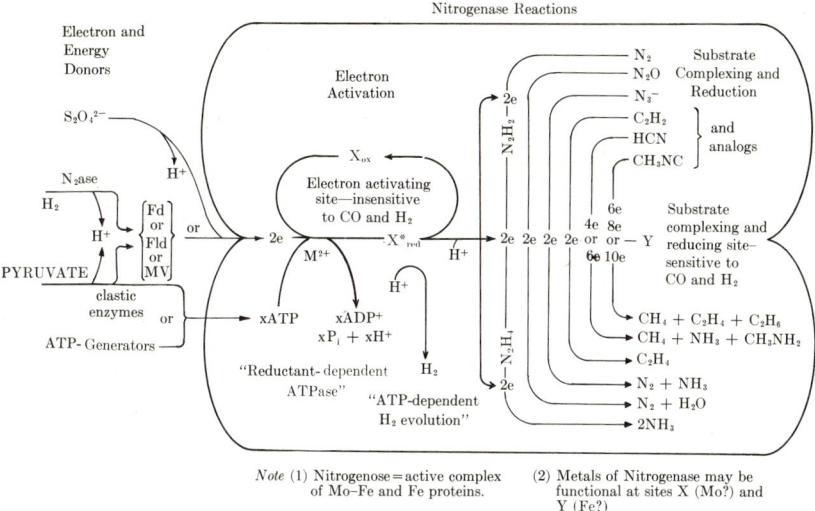

Figure 1. Scheme for nitrogenase and its reactions based on the electron-activation, two-site hypothesis according to Hardy and Burns (3)

Table 1. Requirements for Light-Dependent N_2 Fixation and Acetylene Reduction by a Cell-Free Extract of *Chromatium*[a]

Treatment[b]	Nmoles Formed per Mg Protein per Min	
	Ammonia	Ethylene
Complete	4.8	8.2
Complete, dark	0	0
Minus ferredoxin	0.1	0.01
Minus chloroplasts	0.5	0.01
Minus (creatine phosphate plus ATP)	0.2	0
Minus (ascorbate plus DCIP)	0	0.01
Minus Mg^{2+}	2.0	–
Complete but N_2 or acetylene replaced by argon	0	0

[a] According to D. C. Yoch and D. I. Arnon, Ref. *4*.
[b] The complete system contained in a total volume of 1.5 ml: heated spinach chloroplast fragments (equivalent to 185 µg chlorophyll); 100 µmoles ascorbate, 0.05 µmole DCIP, 10 µmoles $MgCl_2$, 5 µmoles ATP, 50 µmoles creatine phosphate, 0.2 mg creatine phosphokinase, 200 µg ferredoxin from *C. pasteurianum* and 6.6 mg Chromatium extract. Gas phase was N_2 or 0.1 atm acetylene and 0.9 atm argon, depending on type of nitrogenase assay used. Temp.: 30°; light intensity: 20,000 lux.

to maintain a homogeneous catalytic system. Since none of the metals displayed any particular tendency to reduce acetylene to ethylene rapidly at this early stage of complexity of our model system, we have tested various coordinating ligands as additional components. In these systematic experiments, nitrogen- or phosphorus-containing ligands were not particularly effective. However, when certain thiols were supplied to solutions containing molybdenum, a very high activity was observed in the reduction of acetylene to ethylene. Of all elements investigated, molybdenum exhibits the highest activity (Table II) while iron, even though it is present in nitrogenase, is inactive under similar reaction conditions. We therefore conclude that molybdenum is the site of acetylene reduction in the enzyme. In analogy with the other substrates reduced by nitrogenase, we also find that methylacetylene is reduced to propylene, while 2-butyne, under certain conditions (with sodium dithionite as reducing agent), is not acted upon (Table III). These results demonstrate that essential features of the substrate specificity of nitrogenase can be

Table II. Relative Activities of Transition Metals in the Reduction of Acetylene to Ethylene at 27°[a]

Metal	Relative Activity[b]	Metal	Relative Activity[b]	Metal	Relative Activity[b]
Ti	0	Y	0	Hf	[c]
V	0	Zr	0	Ta	0.1
Cr	0	Nb	0.1	W	0
Mn	0	Mo	100.0	Re	0.1
Fe	0	Tc	[c]	Os	0.1
Co	0	Ru	2.9	Ir	15.5
Ni	0	Rh	0.4	Pt	0.1
Cu	0	Pd	1.3	Au	0
Zn	0	Ag	0	Hg	0

[a] Reaction conditions: Reaction solutions containing 1 mmole of transition metal salt (mostly chloride), 1 mmole of 1-thioglycerol, 2 mmoles of $Na_2S_2O_4$, and 10 mmoles of NaOH in 10 ml of H_2O were placed into glass vials of 15-ml volume and were sealed with rubber serum caps. The air was then replaced by water-washed acetylene at 1 atm of pressure. Relative rates are reported for 16-hour reaction periods.
[b] Relative rate of Mo = 100.0.
[c] Not determined.

Table III. Reduction of Substrates by Nitrogenase and the Nonenzymatic Molybdenum–Thiol Systems

	Products	
Substrate	Enzymatic	Nonenzymatic
Acetylene	Ethylene	Ethylene
Methylacetylene	Propylene	Propylene
2-Butyne	–	– (with $Na_2S_2O_4$)
		2-Butene (with $NaBH_4$)
Azide	N_2 + NH_3	N_2 + NH_3
Alkylisocyanide	CH_4, C_2H_4, C_2H_6, C_3, C_4 hydrocarbons	CH_4, C_2H_4, C_2H_6, C_3, C_4 hydrocarbons

mimicked at a very early stage of the chemical evolution of our model system.

The evidence accumulated in the course of these experiments led us to try molecular nitrogen as the substrate. Our present results indicate that little if any nitrogen is fixed by molybdenum–thiol catalysts at normal pressure. Detectable amounts of ammonia have been obtained at elevated pressures, however, in the presence of iron as the cocatalyst. We are presently establishing the optimal reaction conditions. The specific and high catalytic activity of molybdenum in the model system and the established requirement for a sulfur ligand strongly suggest that most substrates of nitrogenase are reduced at a molybdenum- and sulfur-containing active site (5).

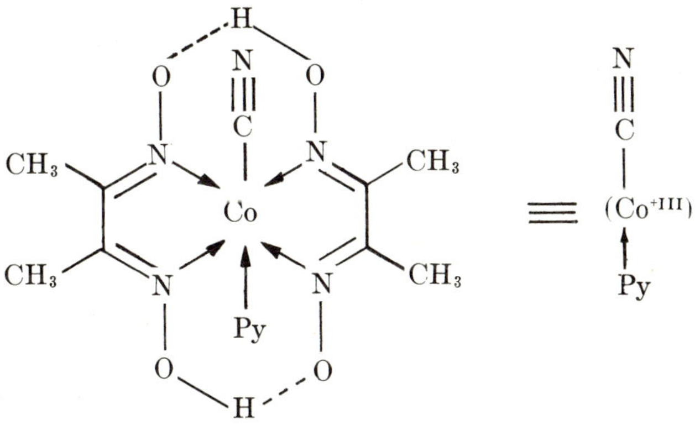

Figure 2. Cyano-pyridinato-cobaloxime, a model of cyanocobalamin

Models in Cobalamin Chemistry

In turning our attention to vitamin B_{12} and its coenzymes, it should be recognized that present differences in opinions and in the interpretations offered on the mechanism of coenzyme B_{12}-catalyzed processes are largely a consequence of the complexity of these reactions and the difficulty of interpreting the results of enzymological studies in the absence of empirical data derived from model experiments. The situation is best exemplified by first considering an early mechanism postulated for the function of coenzyme B_{12} in diol dehydratase, not supported by model studies. Ref. 6 contains a detailed discussion.

$$\begin{array}{c}\text{H}\rightarrow\text{CoB}_{12}\\|\\ \text{HC—OH}\\|\\ \text{HC—OH}\\|\\ \text{CH}_3\end{array} \longrightarrow \begin{array}{c}\text{HC—OH}\\|\diagdown\text{O}\\ \text{HC}\diagup\\|\diagdown\text{H—CoB}_{12}\\ \text{CH}_3\end{array} \longrightarrow \begin{array}{c}\text{OH}\\|\\ \text{HC—OH}\\|\\ \text{HCH}\\|\\ \text{CH}_3\end{array} \xrightarrow{-\text{H}_2\text{O}} \begin{array}{c}\text{CHO}\\|\\ \text{CH}_2\\|\\ \text{CH}_3\end{array} \quad (1)$$

Equation 1 has been derived entirely from studies with the enzyme and is in accord with the available experimental evidence (7). It is unsatisfactory, however, in that virtually no step postulated is in accord with known chemistry or otherwise inductively supported by model experiments.

In order to develop a model approach similar to that exemplified for nitrogenase, we have used simple model compounds of vitamin B_{12}, the cobaloximes (Figure 2), details of which will not be discussed here since they are well documented in the literature (8). However, let us turn our attention to coenzyme B_{12}, the remarkable organometallic complex shown in Figure 3. How could this compound function in the enzyme? Coenzyme B_{12} is a substituted alkylcobalamin in which the cobalt ion is inaccessible for any substrate. The compound also does not show oxidizing or reducing properties and hence is unlikely to function as such in any enzyme. Most workers in the field now agree that the cobalt–carbon bond in the coenzyme must be broken in some way for the coenzyme to become active. Nonenzymatically, this could be achieved either on reaction with acids, homolytically, or on interaction with bases (Figure 4). The reaction with acids is irreversible and hence cannot be incorporated into a catalytic scheme in which the coenzyme is regenerated after completion of a reaction cycle. A homolytical mechanism of coenzyme B_{12} activation would of necessity lead to the postulate of a free radical mechanism of diol dehydration, which is chemically out of the question. The reaction with bases occurs via β-elimination and hence is a reversible process. This remarkable reaction was discovered in our laboratory by studying the properties of the cobaloxime model of coenzyme B_{12} (Figure 5) (6). The products of the cleavage of coenzyme B_{12} and of the cobaloxime model with base are 4′,5′-didehydro-5′-deoxyadenosine and the powerfully nucleophilic Co(I) derivatives of the cobalamin or the cobaloxime (Figure 6). Since β-eliminations involving cobalamins and cobaloximes have previously been shown to be reversible, an attractive mechanism for the activation of coenzyme B_{12} is available by which the most reactive reduced cobalamin derivative (vitamin B_{12s}) may be generated even in the absence of a reducing agent. All that would be required for such a reaction to occur (disregarding, for simplicity, secondary Co–C bond labilizing effects of the enzyme protein) is a basic center in the vicinity of the coenzyme, whose interaction with the 4′ hydrogen atom of the coenzyme could initiate Co–C bond cleavage via β-elimination. There is some evidence that the cobalt–carbon bond of the coenzyme is cleaved in the diol dehydratase holoenzyme. Although it is difficult to prove that the cobalamin is pres-

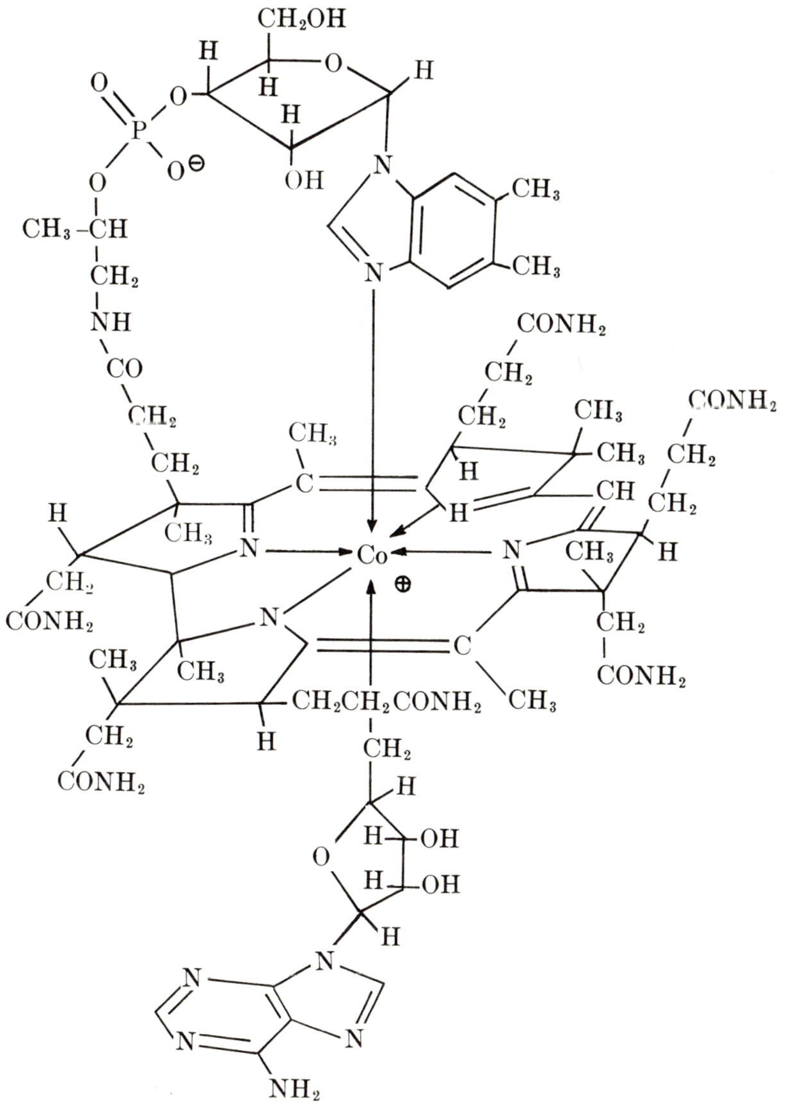

Figure 3. Coenzyme B_{12}

ent in the Co(I) state, for reasons that we shall discuss, there is indication that the Co–C bond is broken in the holoenzyme since it is oxygen-sensitive whereas coenzyme B_{12} itself is not. The model compound also shown in Figure 6 undergoes Co–C bond cleavage with bases. The olefinic elimination product 2-methylenefurane, furthermore, reacts with vitamin B_{12s} to regenerate the organocobalamin, thus establishing the reversibility of the elimination reaction in a model situation (6).

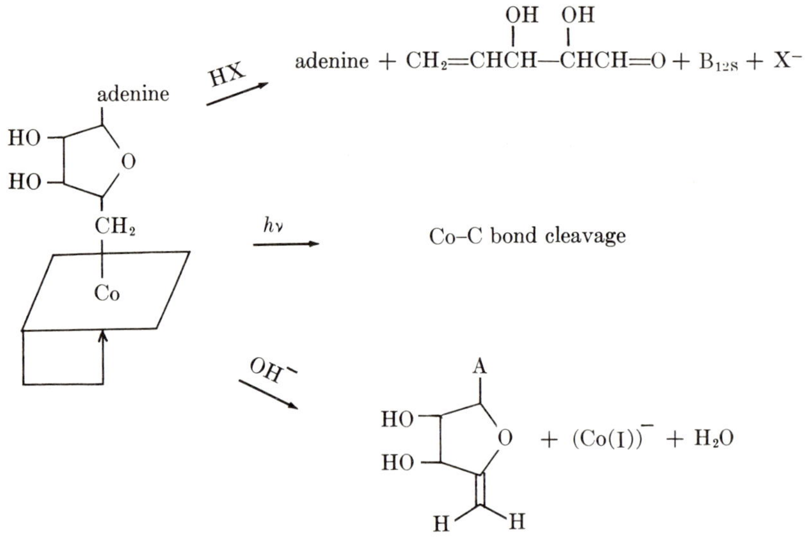

$$\text{adenine} + CH_2=CHCH-CHCH=O + B_{12s} + X^-$$
(with OH OH on the middle carbons)

Figure 4. Reactions of coenzyme B_{12} with acids, light, and bases

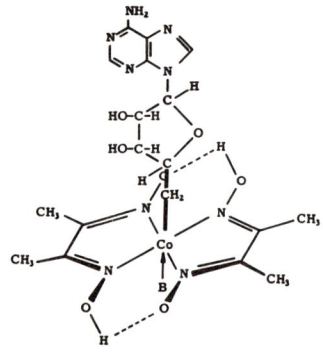

Figure 5. Cobaloxime model of coenzyme B_{12}

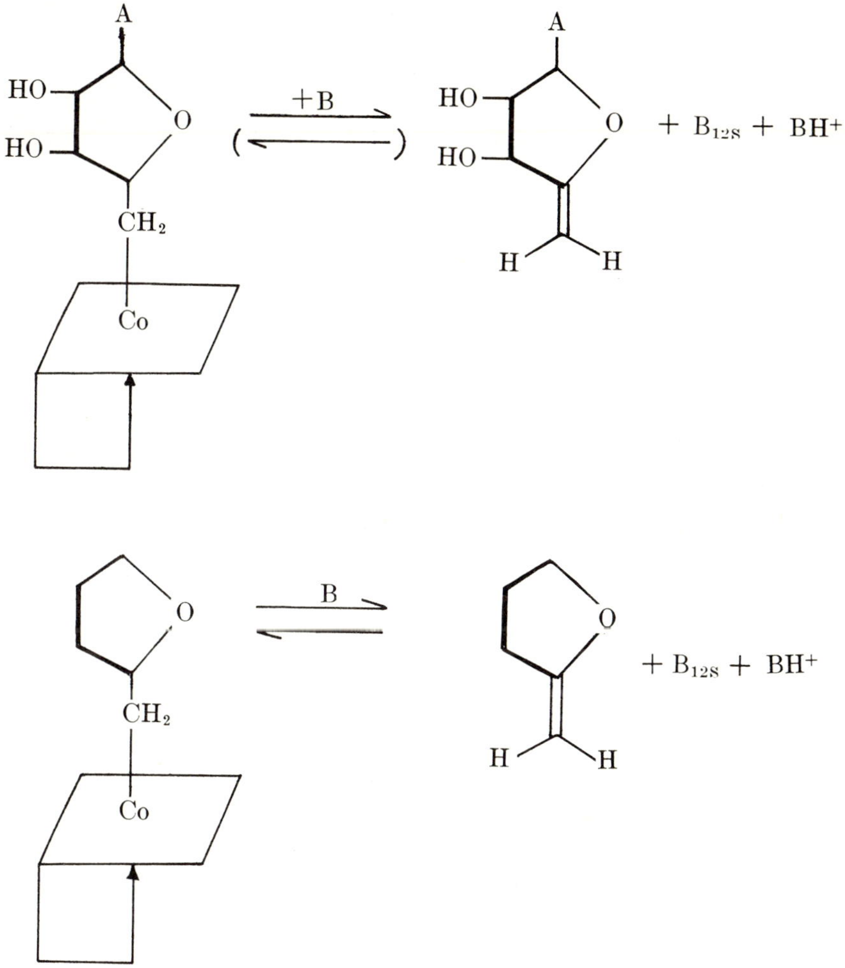

Figure 6. Cleavage of coenzyme B_{12} and of the cobaloxime model with base

Mechanism of Elimination Reactions of Substituted Alkylcobaloximes and Alkylcobalamins

In view of the reversibility of β-elimination reactions in the chemistry of substituted alkylcobalamins and -cobaloximes and the potential biochemical importance of such reactions, we will now discuss the mechanism of these remarkable reactions (9).

Using β-substituted ethylcobaloximes such as β-cyanoethyl- and β-carbethoxyethylcobaloxime, we have found that these elimination reactions bear little if any similarity with common organic β-elimination reactions. Upon the addition of alkali to aqueous–alcoholic solutions of the complexes, the Co(I) nucleophiles and olefinic elimination products are not formed directly. The initial products are π-complexes of the Co(I) nucleophiles with the olefins (Figure 7). These new complexes

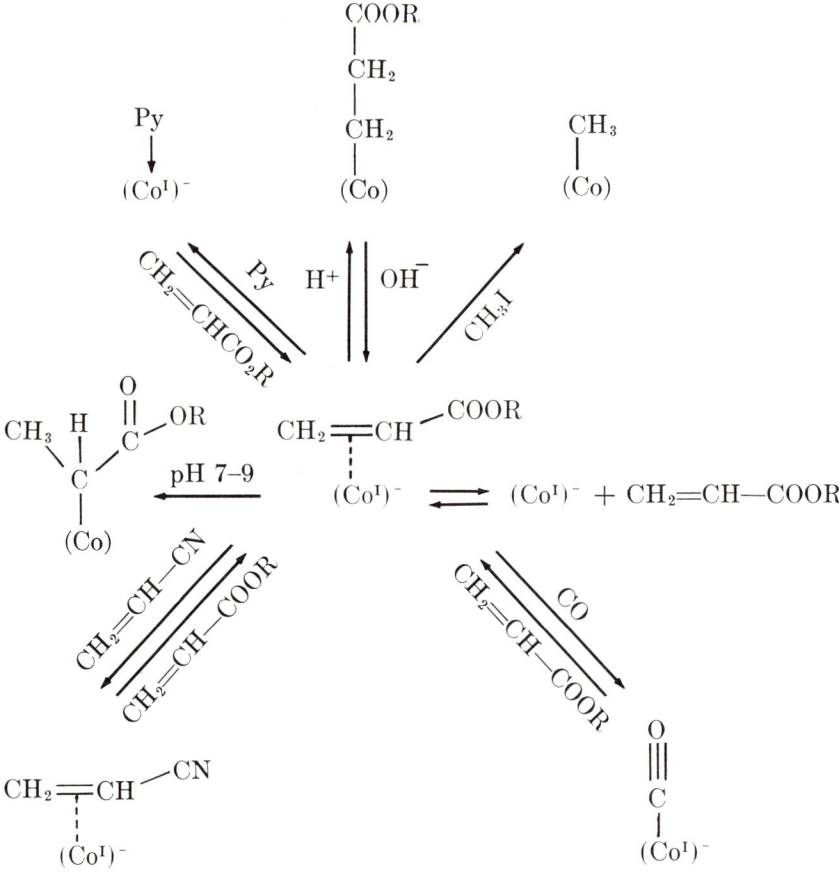

Figure 7. Reactions of the cobaloxime(I)–acrylic ester π-complex; (Co) represents the Co(Dmg)$_2$ part of the cobaloxime (axial base not shown)

possess spectroscopic properties substantially different from those of the free Co(I) nucleophiles. The characteristic low-energy ligand field transitions are shifted to higher energies, to a degree depending on the electron donor–acceptor properties of the olefinic ligands. Experimental term-level schemes for complexes with various ligands are shown in

Figure 8. The structural possibilities of the complex with acrylonitrile are given in Figure 9.

The π-complexes of cobaloximes with olefinic ligands are at equilibrium with the free nucleophile and the olefin. The equilibrium concentrations of the free nucleophile are established sufficiently rapidly to permit the displacement of the π-bonded olefin by other ligands, or to remove the volatile olefinic component simply by passing a stream of

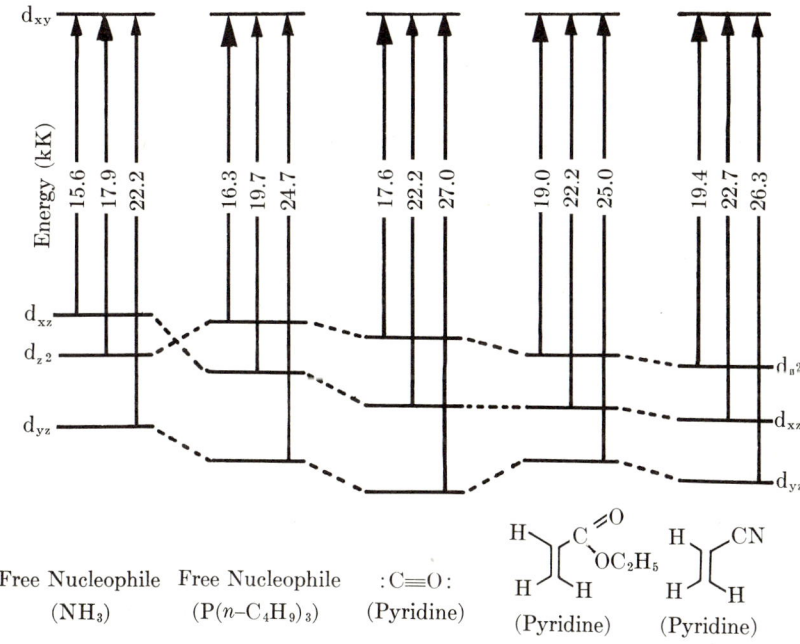

Figure 8. *Experimental term level diagrams for complexes of Co(I) nucleophiles with various ligands; the transitions $3d_{x^2-y^2}$-$3d_{xy}$ are not resolved and are not included in the diagrams*

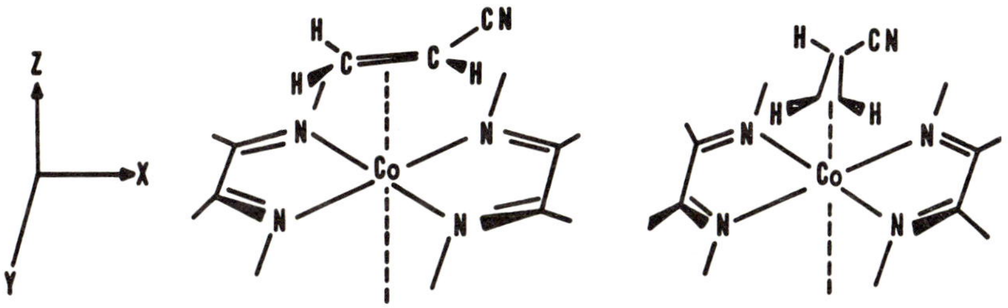

Figure 9. *Possible structures of the cobaloxime(I)-π-complex with acrylonitrile*

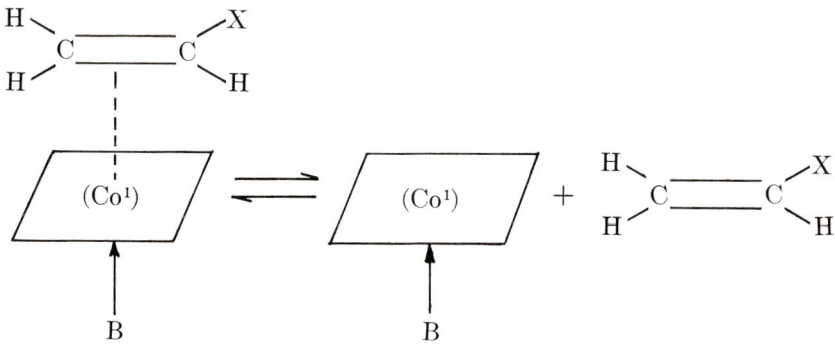

Figure 10. Stabilization of the Co(I) nucleophile through π-complex formation; X is an inductively electron-withdrawing substituent, e.g., —CN^- or —COOR

Table IV. Rates of π-Complex Formation From Substituted Alkylcobaloximes and Alkylcobalamins at 25°, and Co(I) Pearson Nucleophilicities[a]

π-Bonded Olefin Generated	In-Plane Ligand	Axial Base	k^{2nd}	Co(I) Nucleophilicity[b]
$CH_2=CH—CN$	Dmg	H_2O	0.48	14.3
$CH_2=CH—CN$	Dmg	Pyridine	0.32	13.8
$CH_2=CH—CN$	Dmg	$P(n-C_4H_9)_3$	0.12	13.3
$CH_2=CH—CN$	Dmg	Pyridine	0.08	13.1
$CH_2=CH—CN$	$DmgB_2F_4$	Pyridine	0.046	11.7
$CH_2=CH—CO_2C_2H_5$	Dmg	Pyridine	0.16	13.8
$C_6H_5CH=CH—CN$	Dmg	Pyridine	0.030	13.8
$H_5C_2O_2C—CH=CH—CO_2C_2H_5$	Dmg	Pyridine	0.0080	13.8
$NC—CH=CH—CN$	Dmg	Pyridine	0.12	13.8
$CH_2=CH—CN$[c]	Corrin	[d]	3.8^e	14.4
$CH_2=CH—CO_2C_2H_5$[c]	Corrin	[d]	0.32^e	14.4

[a] Most rates measured in 0.1 M to 1.0 M NaOH in methanol; k^{2nd} in 1 mole^{-1} sec^{-1}.
[b] Reference 8.
[c] Rate measured in NaOH in water.
[d] 5,6-Dimethylbenzimidazole.
[e] π-Complexes unstable.

argon through the solutions of the π-complexes. It is also possible to alkylate the Co(I) nucleophile present in equilibrium amounts. The π-complexes may for this reason be regarded as stabilized forms of the Co(I) nucleophile (Figure 10). The subsequent study of the mechanism of these elimination reactions provided us with a striking example for a d-orbital–assisted process. Since the π-complexes are the initial reaction products, it is possible to measure the rate of the σ–π conversion reaction in the presence of alkali separately. The results show clearly that the rate of π-complex formation increases with increasing Co(I) nucleophilicity (Table IV). Improving the leaving group properties of the Co(I) nucleophile by attaching electron-withdrawing bases to cobalt in the axial coordination site actually lowers both the rate of the σ–π conversion and the stability of the resulting complex. The role of the metal in the process is schematically represented in Figure 11. The hydroxyl

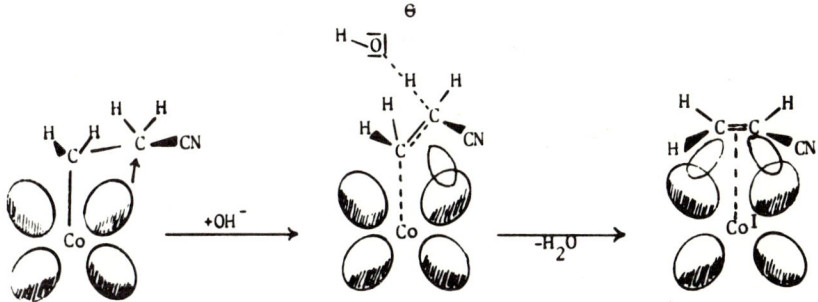

Figure 11. Metal-d-orbital assistance in the σ–π conversion reaction of substituted alkylcobaloximes

Table V. Relative Rates of Alkylation in the Presence of π-Bonded Olefins at 25°[a,b]

π-Bonded Olefin	Olefin–Cobalt Ratio	In-Plane Ligand	Relative Rate
None	0	Dmg	1.0
$CH_2=CH-CN$	1:1	Dmg	0.020
$CH_2=CH-CN$	2000:1	Dmg	0.00071
None	0	$DmgB_2F_4$	1.0
$CH_2=CH-CN$	1:1	$DmgB_2F_4$	0.82
$CH_2=CH-CN$	500:1	$DmgB_2F_4$	0.024
$CH_2=CH-CN$	2000:1	$DmgB_2F_4$	0.0082
None	0	Cobalamin	1.0
$CH_2=CH-CO_2C_2H_5$	1000:1	Cobalamin	1.0
$CH_2=CH-CN$	1000:1	Cobalamin	0.35
$NCCH=CHCN$	1000:1	Cobalamin	0.018

[a] Axial base is pyridine for cobaloximes and 5,6-dimethylbenzimidazole for cobalamins.
[b] Alkylating agent: ethyl bromide.

ion, on the other hand, labilizes the C–H bond in the β-position as indicated. The comparatively small deuterium–hydrogen isotope effect of 1.7 associated with the σ–π conversion of β-cyanoethylcobaloxime indicates that the cleavage of the carbon–hydrogen bond is a late event in the activation profile. Parallel studies with β-substituted alkylcobalamins and -cobinamides indicate that the β-eliminations in these compounds are also d-orbital–assisted. The π-complexes of the Co(I) form of vitamin B_{12} (vitamin B_{12s}) with the olefinic elimination products are less stable than those of cobaloximes(I), presumably for steric reasons. The tendency of the Co(I) ion in vitamin B_{12s} to form complexes with olefins follows from the observed diminished rates of alkylation of vitamin B_{12s} in the presence of excess of olefinic ligand (Table V). The strong π-acceptor fumaronitrile forms a more stable π-complex and causes the spectrum of vitamin B_{12s} to look more like vitamin B_{12r} (Figure 12).

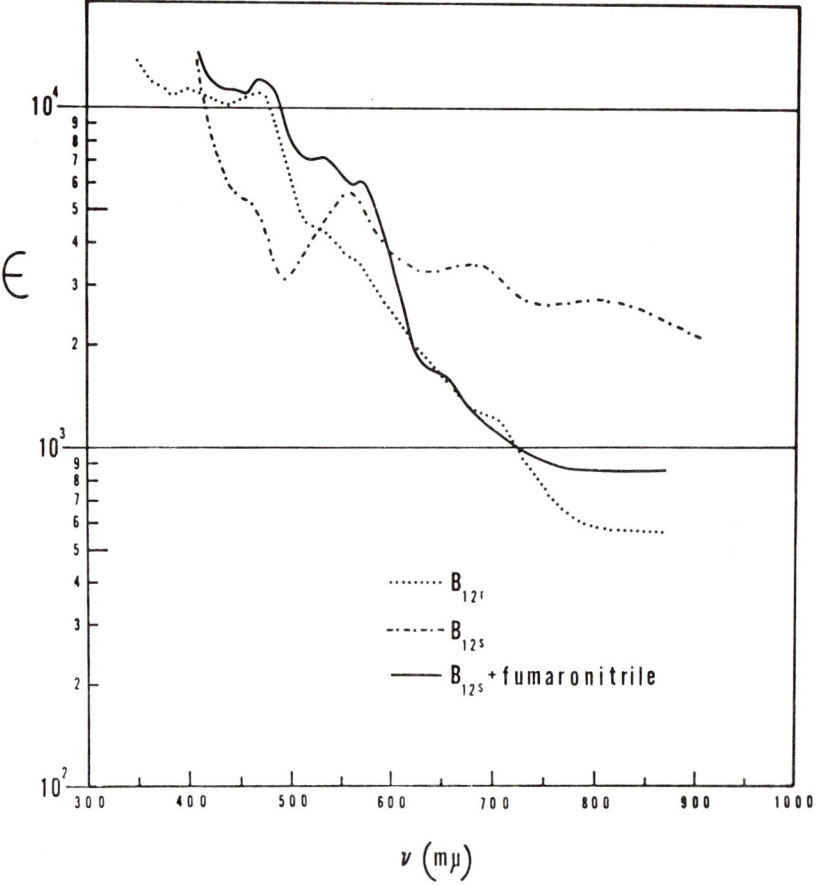

Figure 12. *Optical absorption spectra of vitamin B_{12r}, vitamin B_{12s}, and of vitamin B_{12s} in the presence of excess fumaronitrile*

The demonstrated ability of vitamin B_{12s} to form π-complexes with electron-accepting olefinic ligands shows that it may be difficult to detect the presence of the Co(I) nucleophile in coenzyme B_{12}-dependent enzymes. Apart from the fact that the stationary concentrations of vitamin B_{12s} are expected to be altogether too low for direct observation, the spectroscopic properties of enzyme-bound vitamin B_{12s} need not be necessarily identical with those of the free compound.

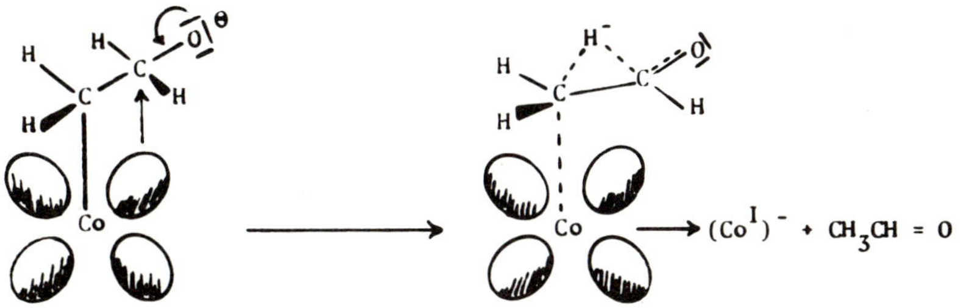

Figure 13. Decomposition of β-hydroxyethylcobaloxime via hydride transfer, yielding the Co(I) nucleophile plus acetaldehyde

Studies on the Mechanism of Diol Dehydratase

I interject at this point the Co–C bond cleavage reaction of 2-hydroxyethylcobaloximes and -cobalamins which yields the Co(I) nucleophile plus acetaldehyde according to the reaction scheme in Figure 13. This reaction represents the only empirically established conversion of glycol derivatives to aldehydes which occurs with cobalamins or cobaloximes as catalysts and provides the basis for our mechanism of the enzymatic dehydration of glycols to aldehydes by *Aerobacter aerogenes*. Our mechanism (6) for the essential steps in the enzymatic dehydration of propanediol is summarized in Figure 14. The interpretation offered for the substrate–coenzyme hydrogen exchange reactions is supported by model experiments and is in agreement with the available evidence from enzymological studies (10, 11). In particular, it offers an explanation for the observed transfer of tritium from the 5′-position in the enzyme-bound coenzyme to added propionaldehyde.

Very recently, Abeles *et al.* (12) reported the results of tritium labeling experiments of enzyme-bound coenzyme by ^{3}H-labeled propanediol which they feel contradict our mechanism. Their data may be readily interpreted in terms of our mechanism. In our view, the tritiation of the coenzyme occurs as the result of the stereospecific removal of the ^{3}H ion

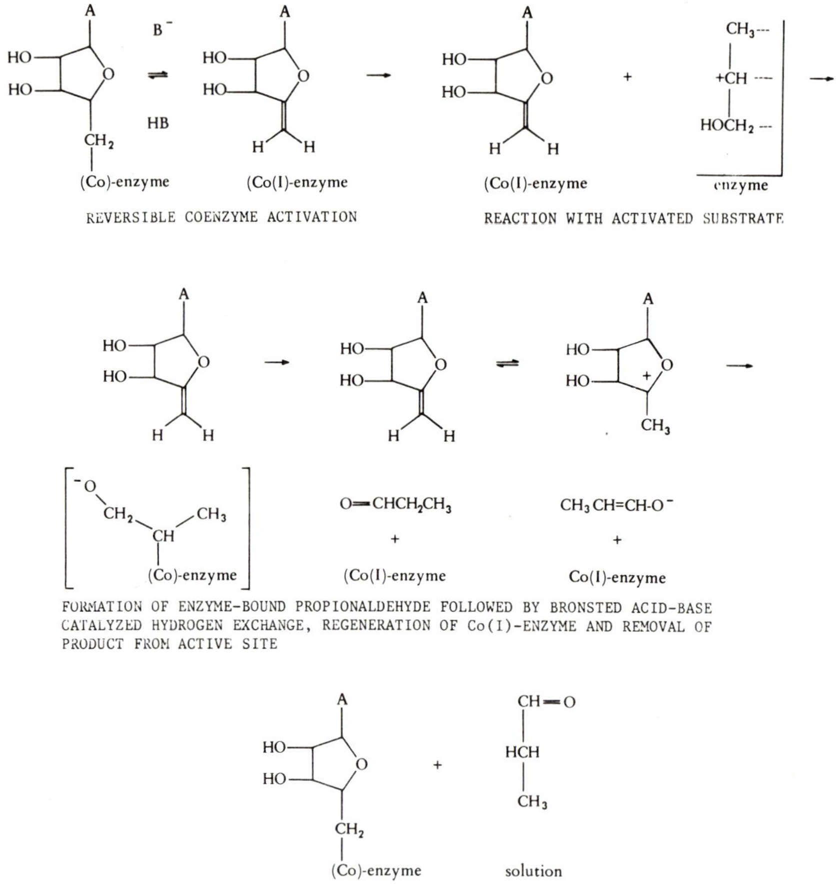

Figure 14. *Proposed mechanism of the enzymatic dehydration of propanediol*

from the α-position of enzyme-bound propionaldehyde as it is generated at the active site. The ^{3}H ion is subsequently transferred to the 5′ carbon atom of the enzyme-bound 4′,5′-didehydro-5′-deoxyadenosine, yielding a carbonium ion in which the labeled and unlabeled hydrogen atoms at C-5′ are equivalent. This process can occur as a general Bronsted acid–base catalyzed reaction and has been demonstrated by model experiments. The H/T effect associated with this removal and the statistical factor of 1/3 owing to the presence of the two other hydrogen atoms are jointly responsible for the accumulation of tritium in the coenzyme. Assuming plausible values for the H/T effect, a theoretical tritium uptake curve has been calculated which agrees with the data reported in Ref. *12* (Figure 15). The essential point of our interpretation of the ^{3}H uptake is that it occurs at the active site, prior to equilibration of the enzyme-bound aldehyde with aldehyde outside the enzyme. This assumption appears to be reasonable, since all reactions of diol dehydratase also occur without hydrogen exchange with the solvent. The hydrogen exchange reactions are interpreted to occur in the final stages of the enzymatic reactions and are considered to be associated with the removal of the enzyme-bound product from the active site. The mechanism thus is supported by established reactions of glycol derivatives and of the coenzyme. It also ascribes the catalytic effect of the coenzyme to the corrin cobalt ion in its most reactive univalent state, and furnishes a chemically plausible explanation for the involvement of a cobalt-containing coenzyme in the process. Alternative interpretations of the mechanism all invoke substrate–coenzyme hydrogen exchange processes. In view of the non-lability of glycol hydrogen atoms, any intermediate substrate (glycol) dehydrogenation step is very difficult to envisage chemically and is unlikely energetically (6).

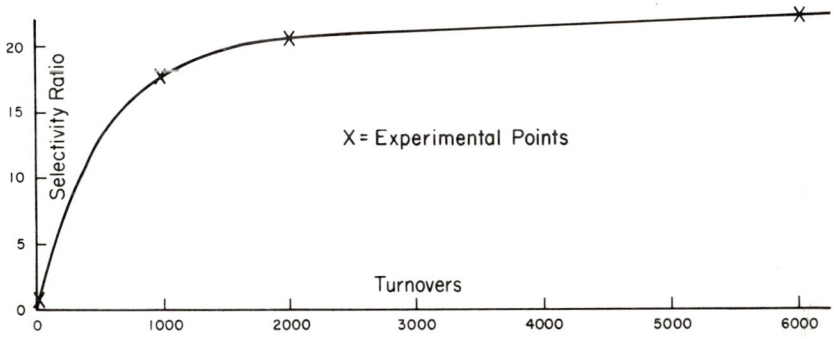

Figure 15. Calculated tritium uptake curve for coenzyme B_{12} in diol dehydratase; selectivity ratio is defined as the quotient of the radioactivity of coenzyme and the activity of substrate; assumed over-all H–T effect is 25. Experimental data from Ref. 12

The Co–C bond cleavage of 2-hydroxyalkylcobalt complexes does not take place via intermediate π-complexes between the Co(I) nucleophile and vinyl alcohol. There is evidence that the reaction occurs with a 1,2-hydride shift rather than as a β-elimination (6) which is associated with a H/D effect of 5.5. This is larger than observed in normal 1,2-hydride shift reactions. We presently favor the view that the migrating hydrogen attains a greater degree of "hydride ion character" in the transition state than in conventional 1,2-hydride shift reactions (this is schematically indicated by the dotted lines in Figure 13).

Concerning the Mechanism of Coenzyme B_{12} Dependent Mutase Reactions

The catalytic action of vitamin B_{12} coenzyme in the enzymatic conversion of methylmalonyl–coenzyme A to succinyl–coenzyme A and of related mutases is as yet unexplained and poses fascinating problems (Equations 2–5).

Succinyl–methylmalonyl–coenzyme A mutase:

$$\text{HOOC—CH}_2\text{—CH}_2\text{—COS—CoA} \rightleftarrows \text{HOOC—CH(CH}_3\text{)—COS—CoA}$$

$$\text{CoA} = \text{Coenzyme A} \quad (2)$$

Glutamate–methylaspartate mutase:

$$\text{HOOC—CH}_2\text{—CH}_2\text{—CH(NH}_2\text{)—COOH} \rightleftarrows$$
$$\text{HOOC—CH(CH}_3\text{)—CH(NH}_2\text{)—COOH} \quad (3)$$

Diol dehydratase:

$$\text{HO—CH}_2\text{—CH(R)—OH} \xrightarrow{-H_2O} \text{O=CH—CH}_2\text{(R)}$$

$$R = H, CH_3, CH_2OH \quad (4)$$

Ethanolamine deaminase:

$$\text{HO—CH}_2\text{—CH}_2\text{—NH}_2 \longrightarrow \text{CH}_3\text{—CH=O} + \text{NH}_3 \quad (5)$$

We are actively pursuing all possible mechanistic alternatives for these unusual reactions. The difficulty of finding model reactions for these molecular rearrangement processes is that certain fundamental reactions of the substrates have not yet been discovered. Although a successful rearrangement reaction in a model system is not now available, we have experimental data which seem to direct us toward our aim of mimicking

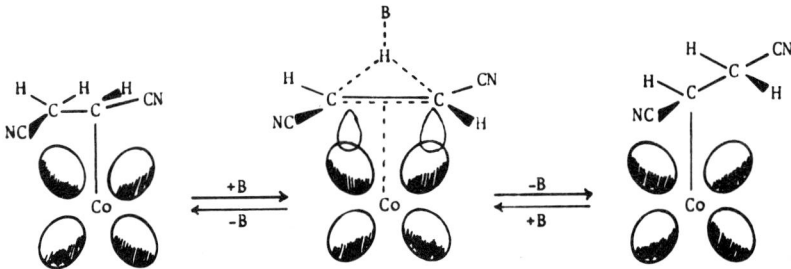

Figure 16. *The base-catalyzed valence tautomerization of 1,2-dicyanoethylcobaloxime via a π-complex intermediate*

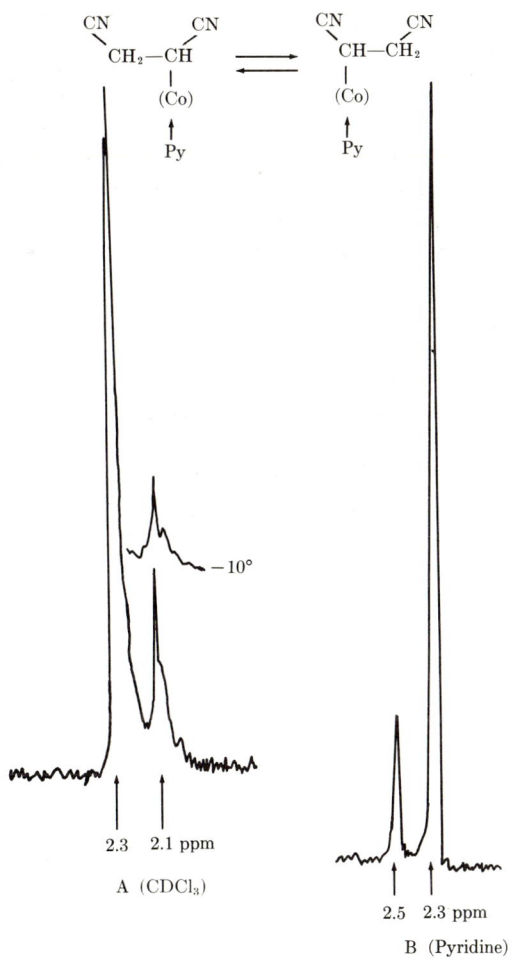

Figure 17. Partial H^1–NMR spectrum of 1,2-dicyanoethyl(pyridine)cobaloxime demonstrating the equivalence of the three protons of the dicyanoethyl moiety at room temperature

such a process. The reaction of 1,2-dicyanoethyl(pyridine)cobaloxime with bases produces a π-complex in analogy with the elimination reaction of β-substituted ethylcobaloximes. However, in this case, the π-complex is symmetrical (Figure 16) and hence should give rise to a reversible valence tautomerization. NMR measurements on several 1,2-dicyanoethylcobaloxime derivatives provide strong support for the occurrence of this process. In Figure 17 a portion of the NMR spectrum of 1,2-dicyanoethyl(pyridine)cobaloxime is shown. The three dicyanoethyl protons give rise to a singlet whose chemical shift depends on the polarity of the solvent. This is anomalous, inasmuch as the three protons should produce a typical AB_2 or ABB' pattern, as actually observed in 1,2-dicyanoethyl(tributylphosphine)cobaloxime. When solutions of 1,2-dicyanoethyl(pyridine)cobaloxime in $CDCl_3$/pyridine are cooled to $-45°$, the singlet of the dicyanoethyl protons collapses to the expected AB_2 pattern, indicating that the complex in solution at room temperature indeed undergoes reversible valence tautomerization *via* π-complex intermediates.

Concluding Remarks

This presentation was of necessity brief. I sincerely believe, however, that carefully selected models intelligently applied will bring us closer to essential aspects of nitrogen fixation and the function of coenzyme B_{12} in enzymes, and I also believe that a promising start has been made toward the solution of these problems.

Acknowledgment

The work on nitrogenase models was carried out in collaboration with my students, Gordon Schlesinger and Paul Doemeny. The studies on cobaloximes and cobalamins were carried out together with T. M. Beckham, J. W. Sibert, and J. H. Weber. The support of this work through the National Science Foundation is gratefully acknowledged.

Literature Cited

(1) Hesse, M. B., "Models and Analogies in Science," University of Notre Dame Press, Notre Dame, Ind., 1966.
(2) Campbell, N. R., "Physics, the Elements," Ch. VI, Cambridge, 1920.
(3) Hardy, R. W. F., Burns, R. C., *Ann. Rev. Biochem.* (1968) **37**, 331 and references cited therein.
(4) Yoch, D. C., Arnon, D. I., *Biochem. Biophys. Acta* (1970) **197**, 180.
(5) Schrauzer, G. N., Schlesinger, G., *J. Am. Chem. Soc.* (1970) **92**, 1808.
(6) Hogenkamp, H. P. C., *Ann. Rev. Biochem.* (1968) **37**, 225.
(7) Schrauzer, G. N., *Accounts Chem. Res.* (1968) **1**, 97.
(8) Schrauzer, G. N., Sibert, J. W., *J. Am. Chem. Soc.* (1970) **92**, 1022.
(9) Schrauzer, G. N., Weber, J. H., Beckham, T. M., *J. Am. Chem. Soc.*, in press.
(10) Abeles, R. H., Zagalak, B., *J. Biol. Chem.* (1967) **242**, 5369 and references cited therein.
(11) Abeles, R. H., Zagalak, B., *Ibid.* (1966) **241**, 1246.
(12) Frey, P. A., Abeles, R. H., Essenberg, M. K., Kerwar, S. S., *J. Am. Chem. Soc.* (1970) **92**, 4488.

RECEIVED June 26, 1970.

2

Studies on Enzyme Models

RONALD BRESLOW

Columbia University, New York, N. Y. 10027

> *Model enzymes have been produced which bind the substrate into an oriented mixed complex and then attack with appropriately placed functional groups. Metal complexing produces such systems, with accelerations of hydrolysis by up to 10^9. Hydrophobic binding into a cyclodextrin cavity complexes the substrate to a functionalized cyclodextrin, which catalyzes hydrolysis within the complex. The high positional selectivity of enzymatic attack on a particular molecule has been duplicated by forming oriented complexes of a substrate with a hydrogen-abstracting reagent, leading to selective functionalization of particular unactivated carbons. In another example, cyclodextrin binding of an aromatic substrate has also been used to promote selective substitution. These model studies have led to the development of useful new synthetic reactions.*

There are several purposes of biochemical model studies. The first is to help elucidate biochemical mechanisms. The common proposition that you explain something by relating it to something else is the argument for doing biochemical mechanism work in models—that is, you produce the "known" facts and relate the enzyme to them.

Secondly, there are points about biochemical model studies which are unique to models and which elevate the field somewhat above this simple handmaiden role. For instance, one can explore structure–activity relationships and generalize the chemistry in models in a way that often cannot be done in the enzyme itself; you can explore exactly what the significance is of some particular feature of the enzyme. How important is it that a carboxylate ion be located exactly at some particular spot? This can be investigated better in models than in the enzyme because the enzyme imposes a number of other special requirements related to binding.

The third point of model studies is the discovery of new chemical facts and principles. By developing a branch of chemistry that must contain new types of chemistry, because we do not yet understand how enzymes work, we certainly can expect to discover new facts as worthwhile objects in themselves.

Finally, the fourth point is that we might even do something useful out of all this; we might invent new chemical reactions or new processes, inspired in a sense by the biochemical system.

There are two general types of model systems (Table I). Much work in this field is concerned with models for what I call "special chemistry," models for co-enzymes, oxidative phosphorylation, steroid biosynthesis, and that type of thing.

Table I. Types of Model Systems

A. *Models for "Special Chemistry"*

(1) Coenzymes: pyridoxal phosphate, thiamine pyrophosphate, nicotinamide coenzymes, tetrahydrofolic acid, biotin, coenzyme B_{12}, etc.

(2) Oxidative phosphorylation, terpene cyclization, and other "biogenetic" syntheses, etc.

B. *Models for Enzyme Catalysis*

(1) Chemistry of the functional groups of proteins
(2) Intramolecular reactions and catalyses
(3) Catalyses and reactions in mixed complexes

The second area is models for enzyme catalysis itself. Here, as a first approximation, one will deny the idea that only a macromolecule can have enzyme-like properties. It is an open question whether or not a macromolecule is required, but this paper indicates some of the approaches one can make to this type of thing. In other words, what can one do about a model for the enzyme part of biochemical catalysis?

One approach concerns the chemistry of functional groups of the protein, such as imidazole. A second approach involves work on intramolecular reactions as a model for reactions within the enzyme substrate complex. One can look at an intramolecular model to determine rates and properties. Finally, the third area is catalysis and reactions in mixed complexes. This is the area under discussion: chemistry in complexes, hopefully of well-defined geometry, rather than by random collision in solution.

Figure 1 shows mechanisms of two enzymes which use a metal ion catalytically. We are going to consider model reactions involving a metal ion and some other possible functional groups. Carboxypeptidase has been described as having a simple set of functional groups required for the activity of the enzyme: there is a zinc ion which is coordinated to the

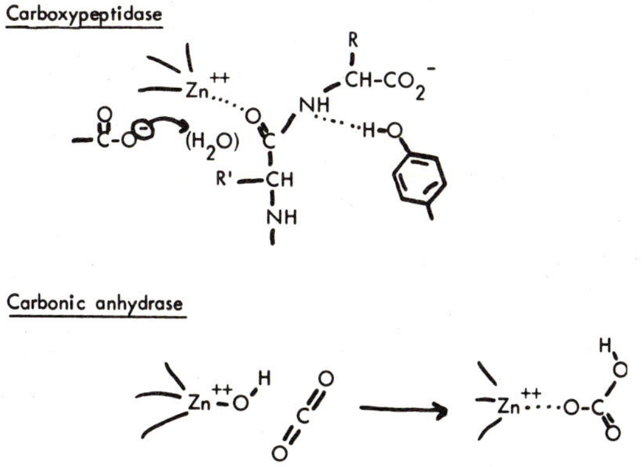

Figure 1. Two different functions of a metal ion in proposed enzyme mechanisms

oxygen; a tyrosine phenolic hydroxyl is in position to assist the departure of the leaving nitrogen when the amide is hydrolyzed; and the hydrolysis is promoted by a glutamate carboxylate ion. This is either directly attacking the carbonyl to form an anhydride intermediate or attacking a water molecule to act as a general base, and it is apparently not yet possible to decide between these two mechanisms. This is one way in which a metal ion can function, to act as a Lewis acid to stabilize the negative charge on the carbonyl oxygen. Another type of function is the one in the illustrated mechanism for carbonic anhydrase, in which the zinc ion in the carbonic anhydrase interacts with carbon dioxide bound into the hydrophobic cavity to act as a donor of a zinc ligand group (hydroxide) rather than as a Lewis acid.

I now want to consider some metal-catalyzed model reactions, with and without extra catalytic groups. Many people have been interested in metal-catalyzed hydrolyses of coordinated ligands (1). We chose some years ago (2, 3) to look at the hydrolysis of groups attached to the 1,10-phenanthroline nucleus because this can be prepared very conveniently as a one-to-one complex with a variety of metal ions with no danger of metal ion equilibria complicating all the rest of the hydrolysis. Table II shows a typical example of hydrolysis. The carbonyl oxygen of the amide is coordinated to the metal ion and so one can get Lewis-acid-assisted metal ion help by nickel. Phenanthroline carboxamide hydrolyzes with a second order rate constant (there is a kinetic term in hydroxide and one in substrate) of 7.26×10^{-2}. If one puts a nickel ion into the phenanthroline, the kinetics stays the same (it is still hydroxide times the complex) but now there's an increase of the rate constant by 400-fold. This is still a very sluggish hydrolysis, owing to the metal ion alone, and is cer-

Table II. Metal-Promoted Hydrolysis of Phenanthroline 2-Carboxamide

$$R-\underset{\|}{\overset{O}{C}}-NH_2 + OH^- \xrightarrow[H_2O]{k_2} R-\underset{\|}{\overset{O}{C}}-O^-$$

Substrate	$k_2(50°, M^{-1} Min^{-1})$
Phenanthroline amide	7.26×10^{-2}
Ni^{2+}–Phenanthroline amide	33

Table III. Metal-Promoted Hydration of Phenanthroline 2-Carbonitrile

$$R-C\equiv N + OH^- \xrightarrow[H_2O]{k_2} R-\underset{\|}{\overset{O}{C}}-NH_2$$

Substrate	$k_2(25°, M^{-1} Min^{-1})$	$H^{\neq}$, Kcal/Mole	$S^{\neq}$, e.u.
Phenanthroline nitrile	0.154	15.1	−20
Ni^{2+}–Phenanthroline nitrile	0.142×10^7	15.7	+14
Cu^{2+}–Phenanthroline nitrile	10^8	–	–
Zn^{2+}–Phenanthroline nitrile	0.22×10^4	–	–

tainly not enough to approximate an enzyme. However, if the group which is being hydrated is a cyano group instead of an amide group, there is a very much larger effect of metal ions. Table III shows the phenanthroline nitrile with the cyano group, and in this case nickel ion gives an increase of 10^7, not 10^2. Copper gives an increase of the order of 10^9 in the second order rate constant under these conditions, and even zinc, which is a relatively poor Lewis acid, gives an increase of the order of 10^4. As Table III shows, this increase is all owing to the entropy of activation term; the enthalpy of activation is essentially the same, and, in fact, is slightly more favorable for the nonmetal case.

There are two possible mechanisms for the nitrile hydrolysis. One is that the nickel ion, with a ligand hydroxide ion, holds the ligand in such a position that it is directly opposite the cyano group in the complex and is in a position to attack it. The other mechanism is that hydroxide comes in externally, attacking the carbon as the nitrogen anion swings in towards the nickel. Since this is one of the largest metal ion catalyses known, we did some work on this question. Our conclusion is that the second mechanism is the correct one—*i.e.*, hydroxide is attacking from the outside as the nitrogen anion swings in. The reason for the very large rate advantage in the case of this cyano group over the amide group is that the cyano group initially starts off uncoordinated to the metal but near it, and as one adds to the cyano group, it can swing in and become coordinated. There is a gain in coordination, and this is responsible for the very large effect in the cyano compound as compared with the amide.

$$\text{[Ni}^{2+}\text{-phenanthroline-C}\equiv\text{N]} \xrightarrow{\text{ROH}} \xrightarrow{\text{H}_2\text{O}} \text{[Ni}^{2+}\text{-phenanthroline-C(=O)-OR]} \quad (1)$$

R = Et—, $NH_2CH_2CH_2$—, $NH_2CH_2CH_2NHCH_2CH_2$—

Evidence for the mechanism was obtained by looking at a variety of alcohols. The addition of alcohols to the cyano group is also catalyzed by metals (Equation 1). Ethanol had the normal reactivity of free ethanol in solution. That is, the ratio of reactivity of alcohol to water in mixed reactions was of the order of ten to one. This is characteristic of attack of external ethanol in many acylation reactions and consistent with the second mechanism. Ethanol in the coordination sphere of nickel, as in the first mechanism, would be selected against. We also tried to put ligands into the coordination sphere of nickel, as in the case of ethanolamine or hydroxyethylethylenediamine, and these did not show any selective reactivity, excluding the possibility that it was a nickel-bound ligand which was attacking.

Metal-catalyzed reactions of 2-picolinonitrile have an even larger effect, 10^9 acceleration of the second order hydration rate (Table IV).

Table IV. Metal-Promoted Hydration of 2-Picolinonitrile

$$R-C\equiv N + OH^- \xrightarrow[H_2O]{k_2} R-CO-NH_2$$

Substrate	k_2, 30°, M^{-1} Min^{-1}	$H^{\neq}$, Kcal/Mole	$S^{\neq}$, e.u.
2–Picolinonitrile	1.08×10^{-1}	15.6	−20
Ni^{2+}–Picolinonitrile	6×10^8	—	—

This transition state is a little more flexible, and in the pyridine nitrile case it is possible for a ligand (Equation 2) on the copper to reach out-

$$\text{[pyridine-C}\equiv\text{N···Cu}^{2+}\text{(H}_2\text{N-CH(CH}_2\text{OH)}_2\text{-OH)]} \longrightarrow \text{[pyridine-C(=N-O)-C(CH}_2\text{OH)}_2\text{-CH}_2\text{]} \quad (2)$$

side and get at the external position in the cyano group. Even though with phenanthroline other ligands on the copper are not able to attack the cyano group, in the case of the pyridine nitrile, one of the hydroxyls of coordinated tris(hydroxymethyl)methylamine can get to the outside, with very slight strain in models, and attack. An intramolecular reaction which is at all sterically reasonable always seems to be favored over an intermolecular analog. Thus very selectively (4) the one-to-one tris-copper complex in water forms the oxazoline by attack of the hydroxyl and then Schiff base formation. This is the exclusive product in competition with 55 M water when the tris-copper complex is the order of 10^{-3} M, so this is a very selective intracomplex reaction—attack of one ligand on another.

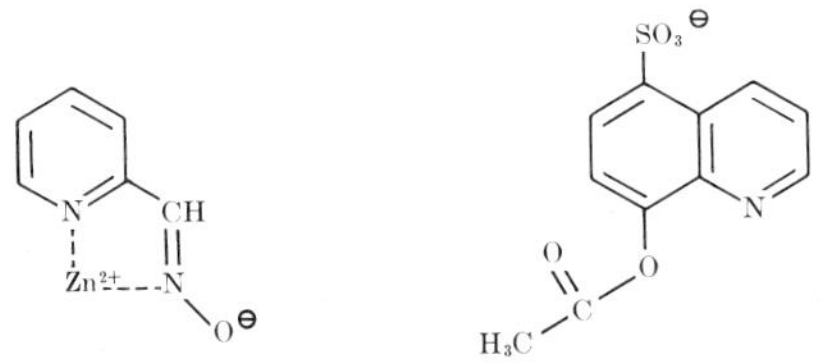

Figure 2. Components of a nucleophilic reaction in a mixed-metal complex

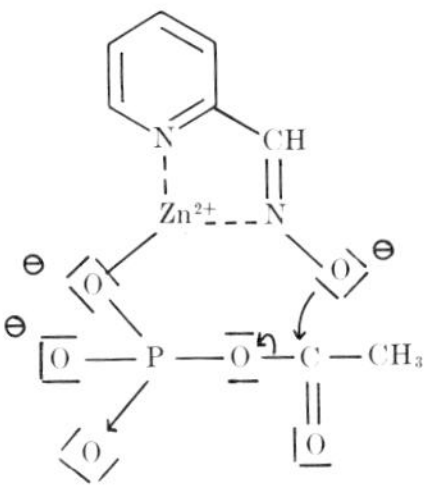

Figure 3. Intracomplex attack on acetyl phosphate by a metal-bound nucleophile

Some years ago (5, 6) we looked at the possibility of doing intracomplex reactions with a system (Figure 2) in which nucleophilic catalytic ligand and a substrate ligand were bound by metal binding. The nucleophile–metal combination was particularly good as a catalyst for the deacetylation of acetoxyquinolinesulfonate. In the mixed complex, the oxide ion is in position to attack the acetyl group. There is an acetyl transfer and an intermediate oxime acetate formed. Oxime acetate is hydrolyzed in a fairly fast metal-catalyzed step, also. The metal–nucleophile combination is especially reactive toward a coordinating substrate, as compared with simple substrates such as *p*-nitrophenyl acetate.

Figure 3 is a picture of the kind of transition state we are talking about, in which the substrate is acetyl phosphate. This is some work in our laboratory (7) in which the metal–nucleophile combination acts as a good acetyl phosphatase, although it is certainly a very small molecule. At modest concentrations, it gives several thousand-fold acceleration of the hydrolysis of acetyl phosphate over just the aqueous solution at neutrality. Again by coordination and attack of the oxygen on the acetyl group, an intermediate acetate is formed which then hydrolyzes.

All of these processes have involved using the metal ion to bind the substrate to the catalytic group and, of course, nature does not use metal binding as the principal means of binding the substrates to an enzyme. This is only one of the forces and probably a minor one; thus, we were also interested in the question of whether we could extend this type of work to a model system in which we bound the substrate with hydrophobic rather than metal-binding forces.

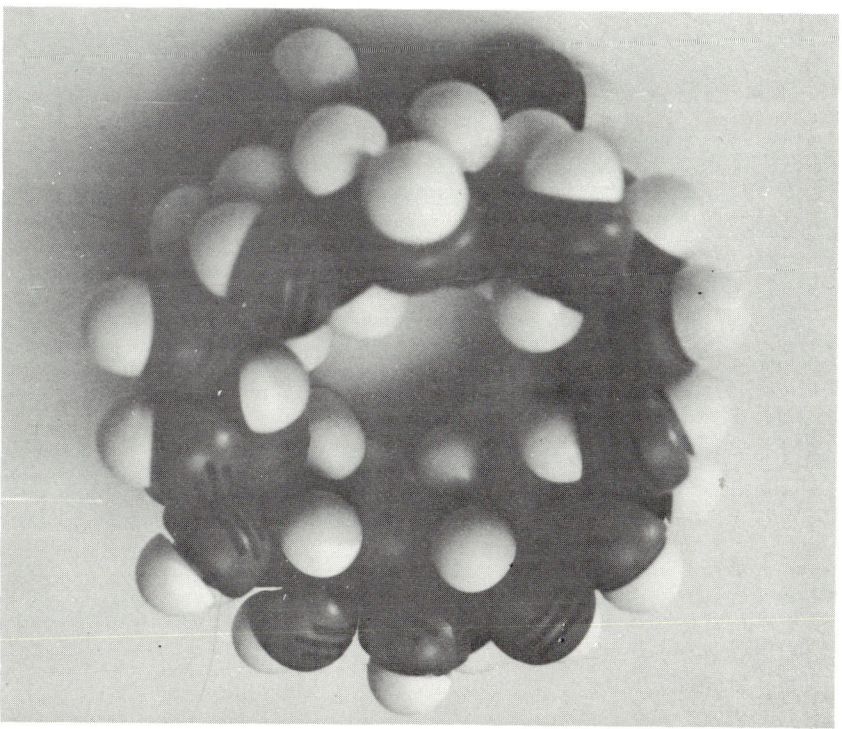

Figure 4.   α-Cyclodextrin (cyclohexaamylose)

For that, we went to a substance called α-cyclodextrin (Figure 4). A set of cyclodextrins is available by biological digestion of starch, and the one shown here has six glucoses in a ring. The top of the molecule has a set of 12 hydroxyl groups on glucose attached to carbons 2 and 3. On the bottom of the molecule there are six primary hydroxyls. None of the hydroxyls point into the cavity. The cavity in fact has CH bonds, and ether oxygens, and the cavity is therefore hydrophobic. It had been known for a long time that the cyclodextrins will bind aromatic substrates.

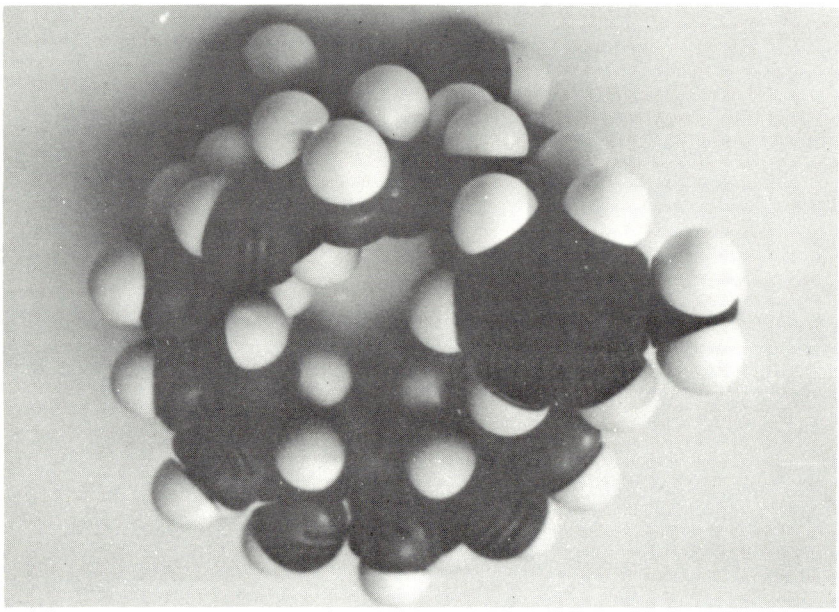

Figure 5. α-Cyclodextrin and toluene

Figure 5 shows a toluene molecule approaching the cavity, to indicate the relative sizes. Even in the model (Figure 6) the toluene has no difficulty in getting into the cavity. Dissociation constants depend on the exact aromatic substrate, but these are bound well enough that it's frequently possible to saturate the cavity with available concentrations of substrates. We decided that this kind of molecule would be useful in trying to construct enzyme models which would bind substrates hydrophobically; of course, this idea is not unique with us.

Friedrich Cramer in Germany (8) has worked for many years on the properties of cyclodextrins, and Myron Bender in this country has recently done some work (9) which was particularly useful to us. Bender has found that if a para-substituted phenylacetate is put into solution, it would bind into the cavity in such a way (Figure 7) that the acetyl group lies somewhere near the axis of the system and is not accessible to the secondary hydroxyl groups off on the sides. However, a meta-substituted phenyl acetate—for instance, *m*-nitrophenyl acetate or *m*-chlorophenyl acetate—binds into the cavity in such a way that the acetyl

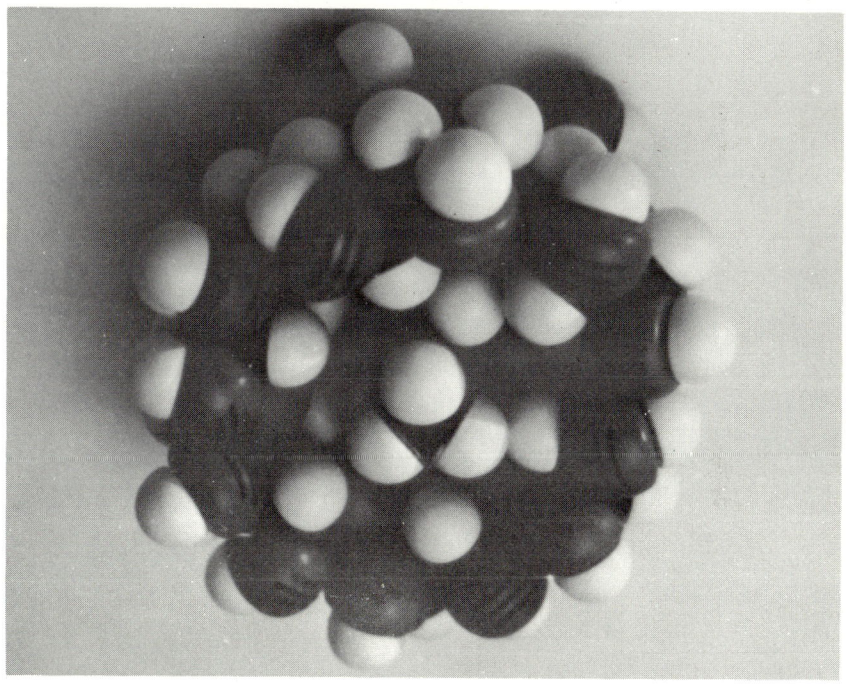

Figure 6. Toluene bound in the cavity of α-cyclodextrin

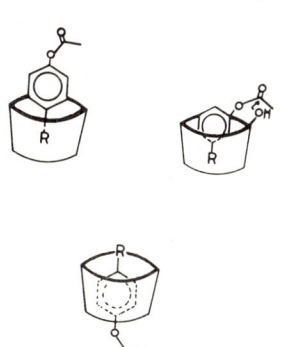

Figure 7. Binding of p- and m-substituted phenyl acetates by a cyclodextrin

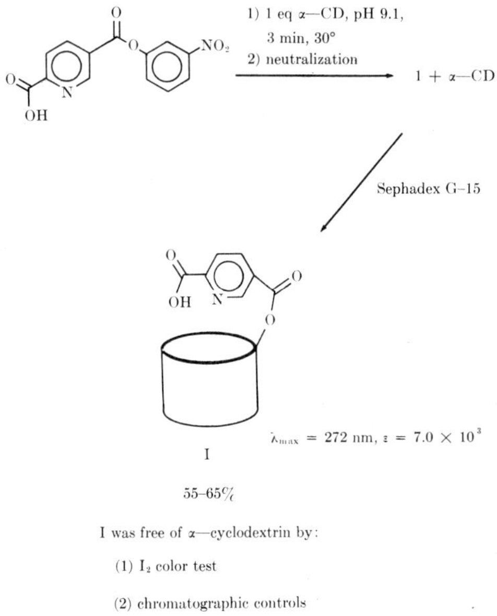

Figure 8. Preparation of the α-cyclodextrin ester of pyridine-2,5-dicarboxylic acid

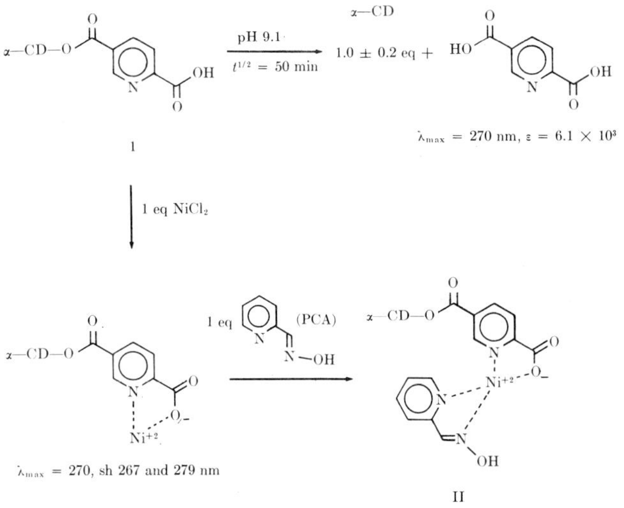

Figure 9. Formation of catalysts combining cyclodextrin with metal-containing functional groups

group can be transferred to the cyclodextrin hydroxyl; at fairly high pH ($\sim$ 10.5 or 11.0) there is a reasonably fast transfer of this acetyl group to the hydroxyl. One gets an intermediate acetate, and this then will cleave hydrolytically. Thus one has something that is short of an enzyme, operating at fairly high pH, and with the second step being rather slow. We decided to use this in order to construct a system with both the metal ion and a better binding site, to try to extend metal-catalyzed reactions to the cases in which the substrate is not a good metal ligand; metal-binding is certainly restricted in terms of its possibility for forming strong complexes. We prepared (*10*) the *m*-nitrophenyl ester of pyridine-2,5-dicarboxylic acid (Figure 8). When this is incubated with cyclodextrin at pH 9 for three minutes, the acyl group is attacked by one of the secondary hydroxyls and one ends up with a molecule with the picolinate attached to the cyclodextrin. This material is easily isolated as a pure substance in good yield and in large quantity; we have several grams of it at a time if we so desire. Figure 9 indicates that this material will bind nickel or other metal ions to form a metal complex. In the first work to be discussed, we constructed a somewhat more elaborate catalyst by using the pyridine carboxaldoxime as a second ligand to produce a very powerful nucleophile. This is our molecule II, a molecule which we will refer to subsequently as "enzyme." In contrast to the simple cyclodextrin itself, this complex is able to reach the acetyl group of *p*-nitrophenyl acetate (Figure 10). That is, *para*-nitrophenyl acetate will bind so that the acetyl group lies on the axis of the cavity and is not accessible to the cyclodextrin hydroxyls, but the model shows that this system can easily reach the acetyl group. However, it reaches it by the restriction of a number of free rotations, and that, as we shall see, poses a problem. Figure 11 is a picture of the molecule cyclodextrin with the large appended group hanging on the side so it can be turned away, leaving the cavity perfectly accessible. Within the cavity one can bind *p*-nitrophenyl acetate (Figure 12) in such a way that by rotation around a couple of single bonds the large appended group can swing around and can bring itself to the point

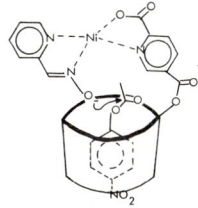

Figure 10. Attack by the artificial enzyme on bound p-nitrophenyl acetate

Figure 11. Molecular model of the artificial enzyme

Figure 12. Molecular model of the artificial enzyme–substrate complex

where the nucleophilic oxygen of the oxime can rest directly on the carbon of the acetyl carbonyl group (Figure 13). The prediction was that this very complicated functional group should be able to attack *p*-nitrophenyl acetate bound into the cavity.

This turns out to be the case, but what we are really looking for in a good enzyme is a cooperative effect. We already know that the nucleophilic functional group is a good catalyst for the deacylation of *p*-nitrophenyl acetate, but what we want to see is how much help we get from the hydrophobic binding.

Figure 13. *Molecular model of the attack at the active site*

Table V. Hydrolysis of *p*-Nitrophenyl Acetate, pH = 5.17

Catalyst, mM	k, Min^{-1}
—	7.1×10^{-5}
"Enzyme" (5.11)	9.9×10^{-2}
"Enzyme" (10.0)	19.3×10^{-2}
Ni–PCA (5.01)	2.6×10^{-2}
Ni–PCA (10.0)	5.06×10^{-2}
Enzyme (5.09) + Cyclohexanol (50)	5.99×10^{-2}
Enzyme (5.09) + Cyclohexanol (134)	4.78×10^{-2}
Ni–PCA + Cyclohexanol (50 or 134)	2.6×10^{-2}

As Table V shows, this "enzyme" at 5 mM increases the hydrolysis rate of p-nitrophenyl acetate by a factor of over 1000; at 10 mM the rate goes up, and it's nicely linear. This is not the region of saturation. The data also show that most of the catalytic effect originates in the ligand itself—i.e., in this very good catalytic nucleophilic group—and only a modest extra effect of the order of four or slightly less can be attributed to the binding. This binding, far from turning this material into a good "enzyme," has only increased its catalytic effectiveness in a modest way, probably because a cooperative interaction would require freezing out a number of single-bond degrees of rotational freedom. On the other hand, this increase is the first bit of evidence that the catalyst is binding p-nitrophenyl acetate into the cavity and then attacking the material within the cavity. A second piece of evidence is that competitive inhibition is observed with molecules like cyclohexanol which can also bind into the cavity, so cyclohexanol will lower the rate to a region near that of the simple ligand. It is not quite the same as the simple ligand itself, and we think perhaps this inhibited enzyme is catalytically slightly different from the simple ligand itself. Perhaps there's some help from cyclohexanol. Cyclohexanol has no effect at all on the rate with the simple functional group (NiPCA).

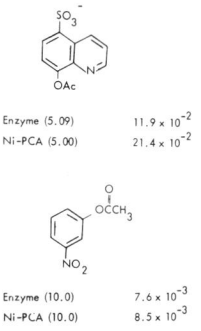

Figure 14. Nonsubstrates for the artificial enzyme

Another piece of evidence that we are dealing with the reaction within this mixed complex involving binding within the hydrophobic cavity is that other substrates which shouldn't fit don't show rate acceleration (Figure 14). The acetoxyquinolinesulfonate is too large and charged in a way which would exclude it from the hydrophobic cavity; it turns out to be only half as reactive towards the "enzyme" as it is towards NiPCA and the rate is not affected by cyclohexanol. In other words, the "enzyme" is a super nucleophile only toward things which can bind in such a way that the geometry is acceptable in the cavity. This is also true for *m*-nitrophenyl acetate which is, at neutrality, essentially inert to ordinary cyclodextrin. It is attacked by NiPCA as well as by the "enzyme," but in the latter case more slowly because again, as a model shows, the *meta*-acetoxy group is not in a position to be reached by the nucleophilic part of the functional group. We have also found other cases which do work (Figure 15). For example, if the *p*-nitrophenyl group is placed in the cavity of the enol acetate of *p*-nitroacetophenone, the acetate is essentially one atom further out from the ring than for *p*-nitrophenyl acetate. It is easily reached and there is an increase in rate with the "enzyme." Dinitrophenyl cinnamate is also a good substrate. So substrates which should fit into the cavity and be reachable are good substrates for the catalyst. Substrates which fit into the cavity, but are then in the wrong position to be attacked, or substrates which should not bind, are inferior substrates. They are attacked more readily by the functional group itself than by this "enzyme."

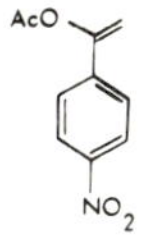

Catalyst	pH = 5.2	k (min^{-1})
Enzyme		10.1×10^{-4}
Ni-PCA		7.2×10^{-4}
Enzyme		7.5×10^{-2}
Ni-PCA		1.67×10^{-2}

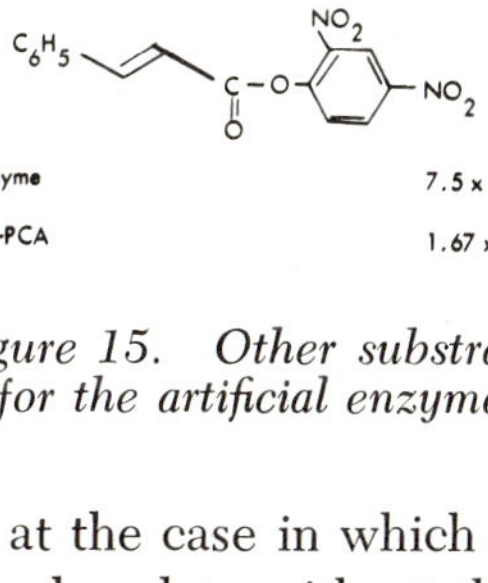

Figure 15. Other substrates for the artificial enzyme

We have also looked at the case in which we have simply the metal attached to the pyridine carboxylate without the oxime—*i.e.*, without the extra nucleophilic ligand; in other words, the case in which we are dealing simply with metal-catalyzed reactions but in which there is not only a metal-binding site but also the hydrophobic-binding site of the cyclodextrin.

Table VI shows the situation. With various concentrations of copper alone, *p*-nitrophenylglycine ethyl ester is hydrolyzed and the copper catalyzes the hydrolysis. The observed catalysis is rather modest for the hydrolysis of this particular amino acid ester, and the hydrolysis rate is

Table VI. Hydrolysis by a Simpler Artificial Enzyme

$$H_2N-CH_2-\overset{O}{\underset{\|}{C}}-O-\underset{}{\bigcirc}-NO_2 \quad 25.0°, \text{pH} = 3.97$$

$Cu^{+2} \times 10^3$	$CDE \times 10^3$	Added	$10^3 \, k, \, Min^{-1}$
–	–	–	6.51
0.625	–	0.625×10^{-3} M each pyridine-2, 5-diCO_2H + α-cyclodextrin	8.35
0.625	–	–	10.3
1.25	–	–	12.0
2.50	–	–	17.8
5.00	–	–	27.3
7.50	–	–	37.2
10.0	–	–	46.5
1.25	1.25	–	56.9
2.50	2.50	–	97.3
5.00	5.00	–	183
7.50	7.50	–	263
2.50	2.75	–	114
2.50	2.75	cyclohexanol 0.129 M	21.3
2.50	2.75	0.0472 M	33.1
2.47	–	0.129 M	17.4

not affected appreciably by adding either pyridine-2,5-dicarboxylic acid or α-cyclodextrin or both. In other words, these materials do not interact with the free system. But if one has the copper and the cyclodextrin ester with a ligand pyridine–carboxylate so that the copper becomes bound to the cyclodextrin ester, then one has a catalyst combining copper and the cavity. Whereas the simple copper-catalyzed reaction, for instance, at 1.25 mM had a superficial pseudo-first-order rate constant of 12, in this case with the cyclodextrin system it goes up to 56, or again by a factor of 4 or slightly more. Again, this is all linear. *para*-Nitrophenyl glycinate is being bound into the cavity in such a way as to increase the rate of hydrolysis by the copper, but the cooperative effect again leads to an increase of only four or so. Again there is inhibition by cyclohexanol, which can compete in the cavity for the *p*-nitrophenyl group. This follows good competitive inhibition kinetics with a reasonable inhibition constant for cyclohexanol.

Cyclohexanol has no effect on the simple copper-catalyzed reaction, so this is all consistent with the mechanism. We are getting binding within the cavity and attack of the bound substrate by the copper, but still the cooperativity here is rather modest.

By contrast, Table VII shows that nickel and zinc will catalyze the deacetylation of pyridine carboxaldoxime acetate, but the cyclodextrin ester does not add anything to the rate. The ester will not increase the metal-catalyzed hydrolysis rate of materials which if bound into the cavity would have the group held in the wrong place. This fits a geometric picture—that is, that the substrate must both bind into the cavity and in addition have the proper orientation with respect to the attacking group within that cavity. But still the cooperativity is rather modest.

Table VII. A Nonsubstrate for the Simpler Artificial Enzyme

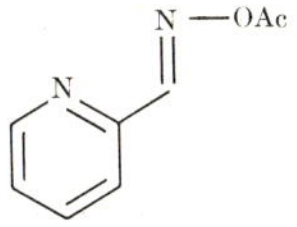

Catalyst (mM)	pH	k, Min^{-1}
—	7.0	2.9×10^{-3}
Ni^{2+} (5.0)	7.0	2.9×10^{-1}
Zn^{2+} (5.0)	7.0	2.1×10^{-2}
Ni^{2+} (5.0)	7.5	8.1×10^{-1}
CDE (5.1) Ni^{2+} (5.0)	7.5	3.5×10^{-1}
Ni^{2+} (5.0)	5.2	2.3×10^{-2}
CDE (5.0) Ni^{2+} (5.0)	5.2	6.9×10^{-3}

We have focused on reactions trying to duplicate the velocities of enzyme-catalyzed processes by putting together a binding site and some functional groups. The evidence so far is that one has not made substantial progress towards the very large rates of enzyme-catalyzed reactions. This certainly leaves open the possibility that one must worry about a macromolecule effect of a special kind, although our models have not really combined all the factors (Table VIII) which are understood to play a role in enzymes.

Table VIII. Enzyme-Catalyzed Chemistry vs. "Organic" Chemistry

Velocity
- Enzyme–substrate complex
- Polyfunctional catalysis
- Proximity
- Medium effects
- Distortion: the rack mechanism

Another thing which is interesting about enzyme-catalyzed processes is their selectivity (Equations 3, 4). We will focus not so much on sub-

Selectivity

$$CH_3(CH_2)_{16}CO_2H \longrightarrow CH_3(CH_2)_7CH=CH(CH_2)_7CO_2H \quad (3)$$
stearic acid   oleic acid

$$\text{α-cholestanol} \longrightarrow \text{(ketone product)} \quad (4)$$

strate selectivity, but rather on selectivity which determines just where in the molecule a given substrate will react under the influence of an enzyme. For instance, stearic acid can be dehydrogenated to oleic acid; α-cholestanol can be oxidized by enzymes in particular spots. This kind of thing would be very useful to duplicate, and it is not out of the reach of chemistry at the present time, even though the velocities of enzyme-catalyzed reactions seem to be pretty far away. Figure 16 is a funny

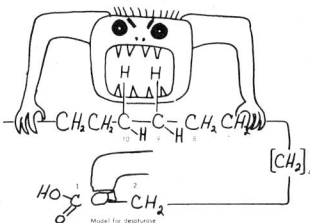

Figure 16. Highly stylized representation of the essential features of selective enzymatic oxidations

sort of picture which was drawn up by Mr. Winnik, one of my graduate students, to illustrate how the enzyme does this. There has to be some mode of attachment at some functional group, and quite far away from the point of attachment there is some group which will remove the hydrogens. I might point out that the kind of thing we're talking about here is a reaction involving an oxidase, and metals are certainly involved in the enzymatic reaction. We looked at this kind of thing to see if it could be duplicated in a simple chemical system—that is, a rigid point of attachment and then remote oxidation far from the attachment site.

Equation 5 illustrates the process (*11, 12*). We attach a rigid reagent

$$\text{I} \xrightarrow[\text{CCl}_4]{h\nu} \quad (5)$$

which one can excite to a radical, in this case photochemically (biologically it presumably is done by an oxidation–reduction process). The oxygen radical is generated in a position remote from the point of attachment (in this case on a long-chain alcohol) in such a way that this would be expected to attack far down the chain from the attachment site. That in fact occurs. It removes a hydrogen very far removed from the point of attachment and then functionalizes by producing a pair of radicals. The radical pair couples, and in one mode of degradation we can dehydrate this and oxidize it so as to put a carbonyl group in. The net change (Equation 6) is to have gone from a straight chain alcohol derivative to a keto-alcohol, by oxidation well down in the chain.

$$(6)$$

Figure 17 shows that though there is some distribution, this is a rather selective process. It involves primarily carbon number 14, way down the chain just two carbons from the end. It is possible to reach far down this saturated chain and attack relatively selectively. We have also looked at this kind of thing with steroids (*13*), which is a case where perhaps more interest lies. We have tried to attach reagents onto steroids in order to attack particular spots. While many reagents don't work,

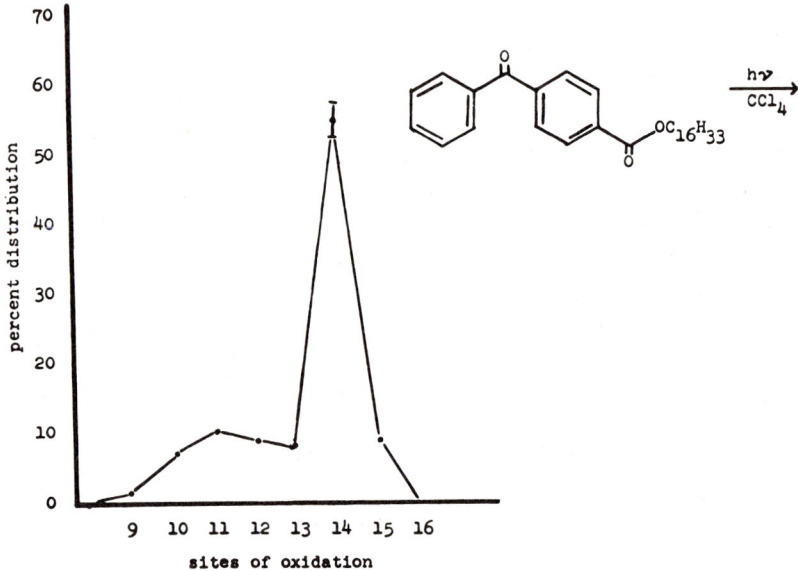

Figure 17. Photo-oxidation of the C-16 ester in carbon tetrachloride

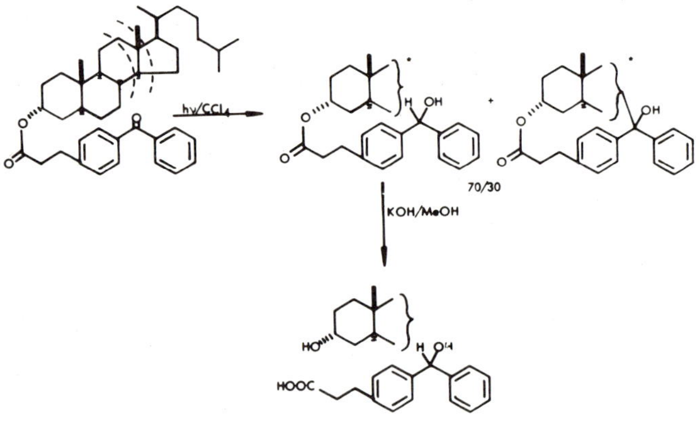

Figure 18. Attack on a steroid by an attached benzophenone

the benzophenone derivative on Figure 18 can be photoexcited to produce an oxygen radical which swings under the steroid in such a way that it is expected to be able to remove one of the hydrogens on carbon-12 or carbon-14 and produce a functionalized steroid derivative. After a degradation scheme which involved just some simple chemical transformations it is possible to find out where the oxidation has occurred. The only ketone which is produced is 12-ketocholestanol. Again this rigid oxidizing reagent has swung far from the point of attachment and attacked a particular spot.

There are other kinds of products as the reaction conditions change. We can get attack on carbon-14 (Equation 7), and this can lead to olefin formation (Equation 8). There is an oxidation (Equation 9) in which

$$\text{(7)}$$

$$\text{(8)}$$

the steroid is directly, under some conditions, converted to the Δ^{14} steroid in pretty good yield, and this is specifically produced as the only olefin. It is thus possible to duplicate this kind of selectivity of an enzyme for certain atoms by using the fundamental principle that the enzyme uses— attachment to a rigid reagent which can only reach certain spots. All of this is intramolecular.

$$\text{(9)}$$

We are extending these studies and looking at ways in which complexing forces can orient a reagent and its substrate. We published a related work a year ago (14), a study of the effect of α-cyclodextrin on aromatic substitution. Table IX shows that when anisole is chlorinated with aqueous HOCl, it produces a 60/40 ratio of *p*- to *o*-chloroanisole. However, in the presence of α-cyclodextrin this ratio climbs, and it is

Table IX. Product Distribution in the Chlorination of Anisole by Hypochlorous Acid in the Presence of α-Cyclodextrin

Cyclohexaamylose, $M \times 10^3$	p/o Chloroanisole Product Ratio	% Anisole Bound
0	1.48	0
0.933	3.43	20
1.686	5.49	33
2.80	7.42	43
4.68	11.3	56
6.56	15.4	64
9.39	21.6	72

possible with slightly higher concentrations of cyclodextrin to reduce o-chlorination below detectable limits. We expected the ortho positions to be blocked by cyclodextrin complexing, but the data in Table IX indicate that something more is happening. Apparently the para position is being activated, and direct kinetic studies confirm this. Thus, on Figure 19 we indicate that one of the hydroxyl groups ringing the cavity acts as the chlorine donor, to account for the fast attack on bound anisole.

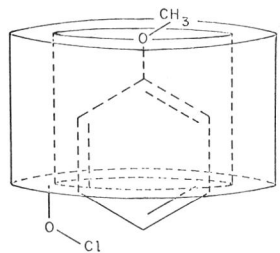

Figure 19. Proposed mechanism for chlorination of anisole catalyzed by α-cyclodextrin

The summary of the situation is that in the approaches to trying to get a model which will work the way an enzyme does, we can in fact put together some catalysts which will bind substrates and attack within the mixed complex. So far, it seems fair to say that none of the velocities that one can produce are close enough to enzymatic velocities to be very exciting, although these studies are still in their infancy. On the other hand, in the other kind of approach to models, the attempt to produce useful chemical reactions by letting the geometry of the "enzyme"–substrate complex be the dominant factor, it does turn out that some interesting new synthetic chemistry has emerged.

Literature Cited

(1) "Reactions of Coordinated Ligands," ADVAN. CHEM. SER. (1963) **37**.
(2) Breslow, R., Fairweather, R., Keana, John, *J. Am. Chem. Soc.* (1967) **89**, 2135.
(3) Fairweather, R., Ph.D. thesis, Columbia University, 1967.
(4) Schmir, M., Ph.D. thesis, Columbia University, 1969.
(5) Breslow, R., Chipman, D., *J. Am. Chem. Soc.* (1965) **87**, 4195.
(6) Chipman, D., Ph.D. thesis, Columbia University, 1965.
(7) Malmin, J., Ph.D. thesis, Columbia University, 1969.
(8) Cramer, F., Saenger, W., Spatz, H.-Ch., *J. Am. Chem. Soc.* (1967) **89**, 14.
(9) van Etten, R. L., Sebastian, J. F., Clowes, G. A., Bender, M. L., *J. Am. Chem. Soc.* (1967) **89**, 3242, 3253.
(10) Breslow, R., Overman, L. E., *J. Am. Chem. Soc.* (1970) **92**, 1075, and unpublished work.
(11) Breslow, R., Winnik, M. A., *J. Am. Chem. Soc.* (1969) **91**, 3083.
(12) Winnik, M., Ph.D. thesis, Columbia University, 1969.
(13) Breslow, R., Baldwin, S. W., *J. Am. Chem. Soc.* (1970) **92**, 732, unpublished work.
(14) Breslow, R., Campbell, P., *J. Am. Chem. Soc.* (1969) **91**, 3085.

RECEIVED June 26, 1970.

3

Chemical Foundations for the Understanding of Natural Macrocyclic Complexes

D. H. BUSCH, K. FARMERY, V. GOEDKEN, V. KATOVIC, A. C. MELNYK, C. R. SPERATI, and N. TOKEL

Ohio State University, Columbus, Ohio 43210

Studies on template effects, condensation reactions, oxidative dehydrogenations, catalytic hydrogenations and ligand deprotonations have made it possible to synthesize almost any tetradentate macrocyclic ligand one might choose. The principle of multiple juxtapositional fixedness is explained and applied to metal binding in enzymes. The inflexibility of the metal ion site in the porphyrin ring is contrasted to those in synthetic macrocycles. TAAB, a related tetraazaannulene, can vary its inner site while other rings fold when the metal ion is too large. Many model compounds for high-spin, five-coordinate deoxyhemoglobin and deoxymyoglobin are presented. The Fe^{III}–O–Fe^{III} bridge is extremely hard to break when iron is bound to a tetraazaannulene (TAAB). Iron is a better catalyst for oxidative dehydrogenation of its own ligand than is nickel.

The sustained development of the new synthetic macrocycle chemistry is traceable mainly to the efforts of those in two laboratories. Neil Curtis first discovered (1) that monocarbonyl compounds condense with diamines to produce cyclic compounds. Table I represents the products of this general category of reaction for such diamines as ethylenediamine, propylenediamine, and trimethylenediamine. The structure of the carbonyl compound must contain an α-methyl group. Simultaneously, macrocyclic complexes were intentionally produced at The Ohio State University in a program then dedicated to the demonstration of the coordination template effects (2, 3). Our first successful cyclization of this class is shown in Figure 1. Though the rapid progress of this field is largely the result of extensive studies by the two groups already mentioned, the excellent contributions of several others must be mentioned even in so

Table I. Macrocycles Prepared from Diamines and Carbonyl Compounds

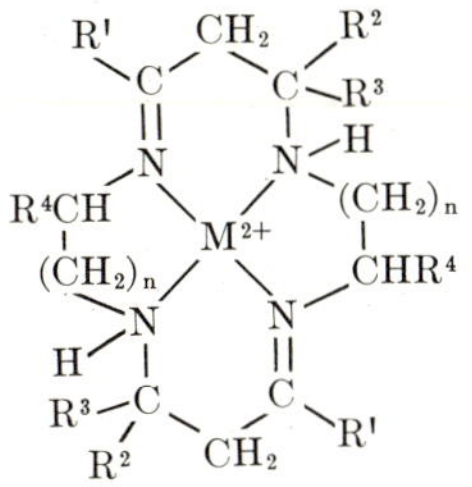

Carbonyl Compound	Amine	Metal	R^1	SR^2	R^3	R^4	n
Acetone	en	Ni	Me	Me	Me	H	1
Acetone	pn	Ni	Me	Me	Me	Me	1
Acetone	en	Cu	Me	Me	Me	H	1
Acetone	pn	Cu	Me	Me	Me	Me	1
Acetone	en	Co	Me	Me	Me	H	1
Methylethyl ketone	en	Cu	Et	Et	Mc	H	1
Acetone	tm	Ni	Me	Me	Me	H	2
Acetone	tm	Cu	Me	Me	Me	H	2

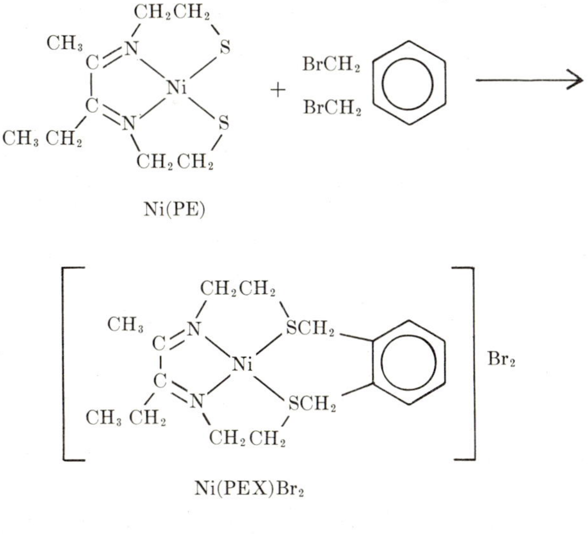

Figure 1. Preparation of s,s'-o-xylyl-2,3-pentanedione-bis(mercaptoethylimine) nickel(II) bromide

cursory a consideration of the subject as is intended here. In 1962, both Schrauzer (4) and Thierig and Umland (5) reported cyclization reactions involving a borate ester and similar linkages. Also, the synthesis of macrocyclic ligands differing only slightly from natural rings was accomplished early by Johnson and his associates (6). An extensive summary on the synthesis of macrocyclic complexes will soon be available (7).

It is particularly true in the case of tetradentate macrocycles that the synthetic techniques are rapidly reaching a stage of maturity such that one can set out to synthesize almost any previously unknown macrocyclic ligand of specific structure with a reasonable expectation of success. The reactions of mono- and dicarbonyl compounds with pairs of adjacent amine groups in the presence of metal ions has provided much of this generality. Figure 2 shows the previously mentioned Curtis reaction. It also shows the reaction of an α-diketone to form the α-diimine chelate ring of a macrocycle. Though the reaction has been known, its successful application to macrocyclization is owing to Baldwin and Rose (8). Jager has made extensive contributions to synthesis of macrocyclic ligands in their complexes. His procedures first incorporated the β-diketone moiety in such structures and he first showed that the expected

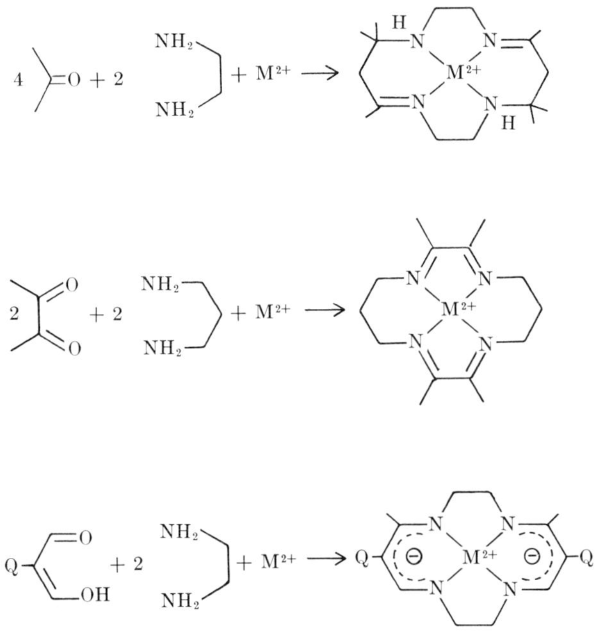

Figure 2. Schiff base condensation in macrocyclization reactions

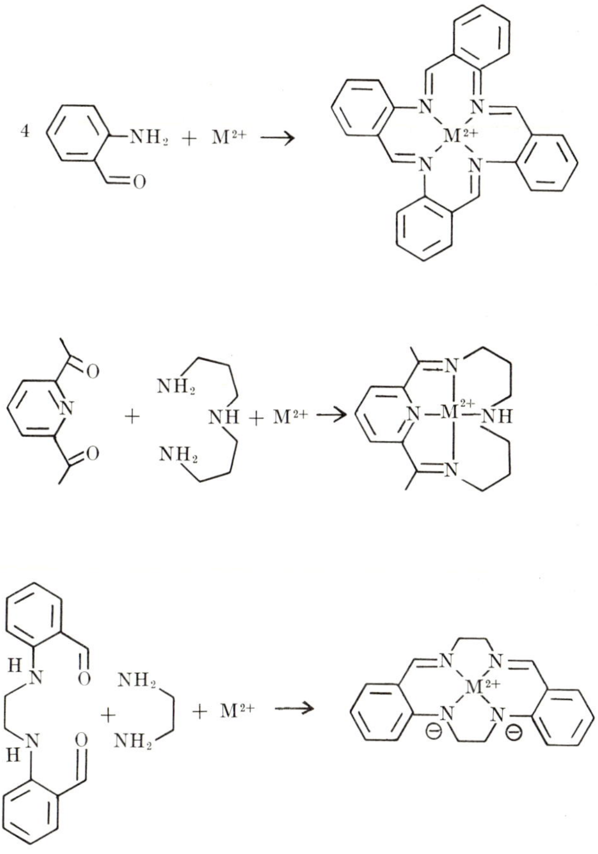

Figure 3. Schiff base condensation in macrocyclization reactions

ionization of a proton occurs in such reactions (9), so that the presence of each such grouping in a structure is accompanied by a unit of negative charge.

Figure 3 shows a number of variations on the use of the Schiff base condensation in macrocyclization reactions. One of the main products of self condensation of *o*-aminobenzaldehyde in the presence of a metal ion is the tetraanhydrotetramer (10). The complexes of this ligand (abbreviated TAAB) are discussed below.

The first example of such reactions using reagents incorporating noncondensing amines as parts of their structures was developed by Curry. He produced a variety of macrocycles starting with 2,6-diacetylpyridine (11). A similar tetrafunctional reagent was developed by Tasker and Green (12), as shown in the last reaction of this figure.

In addition to an almost unlimited array of condensations that might produce rings of the sorts described above, there exist systematic transformations that permit one to generate a variety of products by reactions of a single condensation product. Curtis (1) first produced a hydrogenation-dehydrogenation sequence, as shown in Figure 4. It is generally true that procedures can be found for the reduction of azomethine linkages in complexes containing macrocyclic ligands; however, the oxidative

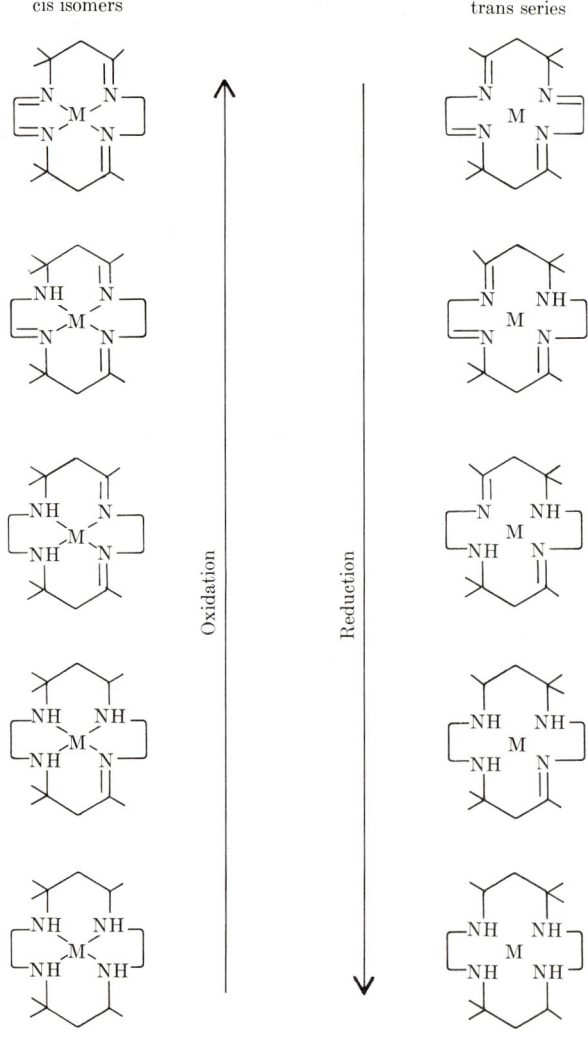

Figure 4. Hydrogenation-dehydrogenation sequences in macrocycles

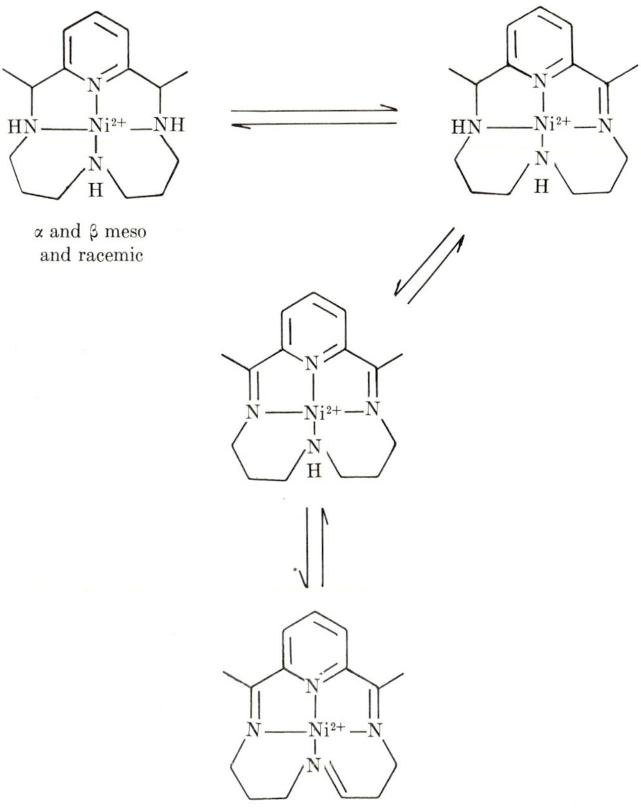

Figure 5. Hydrogenation-dehydrogenation sequence for Ni(II) complexes

dehydrogenation exhibits metal ion–specific effects. In general, the group —CHRNH— can be oxidized to —CR=N— in nickel complexes. Indeed, Curtis' work involved nickel(II), and Barefield's extensions (13) have shown that the nickel complexes are much more susceptible to such reactions than are the corresponding cobalt derivatives (Figure 5). In fact, nickel(III) seems to play a significant role in the mechanisms of such reactions. Goedken has recently found that iron is still more effective at promoting oxidative dehydrogenation of its cyclic ligand (14). We shall return to the latter point shortly.

A third class of reaction derives from the acidity of the coordinated ligands. A variety of the structural members may evidence acidity in complexes. Figure 6 provides a useful example from the studies of Goedken (13). The methylene group that ionizes in this ligand does so in fair measure because of the substantial electron delocalization in the resulting anion. One may couple deprotonations and the concomitant tautomerizations to oxidations to provide still more possibilities in ligand synthesis.

From studies such as these, it is safe to conclude that one is no longer exploring the possibilities but that one can now define synthetic goals and expect to achieve them. In this regard, there are a few structures that must soon become as familiar as those of ethylenediamine and of EDTA to all who are concerned with coordination chemistry (Figure 7). Principal among these are the uncluttered inner great rings of the

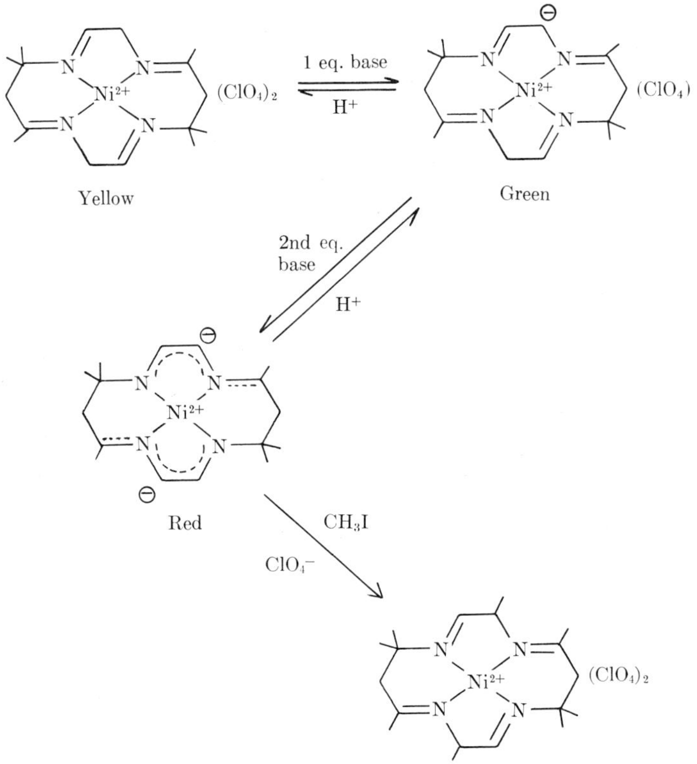

Figure 6. Deprotonation of the nickel(II) complexes of 1,4,8,11-tetraene

two important natural macrocyclic systems. Since the bare 15-membered ring is the core of the Corrin ring, we call it "cor." It follows then that the 16-membered inner ring of porphyrins must be called "por," a name that may be appropriate since there is reason to suspect that it might not be an ideal ring. The remaining structures all represent important variations of the basic structure that serve to test for the importance of the obvious features. For the cor system, is the net ligand charge

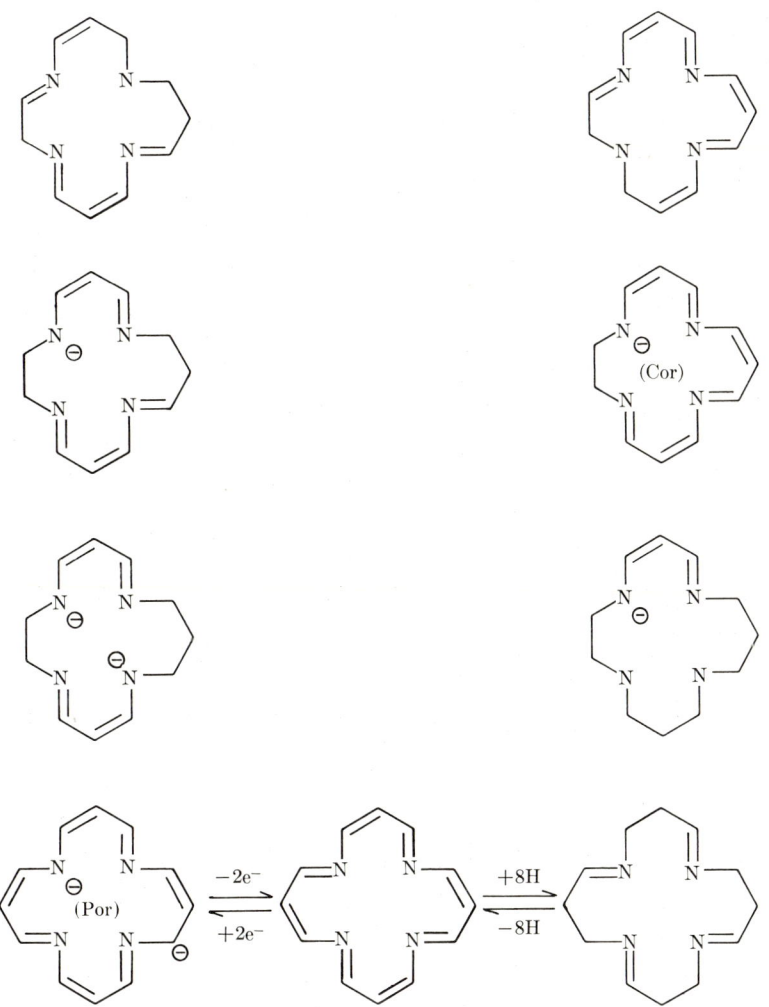

Figure 7. Inner great rings of natural macrocyclic ligand systems

critical? Must the ligand be both minus one in charge and involve full conjugation across all four donors? Is the fact that the conjugation is broken on one side of the ring important? Etc.? With regard to the por system, an important connection already exists between the two-electron oxidation product of porphyrins and the annulenes. This subject is treated at length below.

At this point, it seems valuable to comment on the nature of the contribution to the understanding of biochemistry that can be expected from such fundamental chemical studies as those concerned with the metal complexes of synthetic macrocycles. Whereas the role of the inorganic chemist as model builder may appear obvious, I would contend that the biochemist is best suited to that job. The inorganic chemist could probably most profitably contribute by applying known principles to explain existing biochemical phenomena and by developing new principles that are germane both to his field and to its applications to biochemistry. I shall try to do these things with a few examples that I hope are at least provocative. A few useful generalizations may be derived from the novel aspects of the behavior of macrocycles as ligands. Also, the study of the fundamental chemistry of metal complexes of macrocycles has yielded observations that are of direct use in accounting for the behavior of certain natural systems.

Multiple Juxtapositional Fixedness

This section heading is offered in keeping with the practice of giving incomprehensible, unpronounceable names to straightforward phenomena. Those who have worked with complexes of macrocyclic ligands generally come to realize that the substances are both uncommonly stable in the thermodynamic sense and shockingly inert toward macrocyclic ligand dissociation, even in very strong acids. The nature of the qualitative observations that have so impressed synthetic chemists for some eight years can be made apparent by considering Equations 1–4.

$$Ni(NH_3)_6^{2+} \xrightarrow[H_2O]{H^+} Ni(H_2O)_6^{2+} + 6\ NH_4^+ \qquad (1)$$

$$Ni(en)_3^{2+} \xrightarrow[H_2O]{H^+} Ni(H_2O)_6^{2+} + 3\ enH^+ \qquad (2)$$

$$en = NH_2CH_2CH_2NH_2$$

$$Ni(trien)(H_2O)_2^{2+} \xrightarrow[H_2O]{H^+} Ni(H_2O)_6^{2+} + trien(H^+)_n \qquad (3)$$

$$trien = NH_2CH_2CH_2NHCH_2CH_2NHCH_2CH_2NH_2$$

$$Ni(cyclam)^{2+} \xrightarrow[H_2O]{H^+} \qquad (4)$$

cyclam = (cyclic structure with four NH groups)

Table II. Rates of Formation of Cu^{2+} Complexes

$$Cu_{aq}^{2+} + L \longrightarrow CuL^{2+}$$

L	$k_2(0.5\ OH^-)$	$k_2(pH = 4.7)$
(cyclic tetraamine with 2 NH and 2 NH_2)	$\sim 10^7$	8.9×10^4
(meso tetramethyl cyclic tetraamine)	1.6×10^3	5.8×10^{-2}
hematoporphyrin IX	2.0×10^{-2}	—

Table III. Rates of Dissociation of Cu^{2+} Complexes

$$CuL^{2+} + nH^+ \longrightarrow Cu_{aq}^{2+} + LH_n^{n+}$$

L	$k_1(sec^{-1})$
(cyclic tetraamine with 2 NH and 2 NH_2)	4.1
(meso tetramethyl cyclic tetraamine)	3.6×10^{-7}

The familiar octahedral complexes $Ni(NH_3)_6^{2+}$ and $Ni(en)_3^{2+}$ (en = $NH_2CH_2CH_2NH_2$) are destroyed in fractions of seconds in strongly acidic solutions, showing the existence of a facile mechanism for ligand dissociation. If all four donor nitrogens are made part of one ligand, trien, $NH_2CH_2CH_2NHCH_2CH_2NHCH_2CH_2NH_2$, the rate is diminished, but only just noticeably. In any event, most nitrogen ligands are stripped from nickel(II) in strong acid media in a very short time. In contrast, when the nickel(II) complex of a cyclic ligand, Ni(cyclam)$^{2+}$, is placed in strong acid (6N HCl) nothing happens. That is, the rate of loss of the ligand is unbelievably slow. Cabbiness and Margerum (15) have measured rates of formation and rates of dissociation of copper(II) complexes of linear tetradentate and closely related macrocyclic complexes. Table II shows that the formation rates are retarded by a few orders of magnitude for macrocyclic complexes; however, the greatest effect occurs for the dissociation rate (Table III). The copper complex of the closely related linear tetradentate ligand dissociates some 10^7 times faster than does the macrocyclic complex. The huge effect on the dissociation rate is readily illustrated for the case of the nickel(II) complexes. The usual mechanism of substitution at nickel(II) involves bond breaking in the rate-determining step. As Figure 8 shows, this occurs at an end group in a polydentate derivative. In acidic media, the dissociated groups are protonated quickly and the vacated site in the coordination sphere of the nickel atom is filled rapidly by solvent. A second atom then dissociates and the entire ligand is replaced by solvent in a sequence of S_N1 steps. From the structure of a ring, a simple dissociative step cannot occur because the ring has no end. It is not possible to extend the metal–nitrogen distance sufficiently to constitute bond rupture without additional bond rupture involving the ligand or extensive rearrangement within the co-

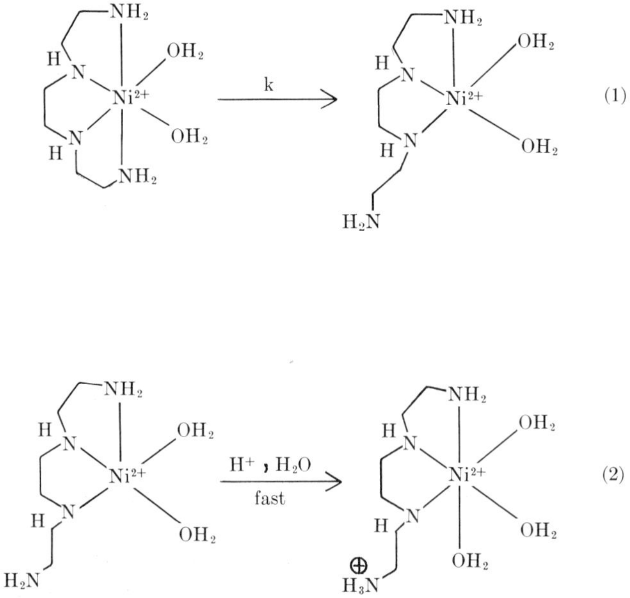

Figure 8. First step of removal of a linear polyamine from Ni^{2+} in acidic media

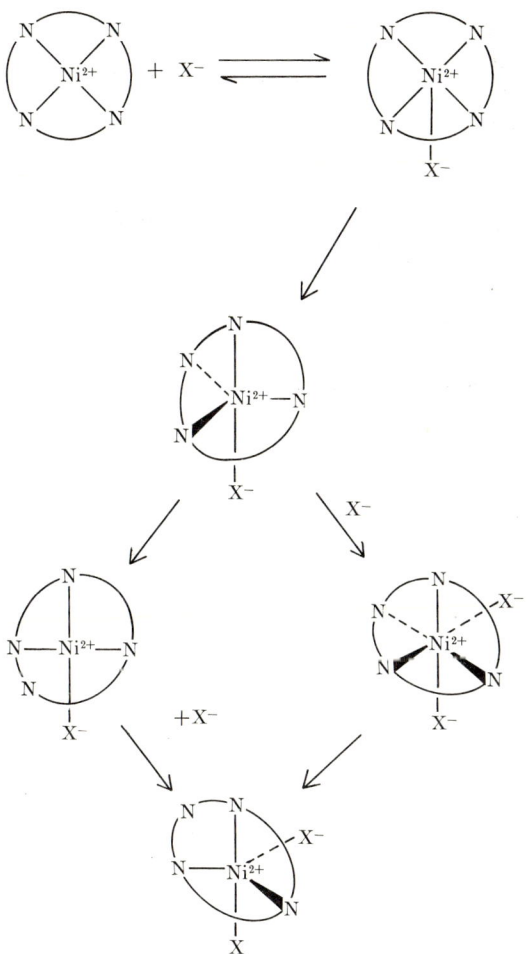

Figure 9. Model for dissociation of macrocyclic complexes

ordination sphere. Figure 9 suggests the general features of displacement of a macrocycle from nickel(II). These processes appear to involve a nucleophile, and therefore their rate processes may resemble bimolecular reactions. The scheme given reflects what is probably the other principal distinguishing feature of such reactions for flexible macrocycles. The ring almost certainly folds before the first bond between itself and the nickel atom is broken. It is relatively simple to move a donor atom away from the metal ion when the ring is in a folded configuration.

Although no detailed model was suggested, Cabbiness and Margerum (15) have referred to the sluggishness of these processes and the attendant stabilization of the macrocyclic complexes as the "macrocyclic effect." It is our conviction that the phenomenon is well illustrated by, but not unique to, macrocyclic ligands. It is a simple matter to illustrate the fact that the metal ion need not be completely encircled by a ring in order to display this phenomenon. The tridentate cyclic ligand trisanhydro-o-aminobenzaldehyde (TRI, Figure 10) coordinates to a single face of an octahedron. The metal ion is not surrounded by the ligand, but wears it like a crown (*16*, *17*). Despite this fact, the complex is extremely inert toward substitution, as witnessed by resolution of salts of the cation Ni(TRI)(H$_2$O)$_3{}^{2+}$ into optical isomers (*18*). Further, these optical isomers show no signs of racemizing, being optically stable in solution for weeks. It is suggested that the inertness occurs when some minimum number of donor atoms, possibly three, is so arrayed that the usual mechanism for stepwise dissociation of the ligand is thwarted. That is, extensive rearrangements must occur, or an entirely different path must be followed for ligand removal. Since the phenomenon arises mainly because of the fixed geometric placement of a set of ligating functions, it is called multiple juxtapositional fixedness.

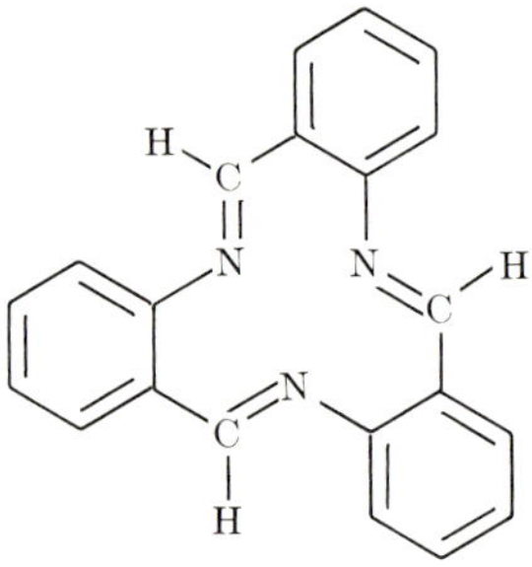

Figure 10. Tridentate ligand, trisanhydro-o-aminobenzaldehyde (TRI)

Since the constraints on mechanism act more severely with regard to dissociation than formation, MJF has strong thermodynamic consequences—the stability constants for macrocyclic and other MJF-affected complexes should be abnormally high. MJF is probably of considerable importance in metal binding to enzymes, for the locations of the donor atoms may be so fixed in space that they cannot be removed from a metal ion in a stepwise fashion without greatly altering the protein conformation; *i.e.*, without denaturing the enzyme.

There is a further consequence of this phenomenon that must be considered. Vallee and Williams (*19*) have suggested an entatic state for metal ions in active sites on enzymes. The state of binding of the metal ion should be such that its energy level contributes at par (or better) to an energetically poised domain. For enzymatic catalysis of nucleophilic processes, this could be accomplished most obviously by binding the metal ion in an extremely electron-poor condition. This suggests weak bonds (poor donor groups or usual donor groups and long bond distances). This seems acceptable but there is a problem. The binding of metal ions to enzymes is not notably weaker than that to simple ligands having comparable ligating groups. In view of the frugality of nature, it is difficult to believe that she has not chosen to use the entatic principle, but it appears that she would have to be doing so without the requisite weakening in the binding of the metal ion to an enzyme. A simple answer can be suggested by invoking MJF and entasis as opposing factors as regards the binding of metal ions. The extra stabilization attributed to MJF may serve to offset the weak bonds required by entasis (Table IV).

Table IV. Possible Role of Multiple Juxtapositional Fixedness in Enzyme Binding of Metal Ions

MJF ⟶ { a) Greatly enhanced binding / b) Normal electron density } ⟶ Strong binding and

Bond strain ⟶ { a) Greatly weakened binding / b) Reduced electron density } ⟶ Great electrophilicity

Effects of Ring Size and Flexibility

Though the porphyrin ring does ruffle a bit around the edges, the size of the site that accepts the metal ion remains very nearly constant for a wide range of compounds (Table V) (20, 21). This inflexibility is associated with the aromatic character of the ring. It is useful to compare the porphyrin ring with a related macrocycle that was introduced above, TAAB, Figure 11. This ligand has a metal ion site that is contained in a sixteen-membered ring that constitutes an alternating nonaromatic structure. It is, in fact, tetrabenzotetraaza-16-annulene, and though the number of members in the inner ring is the same as in porphyrin, a very different set of properties is associated with its structure. Although the metal ion and its donors occupy a single plane, the crystal structure determination reveals a strongly nonplanar arrangement for the rest of the structure (Figure 12) (22). The relative flexibility of this structure can be seen from the data of Table VI. The first two entries

Table V. Porphyrin

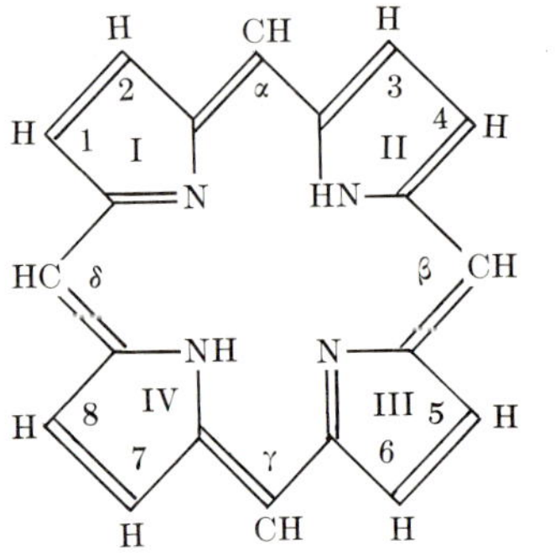

	1	2	3
Aetioporphyrin	CH_3	C_2H_5	CH_3
Deuteroporphyrin	CH_3	H	CH_3
Haematoporphyrin	CH_3	$CH(OH)CH_3$	CH_3
Protoporphyrin[b]	CH_3	$CH=CH_2$	CH_3
Mesoporphyrin	CH_3	C_2H_5	CH_3
Coporphyrin[b]	CH_3	$(CH_2)_2 \cdot COOH$	CH_3
Uroporphyrin[b]	$CH_2 \cdot COOH$	$(CH_2)_2 \cdot COOH$	$CH_2 \cdot COOH$

[a]Taken from "Thorpe's Dictionary of Applied Chemistry," Vol. X, Longmans, Green & Co., London, 1950.
[b]Naturally-occuring porphyrins.

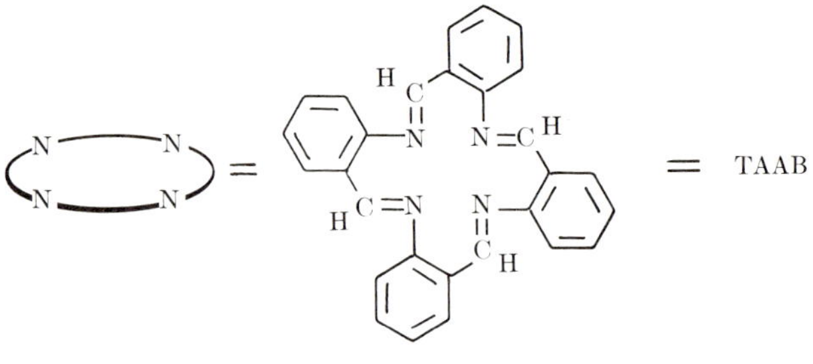

Figure 11. Tetrameric ligand, TAAB

represent examples of high-spin and low-spin nickel(II), both in the center of the TAAB macrocycle. The high-spin nickel ion is 0.16Å, or some 8%, larger than the low-spin nickel ion. In contrast, in the porphyrin complex, the Fe^{III}–N distance of high-spin iron(III) is only some 0.08Å greater than that of low-spin iron(III); further, this increase comes about mainly by expulsion of the high-spin atom about 0.5Å out of the plane of the four porphyrin nitrogens. Indeed, the distance between the center of the porphyrin and the center of one of its nitrogen atoms changes no more than 0.02Å. These differences should be kept in mind

Compounds[a]

	4	5	6	7	8
	C_2H_5	CH_3	C_2H_5	C_2H_5	CH_3
	H	CH_3	$(CH_2)_2 \cdot COOH$	$(CH_2)_2 \cdot COOH$	CH_3
	$CH(OH)CH_3$	CH_3	$(CH_2)_2 \cdot COOH$	$(CH_2)_2 \cdot COOH$	CH_3
	$CH=CH_2$	CH_3	$(CH_2)_2 \cdot COOH$	$(CH_2)_2 \cdot COOH$	CH_3
	C_2H_5	CH_3	$(CH_2)_2 \cdot COOH$	$(CH_2)_2 \cdot COOH$	CH_3
	$(CH_2)_2 \cdot COOH$	CH_3	$(CH_2)_2 \cdot COOH$	$(CH_2)_2 \cdot COOH$	CH_3
	$(CH_2)_2 \cdot COOH$	$CH_2 \cdot COOH$	$(CH_2)_2 \cdot COOH$	$(CH_2)_2 \cdot COOH$	CH_3

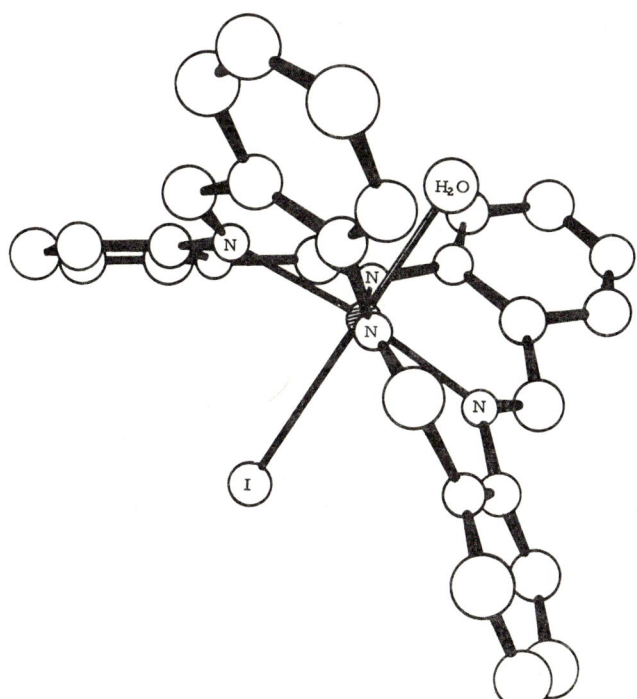

Figure 12. Perspective view of TAAB in $Ni(TAAB)I_2 \cdot H_2O$

Table VI. Metal–Nitrogen Distances in Tetradentate Macrocyclic Complexes

M^{n+}	L	D_{M-N}	Comments
Ni^{2+}	TAAB	1.90	low spin, X = BF_4
Ni^{2+}	TAAB	2.06	high spin, X = I
Ni^{2+}	Porphyrin	1.96	low spin
Pd^{2+}	Porphyrin	2.00	low spin
Fe^{3+}	Porphyrin	2.07	high spin, a = 0.5A
Fe^{3+}	Porphyrin	1.99	low spin (bisimidazole) a = 0

[a] Ref. 20, 23.

for a section to follow on the oxidizability of porphyrins and the probable natures of the oxidation products. As has been indicated above, the two-electron oxidation product of porphyrin should have much in common with TAAB.

Those macrocycles that are quite flexible readily fold when presented with metal ions that are too large for them to encompass. As Figure 13 shows, the critical ring sizes occur at different values for different sets of donor atoms. In the case of four nitrogen donors in a fully saturated ring, 13 members will surround a first row transition element ion while a 12-membered ring must fold (24). However, with the larger sulfur donors, a 14-membered ring is required to encircle the metal ion (25).

Some years ago, Brubaker (26) suggested a possible mechanical constrictive effect on metal ions because of the tight fit of an almost too large metal ion. The data of Table VII provide some support for this view. The Dq value indicating coordination within one plane of all the nitrogen donors of two molecules of dimethylethylenediamine is 1185 cm^{-1}. This may be considered a reasonable value for four such donors. However, 1,7-CTH, a closely related tetradentate macrocycle (30), also binds its four nitrogens in a single plane about the metal ion, and this ligand has been assigned a Dq value of 1426 cm^{-1}. If the values are essentially correct, the ring is exerting a much stronger ligand field strength than is anticipated. This may be the result of mechanical restriction which would shorten the Ni–N distance, for Dq is an inverse function of a large power of that distance.

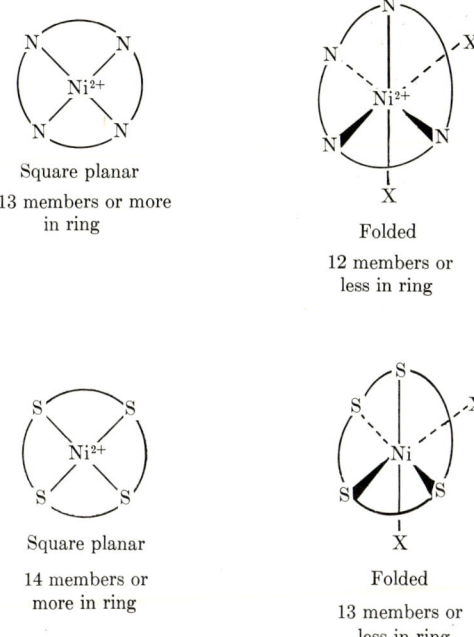

Figure 13. Relationship of ring size to folding for tetradentate macrocyclic complexes

Table VII. Spectral Parameters for Some *trans*-Diacido Complexes, $MA_4X_2^{n+}$

Complex	$\nu_1(kK)$	$\nu_2(kK)$	$Dq^{xy}(cm^{-1})$	$Dq^z(cm^{-1})$
$Co(NH_3)_4Cl_2^+$ [a]	15.90	21.00	2490	1461
$Co(en)_2Br_2^+$ [a]	15.21	21.68	2530	1277
$Ni(Dimen)_2(NO_3)_2$ [b]	7.84	11.85	1185	383
$Ni(Etu)_4Cl_2$ [b]	7.80	8.55	855	705
$Ni(m$-$1,7$-$CTH)Cl_2$ [c]	8.49	14.73	1426	420

[a] Ref. *27*.
[b] Ref. *28*.
[c] Ref. *29*.

The Oxidation-Reduction Properties of Porphyrins and TAAB Complexes

Table VIII summarizes the idealized electron accounting associated with electrochemical reduction of the divalent metal ion complexes of TAAB (Figure 11). For each metal ion, a distinct set of one-electron reduction waves occurs (*31*). In methanol solution, these multiple one-electron processes are followed by a very complex poly-electron wave that is associated with hydrogenation of the azomethine functions of the ligand. From Table VIII, it is clear that the number of one-electron waves is such that the system acquires 42 electrons. This amounts to the eight d-electrons needed to fill the four low-lying d-orbitals on the metal ion

Table VIII. Idealized Electron Accounting for Final Reduction Products of $M^{II}(TAAB)^{2+}$

Complex	M^{II}	TAAB	Electrode	Total
$Cu(TAAB)^+$	9	32	1	42
$Ni(TAAB)^0$	8	32	2	42
$Co(TAAB)^-$	7	32	3	42

$TAAB^{2-}$ requires	$34e^-$	$Cu^{3+} + TAAB^{2-}$
d^8 sq. planar requires	$8e^-$	$Ni^{2+} + TAAB^{2-}$
	$42e^-$	$Co^+ + TAAB^{2-}$

Table IX. Products of the Electrode Reaction $M^{II}(TAAB)^{2+} + ne^- \rightarrow M(TAAB)^{(2-n)+}$

n	Compound	μ_{eff}		C	H	N	X	M
						Analyses		
1	$Cu(TAAB)PF_6$	~0	Calc.	54.15	3.25	9.02	4.99	10.23
			Found	53.53	3.29	9.04	4.84	9.96
1	$Ni(TAAB)ClO_4$	1.94	Calc.	58.9	3.53	9.82	6.21	10.2
			Found	59.2	3.24	9.96	6.50	10.1
2	$Ni(TAAB)$	0.68	Calc.	71.1	4.28	11.9	–	12.5
			Found	70.4	4.13	11.9	–	12.3
1	$Co(TAAB)ClO_4$	0.99	Calc.	58.9	3.53	9.82	6.21	–
			Found	59.6	3.79	9.96	6.71	–
2	$Co(TAAB)CH_3CN$	2.84	Calc.	70.3	4.52	13.7	–	11.5
			Found	69.7	4.43	14.4	–	11.6

Table X. $Co^{II}(TAAB)^{2+}$ and

Compound	I	
	E	Elec. Chg.
$Co^{II}(TAAB)(ClO_4)_2$	+0.58	−1
$Co(TAAB)$	+0.70	−1
$Co(TAAB)CH_3CN$	+0.72	−1

^{a} vs. Ag–AgNO$_3$ (0.1M) in CH$_3$CN.

in D_{4h} (or related) symmetry and a pi electron system containing 34 electrons. That is, the ligand has been reduced to $TAAB^{2-}$, an aromatic dianion.

Most of the electrode products expected to be formed via the one-electron processes have been synthesized by controlled potential electrolysis, isolated, and characterized. Table IX shows evidence for their existence as pure compounds. The one-electron product in the copper system is diamagnetic and has been formulated as $Cu^{III}(TAAB^{2-})^+$. Both products have been isolated from the nickel system. $Ni(TAAB)ClO_4$ contains one unpaired electron and may be either a Ni(III) derivative of $TAAB^{2-}$ or a Ni(II) complex of the radical anion $TAAB^-$. Ni(TAAB) is formulated as a Ni(II) complex of the dianion $TAAB^{2+}$. Only the first two reduction products have been isolated from the cobalt system. These are $Co(TAAB)ClO_4$ (essentially diamagnetic) and $Co(TAAB)CH_3CN$ with a moment that is slightly high for one unpaired electron. Table X presents some of the electrochemical data for the cobalt complexes. It is my purpose at the moment to point out only a single relationship. The oxidation waves for the two reduced products correspond well among the two compounds, but they differ greatly from the oxidation and reduction waves of the starting material. This and other information cause us to formulate this behavior as given in Equations 5–8.

Presumed Process upon Controlled Potential Electrolysis

$$Co^{II}(TAAB)^{2+} + e^- \xrightarrow{\text{electrode reaction}} Co^{I}(TAAB)^+ \quad (5)$$

$$Co^{I}(TAAB)^+ \xrightarrow{\text{rearrangement}} Co^{III}(TAAB^{2-})^+ \quad (6)$$

Probable Oxidation Processes for 1-Electron Reduction Product

$$Co^{III}(TAAB^{2-})^+ \xrightarrow{-e^-} [Co^{III}(TAAB^-)]^{2+} \xrightarrow{-e^-} [Co^{III}(TAAB)]^{3+} \quad (7)$$

Its Reduction Products

Reduction Waves[a]

II		III		IV	
E	Elec. Chg.	E	Elec. Chg.	E	Elec. Chg.
−0.87	+1	−1.23	+1	∼−1.85	∼+1
+0.25	−1	−1.06	+1	−2.00	+1
+0.25	−1	−0.92	−1	−1.78	+1

Probable Reduction Processes for 1-Electron Reduction Product

$$\text{Co}^{III}(\text{TAAB}^{2-})^+ \xrightarrow{+e^-} \text{Co}^{II}(\text{TAAB}^{2-}) \xrightarrow{+e^-} \text{Co}^{I}(\text{TAAB}^{2-})^- \qquad (8)$$

Following the addition of one electron to the cobalt atom in $\text{Co}^{II}\text{TAAB}^{2+}$, the product rearranges *via* intramolecular oxidation-reduction to produce the cobalt(III) complex of the reduced ligand, the dianion TAAB^{2-}. The electrode processes of this substance are then rationalized as shown in the figure. The structure of the dianion is that of an approximately planar aromatic dianion (Figure 14). It is a direct analog of a porphyrin, differing in the locations and kinds of fused rings but having the same inner great 16-membered ring. The most obvious result of the structural

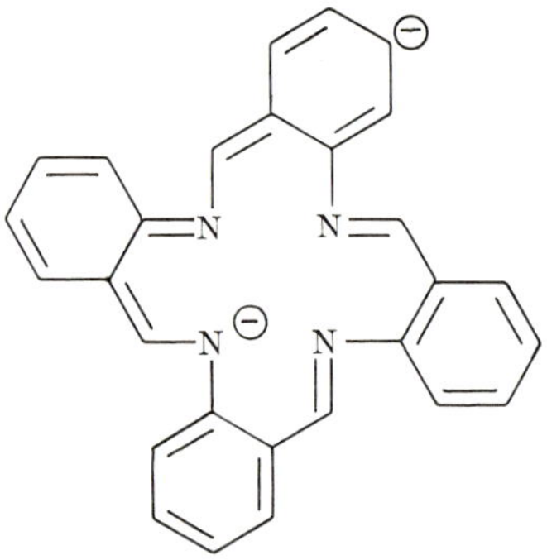

Figure 14. $(TAAB)^{2-}$ *dianion*

Table XI. Oxidation Potentials for Tetraphenylporphyrin Complexes: Benzonitrile vs. SCE, Cyclic Voltammetry[a]

$$[M(II)TPP] \xrightarrow{E_1} [M(II)TPP]^+ \xrightarrow{E_2} [M(II)TPP]^{2+}$$

M	E_1	E_2	μ_{eff}
H_2	1.00	1.20	–
Zn	0.79	1.10	–
Cu	0.99	1.33	2.88(E_1)

$$[M(II)TPP] \xrightarrow{E_1} [M(III)TPP]^+ \xrightarrow{E_2} [M(III)TPP]^{2+} \xrightarrow{E_3} [M(III)TPP]^{3+}$$

M	μ^0	E_1	μ_1	E_2	μ_2	E_3
Ni	–	1.00	–	1.00	–	1.40
Co	–	0.52	–	1.19	–	1.42
Fe	0.0	–0.32	5.10	1.18	2.71	1.50

[a] Ref. *32*.

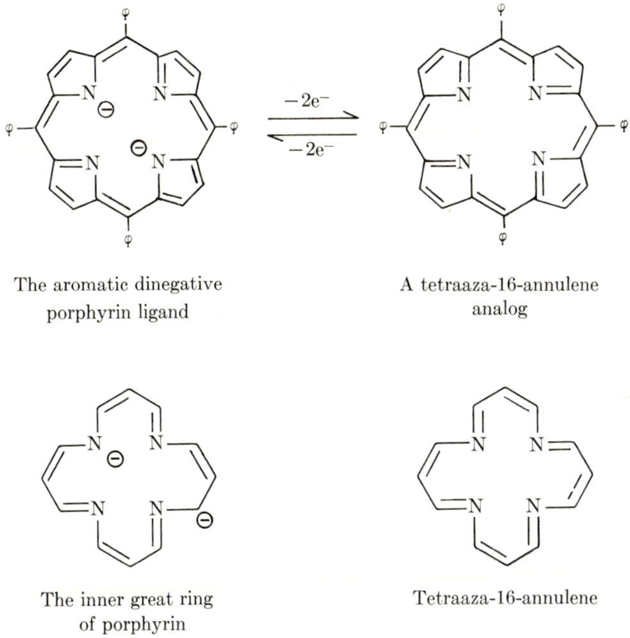

The aromatic dinegative
porphyrin ligand

A tetraaza-16-annulene
analog

The inner great ring
of porphyrin

Tetraaza-16-annulene

Figure 15. Porphyrin and annulene ring systems

differences is the very much greater reducing strength of TAAB^{2-} as compared with the porphyrin dianion.

To continue this comparison, the most recent work (32) on the electrochemical oxidation of complexes of tetraphenylporphyrin is summarized in Table XI. These materials suffer two or three one-electron oxidation steps terminating in the formation of the metal ion complex of the neutral 2-electron oxidation product of the ligand. As shown in Figure 15, the 2-electron oxidation product is closely related to the annulenes; *i.e.*, it contains the same 16-membered nonaromatic alternating heterocyclic hydrocarbon as its great ring as does TAAB. In this state, the ligand should be much more flexible, have a variable-sized metal ion site and, in general, be quite different from the parent dianion. In fact, one should look to the chemistry of TAAB complexes in order to anticipate the chemical behavior of these oxidized products.

A chemical property of M(TAAB)$^{n+}$ that deserves special consideration is the addition of nucleophiles to the carbon atom of the azomethine group (33, 34). Equations 9–13 illustrate these addition reactions while the structure of a typical product is given in Figure 16.

$$\text{Ni(TAAB)(BF}_4)_2 + \text{OR}^- \xrightarrow{\text{ROH}} \text{Ni(TAAB)(OR)}_2 \quad (9)$$

$$\text{Ni(TAAB)(BF}_4)_2 + \text{NH}_4\text{OH} \xrightarrow{\text{H}_2\text{O}} \text{Ni(TAAB)(NH}_2)_2 \quad (10)$$

$$\text{Ni(TAAB)(BF}_4)_2 + \text{OH}^- \xrightarrow{\text{H}_2\text{O}} \text{Ni(TAAB)(OH)}_2 \quad (11)$$

$$\text{Ni(TAAB)(BF}_4)_2 + \text{OH}^- \xrightarrow{dmf} \text{Ni(TAAB)(NMe}_2)_2 \cdot x\ dmf \quad (12)$$

$$\text{Ni(TAAB)(BF}_4)_2 + \text{Me}_2\text{NH} \xrightarrow{\text{H}_2\text{O}} \text{Ni(TAAB)(NMe}_2)_2 \quad (13)$$

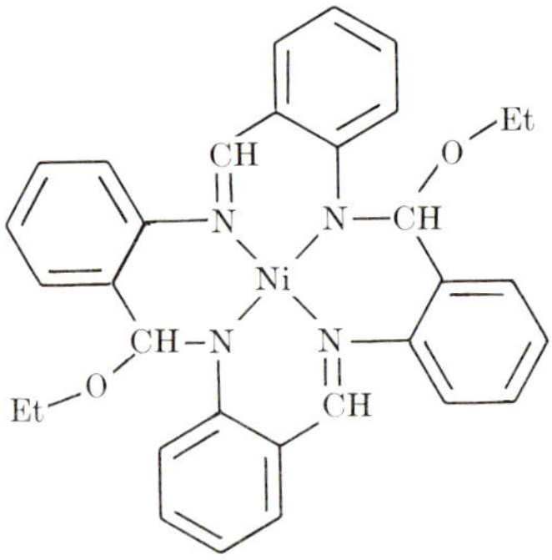

Figure 16. Structure of the product of addition reactions of Ni(TAAB)$^{2+}$ complexes

Very recently the mono-methoxide derivative of 2-electron oxidized zinc tetraphenylporphyrin has been reported (35).

The assignment of structures to the compounds that have been identified in peroxidase and catalase systems presents a strong challenge. Admittedly, the ultimate goal of studying such systems must be the understanding of mechanisms, and it may or may not be necessary to understand the chemistry of the prosthetic group in order to understand the mechanism. From the orientation of the inorganic field, the structure problem would be tackled first since it is the static problem. Mechanisms are difficult enough even when the structural chemistry is clear. In the brief section to follow, we discuss probable structures for the compounds formed by the prosthetic group in peroxidase. No comment is offered at this time on the mechanism of action of these enzymes.

Figure 17 summarizes the compounds derived from peroxidase after Yamazaki and Yokota (36). The iron in peroxidase is trivalent and its

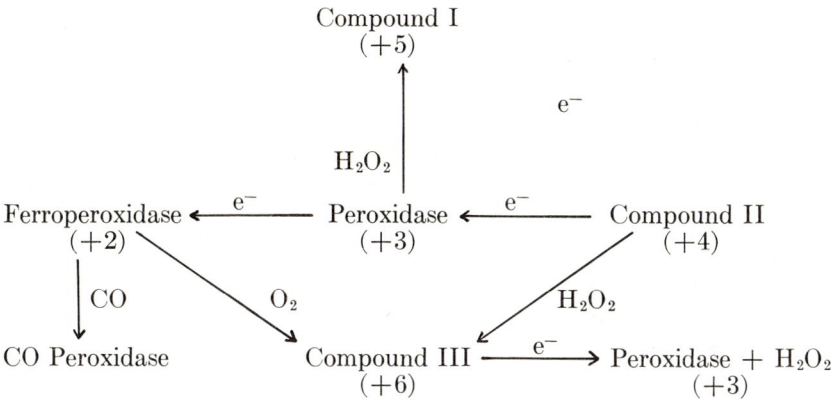

Figure 17. Peroxidase derivatives

interaction with hydrogen peroxide very quickly generates green Compound I, which contains two additional oxidizing equivalents and is written for convenience as (+5). From what has been said above of two-electron oxidation of porphyrins, one should immediately conclude that the ring has experienced 2-electron oxidation, and that in Compound I, the least possible transformation is to the annulene-like state. The spectrum of Compound I is distinctive when compared with Compounds II and III; therefore, the state of either the metal ion or the ligand must differ from those in the reference compounds. Another point that requires emphasis at this stage relates to the comparison of spectra with model compounds. The annulene-like material should be only a weak chromophore as compared with the porphyrin. For this reason and in view of the rather large electrode potentials in these systems, the spectra should depend rather strongly on the metal ion. Though the spectra of Wolberg and Manassen do not strongly support this view, one can only wonder how much sharper the spectra of the purified materials might be. The large extinction coefficients and general characteristics of the absorption bands of Compounds I, II, and III (37) are much more consistent with structures within which nucleophiles have added to the most electrophilic sites of the oxidized ligand (33, 34, 35). The general characteristics of the spectra are far more likely to be preserved in going from metal to metal and from substituent to substituent than are the specific features. It is clear from the spectral properties that addition of nucleophiles to the carbon atoms of these coordinated tetraaza-16-annulenes generates new aromatic systems. Consequently, it is useful to suggest that a nucleophile has added to the oxidized ligand in Compound I. The nucleophile could be a proximal function from the protein in the enzyme.

It might also be the OH⁻ produced by reduction of H_2O_2. A scheme is given (Figure 18) which surprisingly thoroughly accounts for the structures of all the peroxidase compounds and the reactions that convert them. Again it must be emphasized that no attempt is being made to discuss the mechanism of enzyme action and that these structures are presented as deserving equal consideration to that accorded the earlier unsubstantiated suggestions. In the suggested scheme, Compound I has the solvolyzed structure given. It is converted to Compound II merely by one-electron reduction. Compound II is formulated therefore as an iron(II) derivative having the same isoporphyrin ligand as Compound I.

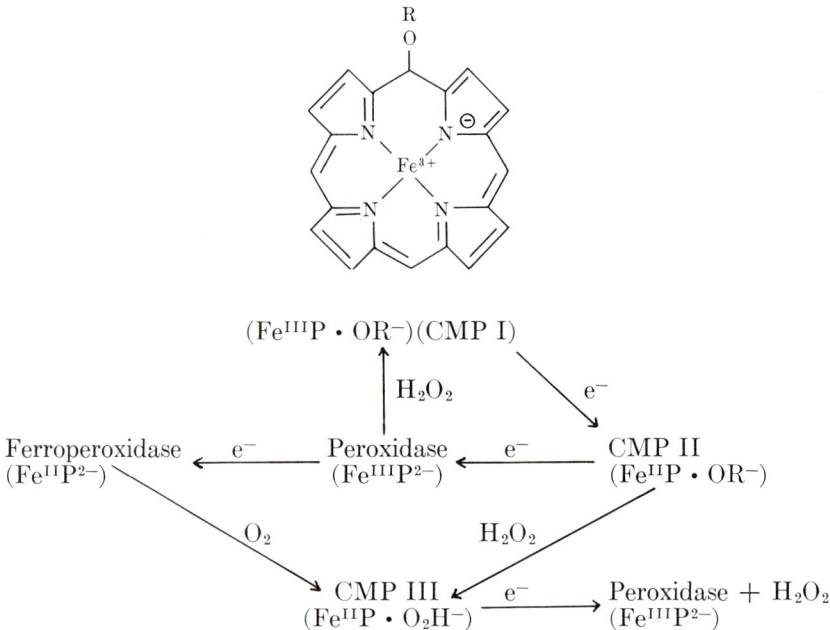

Figure 18. *Reaction scheme for conversion of peroxidase compounds*

This is at variance with the popular view that Compound II is an iron(IV) porphyrin (*38, 39*). This view is not supported by electrochemical studies or model compounds, for example tetraphenylporphyrin (Wolberg and Manassen, *32*), where it is shown that the ligand, not the iron(III), is oxidized. To be sure, the present proposal also deviates from simple use of the electrochemical results; however, in this hypothesis, though the oxidizing equivalents reside on the ring, the addition of nucleophiles would modify the potentials considerably. Having worked a great deal with magnetic susceptibilities, the author is aware of the uncertainties of these values when determined under other than ideal conditions. Consequently, these data are not considered here. Further, the values reported for Compounds I and II are easily adapted to almost any theory.

Certain relationships between Compounds II and III actually provide the main impetus for asking that some thought be given to the structural scheme presented here. Compound III is formed from O_2 and ferroperoxidase ($Fe^{II}P^{2-}$). In the scheme presented here, the ligand undergoes simultaneous 2-electron oxidation and nucleophilic addition of the HO_2^- ion that is an artifact of O_2 reduction. This produces a compound having a structure very close to that of Compound II, differing only in that HO_2^-, and not (presumably) OH^-, is added to a ligand carbon atom. The total oxidation equivalents of O_2 are preserved in the structure without invoking absurd degrees of oxidation of both ligand and metal ion. It is particularly pleasing that Compound III can be prepared from Compound II by addition of H_2O_2. Since HO_2^- is an excellent nucleophile, it would be expected to displace OH^- from the isoporphyrin.

Compounds II and III react very similarly when subjected to 1-electron reduction. Both go to peroxidase; however, Compound III gives up a mole of H_2O_2. In the scheme given here, this is an easily understandable over-all reaction. Both compounds are iron(II) isoporphyrin derivatives. The addition of an electron occurs at the ring and the increased electron density leads to immediate dissociation of the added nucleophile. At this point, the system would contain the iron(II) derivative of the radical monoanion P^-. This is not stable but reverts (if it ever existed) to $Fe^{III}P^{2-}$, peroxidase. It appears that Compound III simply would have to exist as a metal complex of HO_2^- or H_2O_2 if one sticks to schemes based on iron(IV) or on iron(III) combined with radical anions. Perhaps this would not be sufficient change in structure to account for the shifts in wave length of the spectral bands in going from Compound II to Compound III. Since the transitions are mainly owing to the ligand, the incorporation of the electron-releasing HO_2^- into the ligand structure is consistent with observations. Finally, in the event iron(IV) is established as existing in these systems, the probability of nucleophilic ligand addition is high for the oxidized ligand derivative of so highly electrophilic a metal ion as Fe(IV).

Iron Complexes of the Synthetic Macrocycles

Though some difficulties persist, it has been possible to synthesize a large number of iron complexes with synthetic macrocycles. Brief attention will be given first to the derivatives of TAAB (40). As summarized in Figure 19, o-aminobenzaldehyde condenses in the presence of ferrous chloride to form a TAAB complex. Because of the ease of oxida-

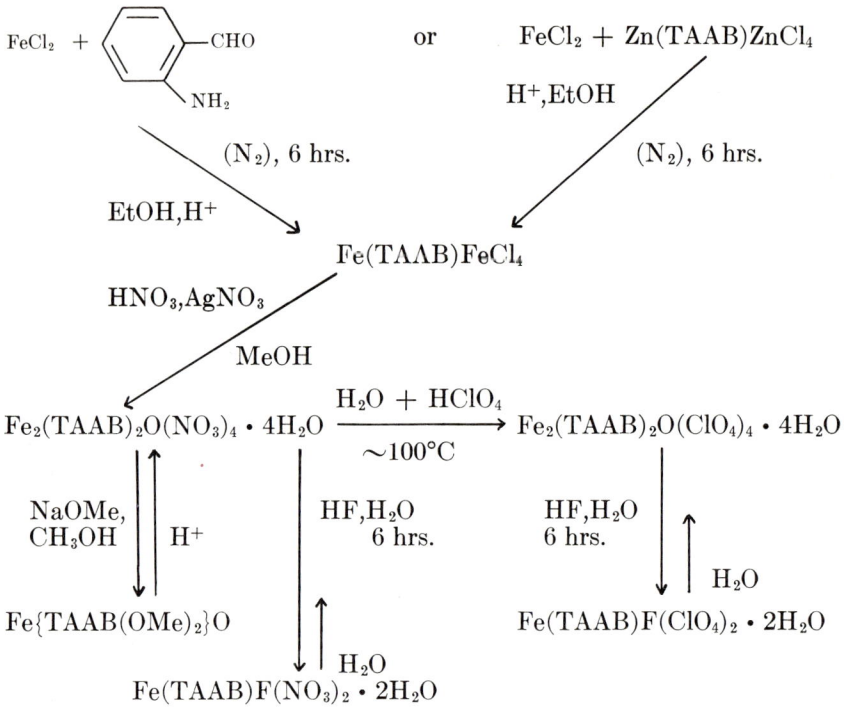

Figure 19. Iron complexes of the tetramer of aminobenzaldehyde

Table XII. Analysis of Iron–TAAB Dimer

$Fe_2[TAAB(OCH_3)_2]_2O$

	C	H	N	O	Fe	OCH_3	Mol Wt
Calcd.	66.91	4.88	10.41	7.43	10.37	11.51	1077
Found	67.05	4.96	10.21	7.30	10.28	10.78	1038, 1080

tion and the very great stability of the iron(III) oxo-bridged dimer $Fe_2(TAAB)_2O^{4+}$, that species is best characterized. Its magnetic properties are typical of spin–spin coupled dimers containing Fe(III) of $S = 5/2$. However, the bridge is quite unusual in another regard. It is exceptionally difficult to break. It has been cleaved only by HF after extended treatment—it is like dissolving glass. The stability of the oxo bridge may be taken as evidence for the great reactivity of the iron(III) complex. From the standpoint of models for the oxidized derivatives of peroxidase, the ethoxide derivative of the dimer is most interesting. Data characterizing this species is presented in Table XII.

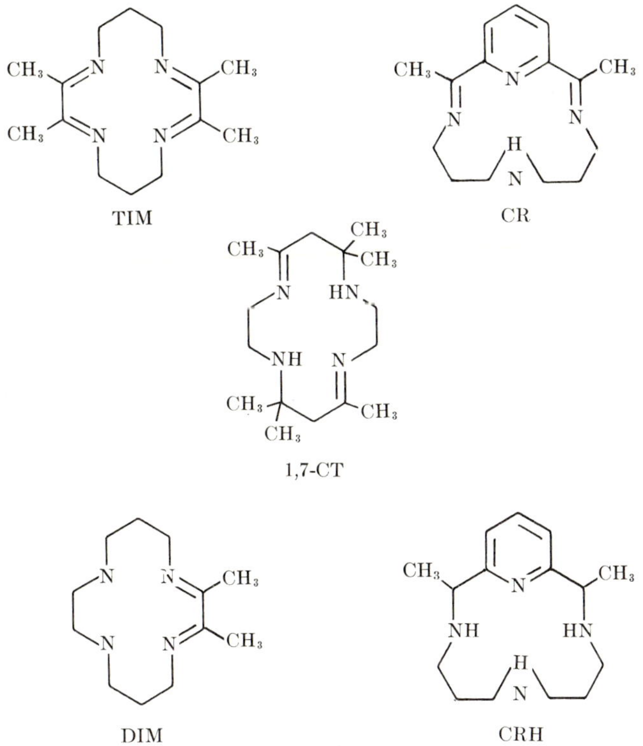

Figure 20. Structures of macrocyclic complexes with iron and other metal ions

A variety of other macrocycles have been used in the synthesis of iron and other transition element compounds. The structures of some of these are presented in Figure 20. The abbreviations presented herein are used in the following discussion. Although iron complexes with several of these ligands have been prepared and characterized, only those of the cyclic diene 1,7-CT will be considered here (41). The several

Table XIII. Iron Complexes of Curtis' Diene

Preparation

a) $Fe(OAc)_2 + 1,7\text{-}CT \cdot 2HX \xrightarrow[24\text{–}36 \text{ hrs.}]{CH_3CN} [Fe(1,7\text{-}CT)(CH_3CN)_2]X_2$

b) $Fe(ClO_4)_2 \cdot 6H_2O + TEOF \xrightarrow[\text{reflux}]{CH_3CN} \xrightarrow[NEt_3]{1,7\text{-}CT\ 2HClO_4}$

$[Fe(1,7\text{-}CT)(CH_3CN)_2](ClO_4)_2$

Other Compounds

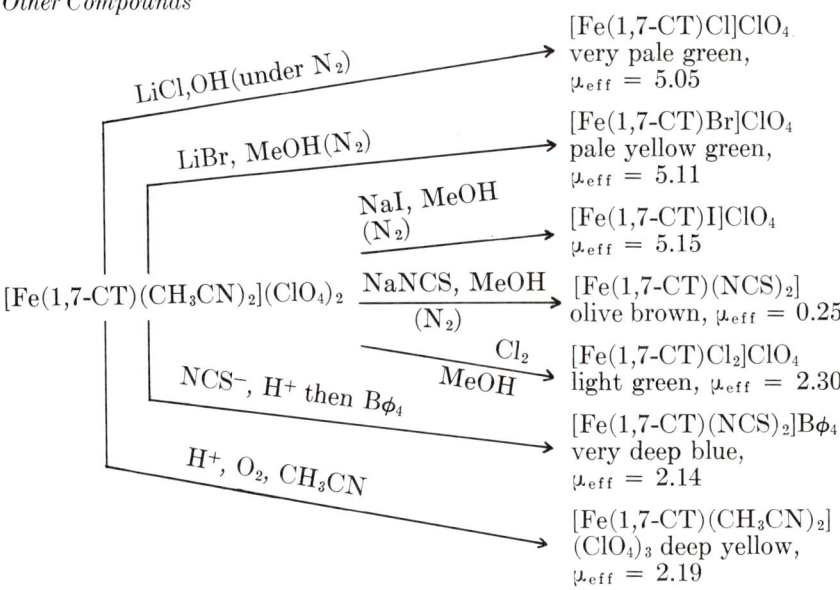

[Fe(1,7-CT)(CH₃CN)₂](ClO₄)₂ reacts as follows:

- LiCl, OH (under N₂) → [Fe(1,7-CT)Cl]ClO₄, very pale green, $\mu_{eff} = 5.05$
- LiBr, MeOH (N₂) → [Fe(1,7-CT)Br]ClO₄, pale yellow green, $\mu_{eff} = 5.11$
- NaI, MeOH (N₂) → [Fe(1,7-CT)I]ClO₄, $\mu_{eff} = 5.15$
- NaNCS, MeOH (N₂) → [Fe(1,7-CT)(NCS)₂] olive brown, $\mu_{eff} = 0.25$
- Cl₂, MeOH → [Fe(1,7-CT)Cl₂]ClO₄ light green, $\mu_{eff} = 2.30$
- NCS⁻, H⁺ then Bφ₄ → [Fe(1,7-CT)(NCS)₂]Bφ₄ very deep blue, $\mu_{eff} = 2.14$
- H⁺, O₂, CH₃CN → [Fe(1,7-CT)(CH₃CN)₂](ClO₄)₃ deep yellow, $\mu_{eff} = 2.19$

classes of complexes formed by this ligand are indicated in Table XIII. The five-coordinate, high-spin complexes Fe(1,7-CT)X⁺ constitute examples of a broad series of such structures that we find to be formed with a variety of macrocycles. They share their structural characteristics with deoxyhemoglobin, for that extremely important iron(II) compound is five-coordinate, high-spin, and surrounded by a tetradentate macrocycle having four nitrogen donors. We suggest that the structure of the natural product should not be considered unusual, for this is as common as any structure we have observed among Fe(II) complexes of tetradentate macrocycles. The thiocyanate derivative of Fe(1,7-CT)²⁺ is low-spin, diamagnetic, and six-coordinate, while all of the Fe(III) complexes are low-spin and six-coordinate. The reaction of Fe(1,7-CT)²⁺ with oxygen is complicated and interesting (Equations 14, 15).

$$[Fe^{II}(1,7\text{-}CT)(CH_3CN)_2]^{2+} \xrightarrow[CH_3CN]{H^+,\ O_2} [Fe^{III}(1,7\text{-}CT)(CH_3CN_2]^{3+} \quad (14)$$

pink   yellow, $\mu_{eff} = 2.2$

$$[Fe^{III}(1,7\text{-}CT)(CH_3CN)_2]^{3+} \xrightarrow[\substack{CH_3CN\\24\ hrs}]{O_2,\ H^+} [Fe^{II}(1,4,7,11\text{-}CT)(CH_3CN)_2]^{2+} \quad (15)$$

                                               red, $\mu_{eff} \sim 0.2$

The Fe(II) complex is very quickly oxidized to the corresponding Fe(III) derivative. However, extended exposure to the air is followed by oxidative dehydrogenation of the diene to a tetraene. There is evidence for formation of an oxygen adduct during the early phases of this process. The reaction has been put to novel use (Equations 16–19).

$$Co^{III}(1,7\text{-}CT)X_2^+ \xrightarrow{(O)} N.R. \quad (16)$$

$$Fe^{II}(1,7\text{-}CT)(CH_3CN)_2^{2+} \xrightarrow{[O]} Fe^{II}(1,4,7,11\text{-}CT)(CH_3CN)_2^{2+} \quad (17)$$

$$Fe^{II}(1,4,7,11\text{-}CT)(CH_3CN)_2^{2+} + 3\ phen \xrightarrow[24\ hrs]{CH_3CN} Fe(phen)_3^{2+} + 1,4,7,11\text{-}CT \quad (18)$$

$$1,4,7,11\text{-}CT + CoBr_2 \xrightarrow{CH_3CN} \xrightarrow[ClO_4^-,\ O_2]{HBr} [Co(1,4,7,11\text{-}CT)Br_2]ClO_4 \quad (19)$$

Attempts to prepare the tetraene complex of cobalt(III) by oxidation of the cobalt complex failed; however, Goedken has succeeded in removing the tetraene from iron and then placing it on cobalt(III).

New Vitamin B_{12} Coenzyme Models

We think of coenzyme B_{12} rather in the schematic way shown on the left in Figure 21—cobalt(III) in a tetradentate nitrogen-donor macrocycle with a Co–alkyl bond and a base in axial sites. Much interest resides in the determination of the characteristics of the macrocycles that lead to the stability of the cobalt–carbon bond. The burgeoning supply of new macrocycles presents the opportunity of studying such relationships, and Table XIV provides the first simple correlation of this kind (42, 43). Two ligands from the list gave stable Co–C bonds. One of these, TIM, contains four imines arranged in α-pairs while the other con-

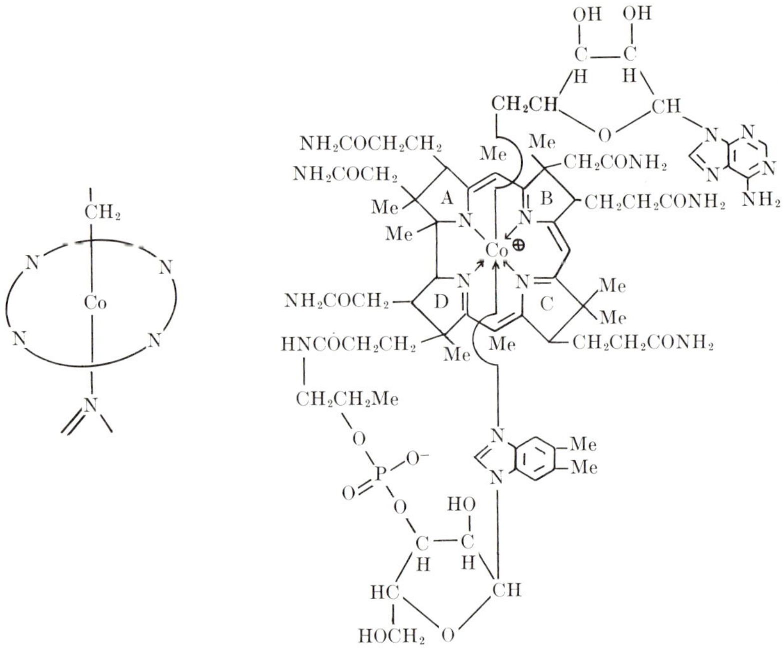

Figure 21. Schematic representation of coenzyme B_{12}

Table XIV. Crystal Field Parameters[a] of Tetragonal Dichloro-Cobalt(III) Complexes

Complex	$Dq^{xy}(cm^{-1})$	$Dt(cm^{-1})$	
[Co(TIM)Cl$_2$]$^+$	2800	766	Form cobalt alkyls
[Co(CR)Cl$_2$]$^+$	2830	785	
[Co(1,7-CT)Cl$_2$]$^+$	2640	674	
[Co(DIM)Cl$_2$]$^+$	2580	640	Do not form cobalt alkyls
[Co(β-CRH)Cl$_2$]$^+$	2580	640	
[Co(cyclam)Cl$_2$]$^+$	2480	594	

[a] Values obtained using the formulae (27)
$Dt = 4/35 (10 Dq^{xy} - \nu_E - C)$
$Dq^z = (Dq^{xy} - 7/4\ Dt)$
assuming
$Dq^z(Cl^-) = 1460\ cm^{-1}\ C = 3800\ cm^{-1}$

Table XV. Formation of *trans*-Dialkylcobalt(III) Complexes of TIM and CR

$$[CH_3Co(TIM)I]BPh_4 \xrightarrow[\text{2) RX}]{\text{1) 2NaBH}_4} [R(CH_3)Co(TIM)]BPh_4$$

$$R = CH_3, C_6H_5CH_2$$

$$[C_6H_5CH_2Co(TIM)Br]BPh_4 \xrightarrow[\text{2) RX}]{\text{1) 2NaBH}_4} [R(C_6H_5CH_2)Co(TIM)]BPh_4$$

$$R = CH_3, C_6H_5CH_2$$

Entirely analogous reactions occur in the CR system.

NMR Changes

Complex	τ $(Co-CH_3)$
$[CH_3Co(TIM)I]BPh_4$	8.85
$[(CH_3)_2Co(TIM)]BPh_4$	9.64
$[CH_3Co(CR)Br]BPh_4$	9.47
$[(CH_3)_2Co(CR)]BPh_4$	9.82

tains two imines and a pyridine ring. The ligands which have failed to yield such derivatives in our hands contain two (DIM and 1,7-CT) or fewer unsaturated nitrogen atoms. The data of Table XIV also suggest that only the ligands of greatest coordinating strength (Dq^{xy}) are effective in this regard, and the ability to pi bond is probably of first importance. Preparation of the monoalkyl compounds proceeds smoothly in accord with classic procedures (Equations 20, 21) (*42, 43*).

$$\left.\begin{array}{l} [Co(CR)Br]Br \cdot H_2O + \\ NaBH_4 \\ \text{or} \\ [Co(TIM)X_2]^+ + \\ 2NaBH_4 \end{array}\right\} \xrightarrow[\text{(acetone)}]{\text{MeOH}} Co(I) \xrightarrow{RY} \begin{array}{l} RCo(CR)Br^+ \\ \text{or} \\ RCo(TIM)X^+ \end{array} \quad (20)$$

$$\left.\begin{array}{l} Co(II) \\ \text{or} \\ Co(III) \end{array}\right\} + Na/Hg\ (1\%) \xrightarrow{CH_3CN} Co(I) \xrightarrow{RY} RCo(MAC)(CH_3CN)^{2+} \quad (21)$$

More interestingly, series of *trans*-dialkyl cobalt(III) complexes are readily synthesized (Table XV). This represents only the second report of such compounds (44). The formation of these substances implies the existence of an intermediate nucleophile containing both cobalt(I) and a Co–C bond, for the second alkyl bond is formed by first reducing the cobalt(III)–monoalkyl complex and then letting it function as a nucleophile in displacing halide from an alkyl halide. Farmery (42) has suc-

Table XVI. Formation of Alkyl-Cobalt(I) Complexes of CR and TIM

(a) $[R\ Co(L)X]Y + 2\ NaBH_4 \xrightarrow{CH_3CN\ +\ MeOH} R\ Co(L)$
   (i) Product difficult to isolate
   (ii) Hydrogenation of ligand often occurs
   (iii) Very useful "*in situ*" synthesis

(b) $[R\ Co(L)X]Y + Na/Hg\ (1\ \%) \xrightarrow{CH_3CN} R\ Co(L)$
   (i) Quantitative reaction
   (ii) Product easily isolated

L = Macrocyclic Ligand   X = halide or neutral ligand

Y = BPh_4^-, ClO_4^-, PF_6^-

ceeded in isolating, purifying, and characterizing the first complexes of this kind $RCo^I(MAC)$ (Table XVI). The crystals of $[CH_3Co(TIM)]$ are red by reflected light but green by transmitted light. The complex is insensitive to light, stable indefinitely in an inert atmosphere, and it does not decompose below its melting point, 109°C. It is soluble in hydrocarbons, benzene, ether, alcohols, and other organic solvents and sublimes without decomposition at reduced pressure. It is notable that these alkyl cobalt(I) compounds are much more stable than those somewhat related Co(I) compounds, such as $CH_3Co(CO)_4$, that have previously been reported. This is a direct consequence of the inertness toward substitution in the planar positions that results from the presence of the macrocyclic ligands and the strong pi electron interactions between the metal ion and the imine nitrogens. Attention has not been directed toward the excellent studies by the groups led by Schrauzer and by Williams, Hill, and Pratt because of the firm conviction that these matters are to be discussed elsewhere in this volume.

Acknowledgment

The studies upon which this report has been based have been generously supported by grants from the National Institute of General

Medical Sciences of the U. S. Public Health Services, by the National Science Foundation, and by the Ohio State University. We are exceedingly grateful.

Literature Cited

(1) Curtis, N. F., *Coord. Chem. Rev.* (1968), 3, 3.
(2) Thompson, M. C., Busch, D. H., *Chem. Eng. News* (1962), Sept. 17, 57.
(3) Busch, D. H., "Alfred Werner Commemoration Volume," p. 174, Verlag Helv. Chimica Acta, Basel, 1967.
(4) Schrauzer, G. N., *Chem. Ber.* (1962), 95, 1438.
(5) Thierig, D., Umland, F., *Angew. Chem.* (1962), 74, 388.
(6) Johnson, A. W., Kay, I. T., *J. Chem. Soc.* (1961), 2418.
(7) Lindoy, L. F., Busch, D. H., "Preparative Inorganic Reactions," Jolly, Ed., Vol. 6, Interscience, New York, in press.
(8) Baldwin, D. A., Rose, N. J., *Abstr. 157th Meeting, ACS, Minneapolis, Minn., 1969*, Inor 020.
(9) Jager, E. G., *Z. Anorg. Allgem. Chem.* (1969), 364, 178.
(10) Melson, G. A., Busch, D. H., *J. Am. Chem. Soc.* (1964), 86, 4834.
(11) Curry, J. D., Busch, D. H., *J. Am. Chem. Soc.* (1964), 86, 592.
(12) Green, M., Tasker, P. A., *Chem. Commun.* (1968), 518.
(13) Barefield, E. K., thesis, the Ohio State University, 1969.
(14) Goedken, V., Busch, D. H., unpublished results.
(15) Cabbiness, D. K., Margerum, D. W., *J. Am. Chem. Soc.* (1970), 92, 2151.
(16) Melson, G. A., Busch, D. H., *J. Am. Chem. Soc.* (1965), 87, 1706.
(17) Fleischer, E. B., Klem, E., *Inorg. Chem.* (1965), 4, 637.
(18) Taylor, L. T., Busch, D. H., *J. Am. Chem. Soc.* (1967), 89, 5372.
(19) Vallee, B. L., Williams, R. J. P., *Proc. Natl. Acad. Sci.* (1968), 59, 498.
(20) Fleischer, E. B., *Accts. Chem. Res.* (1970), 3, 105.
(21) Hoard, J. L., "Hemes and Hemoproteins," Chance, Esterbrook, and Yonetani, Eds., p. 9 ff., Academic, New York, 1966.
(22) Hawkinson, S. W., Fleischer, E. B., *Inorg. Chem.* (1969), 8, 2402.
(23) Countryman, R., Collins, D. M., Hoard, J. L., *J. Am. Chem. Soc.* (1969), 91, 5166.
(24) Collman, J. P., Schneider, P. W., *Inorg. Chem.* (1966), 5, 1380.
(25) Rosen, W., Busch, D. H., *Inorg. Chem.* (1970), 9, 262.
(26) Brubaker, G. R., Busch, D. H., *Inorg. Chem.* (1966), 5, 2114.
(27) Wentworth, R. A. D., Piper, T. S., *Inorg. Chem.* (1965), 4, 709.
(28) Travis, Kenton, thesis, the Ohio State University, 1970.
(29) Lever, A. B. P., ADVAN. CHEM. SER. (1967), 62, 430.
(30) Rowley, D. A., Drago, R. S., *Inorg. Chem.* (1968), 7, 795.
(31) Tokel, N. E., Katovic, V., Farmery, K., Anderson, L. B., Busch, D. H., *J. Am. Chem. Soc.* (1970), 92, 400.
(32) Wolberg, A., Manassen, J., *J. Am. Chem. Soc.* (1970), 92, 2982.
(33) Katovic, V., Taylor, L. T., Busch, D. H., *J. Am. Chem. Soc.* (1969), 91, 2122.
(34) Taylor, L. T., Urbach, F. L., Busch, D. H., *J. Am. Chem. Soc.* (1969), 91, 1072.
(35) Dolphin, D., Felton, R. H., Borg, D. C., Fajer, J., *J. Am. Chem. Soc.* (1970), 92, 743.
(36) Yamazaki, I., Yokota, K., *Biochim. Biophys. Acta* (1965), 105, 301.
(37) Hayaishi, O., "The Oxygenases," p. 280, Academic, New York, 1962.
(38) George, P., *Biochem. J.* (1953), 55, 220.
(39) Brill, A. S., Sandberg, H. E., *Biochemistry* (1968), 7, 4254.

(40) Katovic, V., Busch, D. H., unpublished results.
(41) Goedken, V., Busch, D. H., unpublished results.
(42) Farmery, K., Busch, D. H., 2nd Central Regional Meeting, ACS, Columbus, Ohio, June 3–5, 1970.
(43) Ochiai, E., Long, K. M., Sperati, C. R., Busch, D. H., *J. Am. Chem. Soc.* (1969), **91,** 3201.
(44) Costa, G., Mestroni, G., Licari, T., Mestroni, E., *Inorg. Nucl. Chem. Letters* (1969), **5,** 561.

RECEIVED June 26, 1970.

Developments in Inorganic Models of N_2 Fixation

A. D. ALLEN

Department of Chemistry, University of Toronto, Toronto 5, Canada

> *This review covers most of the major developments reported during the past two years in the chemistry of stable transition metal–N_2 compounds, emphasizing the nature of the N_2 ligand and the remarkable range of compounds containing it, the wide variety of methods and reagents that lead to metal–N_2 compounds, and the properties and reactions of these compounds. The position of the N_2 stretching frequency of the bound N_2 is a good indication of the extent to which the N_2 molecule is modified by complexation, and possibly the extent to which it is activated toward further reaction. Known compounds in which N_2 is bound simultaneously to two transition metals are increasing in number and variety. This development provides the most promising lead so far toward the link between these compounds and biological nitrogen fixation.*

Since this symposium includes authorities on many of the chemical aspects of the fixation of nitrogen and its conversion to ammonia, this paper is restricted to a review of recent developments in the preparation, properties, and reactions of N_2 compounds of the transition metals, compounds that are stable enough to be isolated and characterized.

The connection between these compounds and biological nitrogen fixation is uncertain at the moment. Most people assume that N_2 is coordinated as such to a metal site at some stage in the biological process, and that it is activated, at least partly, by this coordination. Unfortunately, despite early reports (*1, 2, 3, 4*) to the contrary, no N_2 compound that is stable enough to be isolated has yet been reduced, or induced to undergo any reaction that can lead to ammonia or a derivative of ammonia. There is, however, still good reason to believe that an interface between these inorganic systems and the biological systems does exist and that we are slowly reaching the point where an understanding of the one will help in the understanding of the other.

The recent flurry of activity in this area by the inorganic chemists was started by the discovery at Toronto in 1965 of the ruthenium complex $[Ru(NH_3)_5N_2]^{2+}X_2$ (X = halide, BF_4^-, PF_6^-, etc.) (1, 2). The property that is now regarded as diagnostic for all such compounds is a strong, sharp band in the infrared spectrum at somewhere between 1900 and 2200 cm^{-1}. This is regarded as a good, almost sufficient, indication of the presence of an N_2 molecule linearly coordinated to a metal atom or ion (—M—N≡N). It is the one property that is always quoted, since there is some connection between the position of this band and the strength of the interaction between the N_2 and the metal. The N_2 stretching frequency (Raman) of the free N_2 molecule is at 2331 cm^{-1}, and the extent to which this frequency is lowered in the complex may be an indication of the extent of π-bonding from the metal to the ligand and of the degree to which the nitrogen molecule is activated toward further reaction. As is always the case, great care must be taken when comparing compounds of different over-all charge, crystal complexity, and physical state.

The probable unity that underlies this whole area of study is illustrated in Figure 1.

Here we see the group of elements that form "stable" N_2 complexes so far. There is no reason to believe that this is an exclusive group, but its position in the periodic table is significant for the following reasons:

(a) It coincides very closely with the borderline region between Class A and Class B metals according to Ahrland and Chatt (5) and includes those elements that show both Class A and Class B behavior.

(b) It coincides fairly well with the group of metals forming metal carbonyls, indicating a similarity between N_2 and CO.

(c) It contains those elements, Fe and Mo, that are most likely as active sites in biological N_2 fixation.

(d) It contains those elements, principally Fe, that are used as catalysts in the Haber synthesis of ammonia.

The mechanism of the Haber process reaction has been the subject of study and speculation for many years. Since about 1940, a mechanism proposed by Temkin (6, 7) has been most favored. This involves the

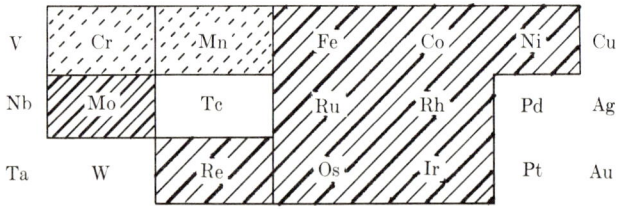

Figure 1. Periodic table of elements forming "stable" nitrogen complexes

adsorption of N_2 as atomic nitrogen on the metal surface. Recently, Carra and Ugo (8) argued that adsorption of hydrogen atoms and N_2 molecules on adjacent metal sites, followed by transfer of hydrogen to the terminal N atom, could explain equally well the kinetic results. This sort of mechanism, involving a double attack on the N_2 molecule, will be familiar to biochemists and is the sort of process that many inorganic chemists are investigating hopefully in their study of inorganic models of nitrogen fixation.

Table I. Some Simple Nitrogen Complexes and their N–N Stretching Frequencies, $\nu(N\equiv N)$

Complexes	$\nu(N\equiv N) cm^{-1}$
$[Os(NH_3)_5N_2]Cl_2$	2010
$[Re\ Cl\ (N_2)(PMe_2Ph)_4]$	1920
$[Mo\ N_2(PPh_3)_2 \cdot PhMe]$	2005
$[Co\ H(N_2)(PPh_3)_3]$	2082
$[Co\ (N_2)(PPh_3)_3]$	2088
$[Ir\ Cl(N_2)(PPh_3)_2]$	2095

Range of "Stable" N_2 Compounds

It was originally thought that the requirements for coordination of a nitrogen molecule would be extremely specific, and that small changes in the electronic state of the metal—*e.g.*, by changing the oxidation state or other ligands present—would destroy its capacity to bind N_2. There is good reason now to believe that this is not so, and that the nitrogen molecule is remarkably flexible in its requirements for the formation of a stable metal–N_2 bond. Table I shows a selection of well-characterized complexes illustrating:

 (a) compounds from four groups (Cr, Mn, Fe, Co) of the periodic table and all three transition series;

 (b) various formal oxidation states from 0 to +2;

 (c) various other ligands, ranging from ammonia to halide to hydride to phosphine;

 (d) various types of stereochemistry, including tetrahedral (?), square planar, trigonal bipyramid, and octahedral.

Table II gives examples of several compounds of Os(II) showing that osmium in a formal oxidation state of +2 can bind N_2 in the presence of a wide range of other ligands, NH_3, $NH_3 + N_2$, halide, alkyl and aryl phosphines. In such compounds, the presence of other ligands that can accept d electrons from the metal reduces the extent of metal-to-ligand π electron transfer in the metal–N_2 bond. This is indicated by an increase in the N_2 stretching frequency, other things being equal. A further example is $cis\text{-}[Os^{II}(NH_3)_4CO.N_2]^{2+}$ in which ν_{N_2} is raised to 2180 cm^{-1}

Table II. Some Nitrogen Complexes of Osmium (II)

Complexes	$\nu(N\equiv N)\,cm^{-1}$
$[Os(NH_3)_5N_2]Cl_2$	2010
$[Os(NH_3)_4(N_2)_2]Cl_2$	2120, 2175
$[Os\,Cl_2(N_2)(PEt_3)_3]$	2064
$[Os\,Br_2(N_2)(PMe_2Ph)_3]$	2090
$[Os\,H_2(N_2)(PEtPh_2)_3]$	2085

from 2010 cm^{-1} in the pentammine, as a result of the strong π electron accepting capacity of carbon monoxide.

In a few cases, two nitrogen compounds have been isolated that are identical except that the metal is in a different formal oxidation state. This loss (or gain) of a whole unit of electronic charge produces a drastic change in the electron density on the metal, but does not, in these cases, destroy the ability of the metal to bind N_2 to itself. Pairs of compounds of this type are shown in Table III. In both the rhenium and osmium cases, we see that an increase in oxidation state of the metal greatly increases the N_2 stretching frequency. This is consistent with a decrease in metal-to-ligand π electron transfer on going from the lower to the

Table III. Nitrogen Complexes of Metals in Different Oxidation States

Complexes	$\nu(N\equiv N)\,cm^{-1}$
$[Re(I)Cl(N_2)(diphos)_2]$	1980
$[Re(II)Cl(N_2)(diphos)_2]^+$	2060
$[Os(II)(NH_3)_5N_2]^{2+}$	2010
$[Os(III)(NH_3)_5N_2]^{3+}$	2140
$[Co(0)(N_2)(PPh_3)_3]$	2093
$[Co(I)H(N_2)(PPh_3)_3]$	2094

higher oxidation state. In the cobalt compounds shown, the change in oxidation state is accompanied by a change in coordination. From the fact that ν_{N_2} is almost the same in both the compounds shown, we can conclude that electronically, so far as the N_2 molecule is concerned, there is little difference between Co(0) and Co(I)H.

Formation of Metal–N_2 Compounds

We shall now consider the types of reaction that lead to the formation of metal–N_2 compounds, with particular emphasis on procedures that have been used to prepare new compounds during the past year or so, since the earlier results were covered in our 1968 review (9).

A surprising variety of reagents is capable of generating metal–N_2 complexes. Within six months of the first report of such a compound, most of the common reagents now used were known to be effective. In chronological order, the early reagents used were hydrazine hydrate (*1*), the azide ion (*2*), acyl azides (*10*), and nitrogen (N_2) (*11*). More recently, substituted hydrazines (*12, 13*) and sulfonyl azide (*14*) have been used, as well as some more complex systems that will be described later.

We shall consider first the reactions using N_2 as a reagent, since these are of obvious interest to this group, but it must be remembered that the more exotic reagents are potentially very important because they provide possible insight into the way bound N_2 might be attacked and brought into reaction—for example, by a simple reversal of the forward reaction path. The simplest reaction is the addition of molecular N_2 with an increase in the coordination number of the metal. Two examples of this type of reaction are (*15, 16*)

$$RuH_2(PPh_3)_3 + N_2 \rightarrow RuH_2(N_2)(PPh_3)_3 \quad \nu_{N_2} = 2147 \text{ cm}^{-1}$$

$$FeH_2(PR_3)_3 + N_2 \rightarrow FeH_2(N_2)(PR_3)_3 \quad \nu_{N_2} \simeq 2060 \text{ cm}^{-1}$$

$$(PR_3 = PEtPh_2, PBuPh_2)$$

Harris and coworkers at Toronto have prepared the dihydridotristriphenylphosphineruthenium (II) used as the starting material in the first example and have found that it reacts rapidly, even in the solid state at room temperature, with gaseous N_2 (and similarly with carbon monoxide) to give the simple addition product, preparation of which had been reported earlier (*17*). The reaction of the iron compound is similar but the analogous osmium–N_2 compound, which is known (*14*), cannot be prepared in this way. We may note in passing that ν_{N_2} for the osmium compound is at 2085 cm⁻¹, between that of the iron and ruthenium compounds. This type of sequence, in which properties of compounds of metals in the second transition series do not lie between those of analogous compounds of the first and third transition series, is not uncommon.

The next type of reaction to be considered is the simple nucleophilic displacement of a coordinated ligand by N_2. Two examples of reactions in which N_2 replaces water on ruthenium (II) are (*11, 18*)

$$[Ru(NH_3)_5H_2O]^{2+} + N_2 \rightarrow [Ru(NH_3)_5N_2]^{2+} + H_2O$$
$$\nu_{N_2} = 2100\text{–}2150 \text{ cm}^{-1}$$

$$[Ru(H_2O)_6]^{2+} + N_2 \rightarrow [Ru(H_2O)_5N_2]^{2+} + H_2O \quad \nu_{N_2} = 2125 \text{ cm}^{-1}$$

The first reaction shown was the first report that N_2 itself can act as a reagent in these reactions. There is no evidence that N_2 can displace more strongly bound ligands, such as ammonia, from ruthenium (II), but it is a more effective reagent towards this center than, for example, halide ions, indicating a remarkable degree of reactivity for so inert a molecule. The penta-aquo compound of ruthenium (II) was recently isolated by Creutz and Taube and details are not yet published, but it appears to be well characterized. We had observed this ion in solution, but were unable to isolate it.

There have been several reports of reactions in which N_2 replaces ligands other than water. The mechanism of these reactions is unknown, so that it is not clear whether we are dealing with an S_N1 or S_N2 type of reaction. In the reaction (19)

$$FeH.Cl(depe)_2 + NaBPh_4 \xrightarrow[\text{acetone}]{+N_2} [FeH(N_2)(depe)_2]^+ BPh_4^- + NaCl$$
$$\nu_{N_2} = 2090 \text{ cm}^{-1}$$

it seems likely that the driving force is the removal of chloride ion as the insoluble sodium chloride and that N_2 occupies the site left vacant by the departing ligand. The reaction (20)

$$CoH_2(PPh_3)_3 + N_2 \rightleftarrows Co(N_2)(PPh_3)_3 + H_2 \quad \nu_{N_2} = 2094 \text{ cm}^{-1}$$

is an example of the direct reversible replacement of hydrogen by nitrogen. It may be that this latest report by Marko and Speier helps to clear up a long-standing problem concerning these cobalt compounds. In our earlier review (9), we described the situation as it appeared in 1968. The existence of the compound $Co(PPh_3)_3H \cdot N_2$ had been confirmed by x-ray analysis (21), but the compound $Co(PPh_3)_3 \cdot N_2$ was in some doubt. The new development here is the preparation and use of $Co(PPh_3)_3 \cdot H_2$ as a starting material. As we shall see later, this N_2 compound has also been prepared by another route. It is reported to be paramagnetic, as required by a simple monomeric structure, and quite stable in air. Final confirmation of the structure of this compound must await x-ray analysis. An interesting feature of this reaction is that the dihydride appears to react preferentially with N_2 rather than O_2 on standing in air. We had observed and reported somewhat similar behavior with aqueous solutions of $[Ru(NH_3)_5H_2O]^{2+}$ (22). It seems that, as would be expected, N_2 as a ligand helps to stabilize low oxidation states, so that once the N_2 complex is formed it may be quite resistant to atmospheric oxidation. If, as is the case with the cobalt dihydride, the starting material is fairly resistant to atmospheric oxidation, there is no reason why the displacement of H_2 by N_2 should not occur even in the presence of oxygen.

Several reactions have been reported in which a reduction is carried out in an atmosphere of nitrogen, the reduced product being stabilized by the coordinated N_2.

$$\text{mer-}[OsX_3(PR_3)_3]^0 \xrightarrow[\text{THF}]{\text{Zn/Hg} + N_2} [OsX_2(N_2)(PR_3)_3]^0$$
$$\nu_{N_2} = 2060\text{–}2090 \text{ cm}^{-1} \quad (23)$$

$$[Co(acac)_2]^0 + AlEt_2(OEt) + PPh_3 \xrightarrow[\text{Et}_2\text{O}]{N_2} [Co(N_2)(PPh_3)_3]^0$$
$$\nu_{N_2} = 2093 \text{ cm}^{-1} \quad (20)$$

$$[Ni(acac)_2]^0 + Al(iBu)_3 + PEt_3 \xrightarrow[\text{Et}_2\text{O}]{N_2} [Ni(N_2)H(PEt_3)_2]$$
$$\nu_{N_2} = 2065 \text{ cm}^{-1} \quad (24)$$

$$[Mo(acac)_3]^0 + Al(iBu)_3 + PPh_3 \xrightarrow[\text{toluene}]{N_2} [Mo(N_2)(PPh_3)_2(tol)]^0$$
$$\nu_{N_2} = 2005 \text{ cm}^{-1} \quad (25)$$

Reducing agents vary from the mild amalgamated zinc in aqueous solution to the strong aluminum alkyls. In their report of the preparation of $[Co(N_2)(PPh_3)_3]$, Speier and Marko emphasize the importance of using cobaltous acetylacetonate rather than the cobaltic complex used by previous workers (26). Ibers (27) points out that it seems unlikely that this would make any difference in the presence of the very strong reducing agents used. This comment seems entirely justified, and it may be that the cobalt story is not yet fully told.

We now come to a number of reactions in which reagents containing an —N—N— grouping, rather than molecular N_2, are used. In the preparation (12, 13) of the rhenium compound, Figure 2, the benzoylhydrazine first acts as a chelate ligand, and on treatment with triphenylphosphine

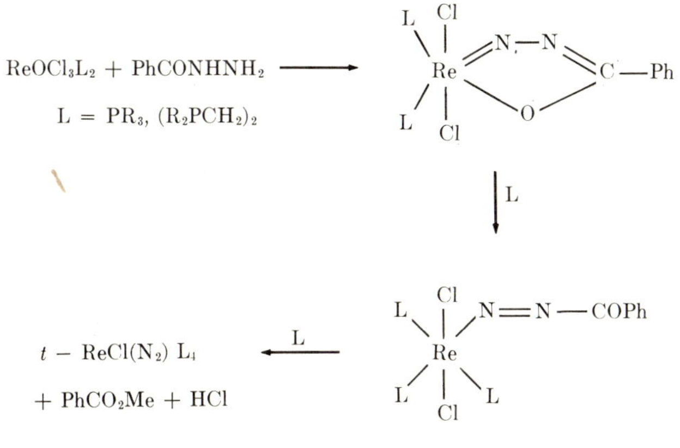

Figure 2. Preparation of rhenium–N_2 compound; data taken from Ref. 12

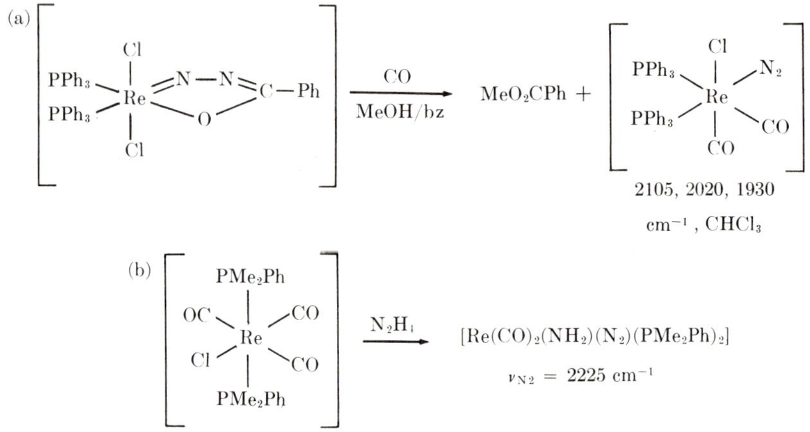

Figure 3. *Preparation of rhenium–N_2 compounds; data taken from Ref. 30 and 31*

in methanol an intermediate benzoylazo complex is produced. On further reaction with the phosphine in methanol, this decomposes to give the product shown. Clearly, if this type of reaction could be reversed, many of our problems would be solved.

The preparation of $[OsH_2(N_2)L_3]^0$ mentioned earlier uses *p*-toluenesulfonyl azide as the reagent (*14*).

$$[OsH_4L_3]^0 + MeC_6H_4SO_2N_3 \rightarrow [OsH_2(N_2)L_3]$$

for $L = PEtPh_2$, $\nu_{N_2} = 2085$ cm^{-1}

It is again assumed that the reagent acts first as a ligand which then decomposes *in situ*, leaving N_2 bonded to the metal.

Only a very few compounds have been prepared that contain more than one coordinated N_2 molecule. Similarly, there are only one or two complexes known so far in which both N_2 and CO are coordinated to the same metal site. Taube (*28*) prepared the first bis-nitrogen compound, *cis*-$[Os^{II}(NH_3)_4(N_2)_2]^{2+}$, by the reaction of nitrous acid with $[Os^{II}(NH_3)_5N_2]^{2+}$. Neither of these two osmium–N_2 compounds can be prepared using N_2 as a reagent, but the carbonyl analogue of one of them, $[Os^{II}(NH_3)_5CO]^{2+}$, has now been prepared (*29*) by the careful reduction of an osmium (III) complex with amalgamated zinc in the presence of carbon monoxide. Treatment of this compound with nitrous acid gives *cis*-$[Os(NH_3)_4CO \cdot N_2]^{2+}$ (*29*).

Two compounds of rhenium that contain one N_2 and two CO groups have been reported, Figure 3. The preparation by Chatt *et al.* (*30*) uses benzoylhydrazine as a reagent in the presence of carbon monoxide; the preparation by Moelwyn-Hughes and Garner (*31*) uses hydrazine on a compound that already contains the carbonyl groups. The band assigned to ν_{N_2} in the infrared spectrum of the second compound (2225 cm^{-1}) is at a surprisingly high frequency, particularly when compared with Chatt's results and bearing in mind that substitution of NH_2^- for Cl^- would be likely to decrease ν_{N_2} rather than increase it. A possible explanation is that this is not an N_2 compound at all, but rather an isocyanate. Further work is needed to clarify this point.

As soon as the first N_2 complex was reported, many workers began looking for compounds of iron and molybdenum because of the biological importance of these metals. Compounds of iron were soon prepared, but it was not until this year that the first molybdenum–N_2 compounds were found. Misono *et al.* have reported (*25*) the composition of the molybdenum complex mentioned earlier, together with the preparation of a second complex containing two molecules of N_2.

$$[Mo(acac)_3]^0 + (diphos) + Al(iBu)_3 + N_2 \rightarrow [Mo(diphos)_2(N_2)_2]^0$$

$$\nu_{N_2} = 2020(w), 1970(vs) \text{ cm}^{-1}$$

From the position and intensities of the infrared bands assigned to ν_{N_2}, it is inferred that the product is the trans compound.

We now come to a type of reaction for producing nitrogen complexes that is important not only in itself but also because it indicates other possible routes back from coordinated nitrogen to a derivative of it. These reactions involve the reactions of a coordinated ligand that contains the —N—N— grouping attached to the metal, using an external reagent that decomposes the ligand, leaving N_2 on the metal. An example (*32*) is

$$[Ru(NH_3)_5(H_2O)]^{2+} + N_2O \rightarrow [Ru(NH_3)_5N_2O]^{2+} + H_2O$$

$$[Ru(NH_3)_5N_2O]^{2+} + Cr^{2+}(aq) \rightarrow [Ru(NH_3)_5N_2]^{2+} + Cr^{3+}(aq)$$

Armor and Taube also found (*33*) that if ^{15}N-labelled N_2O is used, the resulting N_2 complex slowly isomerizes to give a random distribution of ^{15}N in the linearly coordinated N_2 molecule. Since the rate of isomerization is rather faster than the rate of dissociation of N_2 from the complex, they postulate an intra-molecular rearrangement involving a sideways-on π-bonded transition state

$$[(NH_3)_5Ru-^{15}N\equiv N]^{2+} \rightleftharpoons \left[(NH_3)_5 Ru-\overset{^{15}N}{\underset{N}{|||}} \right]^{2+} \rightleftharpoons$$

$$[(NH_3)_5 Ru-N\equiv ^{15}N]^{2+}$$

This proposal is particularly interesting because no compound has yet been found to have this type of bonding, which is, of course, theoretically quite possible.

One of the earliest reported reactions that led to the formation of metal–N_2 compounds was the decomposition of a metal azide. For example, $[Ru^{III}(NH_3)_5N_3]^{2+}$ slowly decomposes on standing at room temperature to give $[Ru^{II}(NH_3)_5N_2]^{2+}$ (2). Reaction of azide complexes with a variety of reagents has now been used successfully in the preparation of N_2 compounds. Basolo et al. (34) prepared the first ruthenium compound containing two coordinated N_2 molecules by the reaction

$$[Ru^{II}(en)_2 N_2 \cdot N_3]^+ + HNO_2(aq) \rightarrow [Ru^{II}(en)_2 (N_2)_2]^{2+} + N_2O$$
$$\nu_{N_2} = 2190, 2220 \text{ cm}^{-1}$$

A similar reaction has been reported by Feltham et al. (35).

$$[Ru^{II}(dias)_2 Cl \cdot N_3]^0 + NOPF_6(\text{methanol}) \rightarrow [Ru^{II}(dias)_2 Cl \cdot N_2]^+ +$$
$$N_2O \quad \nu_{N_2} = 2130 \text{ cm}^{-1}$$

Basolo et al. have also reported (36) their investigations of the acid-catalyzed decomposition of coordinated azide. They find that the primary decomposition product is a protonated nitrene

$$[(NH_3)_5 RuN_3]^{2+} + H^+ \rightleftarrows [(NH_3)_5 Ru\overset{H}{\overset{|}{-}}N-N\equiv N]^{3+} \rightarrow$$
$$[(NH_3)_5 Ru-NH]^{3+} + N_2$$

which reacts further in two ways: (a) by reaction with the original azide

$$[(NH_3)_5 RuNH]^{3+} + [N_3-Ru(NH_3)_5]^{2+} \rightarrow$$
$$[(NH_3)_5 Ru\overset{H}{\overset{|}{-}}N-N-N=N-Ru(NH_3)_5]^{5+} \rightarrow 2[(NH_3)_5 RuN_2]^{2+} + H^+$$

and (b) by dimerization to give the N_2-bridged dimeric ion (to be discussed later).

$$2[(NH_3)_5 RuNH]^{3+} \rightarrow [(NH_3)_5 Ru-N\equiv N-Ru(NH_3)_5]^{4+} + 2H^+$$

These reactions are very reminiscent of the behavior of organic azides when treated with acid or ultraviolet radiation, but there is little obvious chemical resemblance between the N_2-bridged complexes formed and organic azo compounds.

Reactions of Coordinated N_2

We now come to the part of this story that has been most frustrating so far, but which is likely to be the area of greatest activity during the next year or two.

Most attempts to induce the bound N_2 molecule to react further have simply resulted in the displacement of N_2 by a new ligand. This type of reaction provides a useful synthetic route to a variety of compounds (2) but is of little interest to us at this time. There are four obvious ways in which the bound N_2 might be attacked, by reduction, oxidation, photochemical excitation, and further coordination. All have been tried without success so far, but the results of these experiments are important for the information they give about coordinated N_2 and for the possible indication they provide for future work.

The use of strong reducing agents is illustrated by the work of Speier and Marko (37).

$$Co(N_2)(PEt_3)_3 + n\text{-BuLi} \rightarrow Li^+ [Co(N_2)(PEt_3)_2(PEt_2)]^- + C_6H_{14}$$

$\nu_{N_2} = 2059$ cm^{-1} $\quad\quad\quad\quad\quad\quad\quad\quad\quad\quad\quad\quad \nu_{N_2} = 2014$ cm^{-1}

$$[Co(N_2)(PPh_3)_3] \underset{PhI}{\overset{KPPh_2}{\rightleftarrows}} [Co(N_2)(PPh_3)_2(PPh_2)]^- \underset{PhI}{\overset{KPPh_2}{\rightleftarrows}}$$

$\nu_{N_2} = 2093$ cm^{-1} $\quad\quad\quad\quad\quad \nu_{N_2} = 2016$ cm^{-1}

$$[Co(N_2)(PPh_3)(PPh_2)_2]^{2-}$$

$$\nu_{N_2} = 1904 \text{ cm}^{-1}$$

Treatment of the cobalt nitrogen compound with n-butyllithium or potassium diphenylphosphide results in the cleavage of a phosphorus–carbon bond, leaving a phosphide, PR_2^-, as a ligand. The effect of the negatively charged, strongly π-donating ligand is to reduce ν_{N_2} substantially. It is remarkable that the cobalt–N_2 linkage remains intact, even in the dianion, despite the great change in electron density on the metal. The ν_{N_2} in the dianion, at 1904 cm^{-1}, is one of the lowest frequencies so far reported and indicates a degree of activation of the N_2 molecule that should be exploited, for example, by an attempt to coordinate the N_2 to a second metal.

Several workers have studied the oxidation of metal–N_2 compounds. Chatt et al. (38) found that oxidation of $[Os(NH_3)_5N_2]^{2+}$ with Ag^+ and Cu^{2+} ions in ethanol gave clear indications of an osmium (III)–N_2 complex and the possible hint of an osmium (IV)–N_2 complex. Page et al. (39), using electrochemical techniques in aqueous solution, also obtained the osmium (III)–N_2 complex but report that further oxidation leads to decomposition, the rate of which is too fast for any osmium (IV)–N_2 compound to be detected, even by rapid scan techniques. Electrochemical oxidation of the osmium (II)–N_2 to the osmium (III)–N_2 compound is a clean, one-electron, reversible process, the osmium (III)–N_2 compound decomposing to an osmium (III) aquo species at a first-order rate of 2×10^{-2} sec^{-1} at 25°C. Similar experiments by Page et al. (39) with [Ru^{II}(NH$_3$)$_5$N$_2$]$^{2+}$ show that it is oxidized irreversibly to give [Ru^{III}(NH$_3$)$_5$OH$_2$]$^{3+}$. This shows that the ruthenium (III)–N_2 compound, if formed, is very much less stable than the osmium (III)–N_2 analogue.

Chatt et al. (38) have oxidized their robust rhenium (I) compound with metal ions and even with halogens without disturbing the metal–N_2 bond.

$$[\text{Re(diphos)}_2 \text{Cl} \cdot (\text{N}_2)] \xrightarrow[\text{or X}_2]{\text{Ag}^+, \text{Cu}^+, \text{Fe}^{3+}} [\text{Re(diphos)}_2 \text{Cl} \cdot (\text{N}_2)]^+$$
$$\nu_{N_2} = 1980 \text{ cm}^{-1} \qquad\qquad \nu_{N_2} = 2060 \text{ cm}^{-1}$$

Since the ion [Ru^{II}(NH$_3$)$_5$N$_2$]$^{2+}$ has a strong electronic absorption bond at 222 nm, it is obvious that exposure to radiation of this frequency might produce interesting results. When this is done, Sigwart and Spence (40) report that a ruthenium (III) species is formed and gaseous nitrogen is evolved. An important question, however, is the species reduced during the oxidation of ruthenium (II) to ruthenium (III). In experiments with other ruthenium (II) compounds, Ford et al. (41) find that H_2 (from the solvent) is evolved, but Sigwart and Spence do not. It is just possible that some reduction of nitrogen occurs in their experiments, and work with $^{15}N_2$ should be done to clarify this important point.

Strangely enough, the one clear and well-documented reaction that does involve bound N_2 was an early discovery by Harrison and Taube (42).

$$[\text{Ru(NH}_3)_5\text{N}_2]^{2+} + [(\text{H}_2\text{O})\text{Ru(NH}_3)_5]^{2+} \rightleftarrows$$
$$[(\text{NH}_3)_5\text{Ru}-\text{N}_2-\text{Ru(NH}_3)_5]^{4+} + \text{H}_2\text{O}$$

The formation of dimeric N_2-bridged compounds still remains one of the most important discoveries in this field. The largest group of "dimers" of this type contain either two rutheniums or one ruthenium and one osmium. In addition to the above, the list of such compounds (29, 43) now includes:

$$\{[(\text{H}_2\text{O})_5\text{Ru}]_2\text{N}_2\}^{4+}, \qquad \{[(\text{trien})\text{Br} \cdot \text{Ru}]_2\text{N}_2\}^{2+},$$
$$\nu_{N_2} = 2080 \text{ cm}^{-1} \qquad\quad \nu_{N_2} = 2060 \text{ cm}^{-1} \text{ (vw)}$$

$[(NH_3)_5Ru-N_2-Os(NH_3)_5]^{4+}$,
$\nu_{N_2} = 2047$ cm^{-1} (w)

$[(H_2O)(NH_3)_4Ru-N_2-Os(NH_3)_5]^{4+}$, $[(NH_3)_5Ru-N_2-Os(NH_3)_4CO]^{4+}$,
$\nu_{N_2} = 2046$ cm^{-1} (m) $\nu_{N_2} = 2105$ cm^{-1} (m)

$[(NH_3)_5Ru-N_2-Os(NH_3)_5]^{5+}$.
$\nu_{N_2} = 2034$ cm^{-1} (s)

In general, these compounds are prepared by the reaction of a metal–N_2 compound with a ruthenium (II) ion containing a replaceable aquo group. In the symmetrical $\{[(NH_3)_5Ru]_2N_2\}^{4+}$ ion, ν_{N_2} is at best very weakly infrared-active, but it appears in the Raman spectrum at about 2100 cm^{-1} (44), not greatly displaced from its frequency in the parent monomeric ion. The metal—N—N—metal group is essentially linear and the N—N bond distance (1.12Å) is only slightly longer than that in free N_2 (45). The mode of linkage of the ion $[Os(NH_3)_4(N_2)(CO)]^{2+}$ is through the N_2 ligand rather than the carbon monoxide. This illustrates an important difference between the two, since no examples are known of an M—C—O—M bridge between two transition metals. The formation of such bridges between a transition and a nontransition metal, with the latter linked to the oxygen, has been reported recently (46). Presumably, the oxygen of the coordinated carbon monoxide is too π-electron rich to have much affinity for a transition metal ion containing nonbonding d-electrons.

It is clear that in these cases further coordination of the N_2 does little to activate it. Our hope in studying these systems was that the unsymmetrical double coordination of N_2 might lead to increased activation of it. We were particularly interested in the fact that the $[(NH_3)_5Ru-N_2-Os(NH_3)_5]^{4+}$ can be easily and reversibly oxidized to $[(NH_3)_5Ru-N_2-Os(NH_3)_5]^{5+}$. Page et al. (39) have made an intensive study of the electrochemistry of these ions, and find the following. The $\{[(NH_3)_5Ru]_2N_2\}^{4+}$ ion is reversibly oxidized (one faraday at $\sim$0.5v with respect to a saturated calomel electrode) to an ion carrying a 5+ charge, which decomposes ($k_1 = 10 \times 10^{-2}$ sec^{-1}, 25°) to $[(NH_3)_5RuN_2]^{2+} + [(NH_3)_5Ru(OH_2)]^{3+}$. The $[(NH_3)_5Ru-N_2-Os(NH_3)_5]^{4+}$ ion is reversibly oxidized (one faraday, at $\sim$0.1v) to an ion carrying a 5+ charge, which is stable. Further oxidation at higher potential does not give a 6+ ion, but results in oxidative decomposition of the 5+ ion to $[(NH_3)_5OsN_2]^{3+}$ and $[(NH_3)_5Ru(OH_2)]^{3+}$.

Thus, it appears that there is little hope of sufficient activation of coordinated N_2 in systems such as this, but this may be because the two metals involved are not sufficiently different. More promising results along similar lines have recently been reported by Chatt et al. (47). They find that their rhenium complex reacts with a variety of early transition metal halides to give what appear to be deeply colored N_2-bridged compounds with, in some cases, remarkably low N_2 stretching frequencies.

		Product ν_{N_2}
[Re Cl(N_2)(PMe$_2$Ph)$_4$]	+ Ti Cl$_3$ 3THF	1805 cm^{-1}
ν_{N_2} = 1922 cm^{-1}	+ Cr Cl$_3$ 3TFH	1890 cm^{-1}
	+ Mo Cl$_3$ 3THF	1850 cm^{-1}
	+ Mo Cl$_4$ 2Et$_2$O	1795 cm^{-1}
	+ Xs Mo Cl$_4$ 2Et$_2$O	1680 cm^{-1}

The authors confirm the ν_{N_2} band assignments by ^{15}N labelling, so that whether the actual structure involves an N_2 bridge or not, the N_2 bond has been considerably weakened in the product. They point out that only metal–N_2 compounds that already have ν_{N_2} below about 1970 cm^{-1} seem to react in this way; this excludes most of the known metal–N_2 compounds but points clearly in the direction of work to be done. An interesting point here is the fact that the analogous [ReCl(CO)(PMe$_2$Ph)$_4$] does not combine with these reagents. As pointed out earlier, the oxygen of bonded carbon monoxide has little affinity for other transition metal ions, apparently even including those from early in the transition series.

Polymeric N_2-Bridged Compounds

Although it is possibly of little significance to our biological considerations, the existence of stable, bifunctional, monomeric species leads to the obvious possibility of extensive copolymerization.

As a first step on this road, Stevens at Toronto has prepared (29) what is almost certainly the first N_2-bridged trimeric ion, using the stable cis-[Os(NH$_3$)$_4$(N_2)$_2$]$^{2+}$ as a starting material

$$\textit{cis-}[Os(NH_3)_4(N_2)_2]^{2+} + 2[(H_2O)Ru(NH_3)_5]^{2+} \xrightarrow{0.1M \ H_2SO_4}$$
$$\textit{cis-}[(NH_3)_5Ru-N_2-Os(NH_3)_4-N_2-Ru(NH_3)_5]_6^{+}$$

This ion has λ_{max} at 237 and 298 nm, and has one very weak N_2 stretching frequency band at 2097 cm^{-1}. The presence of only one weak infrared band shows that both N_2 groups are forming bridges and that the environment at each bridge is similar.

Page and his group have made a preliminary electrochemical study (43) of this trimeric ion and find that it is oxidized (>0.8 faradays, + 0.44 v) to give [(NH$_3$)$_4$(N_2)Os—N$_2$—Ru(NH$_3$)$_5$]$^{4+}$ + [(NH$_3$)$_5$Ru(OH$_2$)]$^{3+}$, and further oxidized (>0.8 faradays, + 0.65 v) to give

$[Os(NH_3)_4(N_2)_2]^{2+} + [(NH_3)_5RuOH_2]^{3+}$. By rapid cyclic voltammetry, they find that reversible oxidation of the 6+ trimeric ion to a 7+ trimeric ion does occur, but that oxidation of the 7+ ion is irreversible and results in decomposition.

Work is presently in hand using two bifunctional monomeric ions, $[Os(NH_3)_4(N_2)_2]^{2+}$ and $[Ru(NH_3)_4(H_2O)_2]^{2+}$, which possibly may lead to the formation of cyclic or long chain polymeric ions.

Summary

The range of transition metal complexes containing the ligand N_2 is now quite extensive and indicates the versatility of the nitrogen molecule as a ligand. Although no N_2 compounds that are sufficiently stable to permit isolation have yet been shown to be capable of further reaction to produce ammonia, there is still hope that, by the simultaneous coordination of different metal ions onto the same N_2 molecule, sufficient activation of the molecule will be produced to enable a third reagent to break the nitrogen–nitrogen bond. If this is achieved, it will help greatly in our understanding of the biological process of nitrogen fixation.

Literature Cited

(1) Allen, A. D., Senoff, C. V., *Chem. Commun.* (1965) 621.
(2) Allen, A. D., Bottomley, F., Harris, R. O., Reinsalu, V. P., Senoff, C. V., *J. Am. Chem. Soc.* (1967) 89, 5595.
(3) Allen, A. D., Bottomley, F., *J. Am. Chem. Soc.* (1969) 91, 1231.
(4) Chatt, J., Richards, R. L., *Chem. Commun.* (1968) 1522.
(5) Ahrland, S., Chatt, J., Davies, N. R., *Quart. Rev. (London)* (1958) 12, 265.
(6) Temkin, M., Pyzhev, V., *J. Phys. Chem. U.S.S.R.* (1939) 13, 851.
(7) Temkin, M., Pyzhev, V., *Acta Physicochim. U.S.S.R.* (1940) 12, 327.
(8) Carra, S., Ugo, R., *J. Catalysis* (1969) 15, 435.
(9) Allen, A. D., Bottomley, F., *Accounts Chem. Res.* (1968) 360.
(10) Collman, J. P., Kubota, M., Vastine, F. D., Sun, J. Y., Kang, J. W., *J. Am. Chem. Soc.* (1968) 90, 5430.
(11) Harrison, D. E., Taube, H., *J. Am. Chem. Soc.* (1967) 89, 5706.
(12) Chatt, J., Dilworth, J. R., Leigh, G. J., *Chem. Commun.* (1969) 687.
(13) Chatt, J., Dilworth, J. R., Leigh, G. J., Paske, R. J., in "Progress in Coordination Chemistry," M. Cais, Ed., p. 246, Elsevier, Amsterdam, 1968.
(14) Bell, B., Chatt, J., Leigh, G. J., *Chem. Commun.* (1970) 576.
(15) Harris, R. O., private communication.
(16) Sacco, A., Aresta, M., *Chem. Commun.* (1968) 1223.
(17) Knoth, W. H., *J. Am. Chem. Soc.* (1968) 90, 7172.
(18) Creutz, C., Taube, H., private communication.
(19) Bancroft, G. M., Mays, M. J., Prater, B. E., *Chem. Commun.* (1969) 585.
(20) Speier, G., Marko, L., *Inorg. Chim. Acta* (1969) 126.
(21) Davis, B. R., Payne, N. C., Ibers, J. A., *Inorg. Chem.* (1969) 8, 2719.
(22) Allen, A. D., Bottomley, F., *Can. J. Chem.* (1968) 46, 469.
(23) Chatt, J., Leigh, G. J., Richards, R. L., *Chem. Commun.* (1969) 515.

(24) Srivastava, S. C., Bigorgne, M., *J. Organometal. Chem.* (1969) **18,** 30.
(25) Hidai, M., Tominari, K., Uchida, Y., Misono, A., *Chem. Commun.* (1969) 1392.
(26) Yamamoto, A., Kitazume, S., Pu, L. S., Ikeda, S., *Chem. Commun.* (1967) 79.
(27) Ibers, J., oral communication.
(28) Scheidegger, H. A., Armor, J. N., Taube, H., *J. Am. Chem. Soc.* (1968) **90,** 3263.
(29) Allen, A. D., Stevens, J. R., unpublished work.
(30) Chatt, J., Dilworth, J. R., Leigh, G. J., *J. Organometal. Chem.* (1970) **21,** 49.
(31) Moelwyn-Hughes, J. T., Garner, A. W. B., *Chem. Commun.* (1969) 1309.
(32) Armor, J. N., Taube, H., *J. Am. Chem. Soc.* (1969) **91,** 6874.
(33) Armor, J. N., Taube, H., *J. Am. Chem. Soc.* (1970) **92,** 2560.
(34) Kane-Maguire, L. A. P., Sheridan, P. S., Basolo, F., Pearson, R. G., *J. Am. Chem. Soc.* (1968) **90,** 5295.
(35) Douglas, P. G., Feltham, R. D., Metzger, H. G., *Chem. Commun.*, in press.
(36) Kane-Maguire, L. A. P., Basolo, F., Pearson, R. G., *J. Am. Chem. Soc.* (1969) **91,** 4609.
(37) Speier, G., Marko, L., *J. Organometal. Chem.* (1970) **21,** 46.
(38) Chatt, J., Dilworth, J. R., Gunz, H. P., Leigh, G. J., Sanders, J. R., *Chem. Commun.* (1970) 90.
(39) Elson, C. M., Gulens, J., Itzkovitch, I. J., Page, J. A., *Chem. Commun.*, submitted for publication.
(40) Sigwart, C., Spence, J. T., *J. Am. Chem. Soc.* (1969) **91,** 3991.
(41) Ford, P. C., Stuermer, D. H., McDonald, D. P., *J. Am. Chem. Soc.* (1969) **91,** 6209.
(42) Harrison, D. F., Weissberger, E., Taube, H., *Science* (1968) **159,** 320.
(43) Page, J. A., *et al.*, unpublished work.
(44) Chatt, J., Nikolsky, A. B., Richards, R. L., Sanders, J. R., *Chem. Commun.* (1969) 154.
(45) Treitel, I. M., Flood, M. T., Marsh, R. E., Gray, H. B., *J. Am. Chem. Soc.* (1969) **91,** 6512.
(46) Kotz, J. C., Turnipseed, C. D., *Chem. Commun.* (1970) 41.
(47) Chatt, J., Dilworth, J. R., Richards, R. L., Sanders, J. R., *Nature* (1969) **224,** 1201.

RECEIVED June 26, 1970.

Fixation of Molecular Nitrogen Under Mild Conditions

E. E. VAN TAMELEN

Stanford University, Stanford, Calif. 94305

> *Efforts have been directed toward the nonenzymic chemical modification of molecular nitrogen (N_2), reactions being based on titanium(II) as the N_2-coordinating species effective in permitting room temperature and atmospheric pressure phenomena. Accomplishments include: direct NH_3 formation by abstraction of hydrogen from solvent, the development of a cyclic system (over-all catalytic, based on titanium) for NH_3 production, NH_3 formation by direct reaction of N_2 from the air, the catalytic electrochemical synthesis of NH_3, the first case of the preparation and subsequent reductive conversion to NH_3 of a transition metal compound with an N_2 ligand ($C_{20}H_{20}Ti_2 \cdot 2N_2$), elucidation of the titanium(II) (titanium dialkoxide and "titanocene") based N_2 fixation mechanism, controlled reduction of N_2 to hydrazine, and the utilization of N_2 in the synthesis of organic amines and nitriles.*

Our own endeavors in the area of nitrogen fixation had biological origins in more ways than one. Realizing that the process takes place in nature, one might hold out hope that some kind of simulated nitrogen fixation could also be accomplished in the laboratory without the agency of an enzyme *per se*. I would have to admit to entertaining that feeling for a long time, at least ten years. In our own case, the work with titanium had even more specific biochemical origins; the following reasons led us to study certain lower valent titanium species.

In connection with our program dealing with the cyclization of terpenoid terminal epoxides—a program falling in organic and biochemical territory (1)—we had occasion to develop a method for coupling alcohols, especially allylic alcohols (2). Realizing that there was no direct method for converting an alcohol in one step directly to a coupled

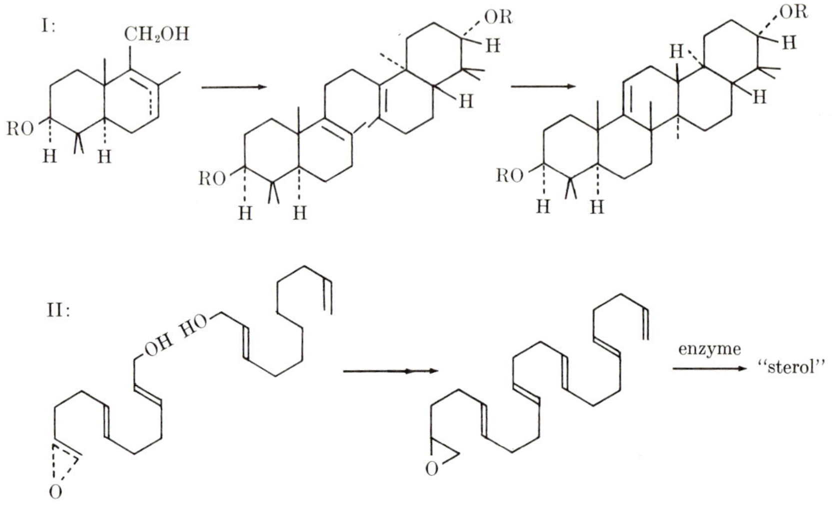

Figure 1. Reaction for converting an alcohol directly to a coupled product

product, we set about to design such a reaction. These needs are symbolized in Figure 1 and are shown in connection with the execution of certain synthetic projects. We wanted to carry out a symmetrical coupling of a bicyclic allyl alcohol (I) to the tetracycle, a coupling reaction of a reductive type. Probably more important, we developed, in connection with enzyme work in this area of sterol biosynthesis, the need for making modified squalene oxides, a structural type written in a highly stylized form under II. The idea was to start off with unsymmetrical halves, one of which would be a potential terminal epoxide, and to carry out some coupling reaction of allyl alcohol types. After the juncture and incorporation of the epoxide moiety, one would utilize this system for enzymic experiments.

The plan evolved for this kind of reaction involved initial conversion of the starting alcohol to the alkoxide of a metal in a lower valent state (II). In this case, we selected a metal which would be considerably more stable in the IV state. We felt that with such a lower valent metal alkoxide in hand, one might be able simply to extrude the metal(IV) oxide and therein produce coupling as shown (Table I). Several experimental methods were developed for carrying out this reaction. They are shown in a, b, and c, and obviously all involve titanium as well as a reducing agent.

The concept for this reaction and the actual mechanism (which was pretty well proved recently) (3) is shown in Table I. With regard to methods (a) and (b), one starts off with the titanium(IV) species. For example, titanium tetrachloride is converted to the mixed titanium di-

Table I. Coupling of Allyl Alcohols

Problem:

2 ROH → R—R + (H$_2$O$_2$)

Basic Plan:

ROH → (RO—(Me)—OR) → R—R + (Me)O$_2$,

where stability of Me(IV) > Me(II)

Experimental:

(a) ROH $\xrightarrow{1, 2, 3, 4}$ R—R   1. NaH   2. 1/2 TiCl$_4$ 3. K°   4. Heat

(b) ROH $\xrightarrow{1, 2, 3}$ R—R   1. Na Naphthalide   2. 1/2 TiCl$_4$ 3. Δ

(c) ROH $\xrightarrow{1, 2}$ R—R   1. 3 CH$_3$Li, 1 TiCl$_3$ 2. Δ

where R = R'—C=C—CH$_2$—;   ArCH$_2$—

Mechanism:

(a) and (b)

RO$^\ominus$ $\xrightarrow{\text{TiCl}_4}$ {(RO)$_2$TiCl$_2$ $\xrightarrow{\text{K}^0 \text{ or Naphthalide}}$ [(RO)$_2$Ti]$_n$ ⇌

(RO)$_2$Ti $\xrightarrow{\Delta}$ (TiO$_2$)2R · } → R—R

(c)

2 ROH + TiCl$_3$ $\xrightarrow{2 \text{ CH}_3\text{Li}}$ {(RO)$_2$TiCl $\xrightarrow{\text{CH}_3\text{Li}}$

(RO)$_2$TiCH$_3$ → CH$_3$· + (RO)$_2$Ti $\xrightarrow{\Delta}$ } R—R

chloride–dialkoxide by reaction with the starting alcohol; and then without isolation, this species is reduced with potassium zero or naphthalene radical anion to the titanium(II) type. Titanium(II) would not exist normally as a monomer; as to whether monomer or polymer actually undergoes the pyrolysis reaction, we are not certain. This reaction was initially developed by a graduate student, Martin Schwartz, and since the reaction turns intensely black at the (II) stage, it should be called the Schwartz reaction! Below 100°, the material—the titanium(II) alkoxide—does pyrolyze, extruding presumably titanium dioxide and forming the coupled product by way of radical type organic residues. Whether these radicals are free, attached to the titanium, or held in a cage circulating around the metal surface, we are not certain. Accordingly, the reaction intermediate is written intentionally in an ambiguous way.

The reaction modification featuring titanium(III) as the starting point was originated somewhat later by Barry Sharpless. The mechanism is shown under (c) and is more speculative than (a) and (b). We hoped that these titanium(II) species could be made, and we are now certain that they are real and available entities. For example, titanium dibenzoxide can be isolated and shown to yield bibenzyl on thermolysis.

It may seem like a "quantum-jump" to proceed from coupling reactions in organic chemistry to nitrogen fixation. On the other hand, the transition is not that discontinuous.

About ten years ago I was involved in the chemistry of diimide, and therein started an interest in molecular nitrogen as well. Just about ten years ago, Professor Burris and I carried out an experiment at the University of Wisconsin designed to reveal the role of intermediates in the enzymic conversion of nitrogen into ammonia. The experiment involved the feeding of a ^{15}N-labelled precursor of diimide (specifically azodicarboxylic acid) to the nitrogen-fixing enzyme system. That experiment, needless to say, was negative.

During the early 1960's, I had occasion to talk with inorganic chemists about the problem of nonenzymic nitrogen fixation. Uniformly, I uncovered the opinion that trying to get N_2 to react under mild conditions was pretty much of a waste of time. In any case, during the early sixties, we were carrying out experiments. Mary Lease attempted reactions based on copper. Alexander Todd checked out tanks of nitrogen and hydrogen and bubbled the gases through media containing iron porphyrins, hoping that some ammonia might be formed. The area opened up during the mid-1960's with the finding, for example, of the first stable N_2 coordination compound by Allen and Senoff (4). Vol'pin's work was pioneering (5), and Collman's contributions followed shortly (6). Taube's work should be mentioned, especially the dramatic example of conversion of N_2 into Allen and Senoff's ion in an aqueous medium (7). At this time, we had initiated our titanium chemistry program and entertained the idea that titanium(II) possibly might resemble a carbene. In connection with this organic parallel, a carbene is very reactive, and one might think that titanium(II) would also have a high reduction potential and therefore in fact react with N_2. We surmised that, in case of reaction, the product could be a titanium analogue of a diazocompound in organic chemistry and that one might be able to reduce this species to ammonia by cleavage of the N–N bond and also perhaps the Ti–N bond.

Table II. Fixation–Reduction of Molecular Nitrogen[a]

(1) $TiCl_4 + 2\ (CH_3)_3CO^-K^+ \xrightarrow[\text{diglyme}]{N_2} TiCl_2[(CH_3)_3CO]_2$

$+ 2\ K^0 \xrightarrow{N_2} 10\text{–}15\%\ NH_3 \uparrow$

(2) $TiCl_4 + 2\ (CH_3)_3CO^-K^+ \xrightarrow{N_2} D_8\text{-tetrahydrofuran}$

$TiCl_2[(CH_3)_3CO]_2 + 2\ K^0 \xrightarrow{N_2} N_mD_n \uparrow$

(3) $(C_5H_5)_2TiCl_2 +$ naphthalene$^{\cdot -}$ $Na\ + \xrightarrow[\text{ethereal solvent}]{N_2} NH_3 \uparrow$

(Na Np)

(4) $Ti[OCH(CH_3)_2]_4 + NaNp \xrightarrow[\substack{\text{ethereal solvent} \\ (30\text{–}60\ \text{min.})}]{N_2} [\text{Nitride}]$

$\xrightarrow{(CH_3)_2CHOH} 65\%\ NH_3 \uparrow$

Argon $\longrightarrow$ Product $\xrightarrow[\text{(days)}]{N_2}$ [Nitride] $\xrightarrow{(CH_3)_2CHOH}$ low yields $NH_3 \uparrow$

[a] Experiments were conducted at room temperature and at atmospheric pressure.

Initially, the nitrogen fixation experiments were run in a fashion very similar to that used for the coupling reaction I have already described (8). For example, one would start with titanium tetrachloride and an alkoxide, perhaps potassium-*tert*-butoxide, and prepare *in situ* the dichloride dialkoxide (Table II). Potassium metal was used in the early experiments, and nitrogen was blown through and run into a trap. In these first experiments, which were carried out by Steve Ela and Bob Fechter, we were running through nitrogen and hydrogen gas simultaneously, hoping that we would make ammonia under the conditions of room temperature and atmospheric pressure. Ammonia was formed; in the course of about two weeks, one could accumulate a maximum of about 10–20% of that simple inorganic substance. Later, we found that one could secure the same results without the use of molecular hydrogen and therefore faced the problem of hydrogen source. We took some pains to exclude water and still observed ammonia as product, and we concluded at this point that perhaps some reactive intermediate was actually abstracting hydrogen from the ethereal solvent. A few years later, this possibility was confirmed by using a deuterated solvent (perdeutero THF), repeating the reaction, and carrying out a mass spectral determination on collected ammonia. A certain amount of deuterated ammonia was in fact detected.

Some years ago, one of my coworkers, Gernot Boche, first utilized a dicyclopentadienyl titanium compound in nitrogen fixation experiments (8). Starting with the well known titanium(IV) dicyclopentadienyl dichloride, nitrogen fixation was observed by the use of the various reducing species, the most interesting modification being the radical anion of naphthalene. Again, in this case one can observe product ammonia, which can be gotten presumably by hydrogen extraction from the solvent.

We quickly learned that a fair amount of material was produced in all of our reactions which represented reduced nitrogen, but in an unprotonated, or bound, state. As we gained experience in this area, we developed a "recipe" which worked fairly well (9). In this modification, one starts with titanium tetraisopropoxide and carries out the reduction with sodium naphthalenide. The whole system is connected to a closed container of nitrogen; and during the course of a half hour to an hour one observes visually a very rapid uptake of nitrogen gas. A highly reactive nitride of the approximate composition $Na_{13}Ti_3N_4(OC_3H_7)_{10}$ is formed. After protonation (for example, with isopropyl alcohol) ammonia is released in yields as high as 65% based on the generation of 2 moles of ammonia per titanium. It is clear that our fixation–reduction method differs fundamentally from the Vol'pin reaction (5), which utilizes reducing agents such as Grignard reagents or metal hydrides.

If the original mixing of the starting titanium species and reducing agent was carried out under argon and then subsequently nitrogen fixation attempted, the reaction rate slows down. In addition, the attainable yields of ammonia are very low. These results suggest that a transient but very reactive titanaceous nitrogen-fixing species probably is generated in the normal fixation reaction taking place in from 30 to 60 minutes.

With the reaction *per se* worked out fairly well, we had occasion to develop this chemistry into what one might call a catalytic process, probably better described at this point as a cycle (9). This process is conceptually and operationally very simple, as demonstrated by Mr. Greeley (Table III) (9). In a normal nitrogen fixation reaction, after liberation of ammonia by the addition of alcohol, it is possible to add more sodium naphthalenide or sodium zero and in effect regenerate the nitrogen-fixing species and carry out a second, reasonably effective nitrogen fixation. After a number of cycles, the operation becomes inefficient but by going through five cycles, one can realize a 170% yield of ammonia, again based on the equation $N_2 \rightarrow 2NH_3$.

Table III. Operation of a Fixation–Reduction Cycle

$$Ti(OR)_4 + NaNp \xrightarrow{N_2} [Product] \xrightarrow{ROH} NH_3 \uparrow$$

$$(NaNp \text{ or } Na^0)$$

Five cycles ≡ 170% NH_3

Fixation of Aerial Nitrogen[a]

(1) $Ti(OR)_4 + (3–12)NaNp \xrightarrow{Air} 10–22\% \ NH_3 \uparrow$

(2) $Ti(OR)_4 + (3–12)NaNp \xrightarrow{Argon} Product \xrightarrow{Air} 1\% \ NH_3 \uparrow$

[a] In 4 equiv. NaNp experiment, no free NaNp is present (ESR).

Another reaction which is very easy to carry out and only requires going over some sort of emotional energy barrier to observe it simply features the use not of pure N_2 but of nitrogen from the air (9). Allen has already described his experience involving preparation of his ion by means of direct reaction under such circumstances (10). Our experiment was carried out apparently at the same time and published concurrently. This reaction works, and in the best example Boche found that a good deal of ammonia is formed in this aerial system. The yield falls in the range 10 to 20% (based on titanium), which is about one-third the yield observed when one uses pure nitrogen.

We worried about the significance of the described result. Sodium naphthalenide is a highly reactive species, and one could imagine that it simply sweeps out the oxygen very rapidly from the atmosphere, leaving some nitrogen, which then reacts with the titanium species present to give some fixation. Whether or not this is true, we are not absolutely certain; but further work tends to substantiate the idea that N_2, under the right circumstances, can compete fairly well with O_2 in a reaction of this type. Under item 2, we see that one can pre-prepare the fixing species under argon in a run using three or four equivalents of sodium naphthalenide which is accordingly essentially used up—and then expose the product of this reaction to the atmosphere and still observe some ammonia formation. The yield is only a few percent, but you will recall that these are the conditions that, even with pure nitrogen, permit only very moderate conversion to ammonia. In the latter case, the conversions were about 3%, whereas here we see about 1%. Again, a 1 to 3 ratio is involved. Finally, in the four equivalent sodium naphthalenide experiments, there is after addition to titanium starting material, no free sodium naphthalenide actually remaining, as indicated by ESR measurements. We feel on this basis that sodium naphthalenide is not complicating the interpretation, but that we may well be seeing true selectivity in terms of partitioning between N_2 and O_2.

So far we have dwelt on the over-all reaction, based on the concept of titanium(II), and the chemistry at this point appears very complicated. One would like to see some clarification of the over-all mechanism. We were pleased to find that one can, in this titanium series, effectively separate the complexation and the reduction steps (*11*). This is as important as the separation of the oxidation and the cyclization steps in the over-all conversion of squalene to lanosterol, a very knotty and complicated problem in the biological province.

At the time we started this mechanism work, one compound was available in the formal titanium(II) state which we thought we would try to use, the so-called "titanocene" (*12*). Although in empirical terms titanocene is dicyclopentadienyl titanium, the substance is isolated as a dimer, $C_{20}H_{20}Ti_2$. In our early work, we actually used a cyclopentadienyltitanium species, $C_{10}H_{10}TiCl_2$, for a fixation reaction. Several years ago at Stanford, Stewart Schneller prepared some titanocene and started to carry out fixation experiments. He quickly secured significant results, which were published early in 1969 (*11*).

Although the conditions for the fixation are exceedingly critical, one can expose titanocene in benzene to an atmosphere of N_2 and observe

Table IV. Mechanism of the Fixation Reactions

I. $N_2 + (C_5H_5)_2Ti \xrightleftharpoons[\text{20–25°}]{\text{benzene}} (C_5H_5)_2TiN_2$

  (1) Product $\xrightarrow{\text{NaNp, H}^+} NH_3 \uparrow$

  (2) Product 1960 cm^{-1} in IR

  (3) Product mol wt 420 in benzene (calc'd for dimer 412)

  (4) Similar N_2 absorption by "Ti[OR(sat.)]$_2$" but not by $(C_6H_5CH_2O)_2Ti$ or $(CH_2=CHCH_2O)_2Ti$

II. Job plot: Maximum yields of NH_3 observed with 1:6 ratio of $[(CH_3)_2CHO]_4Ti:NaNp$, with $\sim$2 moles of NH_3 formed for 1 mole of N_2 utilized.

simple coordination of nitrogen (Table IV). This reaction must be carried out within a very narrow temperature range, 20° to 25°, and over a period of about three weeks. At 30°, one can observe some complexation but the reaction is not a good one. Under ideal conditions one can observe about 93–95% conversion of N_2 to a new product. If one attempts to isolate this material by removing solvent, only titanocene is found. This result points up the ready reversibility of the process.

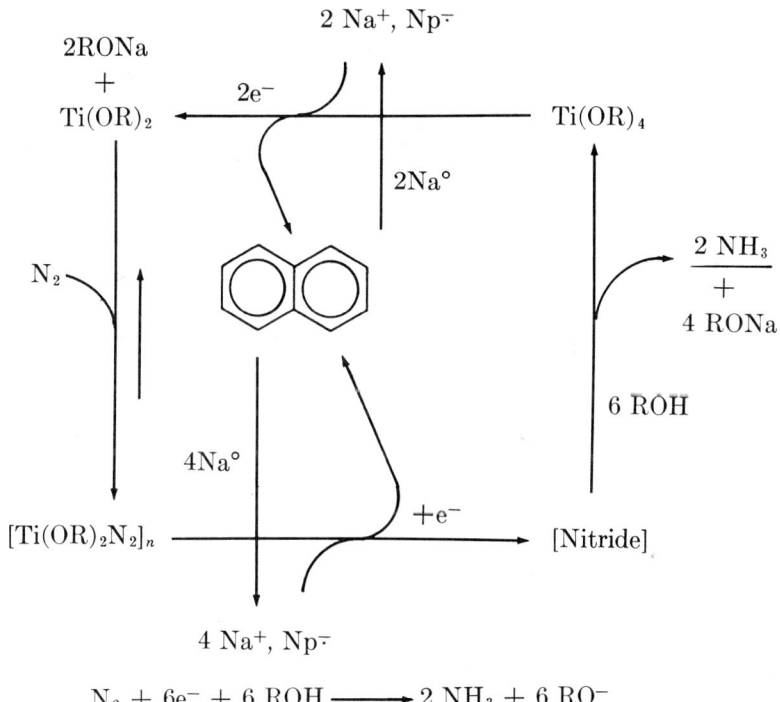

Figure 2. *Cyclic process for reduction of N_2 to NH_3*

Although this new product could not be isolated, in solution it can be reduced quantitatively by sodium naphthalenide to ammonia. On protonation under a variety of circumstances, the N_2-coordination compound does not provide ammonia or any other reduction product, but only releases nitrogen. This is the first species which represents clearcut coordination of nitrogen *per se* and also permits conversion of the ligand to ammonia. Secondly, this N_2 compound in the IR shows a band at 1960 cm^{-1} which is a bit higher wavelength than the usual type of transition metal–N_2 compound. Thirdly, the substance clearly is a dimer in that osmometric determinations show a molecular weight about 420, the calculated value being 412. One sees a similar kind of nitrogen absorption by titanium dialkoxides of the simple saturated alkyl types, but the results then are not clean cut—by no means 100%. In any case, the body of results clearly indicates the function of titanium(II) species in the nitrogen fixation process. VCl_3 and $CrCl_3$ yield V^0 and Cr^0 under conditions used for nitrogen fixation; however, Ti^0 was not similarly implicated (13). The titanium dibenzoxide and the dialkoxide cases do not take up any N_2 at all, the reason probably being that there is internal coordination which prevents nitrogen capture.

A Job plot indicates that maximum yields of ammonia are obtained with about a 1–6 ratio of titanium to reducing agent. Finally, one observes formation of approximately 2 moles of ammonia from 1 mole of N_2.

Now one can incorporate most of this chemistry into a cyclic version, which in a very approximate way might compare to the enzymic process in the sense that one is using a metal species to coordinate nitrogen and is subsequently reducing the ligand in a separate step to ammonia (Figure 2). The functioning of this cycle depends of course on titanium and the presence of a naphthalene pool. Starting with titanium(IV), one uses

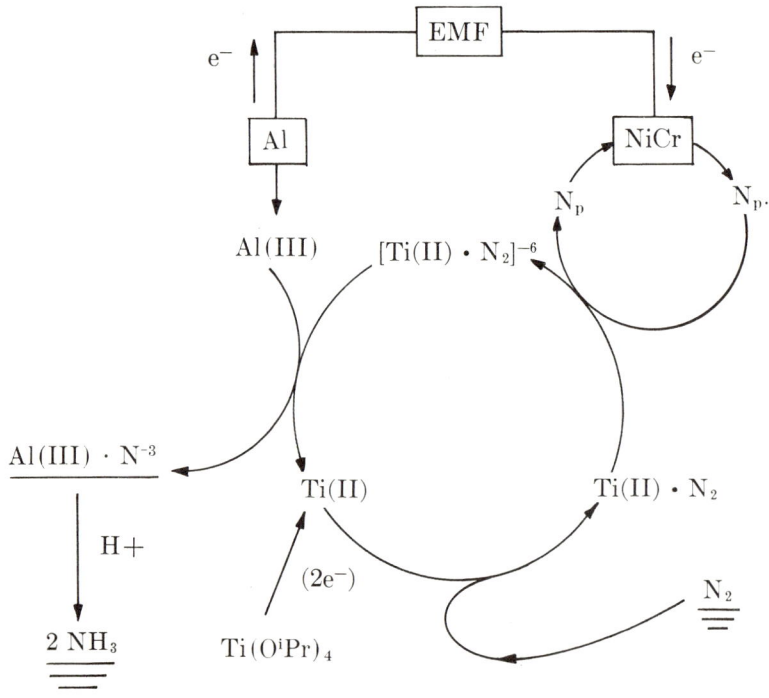

Figure 3. Electrolytic reduction of N_2

sodium naphthalenide in order to reduce to the titanium(II) level while naphthalene is cast off into the pool. The titanium(II) species, once generated, can fix N_2 and proceed to the N_2 ligand compound, which could be monomeric or polymeric. Finally, with more sodium naphthalenide, one can reduce this species to nitride, and again naphthalene falls back in the pool. Protonation finally gives ammonia and what we think is probably again titanium(IV), although we are not sure of this latter. One can go through the cycle again by adding more sodium, drawing on the naphthalene pool, and thus making more sodium naphthalenide available for its dual reduction roles.

One can extend this cycle chemistry by putting electricity to use. In short, it is possible to accomplish electrolytic reduction of nitrogen and to develop the phenomenon into a cycle (Figure 3). Thus, we have achieved true catalytic electrolytic conversion of nitrogen to ammonia, by means of a process based on the behavior of titanium. This system was initiated at Stanford by Akermark (14) and developed by Seeley (15).

Aluminum is used as the anode and nichrome as the cathode. There is also available in the cell a catalytic amount of titanium for fixation purposes, in addition to naphthalene, which serves a purpose again in the reduction stage. The electrolyte is tetrabutylammonium chloride. Aluminum isopropoxide increases the over-all efficiency and turns this process into a catalytic one. This system starts with titanium tetraisopropoxide. Reduction takes place, presumably again to the titanium(II) level. We have evidence from electrochemical experiments that titanium(II) is produced and involved in the fixation. Titanium(II), once formed, picks up N_2 from the atmosphere and forms the complex, which then is available for reduction by sodium naphthalenide. Naphthalene is present in a catalytic amount and is reduced at the cathode to the radical anion. In this experiment, one can actually see it as a greenish color at the cathode. Naphthalenide reduces the N_2 compound, producing the nitride. Normally, the reaction would stop at this point. We believe that in the electrochemical process, aluminum(III) abstracts "nitride" from titanium and forms aluminum nitride. This nitride transfer also can be observed in nonelectrolytic reactions. Thus, aluminum nitride is stored and ammonia is available at any time, merely by protonation. Both titanium and naphthalene are catalytic and permit operation of an over-all catalytic process.

Titanocene chemistry provides an appropriate basis for a discussion of mechanism. Figure 4 shows a recent interpretation of the titanocene fixation–reduction process (13) submitted for publication by the Olives three or four months after our titanocene–N_2 paper appeared (11). The mechanism is a complicated one, based essentially on ESR results meas-

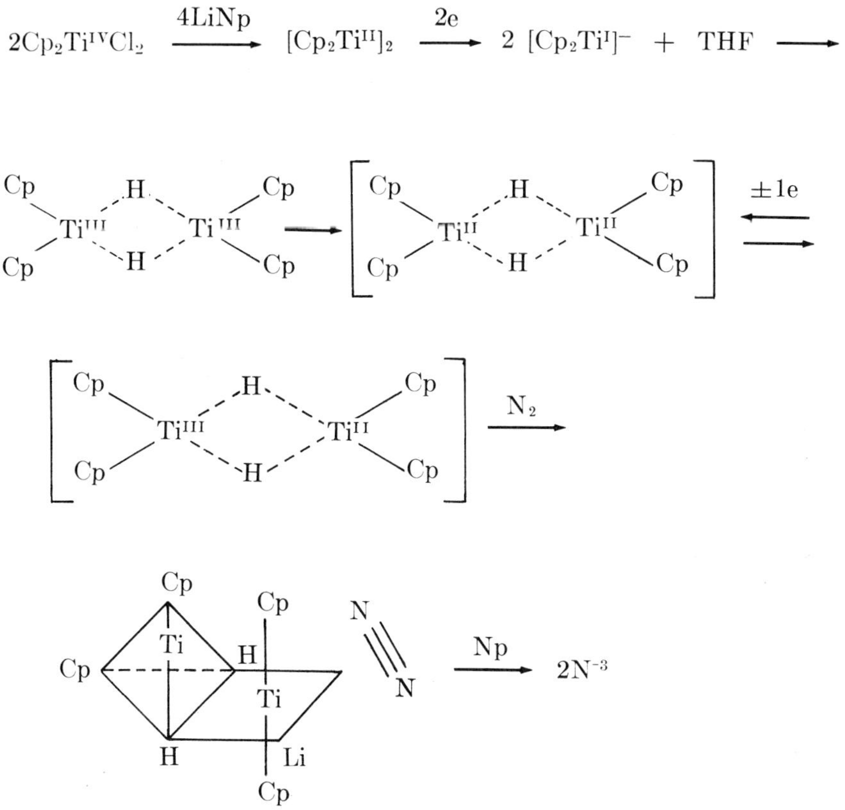

$$2Cp_2Ti^{IV}Cl_2 \xrightarrow{4LiNp} [Cp_2Ti^{II}]_2 \xrightarrow{2e} 2\,[Cp_2Ti^{I}]^- + THF \longrightarrow$$

$$Cp_2Ti^{III}(\mu\text{-}H)_2Ti^{III}Cp_2 \longrightarrow [Cp_2Ti^{III}(\mu\text{-}H)_2Ti^{II}Cp_2] \xrightleftharpoons{\pm 1e}$$

$$[Cp_2Ti^{III}(\mu\text{-}H)_2Ti^{II}Cp_2] \xrightarrow{N_2}$$

[structure: Cp₃Ti₂(H)₂Li with end-on N≡N] $\xrightarrow{Np}$ $2N^{-3}$

Figure 4. *Proposed titanocene fixation–reduction process*

ured in a nitrogen-free system and also on electrometric titrations. In this scheme, the titanium(IV) species is taken down to the (II) level and then further to titanium(I). At this point in the Olive interpretation, proton abstraction from solvent (tetrahydrofuran) takes place to give a binuclear hydrogen bridged titanium(III). In this proposal, it is not this entity which fixes nitrogen but a lower valence species formed by one-electron reduction. Supposedly, there is an equilibrium system featuring as components the titanium(II)–titanium(II) and the titanium(II)–titanium(III) types. The Olives considered the mixed titanium(II)–titanium(III) species to be the actual N_2-fixing agent. After fixation of nitrogen, the coordination number changes, and it is the species shown at the bottom which is assumed to be formed. Figure 4 illustrates end-on bonding of N_2, which is completely arbitrary. Finally, the reduction to nitride occurs.

This concept differs considerably from the one that we published earlier and supported experimentally. In our case, we have observed the facile fixation of N_2 by titanocene itself, a reaction that is not a feature of the Olive scheme. My collaborators, Seeley, Schneller, Rudler, and Cretney, also have been doing some checking of this mechanism, and the results are shown on Figure 5.

First of all, proton removal from the THF implies formation of the anion. One might expect, on addition of tritium oxide to the completed fixation–reduction reaction components, to see formation of tritiated THF. In the experiment, one does not see it at all; moreover, one does not

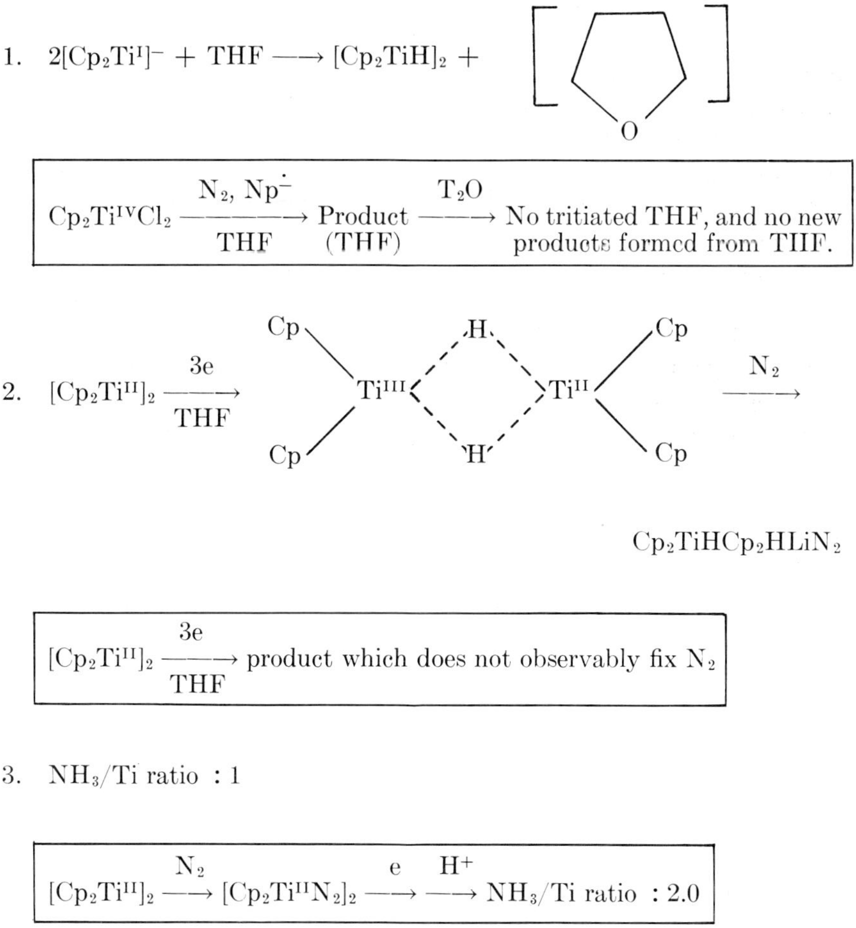

1. $2[Cp_2Ti^I]^- + THF \longrightarrow [Cp_2TiH]_2 + \left[\langle \ \rangle_O \right]$

$Cp_2Ti^{IV}Cl_2 \xrightarrow[THF]{N_2, Np^{\cdot-}} Product \xrightarrow{T_2O} $ No tritiated THF, and no new products formed from THF.
    (THF)

2. $[Cp_2Ti^{II}]_2 \xrightarrow[THF]{3e} Cp\diagdown Ti^{III} \diagup^{H}_{H} \diagdown Ti^{II} \diagup Cp \xrightarrow{N_2}$
   $Cp\diagup \qquad \diagdown Cp$

   $Cp_2TiHCp_2HLiN_2$

$[Cp_2Ti^{II}]_2 \xrightarrow[THF]{3e}$ product which does not observably fix N_2

3. NH_3/Ti ratio : 1

$[Cp_2Ti^{II}]_2 \xrightarrow{N_2} [Cp_2Ti^{II}N_2]_2 \xrightarrow{e} \xrightarrow{H^+} NH_3/Ti$ ratio : 2.0

Figure 5. *Evaluation of proposed titanocene fixation–reduction mechanism*

see any new product of any kind deriving from the THF—for example, the dimer derived from the alpha radical. Of course, these results are in direct conflict with the Olive mechanism.

Secondly, in the Olive mechanism, the fixing species is formed by a three-electron change starting with titanocene dimer. We subjected titanocene to a three-electron naphthalenide reduction and have not been able to observe any nitrogen fixation of any kind with the resulting species.

Finally, the Olive paper demands an ammonia/titanium ratio of 1:1. In our titanocene experiments involving N_2 fixation and reduction, we clearly get an ammonia/titanium ratio of 2:1. The early hydrogen abstraction work which we reported may seem loosely to fit the Olive scheme but is a very slow reaction ($\sim$3 weeks), undoubtedly occurring subsequent to the fixation operation but possibly accounting for the Olive experimental observations. I think the very rapid reaction ($\sim$60 min) that is normally observed does not conform to the Olive interpretation but is actually a titanium(II)–induced fixation without hydrogen removal from solvent.

Our latest simple version of the titanium system, exemplified by titanocene, involves starting with the known titanocene dimer, equilibration with the monomer, and fixation by that latter species. There is good reason to think that the N_2 fixer is the monomer because titanocene dimer absorbs nitrogen very slowly, over a period of three weeks; with "titanocene" generated *in situ* under nitrogen, the reaction with nitrogen is very rapid. Thus, we have evidence that there is a transient, very reactive N_2-fixing species, and we believe this to be the titanium(II) monomer. Brintzinger and Bartell (*16*) recently carried out some Huckel calcula-

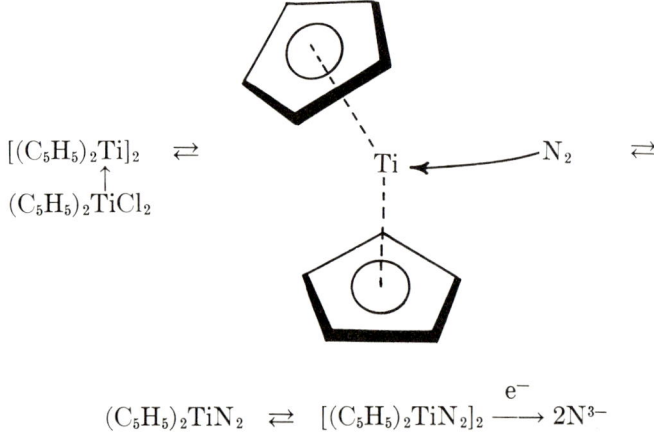

Figure 6. Monomeric titanocene as the N_2-fixing species

tions on this structure and have concluded that it is, in effect, a carbene-like molecule. Moreover, their calculations suggest that monomeric titanocene is not planar but bent, as indicated by the rough pictorial description in Figure 6. In any event, nitrogen absorption proceeds to give the titanocene–N_2 monomer, which can then dimerize to give the more stable, previously described species. Finally, reduction to the nitride level takes place.

Table V. Experiments Involving Conversion of Nitrogen to Hydrazine

$$Ti[OCH(CH_3)_2]_4 + (\text{naphthalene})^{\cdot -} + N_2 \xrightarrow[\text{THF}]{\text{15-90 min}} \xrightarrow{H^+} N_2H_4 (+ NH_3)$$

(1) Yield of N_2H_4 inversely proportional to Ti(IV):NaNp ratio (maximum 1:5–6)
(2) $2NH_3:N_2H_4$ ratio is 3.3–5.0
(3) Overnight, large excess of NaNp increases $2NH_3:N_2H_4$ ratio to 22
(4) $C_{10}H_{10}Ti \cdot N_2$ dimer is not a hydrazine source
(5) No N_2 liberated in protonation step ∴ N_2H_2 not a precursor of N_2H_4
(6) Mechanism:

$$Ti(OR)_4 \xrightarrow{2NaNp} [Ti(OR)_2]_x \xrightarrow{N_2} [Ti(OR)_2 \cdot N_2]_x \xrightarrow{4NaNp}$$

$$[Ti(OR)_2 \cdot N_2]_x^{4-} \xrightarrow{2NaNp} [Ti(OR)_2 \cdot N_2]_x^{6-}$$

$$\downarrow 4H^+ \qquad\qquad\qquad \downarrow 6H^+$$

$$N_2H_4 \qquad\qquad\qquad 2NH_3$$

There is a lot of interest in the conversion of nitrogen to useful products. Ammonia is one. The following is a brief description of some other experiments we have done, again published last year, involving the conversion of the molecular nitrogen to hydrazine (17). The stoichiometric ratio of ammonia to nitrogen was experimentally not exactly two. This is the case because hydrazine is always a product, and if one optimizes the conditions, one can get about a 20% yield of hydrazine from nitrogen (Table V).

Items 1–5 reveal certain observations related to hydrazine formation under these circumstances. In regard to the reaction pathway, in the final protonation step (5), no nitrogen is actually liberated. If N_2 were reduced to the diimide level, then on protonation of bound diimide, one presumably would release diimide, which would undergo disproportionation to give hydrazine and molecular nitrogen. Therefore, we feel that a bound diimide species is not a precursor of the N_2H_4 but that the reduction level must be represented by some form of bound hydrazine. The over-all interpretation features a normal type of complexing by titanium and reduction to two species: one in which nitrogen is held at the hydrazine level, which on protonation gives hydrazine, and a second the known ammonia precursor.

This result is a remarkable one and superficially is against the thermodynamics of the $N_2 \rightarrow NH_3$ change. Of course, the reason for this arrest at the hydrazine level undoubtedly is the fact that there must be formed some peculiarly stable hydrazine-type species, involving titanium and possibly other entities, sufficiently stable to resist to some extent further reduction by naphthalenide.

Literature Cited

(1) van Tamelen, E. E., *Accounts Chem. Res.* (1968) **1**, 111.
(2) van Tamelen, E. E., Schwartz, M. A., *J. Am. Chem. Soc.* (1965) **87**, 3277.
(3) van Tamelen, E. E., Akermark, B., Sharpless, K. B., *J. Am. Chem. Soc.* (1969) **91**, 1552.
(4) Allen, A. D., Senoff, C. V., *Chem. Commun.* (1965) 621.
(5) Vol'pin, M. E., Shur, V. B., *Nature* (1966) **209**, 1236, and references cited therein.
(6) Collman, J. P., Kang, J. W., *J. Am. Chem. Soc.* (1966) **88**, 3459.
(7) Harrison, D. E., Taube, H., *J. Am. Chem. Soc.* (1967) **89**, 5706.
(8) van Tamelen, E. E., Boche, G., Ela, S. W., Fechter, R. B., *J. Am. Chem. Soc.* (1967) **89**, 5707.
(9) van Tamelen, E. E., Boche, G., Greeley, R., *J. Am. Chem. Soc.* (1968) **90**, 1677.
(10) Allen, A. D., Bottomley, F., *Can. J. Chem.* (1968) **46**, 469.
(11) van Tamelen, E. E., Fechter, R. B., Schneller, S. W., Boche, G., Greeley, R. H., Akermark, B., *J. Am. Chem. Soc.* (1969) **91**, 1551.
(12) Watt, G. W., Baye, L. J., Drummond, F. O., *J. Am. Chem. Soc.* (1966) **88**, 1138.
(13) Henrici-Olive, G., Olive, S., *Angew. Chem.* (1969) **81**, 679.
(14) van Tamelen, E. E., Akermark, B., *J. Am. Chem. Soc.* (1968) **90**, 4492.
(15) van Tamelen, E. E., Seeley, D. A., *J. Am. Chem. Soc.* (1969) **91**, 5194.
(16) Brintzinger, H. H., Bartell, L. S., *J. Am. Chem. Soc.* (1970) **92**, 1105.
(17) van Tamelen, E. E., Fechter, R. B., Schneller, S. W., *J. Am. Chem. Soc.* (1969) **91**, 7196.

RECEIVED June 26, 1970.

6

Uptake of Oxygen by Cobalt(II) Complexes in Solution

RALPH G. WILKINS

State University of New York at Buffalo, Buffalo, N. Y. 14214

The general features of the uptake of molecular oxygen by cobalt(II) complexes are reviewed. Two types of interaction are encountered: those in which the $Co:O_2$ stoichiometry is 1:1 and those (2:1) in which bridged complexes are formed. The thermodynamics and kinetics of oxygenation and the spectral, ESR, and structural characteristics of the products are discussed, and a comprehensive table of all characterized cobalt–O_2 carriers is included. A comparison is made of the cobalt(II)–O_2 interaction with the iron(II)– and copper(I)–O_2 systems occurring in the biologically important respiratory pigments.

There are a quite limited number of transition metals which, as complexes, can reversibly interact with molecular oxygen (*1, 2, 3, 4, 5, 6*).

The interaction is reversible in that molecular O_2 can be regenerated from fresh solutions of the adducts by simple means which may include flushing the solution with an inert gas such as nitrogen or argon (sometimes with the addition of acid), increasing the temperature, or shaking the solution *in vacuo*. The biologically important complexes include hemoglobin, the commonest respiratory pigment present in the bloods of vertebrates and invertebrates, the relatively rare hemerythrin and chlorocruorin which are all iron-containing species, and the copper protein hemocyanin, the constituent of numerous *Mollusca* and *Crustacea* (*7*). It is annoying that iron(II) and copper(I), which function so well as reversible oxygen carriers in these respiratory proteins, hardly display this property in simpler complexes, which could then act as model systems. Copper(I) complexes which have been studied so far are rapidly oxidized by O_2 to copper(II) species. These include Cu_{aq}^+ (*8*; second order rate constant $> 10^5 M^{-1}$ sec^{-1} at 25°), $Cu(CH_3CN)_2^+$ (*9*), $Cu(bipy)_2^+$

(*10*), $Cu(NH_3)_2^+$ (*11*), and $Cu(imidazole)_2^+$ (*11*). The over-all rate of the reactions is likely to make difficult the experimental observation of appreciable amounts of oxygen adducts as intermediates, although mononuclear and copper-bridged oxygen species are postulated in suggested mechanisms (*9, 10, 11, 12*).

Even fewer detailed studies of the reaction of O_2 with iron(II) complexes appear to have been made. The rates range from the very slow [oxidation of Fe_{aq}^{2+} in acid perchlorate (*13*)] to the rapid [interaction with oxygen of, *e.g.*, the Fe(II) complex of cysteine (*14*)]. Once again the incorporation of O_2 in the coordination sphere of the metal appears likely from considerations of the rate law (*13, 14*) as well as in certain cases from the nature of the product. The reversible interaction of O_2 with the iron(II) phthalocyanine tetrasulfonate complex has been observed both in solution (*15*) and in the solid state (*16*) but no kinetic studies have been reported.

The most outstanding of the transition metals in its ability to act as an oxygen carrier is cobalt(II). However, it apparently does not function as such in the biological world, although there are reports of cobalt protection of brain respiration against high oxygen pressure and inhibition of tissue respiration (*17*). The study of the reversible reactions of cobalt(II) complexes with O_2 is intrinsically interesting and important in view of the rarity of the phenomenon. From the point of view of its relevance to the subject of this symposium, the cobalt(II)–oxygen interaction may be considered as a model for the metal centers of the biological systems, although of course the protein portion cannot be so simulated and is something else again. There are a number of similarities in the behavior and properties of the three metal (Fe, Cu, and Co) systems.

Oxygen adducts of cobalt complexes with both 1:1 and 1:2 (O_2:metal) stoichiometry are known, resembling the behavior of the myoglobin and hemoglobin (1:1) and the hemerythrin and hemocyanin (1:2) systems. The interactions are strongly exothermic, with ΔH values astonishingly high, for example, -38 kcal mole^{-1} for the bis(histidinato)-cobalt(II)–oxygen system (*17, 18*) and somewhat smaller values for the myoglobin (*ca.* -14) (*19*), hemoglobin [*ca.* -15 per metal center (*19*), depending on the model chosen (*20*)], hemerythrins (-12 to -20, depending on origin), and hemocyanins (*ca.* -13) (*21*). Oxygen uptake appears rapid in all cases and requires flow or relaxation techniques for kinetic measurements. The relatively straightforward kinetics of reaction of O_2 with myoglobin (*22*), the α- and β-subunits of hemoglobin (*22*), and the cobalt(II) complexes (*18, 23*) contrasts markedly with the biphasic reactions (two relaxation times) of octopus hemocyanin [Brunori (*24*), in the only reported kinetic study of hemocyanin–O_2 reaction, shows

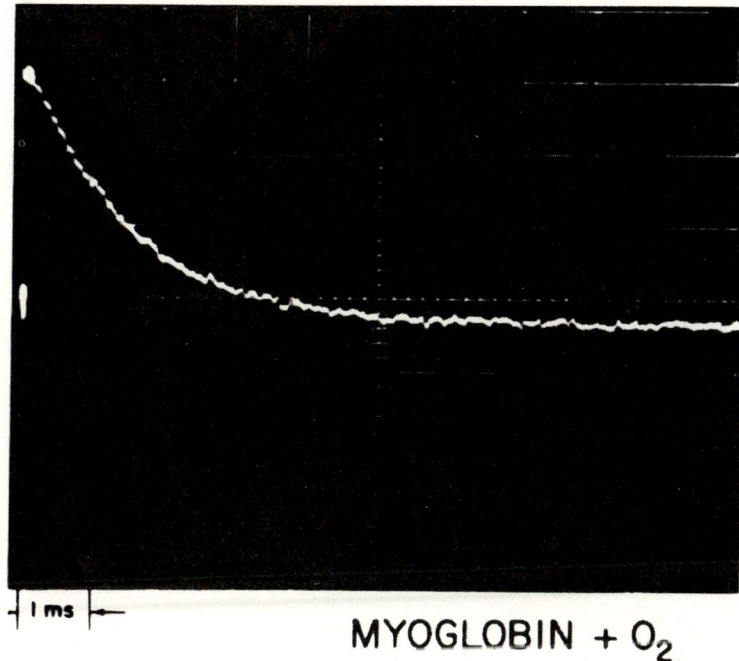

Journal of Biological Chemistry

Figure 1. Oscilloscope trace for the reaction of sperm whale myoglobin with O_2 at pH 7 in 0.1M phosphate + 0.1M KNO_3 at 20°; sweep time was 1 msec per cm; $\lambda = 436$ nm (22).

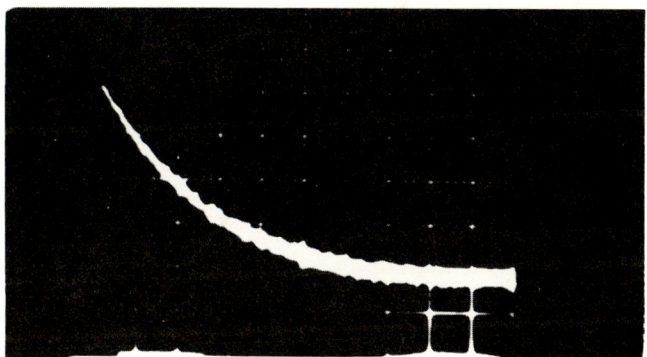

Figure 2. Oscilloscope trace for the reaction $2\ Co(\text{L-histidine})_2 + O_2 \rightleftharpoons Co_2(\text{L-histidine})_4 O_2$; total $Co(\text{L-histidine})_2 = 5.3 \times 10^{-3}$M; temperature = 25°; pH = 9.8; time scale = 50 msec cm^{-1} (54, 90)

by T-jump a bimolecular uptake of O_2 ($k = 1.5 \times 10^7 M^{-1}$ sec^{-1} at 25°, pH = 7) with a slower, possibly conformational change.] and hemerythrins (25) and the extremely complex polyphasic reaction of hemoglobin with oxygen, which still defies accurate description (20, 26) (Figures 1, 2, 3). The oxidized forms—i.e., cobalt(III) complexes (27) and methemoglobin, methemerythrin, and methemocyanin—are all oxygen-insensitive.

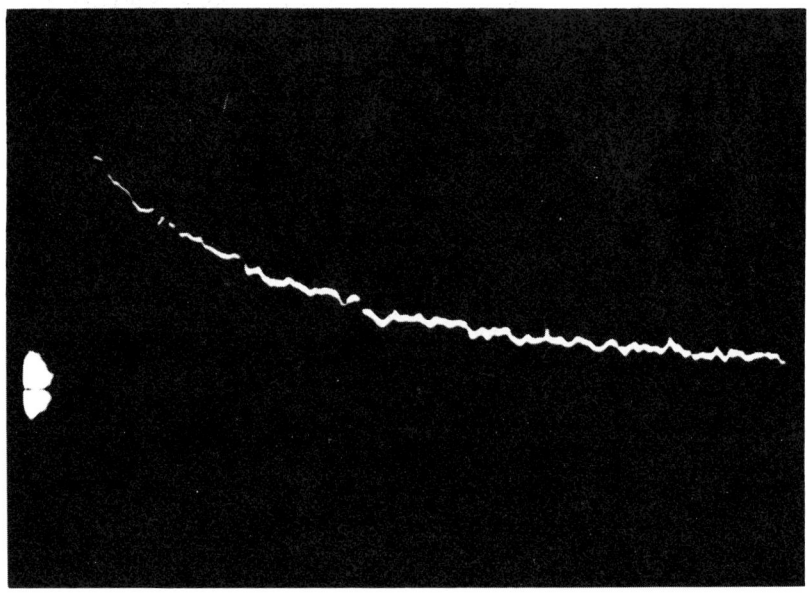

Journal of Molecular Biology

Figure 3. Oscilloscope trace of a temperature-jump experiment on Octopus hemocyanin reacting with oxygen. Potassium phosphate buffer, 0.2M, pH 7, and 20°C (before the jump). Discharge 30 kv yielding a temperature increase of 4° to 5°C. Protein concentration = 4.5 × 10⁻⁵ binding equivalent l.; fractional saturation with oxygen = 0.53; free oxygen concentration = 3.4 × 10⁻⁴M. Sweep time = 100 μsec per large screen division; observation wavelength = 348 nm (24).

The elucidation of the nature of the metal–oxygen bonding appears a more tractable problem with the cobalt than the naturally-occurring systems. Electronic spectral, ESR, and x-ray crystallographic methods have been used with varying success. Information on the systems are interrelated—e.g., the bonding in oxyhemoglobin may be comparable with that in the Co:O_2 (1:1) systems (28, 29) while that in oxyhemocyanins may resemble the cobalt(III)-bridged (1:2) peroxo products (30, 31; cf. 32).

The structure of the ligand surrounding the cobalt has a strong influence on the stability of the oxygen adducts towards further change (*32*). The protein portion of the globins (and hemocyanins?) is believed to protect the metal from further interaction with the oxygen moiety once the adduct has been formed (*33*). In a related way, the oxygenation of cob(II)alamin (Vitamin B_{12r}) stops at the (unusual) 1:1 stage, probably because the ligand shielding of the cobaltous ion prevents binuclear bridging (*34*). All adducts are metastable, however, eventually going to further products, although in some cases this may take many days or weeks (*19, 32, 35*). Thus, the ability of the metal to bring substrate (ligand) and oxygen together in a ternary complex and promote metal-catalyzed autoxidation is a recognized possibility in all cases (*6, 30, 36*). Finally, the study of the oxygenation of cobalt(II) complexes has assumed added biological relevance in view of the fascinating finding that coboglobin (the Co(II) analogue of hemoglobin) also binds O_2 in a cooperative manner (*37*).

General Features of Cobalt(II) Complexes–O_2 Interaction

A large number of cobalt(II) complexes with a wide variety of types of ligands are able to interact reversibly with O_2 in solution (Table I). For this reason, it is important that spectral (*38*), thermodynamic (*39*), kinetic (*40*), and other studies of these complexes should be carried out with anaerobic conditions, a point which has not always been appreciated. The process of oxygenation and regeneration may be possible for only a few cycles since all oxygen carriers eventually break down to products which can no longer easily give up molecular oxygen. These further reactions may take place so readily—as in the cobalt(II)–cyanide (*41, 42*) and cobalt(II)–glycylglycineamide–O_2 (*32*) systems—that they become important as soon as the initial uptake has occurred; or they may occupy the space of several days and sometimes involve further irreversible O_2 uptake as with the cobalt(II)–histidine–O_2 case (*17, 18, 43*). The oxygen adducts are usually dark brown in solution and brown, red, or nearly black solids, with (analytically useful) strong ($\epsilon = 10^3$–10^4) charge transfer bands around 350 nm. There are two types of products encountered on the basis of the oxygen:cobalt stoichiometry. In both cases, it appears that the cobalt is invariably octahedrally coordinated, with one position occupied by the O_2 unit. This may require that a ligand be picked up from solution as with the Co(II)salen complexes (*14* and *19*, Table I) which are 4-coordinate originally. The extraplanar ligands in the Co(II)–phthalocyanine tetrasulfonate-containing octahedral complex have an important bearing on the interaction of the complex with O_2. When these are NO_2^-, SCN^-, ClO_4^-, or H_2O alone, no O_2 uptake occurs; but with imidazole or cyanide occupying the extraplanar positions, reaction with O_2 is smooth (*16*).

Table I. Oxygen Carriers Prepared from Direct Oxygenation of Cobalt(II) Complexes in Solution[a]

	1:2 Species	Refs.
1	$[(NC)_5Co]_2O_2^{6-}$ [b]	*41, 42, 75*
2	$[(H_3N)_5Co]_2O_2^{4+}$	*27, 44, 57*
3	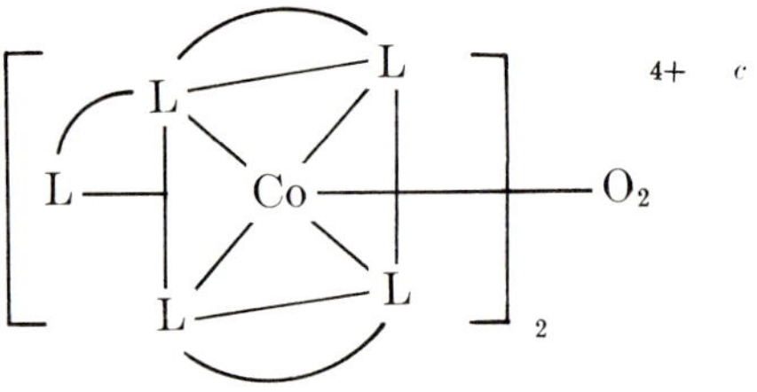 $\left[\begin{array}{c} L_4Co-O_2 \end{array} \right]_2^{4+}$ [c]	*76*

$\overset{\frown}{L\ L}\overset{\frown}{\ L}$

$H_2N(CH_2)_2NH(CH_2)_2NH_2$ or

$H_2N(CH_2)_3NH(CH_2)_3NH_2$

$\overset{\frown}{L\ L}$

$H_2N(CH_2)_2NH_2$ or

$H_2N(CH_2)_3NH_2$

4	$\left[\begin{array}{c} (HN)(NH)(NH_2)(NH_2)Co-O_2 \end{array} \right]_2^{4+}$	*23, 76*

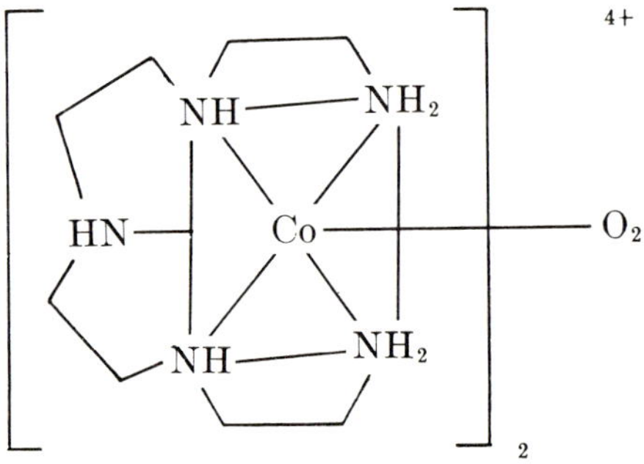

[a] Confined to systems in which 1:1 or 1:2 stoichiometry is established in solution by direct measurements or from which definite compounds are isolated. Aqueous medium unless otherwise specified.

[b] There is evidence for a 1:1 species also in solution, but not as a result of direct oxygenation.

[c] Solids only, probably reversible carriers in solution.

Table I. Continued

1:2 Species   *Refs.*

5   [Co complex with tetradentate amine ligand]—O_2 *49*

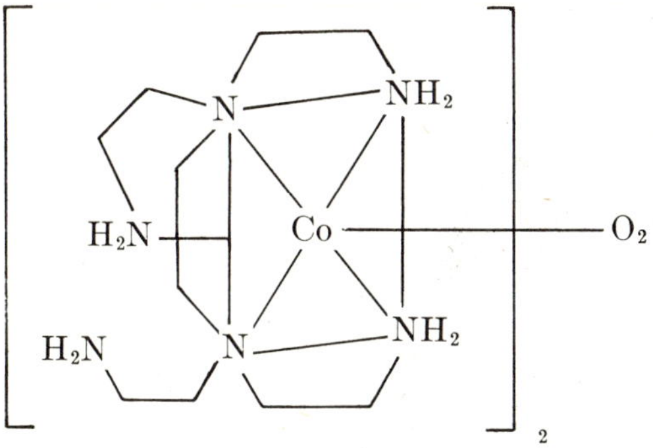

[Co complex with L ligands]—O_2

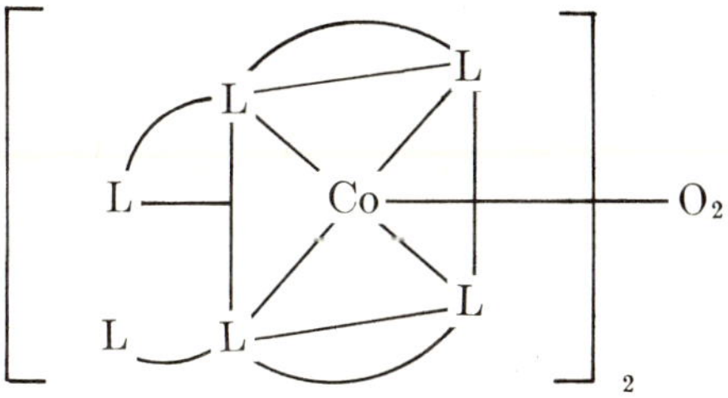

L⌒L⌒L[d]

6	$H_2N(CH_2)_2NH(CH_2)_2NH_2$	*23, 77*
7	$H_2NCH_2CH(NH_2)COO^-$	*32*
8	$H_2NCH_2CO\bar{N}CH_2COO^-$ [e,f]	*32*
9	$H_2NCH_2CO\bar{N}CH_2CO\bar{N}H$ [e]	*32*

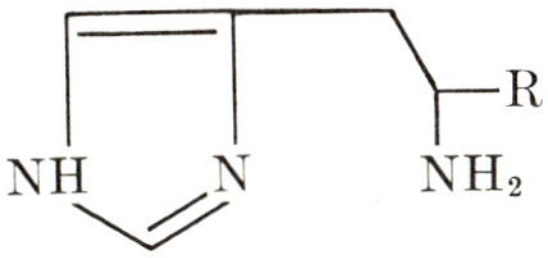

[imidazole-NH$_2$-R structure]

10   R = COO^- (hist)   *17, 18*

---

[d] In this series of compounds, **6–11**, it is believed that one of the terdentate ligands is fully coordinated, and one of the ligands is attached at only two centers, for each Co.

[e] Peptide or amide H's ionize on formation of the adduct (*32*).

[f] Other glycyl peptides also investigated.

Table I. Continued

1:2 Species   *Refs.*

11   R = $CO\bar{N}H$ or $CO\bar{N}CH_2COO^-$ [e]   32

12   L = py, Me_2S, $N(C_2H_5)_3$,
          $P(C_6H_5)_3$, $As(C_6H_5)_3$ [b,g]   58

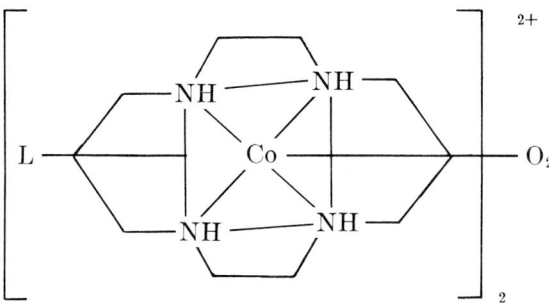

13   L = Cl, NO_2, NCS, $H_2O(4+)$ [c]   48

[g] CH_2Cl_2 or C_6H_6 solvent.
[h] DMF or DMSO solvent unless L present in solution, then wide variety of solvents.

Table I. Continued

1:2 Species		Refs.
14	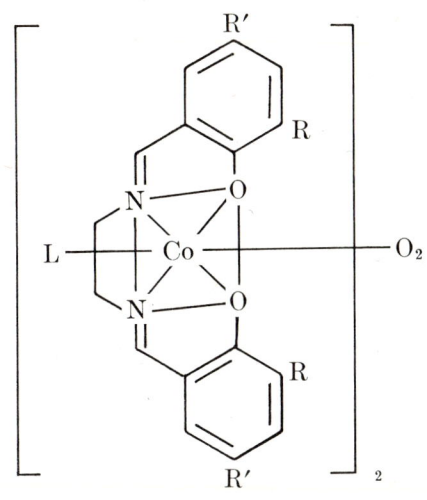	45, 46, 47

R = R' = H(salen), L = DMF, DMSO, py, pyo; SCN⁻, N₃⁻ or CH₃COO⁻ [h]

R = OCH₃, R' = H, L = DMSO;

R = H, R' = Cl, L = DMSO, py

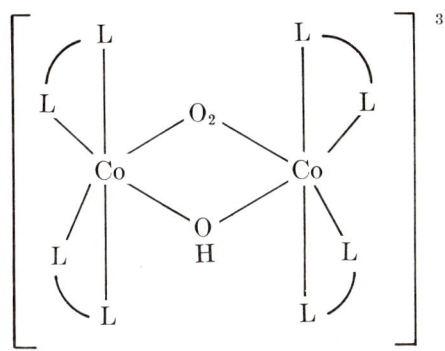

L⌒L

| 15 | en, histamine, NH_2CH_2CONH [e] | 23, 32, 77, 78 |
| 16 | 1/2 trien | 55, 56, 77 |

Table I. Continued

1:1 Species   *Refs.*

17   Vitamin $B_{12r} \cdot O_2$ *34, 58*

18   L = DMF, py, 4-NH₂py, 4-CH₃py, and 4-CNpy[i]   *60, 61, 62*

19   L = py, DMF, DMSO, R = OCH₃[j]   *47, 61,*

                                                                                         62, 63

         L = py, R = H   *61, 62*

[i] DMF, py, or toluene solvent at temperatures < 0°.
[j] L = solvent.

Table I. Continued

1:1 Species   *Refs.*

20 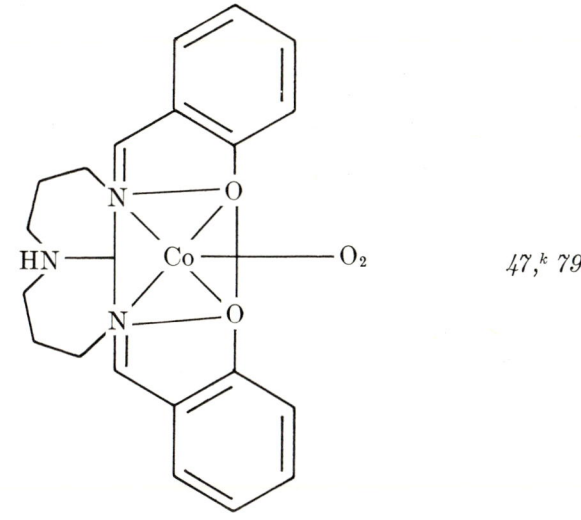   47,[k] 79

21   Cobaltoglobin (O$_2$)$_4$ 37
22   Co(X-TPP).L.O$_2$[l]   64

[k] THF or CHCl$_3$ solvent; evidence for 1:2 species in toluene.
[l] X-TPP is α, β, γ, δ-tetra(X-phenyl)porphinato cobalt(II), with X = p-OCH$_3$ and others, L = py and other amines in toluene and other solvents.

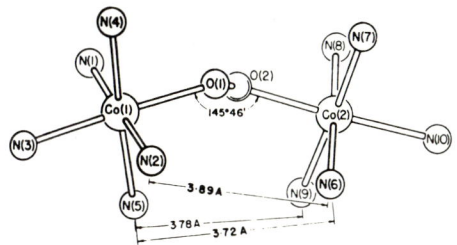

Inorganic Chemistry

Figure 4. The decaammine-μ-peroxo-dicobalt (+4) cation (44)

1:2 Species. This is the most common class, in which two molecules of the cobalt(II) species combine with one molecule of O$_2$. Most of the work relates to aqueous solution. The adducts are known in two cases (Figures 4 and 5) [x-ray structural determinations (44, 45, 46)] and believed in the others [from infrared (47) and visible (48) spectral evidence] to contain an O_2 unit bridging the two octahedral cobalt centers. Reinforcement for this idea comes from the ESR of certain one-electron oxidation products which also indicates a Co–O_2–Co bridge. Further elaboration of the bonding details is deferred until later. If the original cobalt(II) complex has all its coordination positions occupied by chelated ligands as in Co(dien)$_2^{2+}$ (23), Co(histidine)$_2$ (50), or Co(penten)$^{2+}$ (49) then part of the chelate ring must be pried open so that the initial Co–O_2 bond can be made. This will probably be at the weakest bond. There is some evidence for one noncoordinated carboxylate per Co residue in the Co$_2$(histidine)$_4$O$_2$ product (50) (Figure 6)

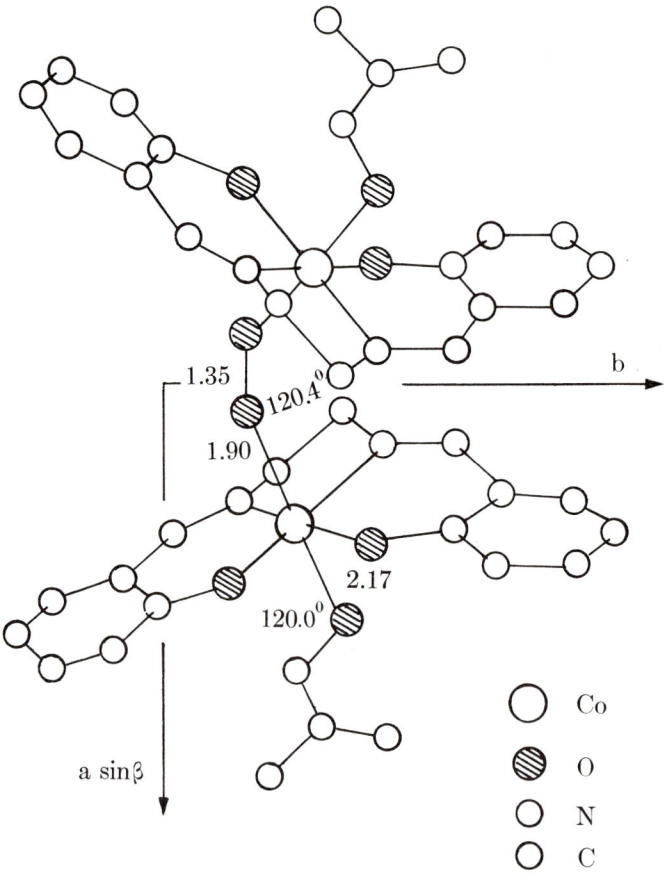

Chemical Communications

Figure 5. Structure of 1:2 adduct of O_2 with ethylenebis(salicylaldehydeiminato) cobalt(II) containing DMF (45)

but unfortunately (*51*) no x-ray structural data are available. In the event that it is difficult to remove part of the chelate as in the stereochemically rigid 1,3,5-tris(*a*-pyridyl methyleneamino)–cyclohexane–cobalt(II) perchlorate complex (*52*) or if one of the ligands dissociates very slowly from the highest complex, *e.g.*, Co(phen)$_3^{2+}$ ion (*53*), then O_2 uptake appears not to occur. The bis-(phenanthroline)cobalt(II) complex does, however, take up O_2 reversibly at pH ~ 8 (*54*).

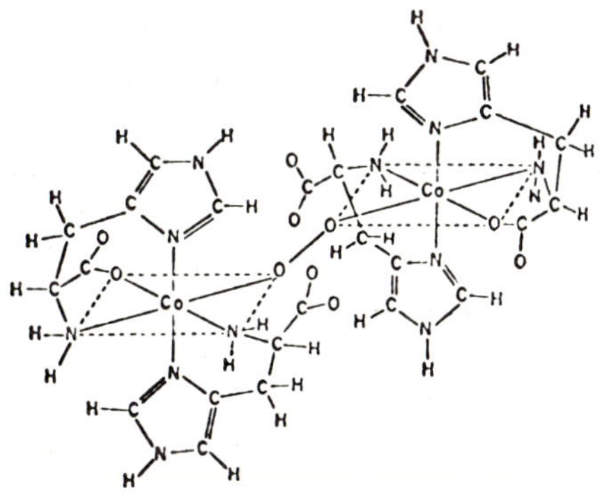

Journal of Inorganic Nuclear Chemistry

Figure 6. Possible structure of Co_2(histidine)$_4O_2$ (*50*)

If there are coordinated water molecules cis to the O_2 bridge, then the simple adduct can undergo further bridging to give the dibridged (O_2, OH) complex. This appears to happen, for example, with the reactions of Coen$_2$(H$_2$O)$_2^{2+}$ (*32*), Co(histamine)$_2$(H$_2$O)$_2^{2+}$ (*32*), and Co(trien)(H$_2$O)$_2^{2+}$ (*55, 56*) (**15** and **16**, Table I). If, however, the ligand geometry is such as to preclude this possibility, as in the trans-Co(II)–dmg and cyclam complexes (**12** and **13**, Table I) and salen condensates (**14**, Table I), then the single O_2 bridge remains. In these cases, a variety of unidentate ligands can occupy the positions trans to the O_2 bridge (*47, 48*).

All the complexes are diamagnetic or only slightly paramagnetic. The O–O bond distance (1.48Å) and geometry in [Co$_2$(NH$_3$)$_{10}$O$_2$](SO$_4$)$_2 \cdot$ 4H$_2$O (*44*) (Figure 4) strongly supports a cobalt(III)–O$_2^{2-}$–cobalt(III) formulation for this species but the substantially shorter O–O bond distance of 1.35Å in **14**, L = DMF (*45, 46*) indicates only partial transfer of electrons from Co to O_2. This is also considered (*58*) to be the case with the dmg complex (**12**) which is purple and has its charge transfer band in the visible rather than in the more usual near ultraviolet region, *e.g.*, in Co$_2$(CN)$_{10}$O$_2$ in which, once again, charge transfer appears complete. The σ-donor strength of the ligand has been related to the energy of the charge transfer transition, through increasing the charge density on the cobalt (58). It is sometimes difficult to reconcile the chemistry of these compounds with their formulation as cobalt(III) and peroxide compounds. The lability of the compounds towards decomposition (23, 57) appears inconsistent with the known inert character of cobalt(III), while strong reducing agents such as Cr^{2+} and V^{2+}, which attack peroxides readily, only react with $[Co_2(NH_3)_{10}O_2]$ via the oxygen in labile equilibrium with it (59). It would be very interesting to examine the reduction of less labile species—e.g., Co_2(histidine)$_4O_2$ (18)—to see whether direct reduction now occurs. On the other hand, the visible spectra (48) or the circular dichroism peaks (32) of these adducts resemble those of classical cobalt(III) mononuclear complexes, and the O_2 unit has a basicity consistent with its formulation as a bridged peroxo group (27, 57). A number of the 1:2 species react with strong one-electron oxidants without breakdown to form superoxide bridged species where there is good evidence for a basic $Co(III)-O_2^--Co(II)$ structure. These apparently contradictory facts can be reconciled by understanding that the inert cobalt(III) complex can become a labile cobalt(II) complex via a redox pathway (the oxygenation equilibrium) that is not normally available to cobalt(III) complexes (33). The peroxo species are believed (33) to be mandatory intermediates in the preparation of mononuclear Co(III) complexes by aerial oxidation methods, so effectively employed by Werner (see also Ref. 91). Some comments follow on specific complexes in Table I.

Complex 1 is formed very rapidly in solution, second-order rate constant > $10^5 M^{-1}$ sec^{-1} at 5°, which may partly result from the supposed structure of $Co(CN)_5^{3-}$ in that no solvent has to be displaced by entering O_2 (54). The oxygenation may be irreversible since $Co(CN)_5^{3-}$ does not appear on subjecting oxygenated solution to a vacuum (41). The adduct is only stable in high alkalinity >1M OH$^-$ (41), it has pK ~ 12 ascribed to protonation of bridged O_2 (42) but this may indicate the latter is
$$Co-\overset{O}{O}-Co \quad (27).$$ Some evidence exists for $Co(CN)_5O_2^{3-}$ from ESR of solutions of $Co_2(CN)_{10}O_2^{6-}$ and O_2 in high [OH$^-$] (65).

Aqueous ammonia solutions containing cobalt(II) salts have been long known to turn brown on exposure to oxygen (for short history, see Refs. 27 and 57). Complex 2 should be regarded as an O_2 carrier in solution (57) although texts do not always recognize this (2, 71).

Formed very rapidly in solution (23), complex 4 can be isolated as a solid perchlorate (76). The geometry of the polyamine in the complex is unknown.

Some recent papers (32, 80, 81) have cleared up a number of misunderstandings and confusion which existed with the frequently studied

Co(II)–glycylglycine–O_2 system. A red or purple product isolated from the reaction mixture is a mononuclear cobalt(III) species, Co($NH_2CH_2CONCH_2COO$)$_2^-$, in which the peptide acts as a terdentate ligand *via* NH_2, deprotonated peptide N, and carboxylate groups. The red product does not, as was often assumed (Refs. *32* and *81* for background), contain an O_2 unit, nor is it an hydroxylated species. The initial (reversible) product of the oxygenation is a brown species which is believed to have structure **8** since its formation is accompanied by the appropriate pH decrease. A recent thorough study of the system (*32*) has shown that cobalt(II), like copper(II) and nickel(II), promotes ionization of the peptide H's, although only at pH $\sim$ 11. It is the deprotonated cobalt(II) complex that is the reactive species towards O_2 (*82, 83*). This was indicated by earlier studies on the kinetics of the oxygen–Co(II) glycylglycine interaction at lower pH (*82*) and confirmed by recent flow studies on the direct formation of the brown species at higher pH (*see* Table III) (*83*). Gillard and Spencer (*80*) have indicated that there is a (red) intermediate between the initial brown and final purple-red product but its nature is uncertain. The oxygenation process occurs with a number of amides and peptide complexes with concomitant H^+ loss (*32*) but the subsequent reactions may differ markedly in character with different dipeptides (*80*). Significantly, complexes of glycylsarcosine and glycylproline, without peptide H's, do not interact with O_2 (*84*).

The uptake of oxygen by bis(histidinato)cobalt(II) in aqueous solution has been carefully studied since its first observation and detailed characterization (*17, 85*). The amber-colored diamagnetic product **10**

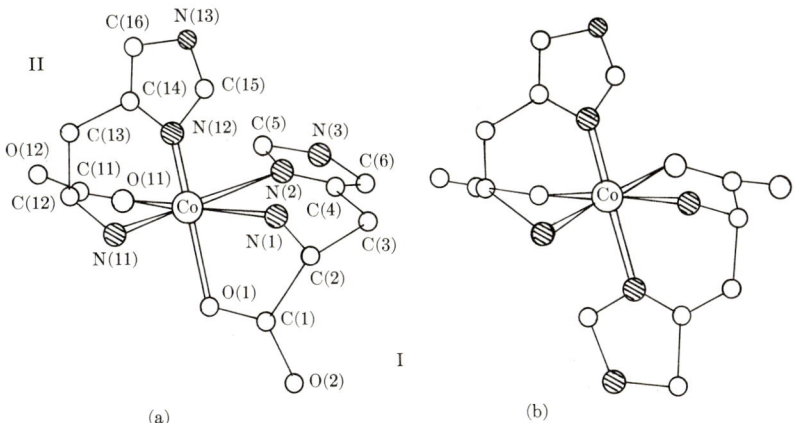

Journal of the Chemical Society

Figure 7. Structures of (a) Co(L-histidine)(D-histidine) and (b) Co(L-histidine)$_2$ (86, 87)

probably has one carboxylate free on each cobalt (50) but it would be extremely helpful to have a definitive structure determination. This interest is particularly heightened by the observation that the structures of Co(L-histidine)$_2$ (86) and Co(L-histidine)(D-histidine) (87) are radically different, with the imidazole rings trans and cis to one another, Figure 7(b) and (a), respectively. These species react at slightly different rates with O$_2$, and the decomposition behavior of the adducts (k_{-2} in scheme) are quite different (18) (Table IV). Co(L-histidine)$_2$ reacts with base (> 0.1M OH$^-$) to give a strongly colored species, which is believed to be tetrahedral and to have lost imidazole H$^+$ ions (54, 88, 89). This species also picks up O$_2$ more rapidly (90) (Table IV) and with greater avidity (91) than the octahedral Co(L-histidine)$_2$. At least two rapid steps accompany the O$_2$ uptake by the tetrahedral species (83, 90) and the product is possibly a dibridged (O$_2$, OH) species (88) which does not revert to the "normal" O$_2$ species on adjusting the pH to 9.5, although the reverse does take place (91). The octahedral Co(L-histidine)$_2$ species reacts with NO (92) but not CO (17) and in this respect resembles the behavior of hemerthyrin and coboglobin (37) but not hemoglobin or myoglobin (21). There has been very limited study of the oxygen affinity of amino acid complexes of cobalt(II). Aerial oxidation of cobalt(II) glycine (93) and cysteine (94) complexes probably go through brown binuclear peroxo species, to give respectively *cis/trans* Co(gly)$_3$ and mononuclear cobalt(III) cysteine complexes, the exact structures of which are quite debatable (94). The second order rate constant for the reaction of Co(cysteine)$_2^{2-}$ with O$_2$, $2.5 \times 10^5 M^{-1}$ sec^{-1} at 25°, falls in the generally observed range for such reactions. Although there are large spectral differences between the cobalt(II) cysteine species at pH 7.5 and 10.5 (94), the similar rate constant for the oxygenation at these pH values suggests that these are not the result of large structural differences (83). Within seconds after the initial oxygen uptake, further changes occur so that any oxygen-carrying capacity of the system is transient.

Structures **14** and **19** include the first synthetic complexes recognized as O$_2$ absorbers in the solid state (95). The class was the object of exhaustive studies of reversible oxygen uptake by cobalt(II) complexes in the solid state by Diehl, Calvin, and their coworkers (1). Recently they have been examined in solution (45, 46, 47, 63, 64) and represent an extremely interesting group in that slight changes in ligand structure (substitution on a phenyl ring, for example) or solvent can modify the oxygen:metal stoichiometry.

1:1 Species. This behavior is shown by a relatively small class of (mainly) 4- or 5-coordinated cobalt(II) complexes of quadridentate imine condensates (47, 60, 61, 62, 63), dimethylglyoxime (58), α,β,γ,δ-tetra-(p-methoxyphenyl)porphyrin (*64*), as well as Vitamin B_{12r} (*34*) and coboglobin (*37*). Most of the work has been carried out in non-aqueous solvents but this is probably, as much as anything (see Ref. *58*), a consequence of the solubility characteristics of the zero-charged cobalt(II) complexes. Furthermore, solutions of these complexes are rarely stable above 0°C. No kinetics of oxygen uptake in solution or x-ray structural determination of isolated solids have so far been reported.

The paramagnetism ($\mu = 1.5–2.2$) of the 1:1 adducts allows some useful ESR studies. These are characterized by 8-line (hyperfine splitting from a single (*61*, *62*) Co nucleus) spectra (*58*, *61*, *62*, *63*, *64*, *65*) which, with molecular weight determinations, indicate monomeric character of the complexes in solution and the solid state. Detailed infrared (*60*) and, particularly, ESR analysis (*34*, *61*, *62*, *63*) indicate the presence of a Co(III)–O_2^- moiety, essentially a superoxide ion coordinated to formally cobalt(III). Angular disposition of the O–O unit (see II in Figure 8)

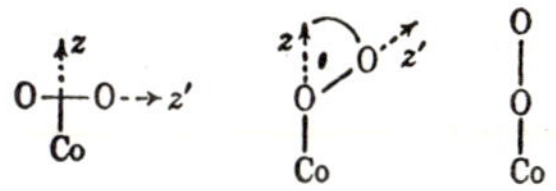

Figure 8. Disposition of O–O unit with respect to Co (*61*)

Journal of the American Chemical Society

is favored in the 1:1 species depicted in **18** and **19** (*61*, *62*). It has been suggested that **19** (R = OCH_3) has O_2 symmetrically disposed with respect to the metal (*47*), but this view is not shared by others (*61*, *62*, *63*). In keeping with this idea, good electron donors present in L appear to stabilize the Co–O_2 bond by compensating for the charge drift from Co to O_2. Increasing base strength of L parallels increasing stability of the adduct (*60*, *64*). The presence of the trans L ligand in the O_2 adduct appears essential (*47*, *58*, *60*, *63*), although Co(acacen)O_2 may be 5-coordinate in solution (*61*, *62*). Once again, the extreme behavior of complete electron transfer from cobalt to oxygen is unlikely always to be displayed.

Kinetic Aspects. The formation of all the adducts described above is fast, even at lowered temperatures, and this may have delayed the investigation of their reaction rates. So far only a number of the 1:2 species have been examined in a series of investigations using the stopped-flow technique (*18*, *23*, *55*, *57*). Kinetic data strongly support a two-step mechanism

$$CoL_2^{n+} + O_2 \rightleftharpoons CoL_2O_2^{n+} \quad (k_1, k_{-1})$$
$$CoL_2^{n+} + CoL_2O_2^{n+} \rightleftharpoons Co_2L_4O_2^{2n+} \quad (k_2, k_{-2})$$

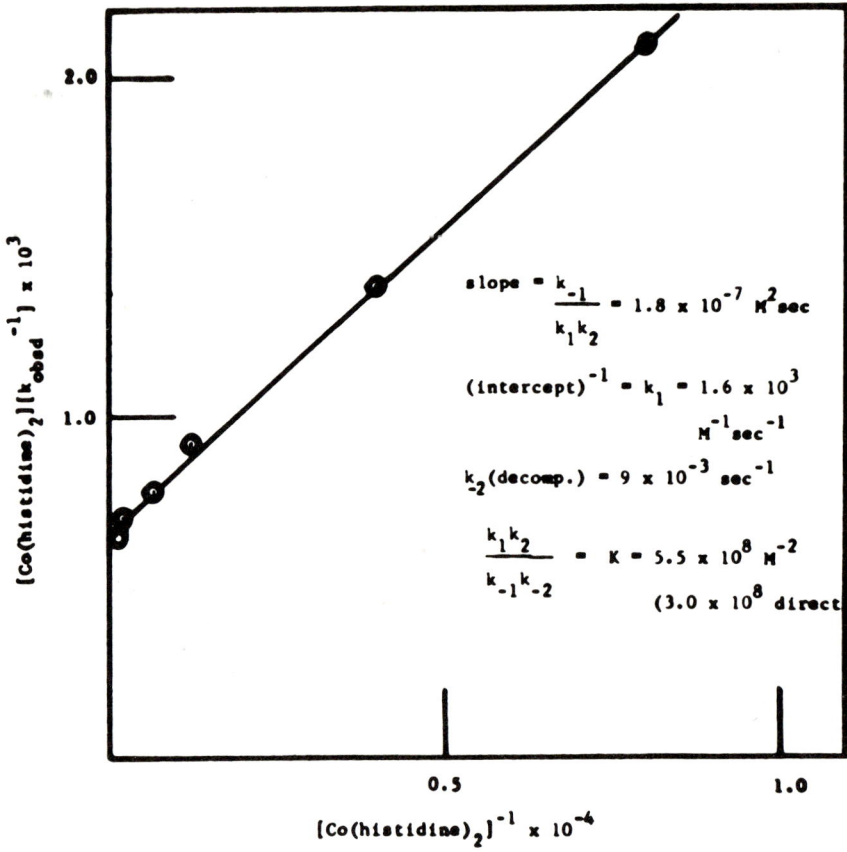

Figure 9. Formation of $[Co(histidine)_2]_2O_2$ at $4°C$ (18)

where L = histidine ($n = 0$) dien or histamine ($n = 2$) with any coordinated water omitted. With excess CoL_2^{n+}, pseudo first-order kinetics were observed, $d/dt(\text{peroxo species}) = \dfrac{k_1 k_2 [CoL_2^{n+}]^2 [O_2]}{k_{-1} + k_2 [CoL_2^{n+}]} = k_{\text{obsd}}[O_2]$ with rate constant = k_{obsd}. A plot of $[CoL_2^{n+}]/k_{\text{obsd}}$ against $[CoL_2^{n+}]^{-1}$ was linear (Figure 9) from which values for k_1 and k_2/k_{-1} could be obtained. In conjunction with studies of $EDTA^{4-}$ or H^+, decomposition of $Co_2L_4O_2^{2n+}$ which gave k_{-2}, $k_1 k_2/k_{-1} k_{-2}$ values could be calculated. These were in good agreement with the over-all equilibrium constant obtained from probe measurements of O_2 concentration in cobalt(II) complex solutions of various compositions. In the case of other cobalt(II)–ligand systems, only the limiting second-order rate constants k_1 were measured ($k_2[CoL_2^{n+}] \gg k_{-1}$) for comparative purposes and some of the results are collected in Tables II and III. The parent cobalt(II)–ammonia–O_2

Table II. Kinetic Data for Second-Order Cobalt(II) Complex–O_2 Interaction at 25°

Reactant	k_2, $M^{-1}Sec^{-1}$	ΔH^*, $Kcal/Mol$	ΔS^*, eu
$Co(trien)(H_2O)_2^{2+}$	2.5×10^4 [a]	7	-15
$Co(trien)(OH)(H_2O)^+$	2.8×10^5 [a]	8	-7
$Co(tetraen)(H_2O)^{2+}$	$\sim 10^5$ [b]		
$Co(histamine)_2(H_2O)_2^{2+}$	1.8×10^4 [b]	5	-23
$Co(en)_2(H_2O)_2^{2+}$	4.7×10^5 [b]	15	$+19$

[a] From Ref. 55.
[b] From Ref. 23.

Table III. Second-Order Rate Constants for O_2 Reaction with Co(II)–Amino Acid Type Complexes at 25°

Complex	k, $M^{-1} Sec^{-1}$
Co(L-histidine)$_2$ octahedral	3.5×10^3 [a]
Co(L-histidine-H)$_2^{2-}$ tetrahedral[b]	2.6×10^4 [b]
$Co(NH_2CH_2CONCH_2COO)_2^{2-}$	7×10^3 [c]
$Co(NH_2CH_2CONCH_2CONH)_2^{2-}$	1×10^5 [c]
Co(L-cysteine)$_3^-$	2×10^5 [d]

[a] Ref. 18.
[b] Ref. 90, 0.8M [OH$^-$].
[c] Ref. 83, pH 11–12 and high ligand/metal ratio.
[d] Ref. 83, same value at pH 7.5 and 10.5.

Table IV. Kinetic Data for the Cobalt(II)–Histidine–O_2 System at 25°[a]

System	$k_1(k_{-2}, Sec^{-1})$, $M^{-1}Sec^{-1}$	$\Delta H_1^*(\Delta H_{-2}^*)$, $Kcal\ Mole^{-1}$	$\Delta S_1^*(\Delta S_{-2}^*)$, eu
L-Histidine	$3.5 \pm 0.4 \times 10^3$ (0.47 ± 0.03)	5 ± 1 (30 ± 2)	-25 ± 3 $(+38 \pm 6)$
DL-Histidine	$2.6 \pm 0.3 \times 10^3$ (0.043 ± 0.004)	6 ± 1 (16 ± 2)	-23 ± 3 (-9 ± 6)

[a] From Journal of the American Chemical Society (18).

system was also investigated (57). The pentamine reacts much more rapidly with O_2 than the hexamine, while the tetramine and lower species do not pick up O_2 over relatively long periods of time. The decomposition of the oxygenated species generated *in situ* yields identical activation parameters to those exhibited by the well-characterized solid $[(NH_3)_5CoO_2Co(NH_3)_5](NO_3)_4$ dissolved in aqueous ammonia, thus strongly supporting the formulation of the oxygenated adduct as the bridged species (57). One is struck by the remarkably small spread of k_1 values (and activation parameters where available) for such a variety of complexes in Tables II, III, and IV. This constancy has been interpreted

$$2\text{ (trien)Co}(H_2O)_2{}^{2+} + O_2 \xrightleftharpoons{2.2 \times 10^4\ M^{-1}\ \text{sec}^{-1}}$$

$$\text{p}K = 11.0 \updownarrow \text{ fast}$$

$$2\text{ (trien)Co}(H_2O)(OH)^+ + O_2 \xrightleftharpoons{2.7 \times 10^5\ M^{-1}\ \text{sec}^{-1}}$$

Figure 10. Equilibria and rate constants

(23) as arising from a similar process of H_2O replacement by O_2 in all these complexes, dominated by H_2O exchange. Comparison with the larger rate constant for the reaction of $Co(NH_3)_5OH_2{}^{2+}$ with NH_3 ($\sim 5 \times 10^6 M^{-1}\ \text{sec}^{-1}$ from data in Ref. 66) indicates that there may be some strict orientation requirements for the O–O grouping with respect to the cobalt complex in the transition state, this being reflected in the large negative ΔS^* values. This relative constancy for the formation rate constant is duplicated in the reactions of other octahedral complexes with small neutral molecules, *e.g.*, in the reactions of $Ru(NH_3)_5H_2O^{2+}$ with N_2, N_2O, and CO (67) and of the aquated iron(II) ion with a variety of ligands (68), including NO (69) but not apparently in the reaction of hemoglobins or myoglobins with NO, O_2, CO, and other ligands for which a wide range of rate constants (10^3–$10^7 M^{-1}\ \text{sec}^{-1}$) is encountered (70).

Finally, with certain cobalt(II) complexes, the rapid oxygenation process is followed by an only slightly slower reaction. This has been fully investigated in the cobalt(II)–trien system and a scheme (Figure 10) has been suggested on the basis of kinetic, pH changes, and other behavior.

Conclusion

Taube has outlined some of the criteria for a successful oxygen carrier and indicated ways in which this might be achieved (71). The frequency with which cobalt(III) complexes are prepared by aerial oxidation of solutions containing cobalt(II) salts and the ligand suggests that a very large number of oxygen-carrying adducts exist. Even organometallic compounds such as $Co(\pi - C_5H_5)_2$ are sensitive to O_2 and form

```
                    O₂
                   ╱ ╲
       (trien)Co     Co(trien)³⁺                1.8 sec⁻¹
                ╲ ╱                      ⇌                    O₂
                OH₂ HO                                       ╱ ╲
                                             (trien)Co         Co(trien)³⁺
       pK ~ 10.8  ⇅ fast                              ╲     ╱
                                                       O
                    O₂                   ≤ 0.08 sec⁻¹   ╲
                   ╱ ╲                   ⇌              H
       (trien)Co     Co(trien)²⁺
                ╲ ╱
                OH  HO
```

(25.0°C, I = 0.2M) in Co(II)–trien–O_2 system (55)

red and brown oxygenated species (in which the cobalt/O_2 ratio is 4:1 and 2:1) at 0° and 25°, respectively (72). The structures of these compounds are unknown. There have been a limited number of systematic studies of this interesting phenomenon of oxygenation, however. It is obvious in reading accounts of the subject of oxygen carriers (2, 3, 6, 71) that the factors which influence the ability of a cobalt(II) complex to interact reversibly with O_2 are not really understood. It was suggested (3) that cobalt(II) species in a certain redox potential range could act as carriers. This range would not permit sufficient donation of electrons to the O_2 to cause irreversible oxidation but this concept is difficult to reconcile with the apparent complete transfer of electrons from cobalt to oxygen in the cobalt(III) amine peroxo product. From examination of the behavior of a number of complexes with ligands containing only N or N and O donor atoms, it appeared that the ligand must contain at least three N's (6), but even this empirical rule ought to be abandoned with the finding that Co(EDDA)(H_2O)₂ picks up O_2 in basic media (73) and that there is evidence for a species (CoEDTA)$_2O_2^{4-}$ as a reasonably stable intermediate in the reaction of CoEDTA^{2-} with H_2O_2 (74). Groups which tend to donate electrons appear to promote oxygenation; *i.e.*, primary NH_2 groups. There is sufficient subtlety in the phenomenon of oxygenation of cobalt(II) complexes to warrant their examination and to hope that their behavior may shed light on the factors which govern the reversible uptake of oxygen by transition metal complexes.

Acknowledgment

Some of the work reported in this account was supported by N.S.F. Grants GP 5671 and 8099, and these are gratefully acknowledged.

Literature Cited

(1) Martell, A. E., Calvin, M., "Chemistry of the Metal Chelate Compounds," p. 337–390, Prentice-Hall, New York, 1952.
(2) Basolo, F., Pearson, R. G., "Mechanisms of Inorganic Reactions," p. 641, Wiley, New York, 1967.
(3) Vogt, L. H., Jr., Faigenbaum, H. M., Wiberly, S. E., Chem. Rev. (1963) 63, 269.
(4) Connor, J. A., Ebsworth, E. A. V., Advan. Inorg. Chem. Radio Chem. (1964) 6, 279.
(5) Bayer, E., Schretzmann, P., Struct. Bonding (1967) 2, 181.
(6) Fallab, S., Angew. Chem. (1967) 79, 500.
(7) Ghiretti, F., Ed. "Physiology and Biochemistry of Hemocyanins," Academic, New York, 1968.
(8) Laski, F., Wilkins, R. G., unpublished results.
(9) Gray, R. D., J.Am. Chem. Soc. (1969) 91, 56.
(10) Pecht, I., Anbar, M., J. Chem. Soc. (1968) (A) 1902.
(11) Zuberbuhler, A., Helv. Chim. Acta (1967) 50, 466.
(12) Henry, P. M., Inorg. Chem. (1966) 5, 688.
(13) George, P., J. Chem. Soc. (1954) 4349.
(14) Gilmour, A. D., McAuley, A., J. Chem. Soc. (1970) (A) 1006.
(15) Vonderschmitt, D., Bernauer, K., Fallab, S., Helv. Chim. Acta (1965) 48, 951.
(16) Weber, J. H., Busch, D. H., Inorg. Chem. (1965) 4, 469, 472.
(17) Hearon, J. Z., Burk, D., Schade, A. L., J. Natl. Cancer Inst. (1949) 9, 337.
(18) Simplicio, J., Wilkins, R. G., J. Am. Chem. Soc. (1967) 89, 6092.
(19) Fannelli, A. R., Antonini, E., Cuputo, A., "Hemoglobin and Myoglobin," Advan. Protein Chem. (1964) 19, 74.
(20) Schuster, T. M., Ilgenfritz, G., in "Symmetry and Function of Biological Systems at the Macromolecular Level," A. Engstrom and B. Strandberg, Eds., p. 181, Interscience, New York, 1969.
(21) Ghiretti, F., "Hemerythrin and Hemocyanin" in "Oxygenases," O. Hayaishi, Ed., p. 517, Academic, New York, 1962.
(22) Brunori, M., Schuster, T. M., J. Biol. Chem. (1969) 244, 4046 and references therein.
(23) Miller, F., Simplicio, J., Wilkins, R. G., J. Am. Chem. Soc. (1969) 91, 1962.
(24) Brunori, M., J. Mol. Biol. (1969) 46, 213.
(25) Bates, G., Brunori, M., Amiconi, G., Antonini, E., Wyman, J., Biochemistry (1968) 7, 3016.
(26) Eigen, M., in "Fast Reactions and Primary Processes in Chemical Kinetics," S. Claesson, Ed., p. 333, Interscience, New York, 1968.
(27) Mori, M., Weil, J. A., Ishiguro, M., J. Am. Chem. Soc. (1968) 90, 615.
(28) Wittenberg, J. B., Inter-American Symp. Hemoglobin, Caracas, Venezuela, December 1969.
(29) Peisach, J., Blumberg, W. E., Wittenberg, B. A., Wittenberg, J. B., J. Biol. Chem. (1968) 243, 1871.
(30) Frieden, E., Osaki, S., Kobayashi, H., J. Gen. Physiol. (1965) 49, 213.
(31) Gray, H., ADVAN. CHEM. SER. (1971) 100, 365.
(32) Michailidis, M. S., Martin, R. B., J. Am. Chem. Soc. (1969) 91, 4683.
(33) Wang, J. H., Accounts Chem. Res. (1970) 3, 90.
(34) Bayston, J. H., King, N. K., Looney, F. D., Winfield, M. E., J. Am. Chem. Soc. (1969) 91, 2775.
(35) Lontie, R., Witters, R., in "The Biochemistry of Copper," p. 455, Academic, New York, 1966.
(36) Beck, M. T., Rec. Chem. Progr. (1966) 27, 37.
(37) Hoffman, B. M., Petering, D., to be published.

(38) Ballhausen, C. J., Jorgensen, C. K., *Acta Chem. Scand.* (1955) **9**, 397.
(39) Bjerrum, J., "Metal Ammine Formation in Aqueous Solution," p. 184, P. Haase and Son, Copenhagen, 1941.
(40) Biradar, N. S., Stranks, D. R., Vaidya, M. S., *Trans. Faraday Soc.* (1962) **58**, 2421.
(41) Bayston, J. H., Beale, R. N., King, N. K., Winfield, M. E., *Australian J. Chem.* (1963) **16**, 954.
(42) Bayston, J. H., Winfield, M. E., *J. Catalysis* (1964) **3**, 123.
(43) Zompa, L. J., Sokol, C. S., Brubaker, C. H., Jr., *Chem. Commun.* (1967) 701.
(44) Schaefer, W. P., *Inorg. Chem.* (1968) **7**, 725.
(45) Calligaris, M., Nardin, G., Randaccio, L., *Chem. Commun.* (1969) 763.
(46) Calligaris, M., Nardin, G., Randaccio, L., Ripamonti, A., *J. Chem. Soc.* (1970) (A) 1069.
(47) Floriani, C., Calderazzo, F., *J. Chem. Soc.* (1969) (A) 946.
(48) Bosnich, B., Poon, C. K., Tobe, M. L., *Inorg. Chem.* (1966) **5**, 1514.
(49) Emmenegger, F. P., Ph.D. Thesis, Juris-Verlag, Zurich, 1963.
(50) Sano, Y., Tanabe, H., *J. Inorg. Nucl. Chem.* (1963) **25**, 11.
(51) Harding, M. M., Long, H. A., *J. Chem. Soc.* (1968) (A) 2554.
(52) Lions, F., Martin, K. V., *J. Am. Chem. Soc.* (1957) **79**, 1572.
(53) Ellis, P., Wilkins, R. G., *J. Chem. Soc.* (1959) 299.
(54) Simplicio, J., Ph.D. Thesis, State University of New York at Buffalo, 1969.
(55) Miller, F., Wilkins, R. G., *J. Am. Chem. Soc.* (1970) **92**, 2687.
(56) Fallab, S., *Chimia* (1969) **23**, 177.
(57) Simplicio, J., Wilkins, R. G., *J. Am. Chem. Soc.* (1969) **91**, 1325.
(58) Schrauzer, G. N., Lee, L. P., *J. Am. Chem. Soc.* (1970) **92**, 1551.
(59) Hoffman, A. B., Taube, H., *Inorg. Chem.* (1968) **7**, 1971.
(60) Crumbliss, A. L., Basolo, F., *J. Am. Chem. Soc.* (1970) **92**, 55.
(61) Hoffman, B. M., Diemente, D. L., Basolo, F., *J. Am. Chem. Soc.* (1970) **92**, 61.
(62) Hoffman, B. M., Diemente, D. L., Basolo, F., *Chem. Commun.* (1970) 467.
(63) Misono, A., Koda, S., *Bull. Chem. Soc. Japan* (1969) **42**, 3048.
(64) Walker, F. A., *J. Am. Chem. Soc.* (1970) **92**, 4235.
(65) Bayston, J. H., Looney, F. D., Winfield, M. E., *Australian J. Chem.* (1963) **16**, 557.
(66) Murray, R., Lincoln, S. F., Glaeser, H. H., Dodgen, H. W., Hunt, J. P., *Inorg. Chem.* (1969) **8**, 554.
(67) Armor, J. N., Taube, H., *J. Am. Chem. Soc.* (1969) **91**, 6874.
(68) Eigen, M., Wilkins, R. G., in "Mechanisms of Inorganic Reactions," ADVAN. CHEM. SER. (1965) **49**, 55.
(69) Kustin, K., Taub, I. A., Weinstock, E., *Inorg. Chem.* (1966) **5**, 1079.
(70) Gibson, Q. H., "Rates of Reaction of Some Haem Compounds," in "Progress in Reaction Kinetics," G. Porter, Ed., p. 319, Pergamon, New York, 1964.
(71) Taube, H., *J. Gen. Physiol.* (1965) **49**, 29.
(72) Weiher, J. F., Katz, S., Voigt, A. F., *Inorg. Chem.* (1962) **1**, 504.
(73) Evans, M., Laski, F., Wilkins, R. G., unpublished results.
(74) Yalman, R. G., *J. Phys. Chem.* (1961) **65**, 556.
(75) Haim, A., Wilmarth, W. K., *J. Am. Chem. Soc.* (1962) **83**, 509.
(76) Gainsford, A. R., House, D. A., *Inorg. Nucl. Chem. Letters* (1968) **4**, 621.
(77) Fallab, S., *Chimia* (1967) **21**, 538.
(78) Foong, S. W., Miller, J. D., Oliver, F. D., *J. Chem. Soc.* (1969) (A) 2846.
(79) Kon, H., Sharpless, N. E., *Spectr. Letters* (1968) 49.
(80) Gillard, R. D., Spencer, A., *J. Chem. Soc.* (1969) (A) 2718.
(81) McKenzie, E. D., *J. Chem. Soc.* (1969) (A) 1655.

(82) Tanford, C., Kirk, D. C., Jr., Chantooni, M. K., Jr., *J. Am. Chem. Soc.* (1954) **76**, 5325.
(83) Watters, K., Wilkins, R. G., unpublished results.
(84) Tang, P., Li, N. C., *J. Am. Chem. Soc.* (1964) **86**, 1293.
(85) Burk, D., Hearon, J., Caroline, L., Schade, A. L., *J. Biol. Chem.* (1946) **165**, 723.
(86) Harding, M. M., Long, H. A., *J. Chem. Soc.* (1968) (A) 2554.
(87) Candlin, R., Harding, M. M., *J. Chem. Soc.* (1970) (A) 384.
(88) Morris, P. J., Martin, R. B., *J. Am. Chem. Soc.* (1970) **92**, 1543.
(89) McDonald, C. C., Phillips, W. D., *J. Am. Chem. Soc.* (1963) **85**, 3736.
(90) Miller, F., Wilkins, R. G., unpublished results.
(91) Bagger, S., *Acta Chem. Scand.* (1969) **23**, 975.
(92) Silvestroni, P., Ceciarelli, L., *J. Am. Chem. Soc.* (1961) **83**, 3905.
(93) Petru, F., Jursik, F., *Coll. Czech. Chem. Commun.* (1969) **34**, 3153.
(94) McCormick, B. J., Gorin, G., *Inorg. Chem.* (1962) **1**, 691.
(95) Tsumaki, T., *Bull. Chem. Soc. Japan* (1938) **13**, 252.

RECEIVED June 26, 1970.

7

The Effect of Metal Ions on the Structure of Nucleic Acids

GUNTHER L. EICHHORN, NATHAN A. BERGER, JAMES J. BUTZOW, PATRICIA CLARK, JOSEPH M. RIFKIND, YONG A. SHIN, and EDWARD TARIEN

Laboratory of Molecular Aging, Gerontology Research Center, National Institutes of Health, National Institute of Child Health and Human Development, Baltimore City Hospitals, Baltimore, Md. 21224

> *Nuclear magnetic resonance line broadening studies have been employed in the determination of metal-binding sites on nucleotides and polynucleotides. Both phosphate and "base" sites are involved, and the specific electron donors on the base sites vary with the position of the phosphate group in isomers of adenosine monophosphate. Metals binding to phosphate stabilize the DNA double helix, whereas metal ions binding to base can unwind and rewind the double helix in a reversible process that mimics steps in the biological replication and transcription of DNA. Metals binding to phosphate bring about the depolymerization of polyribonucleotides like RNA; the tendency to cleave a given phosphodiester bond varies with the nature of the base adjacent to the bond.*

We have been studying the interaction of metal ions with nucleic acids for several years and now present some recent studies, as well as a few highlights of past investigations.

The reason for our interest in metal interactions with nucleic acids is, first of all, the expectation that the nucleoside bases whose sequence constitutes the genetic code will interact preferentially with metal ions, and we have felt from the outset that these reactions would help in the determination of base sequences. Secondly, all of the reactions in which nucleic acids generally participate in biological systems are mediated by metal ions, and we hope that these studies will help us to understand some of the ways in which the metal ions function biologically. Thirdly, from the inorganic chemical point of view, the nucleic acid molecules represent macromolecules which offer a large number of sites to the metal ion. It is interesting to see where the metal ions will go when they are confronted with such a choice.

Other groups have found this topic of interest, but the time limitation does not allow a comprehensive survey of the field.

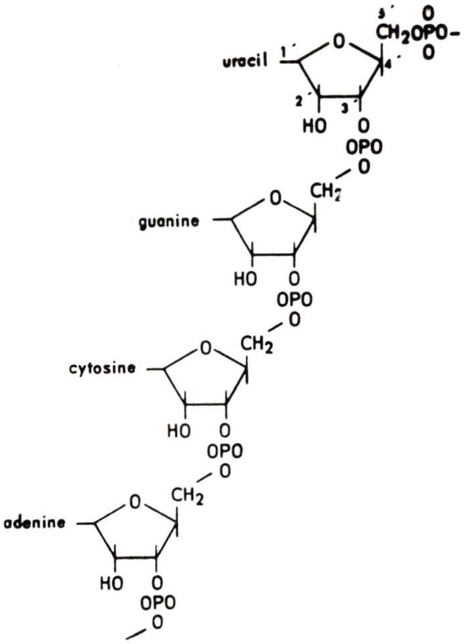

Figure 1.

Figure 1 is a reminder that a nucleic acid molecule consists of a backbone of ribose and phosphate, to which the various heterocyclic bases are attached. The sequence of these bases, of course, constitutes the genetic code. The primes are used in numbered positions of the ribose whereas the unprimed numbers refer to the numbering in the heterocyclic bases. The two different nucleic acid molecules, DNA (deoxyribonucleic acid) and RNA (ribonucleic acid), differ by the presence or absence of the 2'-hydroxyl groups. RNA contains the 2'-hydroxyl groups; DNA does not.

Complexes of Nucleotides

I shall be concerned first with the study of the structure of metal complexes of monomeric residues of the nucleic acid molecule, called the nucleotides.

Phosphate Binding

Figure 2 indicates the broadening of the ^{31}P resonance of the deoxy forms of adenylic acid (5'AMP) and thymidylic acid (5'TMP) as a result of complexation. At the top are the resonance peaks (A and C, respectively) in the absence of metal ions, and at the bottom are the resonance peaks (B and D, respectively) in the presence of added copper ions (1). This type of study indicates binding of copper to the phosphate.

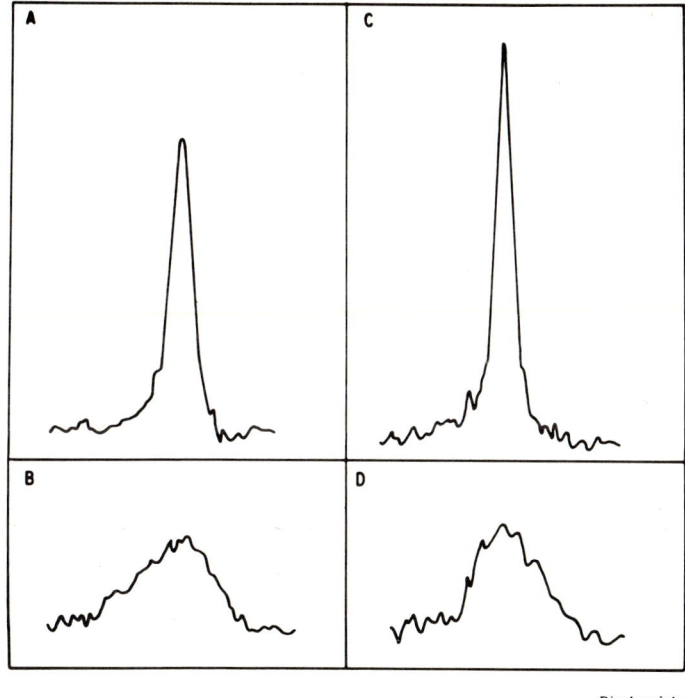

Biochemistry

Figure 2.

Base Binding

The structures of the four most common bases, adenine, cytosine, guanine, and uracil, are depicted in Figure 3; the point of attachment to ribose in Figure 1 is shown by R. Obviously, these molecules contain a large number of electron donor sites, so that it is possible for the metal ions to bind not only to the phosphate but also to various positions on the bases.

Figure 4 indicates work done previously with E. D. Becker on the nuclear magnetic resonance changes of the deoxy forms of 5'-adenylic

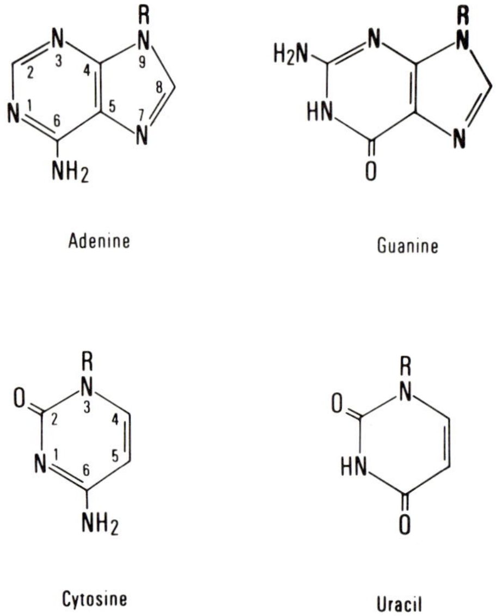

138 BIOINORGANIC CHEMISTRY

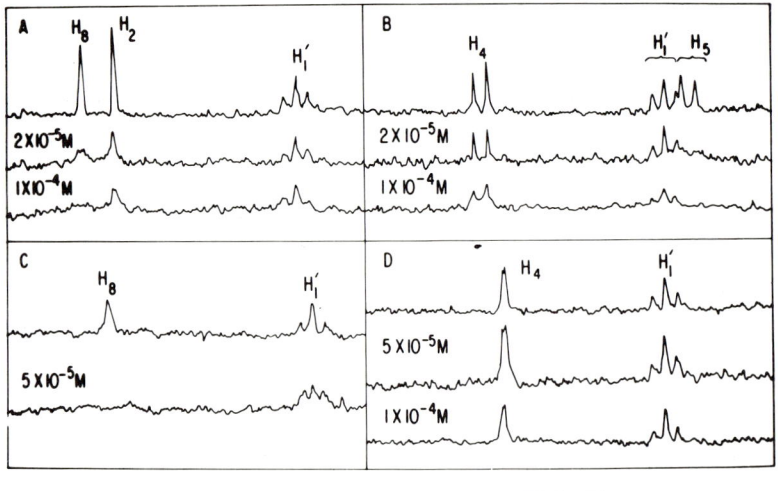

Figure 3.

Figure 4.

acid (Curve A), 5'-cytidylic acid (Curve B), 5'-guanylic acid (Curve C), and 5'-thymidylic acid (Curve D) in the presence of copper ions (1). We see, for example, that in 5'AMP the H_8 is broadened by the addition of $2 \times 10^{-5}M$ copper ions to a very much greater extent than the H_2. Now let me introduce a code that I am going to use in some of the following figures. H_8^+ means that the H_8 is broadened initially; H_2^- means that H_2 is not broadened initially. You will notice that in cytosine, the H_5 peak is plus, the H_4 peak is minus. In guanine, the H_8 peak is plus, and in thymine there is no broadening.

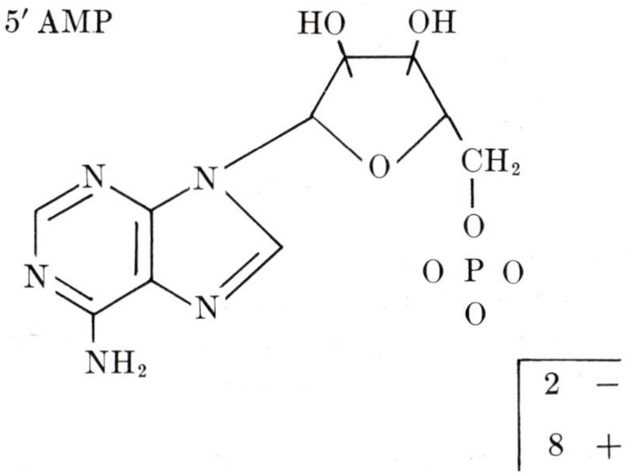

Figure 5.

Figure 5 illustrates, using this code, what has just been demonstrated in the NMR spectra. In 5'AMP, the H_8 is broadened before the H_2, indicating that binding of the copper occurs at N_7, since the effect of the paramagnetic Cu(II) is greatest on the proton closest to it. In the same way, binding of copper is implicated at N_7 in 5'GMP; in 5'CMP, the binding is implicated at N_1; and in 5'TMP no binding is indicated. In dimethyl sulfoxide, it is possible to look at the NMR spectra and to see the proton resonances on the amino groups and on the heterocyclic nitrogens. In every case, no broadening occurred at these sites. We therefore concluded that the 5' phosphates bind copper at the previously indicated sites.

Recently, it has occurred to us that there may be difficulties in the interpretation of such NMR line broadening studies.

The first difficulty arises from some recent x-ray studies on the heterocyclic bases themselves, without any ribose phosphate attached. These studies have indicated that a metal binding on one position of a molecule drastically affects the bond distances at other positions in the heterocyclic ring (2). The question then occurs, is it really possible to use NMR broadening studies to tell where the metal ion is going? To answer this question, we first used a compound in which the nitrogen atom in the 7-position is substituted by a carbon atom. Remember that the metal ion in AMP binds to the N_7 position, thus accounting for the broadening of the H_8. When this nitrogen is substituted by carbon (Figure 6), we get the tubercidin molecule. This molecule, of course, because of the

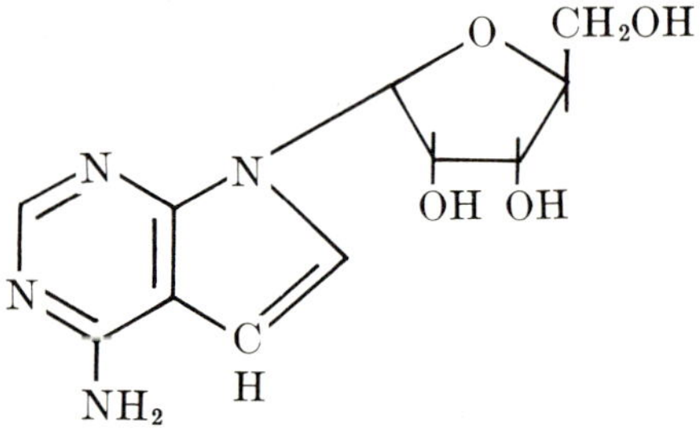

Figure 6. Tubercidin

presence of carbon in the 7-position, cannot bind a copper ion there. If the line broadening NMR studies are reliable indicators of metal binding sites, there should not be any broadening at H_8. This is in fact what happens (3). Copper now binds either at position 1 or 3, and we get broadening at H_2.

The next problem that we felt we had to consider is the fact that copper ions, which bind to the phosphate group in the so-called anti-conformation of the 5'AMP molecule (Figure 5), are very close to the H_8. Studies have been performed on the assumption that it is possible to tell where the H_8 is located on the basis that metal ions binding to the phosphate broaden this position (4). If this assumption is correct, then, of course, broadening at H_8 would not indicate binding at N_7. Consequently, we studied 3'AMP in which the phosphate is bound at the 3' position (Figure 7). You can see that this phosphate—you see it better with models—cannot get anywhere near the H_8 and yet the effect of the copper ions on the NMR broadening is exactly the same as it was with the 5'AMP (3). In other words, there is still preferential broadening of H_8, indicating binding at N_7. We are satisfied that the phosphate group

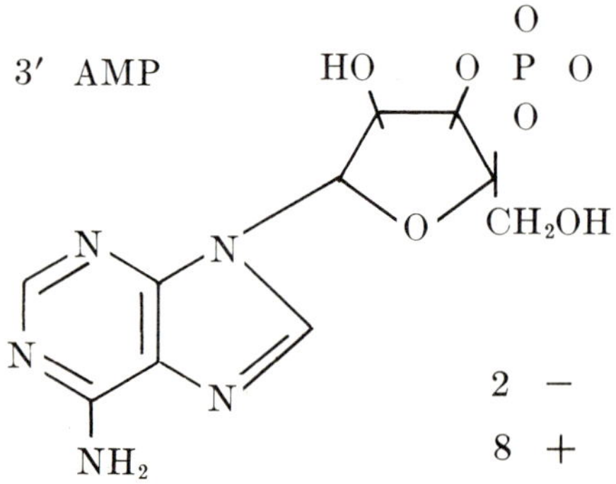

Figure 7.

binding alone cannot be responsible for this broadening effect. However, the matter is even more complicated than we had suspected.

Figure 8 shows that for 2'AMP, where the phosphate is in the 2' position, we get simultaneous broadening at the 2 position and at the 8 position; the same behavior is exhibited by 2'–3' and by 3'–5' cyclic AMP's. Table I reviews the broadening effects on the various AMP isomers: 3' and 5'AMP give preferential broadening at H_8, whereas 2'AMP and the cyclic AMP's do not give preferential broadening at H_8.

2' AMP

2 +
8 +

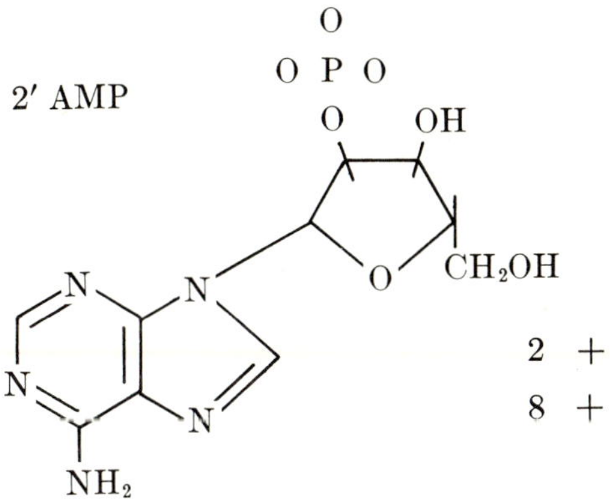

Figure 8.

Table I. Broadening Effects of AMP Isomers

	2	8
3' AMP	−	+
5' AMP	−	+
2' AMP	+	+
2'3' cyc. AMP	+	+
3'5' cyc. AMP	+	+

Sternlicht, Jones, and Kustin (5) have found evidence that the metal ion binds, in such complexes, to the base of one molecule and the phosphate of another molecule. Models demonstrate that our NMR data can be explained if there is a binuclear complex in which each metal ion is bound to the base of one molecule and to the phosphate of another (Figure 9).

Figure 10 represents the structure that we draw for the copper complex of 5'AMP. In this structure, the copper is bound to the 7-nitrogen of one AMP and to the phosphate of the second AMP to form a ring. The 7-nitrogen of the second AMP is bound to a second copper ion, which in turn is bound to the phosphate of the first AMP. We feel that this

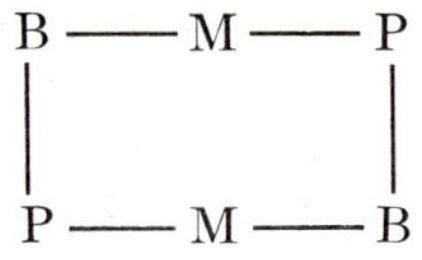

Figure 9.

5' AMP

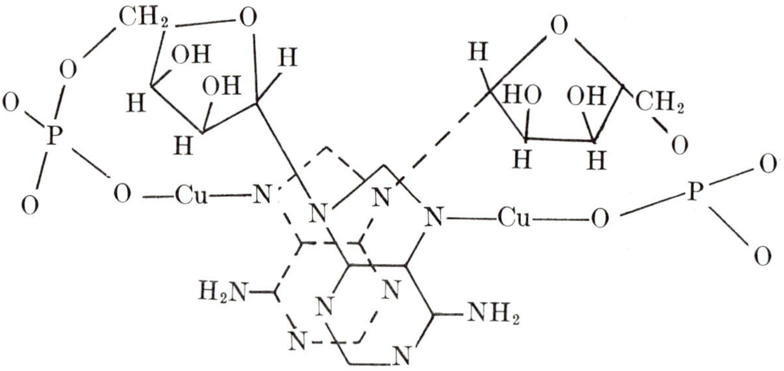

Figure 10.

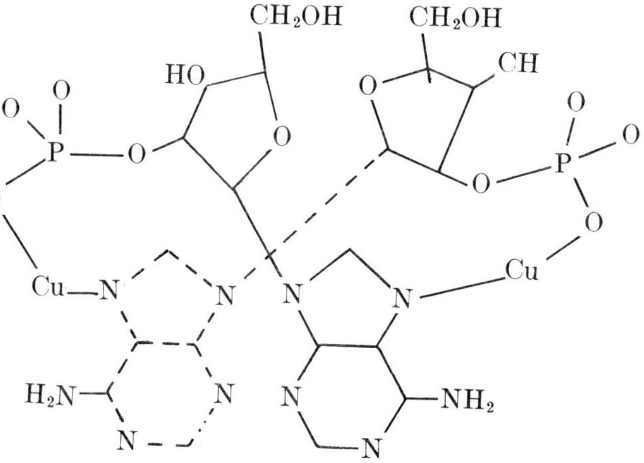

Figure 11.

structure is stabilized with 5′AMP and 3′AMP because it allows these two bases to stack, permitting the kind of π interaction that stabilizes the DNA molecule. Berger has constructed a model of this molecule with the bases stacking when the phosphate is either in the 5′ position or in the 3′ position. A second reason for believing that this structure is correct is that when you look at a model of this molecule, the 2′ and 3′ protons of the ribose are close to the copper, and they are indeed broadened in 5′ and 3′AMP. The 1′ protons are away from the copper and they are not broadened. Figure 11 indicates the same structure, schematically drawn for the 2′AMP, on the basis of model building. The drawing is designed to show that this binuclear complex does not permit base stacking in 2′AMP, and therefore the structure is not stabilized. Similarly, 2′,3′-cyclic and 3′,5′-cyclic AMP cannot form base-stacked binuclear complexes. Thus, the behavior of the various substances, as shown in Table I, is explained (3).

There is evidence in favor of a structure of the copper complex of 2′AMP, involving binding to phosphate and N_3 (Figure 12). This structure is in line with the NMR results, which show a broadening of the 1′ proton. The 1′ proton in this model of the molecule is close to the copper, whereas, e.g., the 3′ proton on the ribose is not.

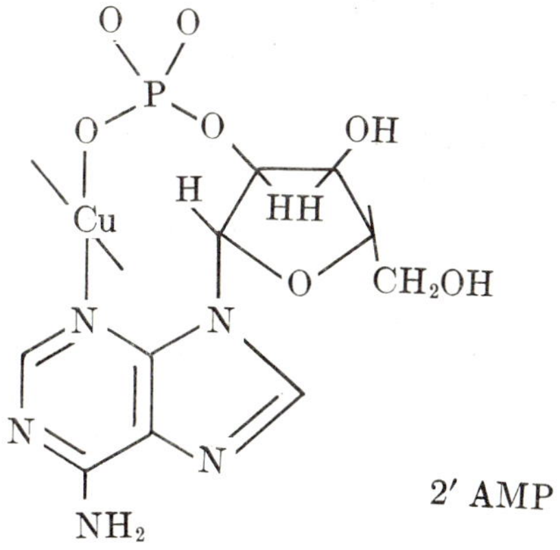

Figure 12.

These results are of interest not only from the point of view of where metals bind in macromolecules, but also from the point of view of where metals bind in the monomeric constituents, i.e., the free nucleotides. They are of some interest in biochemical phenomena because, in various enzymatic reactions, metal ion activators are capable of distinguishing between nucleotide molecules which differ only very slightly—for example, in the position of the phosphate group or in the presence or absence of a 2′ hydroxyl group.

The NMR results show that, in the case of 3′ and 5′ nucleotides, the binding of metal, or at least copper, does occur at N_7 for adenine and guanine and at N_1 for cytosine; very little binding of copper occurs in thymine, the methyl derivative of uracil. NMR studies have also shown us that in the polymeric nucleic acids binding occurs generally in the same position as in these monomers. For example, in polyadenylic acid, the binding is also at N_7, and in polycytidylic acid it is at N_1.

I should like now to discuss some of the consequences for the ligands of metals binding to the bases and to the phosphate groups of the macromolecules.

Effect of Metal Binding on DNA Structure

Let us look at the structural features of the DNA molecule that are important in metal binding (Figure 13). As the previous discussion would indicate, a metal ion can bind to one of two sites, the phosphate groups, which are on the ribose–phosphate backbone on the surface of the molecule, and the bases. When the metal ions bind to the bases, they replace the protons which hold the complementary bases of these two strands together. It is apparent that binding of a metal ion to the phosphate sites and to the base sites in a DNA molecule should produce quite different results. This expectation is borne out by a consideration of the effect of two different metal ions on the DNA structure (Figure 14). Curve A represents the "melting curve" of DNA at low ionic strength; *i.e.*, the conversion of the double helix, which has low ultraviolet absorbance, to unwound, single coils, which have high absorbance. Some metal ions like magnesium (Curve B) increase the melting temperature of the DNA, which indicates stabilization of the double helix through phosphate binding. Other metal ions, such as copper ions (Curve C), decrease the melting temperature of the DNA, indicating destabilization of the double helix through binding to the bases (6).

Stabilization of DNA

The stabilization of DNA by metal binding to phosphate is a phenomenon that has been known for a long time (7, 8). It stems from the fact that DNA, devoid of bound metals, contains an array of negatively charged phosphate ions on its surface that tend to repel each other. Therefore, when DNA is dissolved in distilled water, it spontaneously unwinds because these negative charges want to get as far away from

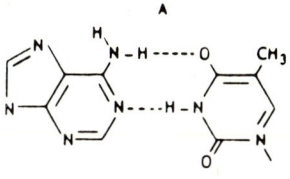

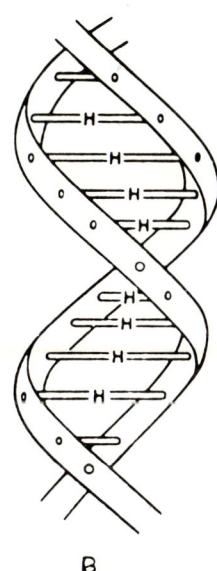

Figure 13.

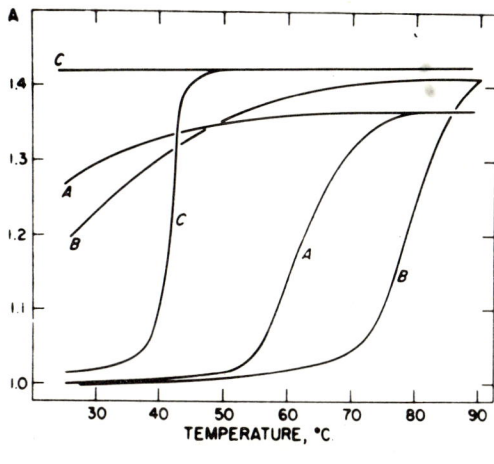

Figure 14.

each other as they can (Figure 15). The addition of counter ions prevents this unwinding phenomenon from taking place, and thus metals binding to phosphate stabilize the DNA structure.

Unwinding and Rewinding of DNA

On the other hand, metal ions binding to bases—*e.g.*, Cu(II)—destabilize the structure. Curve A in Figure 16 is the same as Curve C in Figure 14; it is the melting curve of DNA in the presence of copper ions at low ionic strength. You will notice that when higher concentrations of electrolyte are added to the solution (Curves B–D), the melting temperature increases even in the presence of copper ions. In other words, although copper ions at very low concentrations decrease the melting temperature of the DNA by destabilizing the double helix, this destabilization effect can be counteracted by the addition of a very high concentration of salt (9).

It occurred to us to try the following experiment: We started out with DNA and copper at low ionic strength (Curve A, Figure 16). The solution was heated, resulting in the unwinding of the double helix to form single strands, and the solution was then cooled, resulting in no change from the high absorbance of the heated solution. We then added enough sodium nitrate to make this solution the same ionic strength as solution D. We discovered a gradual decrease in absorbance until the initial absorbance had been regained, indicating that gradually the unwound DNA was being rewound into native DNA.

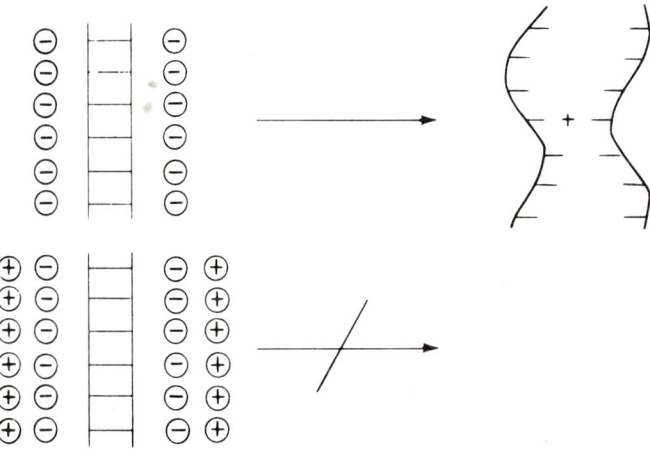

Figure 15. *Stabilization of ordered structure by metals binding to phosphate*

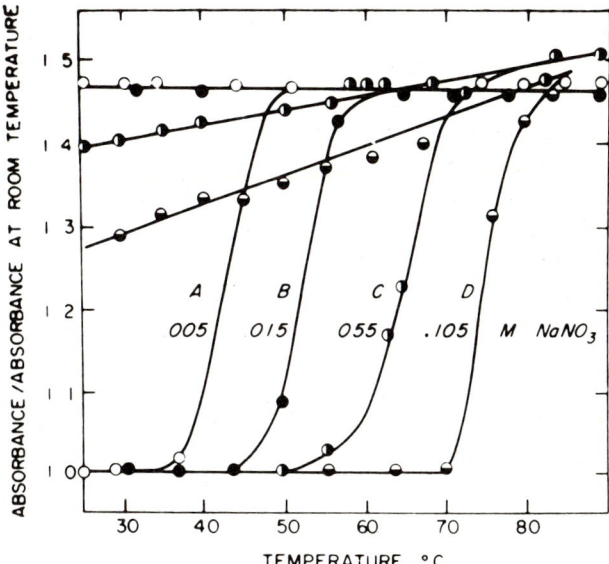

Proceedings of the National Academy of
Sciences of the United States of America

Figure 16.

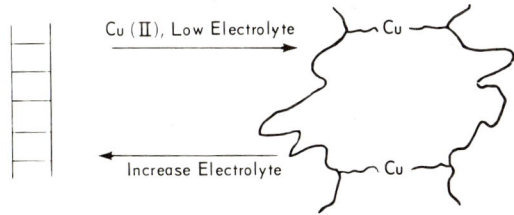

Figure 17. Reaction of DNA with copper (II)

What happened is shown in Figure 17. The native DNA, in the presence of copper and low electrolyte concentration, is unwound upon heating. If there were no copper ions present, it would be very difficult to rewind the DNA molecule because the strands would be tangled up in solution, and the complementary bases simply couldn't find each other (Curve A, Figure 14). However, when a metal ion is present which can form cross links between the DNA strands even after melting, it is possible to regain the double helix when the conditions stabilizing the double helix are again produced. For this reason, we do indeed get a reversible reaction (9).

With zinc ion (Figure 18), this reversibility is perhaps even more striking because we can unwind the DNA molecule by heating and rewind it by cooling (Figure 19). Thus, just by temperature variation alone, it is possible to go back and forth between double- and single-stranded DNA (*10*).

Both zinc and copper ions are capable of promoting the reversible reaction. In the case of copper, cooling the solution alone does not suffice to regenerate the double helix and it was necessary to add electrolyte. In the case of zinc, the double helix regenerates simply by cooling. The reason for the difference between copper and zinc, we believe, is that copper binds more strongly to the bases; it is more difficult to dislodge copper ions from the DNA, and therefore cooling alone does not regenerate the double helix. Zinc is bound less strongly to the bases than copper, hence it is more readily dislodged in the competition between the formation of the double helix and the formation of the metal complex of the single-stranded DNA.

Metal Ions and Polynucleotide Conformation

Let me illustrate another example of the effect of metal ions on the conformation of such macromolecular structures. Figure 20 represents

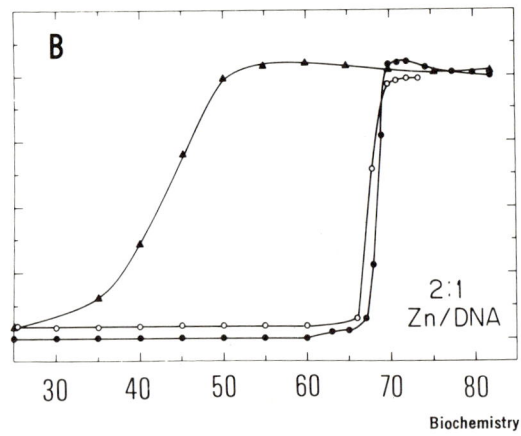

Biochemistry

Figure 18.

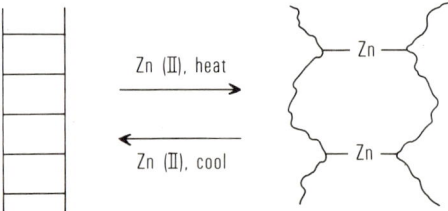

Figure 19.

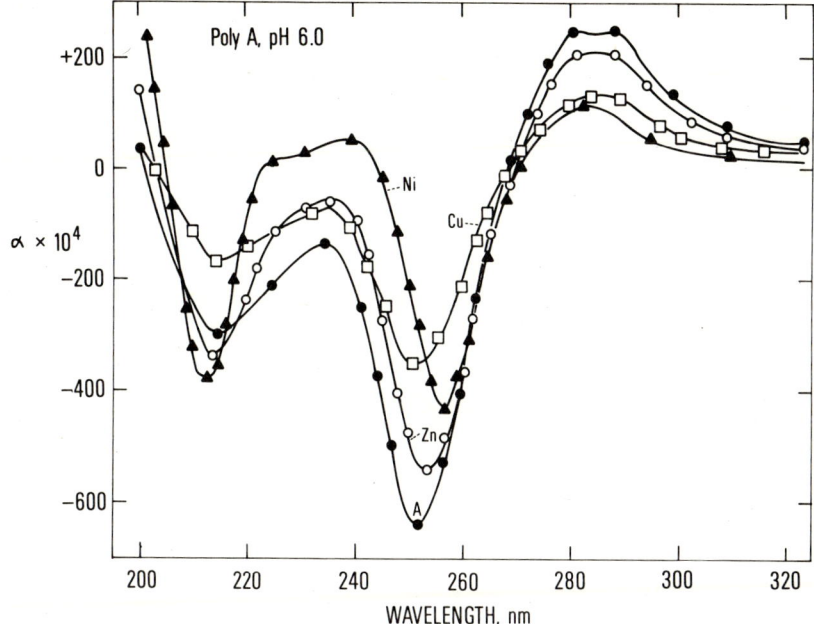

Figure 20.

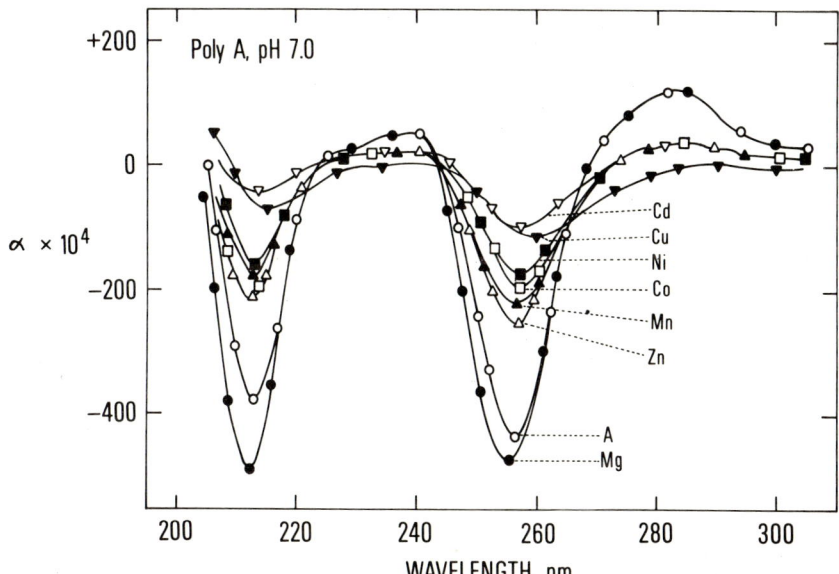

Figure 21.

the effect of various metal ions on the conformation of polyriboadenylic acid at pH 6, where polyriboadenylic acid alone gives an optical rotatory dispersion curve characteristic of a double helical molecule (*11*). In the presence of zinc, we get a similar ORD curve, somewhat displaced. In the presence of nickel, we get a completely different ORD curve, characteristic of a single helical molecule. Copper ions, on the other hand, produce still a third ORD curve, characteristic of a random coil. In other words, it is possible to use different metal ions at room temperature to produce three different conformations of the same macromolecule.

Let me show you another effect that is observed with the same poly A molecule at pH 7 where it exists as a single helical molecule (*11*). When poly A is heated at pH 7, the single helical molecule is gradually denatured into a random coil. Figure 21 illustrates that the addition of various metal ions at room temperature produces the same amount of unwinding as heating to various temperatures. Hence there are two ways in which helical polyadenylic acid can be converted to random coil, by heating to various temperatures or by the addition of various metal ions at room temperature. There are reasons why, under certain conditions, one might not want to heat these molecules; therefore, metal ions can be very helpful in producing the desired helical content.

Phosphodiester Bond Cleavage by Metal Ions

Having considered the unwinding–rewinding phenomenon caused by metal ions binding to the bases of DNA, I should now like to discuss very briefly another reaction which results from metals binding to phosphate. I refer to the ability of metal ions to break the phosphodiester linkages. This reaction works only with RNA; *i.e.*, the 2′ hydroxyl group is necessary. It does not work with DNA; in fact, it is a very good way to separate DNA from RNA if one wants to keep the DNA and is willing to throw the RNA away.

Figure 22 represents the depolymerization by zinc ions of RNA molecules and other polyribonucleotides, in which all of the bases are the same—*i.e.*, polyadenylic, polycytidylic, polyuridylic, and polyinosinic acid (*12*). Notice that the rates of degradation, or depolymerization, are not the same. The ability of the metal ion to cleave a phosphodiester linkage seems to depend on the nature of the base adjacent to that linkage.

This phenomenon can be an interesting one if the cleavage specificity is exhibited not only in homopolyribonucleotides but also in heteropolyribonucleotides like RNA. If so, this kind of metal-catalyzed reaction could be used in the determination of nucleotide sequence. Consequently, we performed some studies to see whether cleavage specificity could also be obtained with RNA.

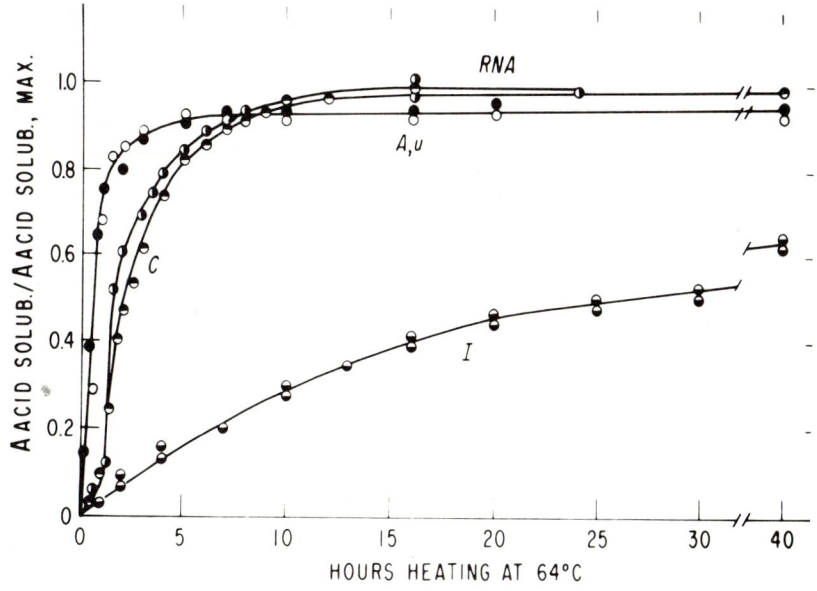

Figure 22.

$$\begin{array}{c} \text{Zn–RNA} \\ N_PN_PN_PN_PN_PN_P \end{array} \xrightarrow{\Delta} \begin{array}{c} \text{Oligonucleotides} \\ N_PN_PN_P \end{array}$$

$$\xrightarrow[\text{Phosphatase}]{\text{Alk.}} N_PN_PN \xrightarrow{\text{KOH}} N_P + N_P + N$$

Figure 23.

Figure 23 is a diagram of the scheme used to determine the cleavage sites in RNA (13). We started out with RNA in the presence of zinc ions. The RNA is schematically shown by $N_PN_PN_PN_PN_PN_P$, with N representing a nucleoside base plus ribose and P representing the phosphate. When this solution is heated, phosphodiester linkages are broken. If the linkage is broken as indicated, then we would get $N_PN_PN_P$. This fragment is treated with an enzyme, alkaline phosphatase, which removes the terminal phosphate group of the cleavage products, yielding N_PN_PN, which then reacts with strong base to cleave all the internucleotide bonds. We then have nucleotides and nucleosides, the latter arising only from the sites at which the original cleavage took place, enabling us to pinpoint the cleavage sites by quantitizing the nucleosides.

Table II shows the composition of guanine, adenine, cytosine, and uracil in the original RNA molecule on the right, compared with the

Table II. Composition in Cleavage Sites and RNA Molecule

pH 8, 64°, 20 Zn

	Cleavage Sites, %	RNA Composition, %
G	7	28
A	20	25
C	30	19
U	42	28

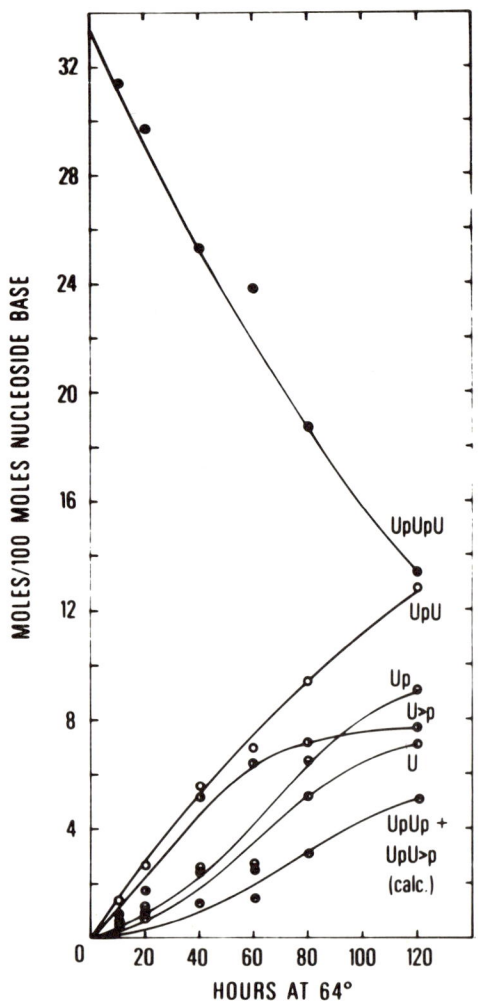

Figure 24.

cleavage sites on the left (*13*). You can see that there is much less guanine at the cleavage sites than in the original RNA, and there is much more uracil at the cleavage sites than in the original RNA. In other words, the metal, when reacting with the phosphate group and breaking the phosphodiester linkage, senses what base is adjacent to the ribose. If it is a uracil, there is a strong tendency to cleave, and if it is a guanine, there is a weak tendency to cleave. Hence there is a specificity even in a heteropolynucleotide—*i.e.*, RNA—which is useful for the determination of base sequences.

We were also interested in finding out something about the mechanism by which this phosphodiester breakage occurs. We first thought, because the 2' hydroxyl group is required (that is, DNA does not undergo the same reaction), that the mechanism involves chelation between the 2' hydroxyl group and the phosphate group (*see* Figure 1). However, the analysis of products of degradation indicated that a good deal of cyclic phosphate intermediate is produced. We then undertook a study of the degradation of a series of trinucleotides (*14*).

The curve at the top of Figure 24 represents the depolymerization of triuridylic acid. The curve representing the formation of a 2' and 3' mononucleotide, uridylic acid (Up), exhibits an initial lag period followed by an accelerated phase. On the other hand, the 2'–3' cyclic uridine phosphate (U > p) is produced very rapidly at first, and then levels off. If this reaction is followed further, the cyclic phosphate (U > p) production falls to zero, while the mononucleotide (Up) levels off. Similar results are obtained with other trinucleotides. These results can be explained as follows: First the metal ion binds to the phosphate group. As a result of this binding, electrons are withdrawn from the phosphate. The phosphorus then has a positive dipole and the oxygen atom, being electronegative, is attracted to it so that we get a pentavalent phosphorus, which then is cleaved at its weakest bond to produce the depolymerization reaction.

These studies illustrate not only that inorganic chemistry is useful in the solution of biological problems, but that the study of the complexes of biological molecules adds to our knowledge of inorganic chemistry. Biological molecules generally, like the nucleic acids, offer a choice of sites to the metal ions and it is interesting to see what varied ligand reactions evolve from binding the metal ions to the different sites.

Literature Cited

(1) Eichhorn, G. L., Clark, P., Becker, E. D., *Biochemistry* (1966) **5**, 245.
(2) Carrabine, J. A., Sundaralingam, M., *J. Am. Chem. Soc.* (1970) **92**, 369.
(3) Berger, N. A., Eichhorn, G. L., manuscript in preparation.
(4) Chan, S. I., Nelson, J. H., *J. Am. Chem. Soc.* (1969) **91**, 168.

(5) Sternlicht, H., Jones, D. E., Kustin, K., *J. Am. Chem. Soc.* (1968) **90**, 7110.
(6) Eichhorn, G. L., *Nature* (1962) **194**, 474.
(7) Shack, J., Jenkins, R. T., Thompsett, J. M., *J. Biol. Chem.* (1953) **203**, 373.
(8) Thomas, R., *Trans. Faraday Soc.* (1954) **50**, 324.
(9) Eichhorn, G. L., Clark, P., *Proc. Natl. Acad. Sci. U.S.* (1965) **53**, 586.
(10) Shin, Y. A., Eichhorn, G. L., *Biochemistry* (1968) **7**, 1026.
(11) Shin, Y. A., Eichhorn, G. L., manuscript in preparation.
(12) Butzow, J. J., Eichhorn, G. L., *Biopolymers* (1965) **3**, 97.
(13) Eichhorn, G. L., Tarien, E. L., Butzow, J. J., manuscript in preparation.
(14) Butzow, J. J., Eichhorn, G. L., manuscript in preparation.

RECEIVED June 26, 1970.

8

Biochemistry of Group IA and IIA Cations

R. J. P. WILLIAMS

Oxford University, Oxford, England

> *The biological function of Group IA and IIA cations of the periodic table is reviewed against the background of their chemistry. Utilization of these cations arises from an ability to form different types of complex compounds, which is dependent upon the radius-ratio effect. If the details of their biochemistry are to be understood, new probe methods for following the cations in biological systems must be devised. Some possibilities based upon the principle of isomorphous replacement are described and tested.*

The chemistry of the Group IA and IIA series of cations is relatively easy to understand as all its features are discernible from an inspection of the ionic model for chemical bonding (*1, 2*). This model shows that in an exchange reaction leading to a complex or an insoluble salt from a simple hydrate

$$M(H_2O)_n + L \rightarrow ML(H_2O)_m + (n-m)H_2O$$

different cation stability sequences can be generated depending on the nature of the ligand, L, even though L is a simple anion or ligand. If L is a very large anion (usually of a strong acid) then the orders are

$$Cs^+ > Rb^+ > K^+ > Na^+ > Li^+ \tag{1}$$

and

$$Ba^{2+} > Sr^{2+} > Ca^{2+} > Mg^{2+} \tag{2}$$

which are sometimes called the lyotropic series. A clear-cut example is provided on association with sulfate anions either in solution or in the solid state but the orders are much more strikingly demonstrated in the solids where cooperative interactions are important. The opposite orders can be generated by simple small anions (usually of necessity of a weak acid) such as hydroxide.

$$Li^+ > Na^+ > K^+ > Rb^+ > Cs^+ \tag{3}$$

and
$$Mg^{2+} > Ca^{2+} > Sr^{2+} > Ba^{2+} \qquad (4)$$

In 1952 we proposed (3, 4, 5) that such changes in the free energy orders arose through a type of radius-ratio effect operating on the associated systems in both solid and solution, salts and complexes, comparable with the effect discussed by Pauling in his book (6) in order to explain the melting points of alkali halide salts. According to the increasing degree of importance of the radius-ratio effect, Orders 3 and 4 change toward Orders 1 and 2 with the changing size of the ligand, and maximum stability can be achieved by any of the cations in the two different groups as Orders 3 and 4 switch to Orders 1 and 2. Our discussion showed this by a consideration of the solubility of both the Group IA and IIA salts (1, 2, 3, 4, 5). Furthermore, when considering binding to multidentate ligands and anions, it is not only the chemical nature of the individual coordinating group which counts but more importantly the packing of the anions around the cations. Table I shows that simple car-

Table I. Some Stability Constants for Group IIA Cations, $\log_{10} K$ [a]

	Mg^{2+}	Ca^{2+}	Sr^{2+}	Ba^{2+}
Acetate	0.82	0.77	0.44	0.41
Oxalate	3.4	2.8	2.5	2.3
Glycine	3.4	1.4	0.9	0.8
Imido diacetate	2.9	2.6	–	1.7
Nitrilotriacetate	5.3	6.4	5.0	4.8
EDTA	8.9	10.7	8.8	7.9
EGTA [b]	5.4	10.7	8.1	8.0
Sulfate (log K)	2.0	2.	–	–
(log S.P.	0.0	5.0	6.5	10.0)
Phosphate (log S.P.	24.0	27.5	27.4	22.5)
ATP	4.2	4.0	3.5	3.3
Carbonate (log S.P.	7.5	8.5	9.0	8.5)
Football ligand	2.0	4.1	13.0	15.0 [c]

[a] Data from Ref. 7.
[b] EGTA is 2,2'-ethylenedioxybisethyliminodi (acetic acid); S.P. is solubility product data; ATP is adenosine triphosphate.
[c] Ref. 8.

boxylates (weak acid anions) give Order 4 but complex carboxylates give an order between 2 and 4. Oxalate in solution gives Order 4 but the insolubility of oxalates follows the order $Ca^{2+} > Sr^{2+} > Ba^{2+} > Mg^{2+}$, showing the increased influence of the radius-ratio effect in the solid state because of packing problems in a continuous lattice. While it was immediately clear that several different orders of the free energy of this exchange reaction could be generated for Group IIA cations both in solution and the solid state and for Group IA in the solid state, it was

Table II. Some Stability Constants, log K, and Other Data for Group IA Cations

	Li^+	Na^+	K^+	Rb^+	Cs^+	Ref.
EDTA (log K)	2.8	1.7	0	–	–	7
$P_2O_7^{2-}$ (log K)	3.1	2.3	2.3	–	2.3	7
Dibenzoylmethane (log K)	5.9	4.2	3.7	3.5	3.4	7
NO_3^- (log K)	−1.0	−0.4	0	–	0.1	7
SO_4^{2-} (log K)	0.6	0.7	0.9	–	–	7
Ring chelate XXXI (log K)	0.0	0.0	2.0	1.5	1.1	12
Football ligand (log K)	–	3.6	5.1	3.7	–	8
Substituted picrylamine anion (log extraction coefficient)	1.0	1.8	3.7	4.2	5.2	13

not possible to demonstrate until recently (9, 10, 11) that several different orders could be generated in solution for Group IA cations by the same radius-ratio effect. In particular, studies using series of organic cyclic ligands, Table II, have redirected attention to this size effect. This has led to the speculation that such cyclic ligands are the only organic ligands which can generate orders different from 3 in solution (9, 10, 11) as most known simple anions give this order. Table II shows that this is not the case, that large inorganic anions give Order 1 and that one large, organic, noncyclic anion can also generate Order 1, at least in the extraction of these cations into organic solvents. Intermediate orders are to be expected with different large anions (9, 10, 11). It is the authors' contention that all the different orders originate as a consequence of the effect of packing upon ionic interactions. The packing problem can be inspected initially through the study of structures.

Since 1952, many crystal structure determinations have been carried out on the alkaline and alkaline earth salts. We have summarized these elsewhere (14); they clearly show that while Na^+ and Mg^{2+} are usually 6-coordinate, K^+ and Ca^{2+} are usually 8-coordinate with the same ligands. In many salts, the cations Na^+ and Mg^{2+} remained relatively highly hydrated compared with the salts of K^+ and Ca^{2+}. These structural changes in the crystals immediately illustrate how packing could produce changes in the stability orders in accord with ionic model considerations. I shall therefore use the words radius-ratio effect to denote that such size factors have caused deviation from Orders 3 and 4 toward 1 and 2. [A different language is used by Eisenman (15) who describes ligands by their "field-strengths" and "effective field-strengths," but I believe that his quite independent approach really refers to the structural property of the ligand as well as the "charge density" (field strength) on a given atom. Eisenman has been extremely successful in using his empirical approach but much of his argument is cyclic, in that the apparent field strength is estimated from the order which it is designed to explain].

More detailed inspection of the ionic model shows that the radius-ratio effect as seen in stability series is not essentially linked to changes in coordination number or to changes in the relative hydration of the cations; such changes are but structural ways in which packing problems manifest themselves. Even with a fixed coordination number, the stability constants of a ligand exchange reaction can take on almost any order in so far that a ligand or group of ligands may most effectively bind to any one of the cations. Steric hindrance will then manifest itself in increased bond lengths over those expected on the ionic model, the increase being the greater the smaller the cation. This is seen clearly in the complexes of ring chelates (9, 10, 11, 12) and must be obviously true in complexes of large molecules such as proteins which have very many possible ways in which they can generate steric fitting and misfitting. As these stabilities are studied against the background of coordination to water—a small ligand which is able to adapt itself to any size of cation—there may be but one coordination number for all the cations with a given ligand but the rather inflexible "hole" which is generated by the ligand will bind certain of them preferentially. Radius-ratio effects can become important in two circumstances, therefore: With large anions when anion–anion and anion–water contact will restrict good packing and will lower stability with small cations and with multichelating agents of rigid structure when the hole size may be such that small cations cannot make good bond distances with all the potential chelating groups or they have to use energy in altering the conformation of the ligand so as to make such contacts.

A further factor enters magnesium chemistry, though it hardly applies to any of the other cations. It is not part of the simple ionic model. The high charge and small size of the magnesium ion allow it to polarize bases. This accounts for the relatively high affinity of magnesium for nitrogen bases such as glycinate (Table I), chlorophyll, and some dyestuffs (*e.g.*, magneson).

In the light of the above observations, we have divided ligands into four major groups as far as biological systems are concerned (Table III).

Table III. Possible Biological Ligands of d^0 Cations

Ligand Type	Example	Order
Strong acid anions	$-OSO_3^-$, $-OP(OR)_2O^-$	1 and 2
Weak acid anions	$-OPO_3^{2-}$, PO_4^{3-} $-CO_2^-$, CO_3^{2-}	All orders are possible, depending upon radius-ratio considerations. Orders closer to 3 and 4 with the simplest ligands
Neutral oxygen groups	Alcohols, ethers, peptides, and esters	
Neutral nitrogen groups	$-NH_2$, imidazole	Mg^{2+} > all others Li^+ > Na^+ > K^+

Table IV. Calcium-Binding Compounds of Cell Walls

Living Systems	Binding Chemicals
Bones	Chondroitin sulfate
	Glycoproteins
Shells and plants	Pectic acids
Celluloses	(Galacturonic acid)
Algae	Alginic acid
	(mururonic acid)
	Fucoidin
	(polyfucose sulfate)
	Carragenin
	(polygalactose sulfate)

It is important to know if such ligands are separated in various biological compartments for then this separation alone would divide the cations functionally. Probably the only important separation is that sulfonate residues are extracellular—*i.e.*, part of cell walls—for much the larger part. Calcium in particular is associated with the sulfonated polysaccharides as expected from Table I, *see also* Table IV, and from its binding to strong acid anion exchange resins (*3, 4, 5, 15*). As has been stressed many times elsewhere, thermodynamic effects such as differential binding of cations to groups inside and outside cells are quite insufficient to explain the ionic distributions generally observed in biological systems.

Ion Distribution in Biology (16)

Very generally, potassium and magnesium are accumulated in cells but sodium and calcium are rejected by them. Thus, the competition for ligands inside a cell is biased compared with that outside. This distribution is essential for life as it stabilizes the cell against osmosis, by rejection of sodium, and against internal precipitation of carbonate and phosphate, etc., by rejection of calcium. Once these two major prerequisites of evolution had been established by the pumping action of membranes, life could develop using all four cations. Evolution has led to the following major systems which utilize the cation gradients.

(1) Use of calcium (a) as an external structure factor and (b) as a cofactor for extracellular enzymes.
(2) Internal use of calcium as a trigger for structural changes.
(3) Internal use of magnesium and potassium as (a) cofactors of enzymes and (b) stabilizers of internal structures.
(4) Transmembrane potentials of especially potassium, sodium, and calcium.

Under (1, a) we include the formation of cell wall structures which became elaborated as shells, bones, and cellulose structures (Table IV).

Table V. Extracellular Calcium Enzymes and Enzyme Precursors

Enzyme or Enzyme Precursor	Function
Trypsinogen	Peptide hydrolysis
Aryl sulfatase	Sulfate ester hydrolysis
B. Sutilis protease	Peptide hydrolysis
Nuclease	Phosphate ester hydrolysis
Amylase	Saccharide hydrolysis
Prothrombin	Blood clotting through peptidase action

Table VI. Possible Probe Ions for Substitution

Native Cation[a]	Substitution[a]
$Na^+(0.95)$	$Li^+(0.60)$ (poor)
$K^+(1.33)$	$Tl^+(1.40)$, $Rb^+(1.48)$, $Cs^+(1.69)$, $NH_4^+(1.45)$
$Mg^{2+}(0.65)$	$Mn^{2+}(0.80)$ to $Zn^{2+}(0.65)$
$Ca^{2+}(0.99)$	$Eu^{2+}(1.12)$, $Mn^{2+}(0.80)$
	$La^{3+}(1.15)$ to $Lu^{3+}(0.93)$
	$UO_2^{2+}(\simeq 1.1)$ etc.

[a] Ionic radii are in parentheses.

Under (1, b) there are many extracellular enzymes of the digestive systems of species ranging from bacteria to animals (Table V). Under (2) we include release mechanisms for hormones and synaptic transmitters; the basic step of nerve transmission may be owing to calcium entry. Muscle contraction, movement of cilia, and many other dynamic structure changes are calcium-induced. (Perhaps cell division is another one of these processes?) Under (3, a) we include the activation of adenosine triphosphatases, the main energy sources of biology, and the possibility of control of a vast range of intracellular enzyme reactions, including those of glycolysis. Section (3, b) refers particularly to the stabilization of ribosomes and therefore to the control by magnesium and potassium of the whole of protein synthesis. Again, without magnesium and potas-

Table VII. The Effect of Thallium on Biological Systems

Diol–dehydratase
Pyruvate kinase
Phosphatases
Na/K ATP–ases (K–function)
Erythrocyte transferases
Muscle excitation

[a] Efficacy is a product of maximum velocity and

sium, the synaptic junction fires spontaneously so that these cations also stabilize the vesicular systems which calcium destabilizes.

Section (4) refers to the restrictions imposed on cation movement by membranes. The relative ease of movement of potassium through a nerve membrane at rest generates a potassium potential. Imposition of a perturbation upon the membrane changes its properties so that it is more permeable to sodium, and the activated membrane shows a sodium potential of reversed sign to the potassium potential. Thereupon a self-propagating spike of depolarization which is rapidly followed by recovery to the rest state flows along the nerve cell and is the nerve message.

It should be clear from the above and from recent summary papers of myself and Wacker (16, 17) that the biochemistry of the Group IA and IIA cations is exceedingly exciting. Inorganic chemists need to think how they can study this chemistry. Here I shall present our views on the possibilities of using series of related cations relying upon our knowledge of their chemistry and isomorphous replacement for K^+, Mg^{2+}, and Ca^{2+} by special probe cations. We are attempting to understand not only the separate actions of the four cations Na^+, K^+, Mg^{2+}, and Ca^{2+} in enzymes but also their selective concentration and competition in membrane transport. A list of possible metal ion substitutions is given in Table VI.

Probes for Group IA Elements

Sodium is unlike any other cation in its charge and radius. Thus, sodium must be followed by its own nuclear properties (18). Potassium can be replaced, in principle, by thallium(I) and cesium. Both are useful as they have suitable nuclei for NMR studies but thallium has additionally an absorption band at 214 nm which is very ligand-dependent, a readily observable fluorescence, and a small temperature-independent paramagnetism which can cause marked shifts in the nuclear resonances of ligand nuclei. We (19) have aimed in the first instance to discover if thallium replaces potassium effectively in enzymes. Table VII shows that it does.

Several Enzymes and Other Biological Systems

Order of Cation Efficacy[a]	Ref.
$Tl^+ > NH_4^+ > K^+ > Rb^+ > Cs^+ > Na^+ > Li^+$	20
$Tl^+ > K^+ > Rb^+ > Cs^+ > Na^+ > Li^+$	21
$Tl^+ > K^+ > Rb^+ > Cs^+ > NH_4^+ > Na^+, Li^+$	22,23
$Tl^+ > K^+ > Rb^+ > Cs^+ > Na^+ > Li^+$	24,25
Tl^+ moves with K^+	26
$Tl^+ > K^+ > Na^+, Li^+$	27

stability of binding in the various processes.

Thallium binds ten times more strongly than potassium to at least four enzymes and is equally effective as a catalyst. We have therefore looked at the stability of binding of thallium as compared with potassium to various ligands. With neutral ligands such as ether–oxygen, thallium and potassium binding are rather similar (R. M. Izatt, personal communication) but an anion such as a single phosphate, $ROPO_3^{2-}$, or a carboxylate, $-CO_2^-$, binds about ten times more strongly, Table VIII. We conjecture, on the basis of this evidence alone, that K^+ (Tl^+) in enzymes may bind at a mono-anion center, sometimes phosphate and sometimes carboxylate, but that the "hole" size closely matches the radius of these two cations, 1.4 Å. In order to prove this assertion, we need crystal structure data on the thallium site, or, failing this, spectroscopic data. In Table VIII we show that carboxylate and phosphate shift the absorption band of thallium(I), and it is known that they quench its fluorescence. Using proton NMR, the Tl(I) ion shifts acetate protons of EDTA by -0.21 ppm and ethylenic protons by -0.17 ppm relative to the parent anion. The examination of Tl(I) phosphate complexes have shown that phosphorus resonances are shifted as follows relative to the respective parent anions: for pyrophosphate, -1.4 ppm; for adenosine triphosphate, α-P -0.5, β-P -2.2, γ-P -1.0 ppm; for adenosine diphosphate, α-P -2.0, β-P -1.3

Table VIII. The Stability Constants and Absorption Spectra of Some Thallium(I) Complexes at Ionic Strength 0.075 [a]

Ligand	$Log_{10}K_{Tl}$	λ_{max}	$Log_{10}K_K$
PO_4^{3-}	2.25	230	
HPO_4^{2-}	0.75	225	
$P_2O_7^{4-}$	3.05	227	1.5
$HP_2O_7^{3-}$	2.35	219	
Ribose–5–phosphate^{2-}	0.90	219	
Adenosine diphosphate^{3-}	1.20		
Adenosine triphosphate^{4-}	2.00		1.0
Ethylenediamine tetracetate^{4-}	5.8	246	1.0
Nitrilotriacetate^{3-}	4.4	243	1.0

[a] Ref. 28.

ppm. These shifts are entirely adequate for NMR studies on complex molecules. Moreover, they show conclusively that Tl(I) (and K) are bound in a polyphosphate chelate in all three cases studied. The binding of Na^+ in ATP crystals would appear to be quite different from Tl(I) binding in complexes, which is to all three phosphates of the ATP. It is equally possible to use the Tl(I) complexes of cyclic ethers, for example, as NMR labels in membranes.

Probes for Group IIA Elements

There are a host of potential probes for Mg^{2+} and Ca^{2+}, as these ions have about the same size as many divalent first row transition metals and divalent lanthanides (Table VI). Trivalent lanthanides have the correct radii although the wrong charge as substitutes for calcium. In isomorphous replacement in minerals, the radius is more important than the charge; this may also be true in biology as Na^+ and Ca^{2+} (radius 1.0 Å) and K^+ and Ba^{2+} (radius 1.3 Å) often compete for sites. This means that a lanthanide may be an excellent competitor for a divalent cation site in biology and there is already some evidence to this effect (Table IX) (29).

Table IX. Lanthanides in Biological Systems

System	Action	Ref.
Lobster axon	Ln^{3+} behaves as "Super–calcium"	30
Bone proteins	Ln^{3+} competes for calcium sites	31
Bacterial nuclease	Ln^{3+} inhibits at calcium site	32
Mitochondria, Muscle, Nerve	Ln^{3+} blocks calcium transport	33,34,35,36

While Shulman (37), Cohn (38), and their coworkers have developed the use in many circumstances of Mn^{2+} as an excellent probe for Mg^{2+} functions, we have been looking for other probes. Ni^{2+} (octahedral by virtue of ligand field effects) may sometimes be the best probe for Mg^{2+} (octahedral on the basis of size) while Co^{2+} and Zn^{2+} (lower coordination numbers often preferred) and Mn^{2+} (too large) may function poorly

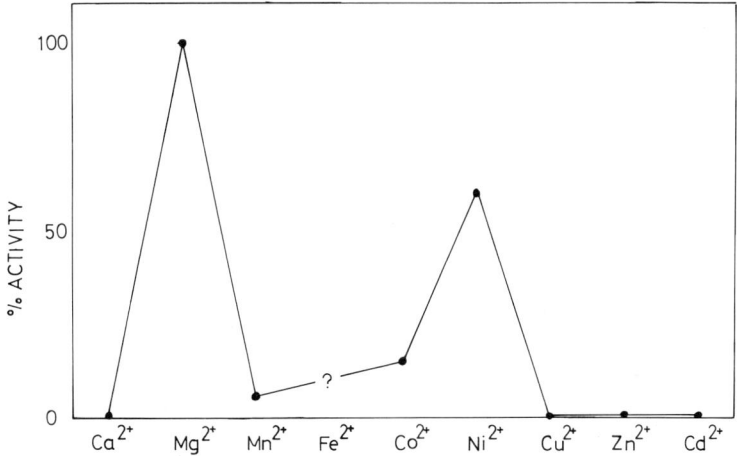

Figure 1. The variation of activity with the metal cation incorporated in the enzyme phosphoglucomutase, data fom Peck and Ray (39)

(Figure 1). This has been established in one enzyme through the work of Ray (39) on phosphoglucomutase. From the nickel(II) ligand field spectra (octahedral) of this protein and the data on metal-binding, we suspect that the metal is bound by a nitrogen and two carboxylates of the protein and that the exact protein geometry is generated by the metal. Cobalt(II) binds in a very different way, its ligand field spectrum suggests five-coordination, and both cobalt(II) and zinc(II) are inactive. Probably none of the metals is at the active site but the metal controls the structure of the active site by remote control, allosteric effect.

Even when manganese(II) can substitute for magnesium(II), two different orders of catalytic efficiency have been observed. Figure 2 shows that these orders arise through the need for closer size-matching when the complexes are of the type enzyme–metal–substrate than when they are of the type enzyme–substrate–metal (38). In general, transition metals fail to activate magnesium enzymes as they bind the wrong groups of a protein—e.g., copper(II) and nickel(II) are often strong inhibitors

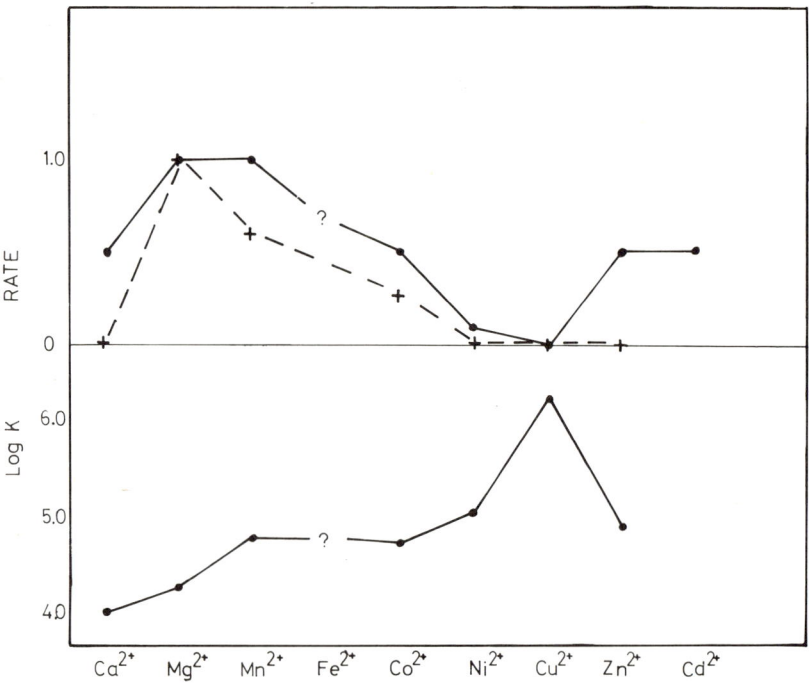

Figure 2. The stability constants for adenosinetriphosphate complexes. bottom figure, which follow the usual order for model complexes, compare with the series of activities produced by the cations in two enzymes, top figure. The full curve in the top figure is for phosphoglycerate kinase and the broken curve for pyruvate kinase. (Note that but for calcium the upper and lower figures are mirror images.)

Table X. Properties of Europium(II), Calcium, and Magnesium[a]

Property	Europium(II)	Calcium	Magnesium
Ion size (Å)	1.12	0.99	0.65
Main coordination No.	8	8	6
Log K (EDTA)[b]	9.6	10.6	8.7
Log K (EGTA)[c]	9.6	11.0	5.2
Log K (PA)[d]	2.8	2.5	2.5
Log K (DPA)[e]	~4.0	4.4	2.4
Solubility of Sulfate	Insoluble	Insoluble	Soluble

[a] Data from Ref. *41*.
[b] Ethylenediamine tetracetate.
[c] 2,2'-Ethylenedioxybisethylimipodi (acetate).
[d] Picolinate.
[e] Pyridine 2,6–dicarboxylate.

(*40*). As ATP is often the substrate in the case of enzyme–substrate–metal complexes, most metals are active for they mostly bind to the triphosphate. Copper(II), mercury(II), and other very strong Lewis acceptors are inhibitors as they bind to the ring nitrogens of ATP and in enzymes they could also block essential sulfhydryls.

A very suitable probe for calcium chemistry is europium(II). This cation has a good NMR nucleus, a good Mossbauer nucleus, useful spectra and fluorescence, and can be studied by EPR or by using its effect on proton (or other nuclear) resonances. We have shown, Table X, that europium(II) has a chemistry closely similar to that of calcium(II) (*42*). We are now studying the interaction between europium(II) and enzymes and proteins. We believe that europium(II) parallels calcium(II) to some degree in the triggering of muscle.

An alternative probe procedure is to use the trivalent lanthanides (*42*). Some lanthanide cations can replace calcium in a functional sense in biological systems. On the other hand, it is generally true that lanthanides inhibit calcium systems in a competitive way and they can be used therefore to probe the calcium sites. The lanthanides, 15 probes of slowly graded ionic size, represent a wonderful experimental tool for the study of biological selectivity and a chance to uncover fine control mechanisms. Almost every physical method can be used in the study of one or the other of these cations.

Once again, crystal structure data are necessary in order to show how the lanthanides bind in proteins. With the help of D. C. Phillips' group (Oxford), we have made a start with lysozyme showing that Gd^{3+} binds between the two carboxylate groups—i.e., exactly at the active site. We (*42*) have done NMR studies on several different M^{3+} cations of the lanthanide series to discover which is most useful. Gadolinium (broadening-probe), europium, and holmium (shift-probes) have much to recommend them, and we have gone on to examine substrate binding to Ln^{3+} proteins. The work will be extended immediately to troponin, the calcium-binding protein of muscle. In order to put the NMR shift work on a firm basis, we have studied the Eu(III) EDTA, EGTA, and NTA complexes in great detail in collaboration with F. J. C. Rossotti. We have followed the changes in conformational equilibrium with temperature in these systems as well as the effect of changing metal:ligand ratio. The systems have also been thoroughly examined by spin–echo methods. All we need state at this moment is that the lanthanides promise to make excellent probes of biological systems.

Possibly an even more important use of the lanthanides could arise in studying the conditions in membranes. Thus lanthanides are not only good NMR and ESR probes but they can be used as fluorescence probes. In this respect, their properties depend upon their further solvation. If we consider the solution of a series of lanthanide chelates then, because of the changing size of the cations (radius-ratio effect), we can expect the coordination number, ligands (6) plus water, to drop from nine at lanthanum to eight at lutecium. Half way along the series, the coordination number will be determined by an equilibrium between the two hydration values. Thus, if we place the chelate in a membrane and there is then an alteration in membrane condition which alters the activity of the water, this will alter the position of the equilibrium. We can then discover if the hydration state of a membrane is a function of the energy state of the membrane by reading out the property of the lanthanide, *e.g.*, the hydration, through NMR studies or through fluorescence.

Summary of Binding Sites

Knowing the causes of the changes in stability and activity sequences in model systems and in enzymes and knowing the sequences of activity which appear in biological systems, we can draw tentative conclusions which are open to test by the above probe studies.

(1) Magnesium is bound preferentially by nitrogen bases plus few anions: $Mg^{2+} > Ca^{2+} > Na^+ > K^+$.

(2) Calcium is bound preferentially by multidentate anions and strong acid anions $Ca^{2+} > Mg^{2+} > Na^+ > K^+$. There is little evidence for nitrogen donors in calcium binding but sulfate ester anions, phosphate ester anions, and carboxylate anions are strongly implicated.

(3) Potassium will be bound in enzymes to a large center composed of neutral donors alone or of neutral donors plus one singly-charged strong-acid donor: $K^+ > Na^+$ and $(Ba^{2+} > Ca^{2+} > Mg^{2+})$.

(4) Sodium will combine with a smaller center of neutral donors and/or of neutral donors plus one singly-charged donor which may be that of a weak acid, $Na^+(>Ca^{2+}) \geqslant K^+(>Mg^{2+})$.

These statements allow us to re-inspect a biological system so as to understand how cation selectivity is used but we must first show how the cation gradients which are found across cell membranes can be generated and if effects other than those in enzymes need to be postulated.

Ion Transport

In the case of sodium and potassium, it is clear from the study of ring ligands (9, 10, 11) and substituted picrylamine anions that potassium in the presence of a large organic reagent or chelating agent can partition into a highly hydrophobic environment but that this is much less readily achieved by sodium. A good transport molecule for potassium, which will not accept sodium, would be one which was generally apolar, though it could carry a single positive charge, so that it can enter the low-polarity membrane. Such a center will not accept magnesium or calcium for they are at least as difficult to dehydrate as sodium. However, weak competition by barium is possible and strong competition from thallium(I), rubidium, cesium, ammonium, and tetraalkylammonium salts is to be expected. Cesium, barium, and especially tetraethylammonium salts act as effective drugs restricting the access of K^+ to the channels through which it moves in membranes. Membranes contain high concentrations of phospholipid which are of the right type of anion and are in a highly hydrophobic environment. By way of contrast, the sodium center (43) must be more highly polar or have a group such as phosphate, $(RO)_2PO_2^-$, or carboxylate, RCO_2^- (9, 10, 11). This would explain the competition by calcium at some sodium sites. Ring chelates alone are not known to separate calcium from sodium.

The transport of calcium would also seem to involve phosphorylated proteins—for example, a phosphorylated protein carries calcium in the blood stream. The binding constant of these centers for Ca^{2+} ($\sim 10^6$) and lanthanides ($\sim 10^9$) indicates that there are probably two (or three) more anionic, carboxylate(?), groups as well as the one phosphate at the binding center. This is certainly the case in the staphylococcus nuclease (44). There are also four carboxylate residues in the terminal sequence of trypsinogen, which is activated by calcium.

The carriers for magnesium in bacterial membranes are more likely to have one nitrogen base, probably imidazole, and two anion groups, e.g., phosphate or carboxylate. Steric restrictions could be built in, as in chlorophyll. The nitrogen base would imply that the carrier would bind transition metals such as cobalt, nickel, and manganese(II) which could make excellent probes. Certain bacteria can be loaded with transition metals and perhaps the mechanism of loading utilizes the magnesium carrier.

An important feature of transport is that it is linked to metabolism. The function of phosphorylated carrier proteins may lie in the ease with which their formation can be linked to energy.

$$\text{Protein} + \text{ATP} \rightarrow \text{Protein-P} + \text{ADP}$$

Hydrolysis at the opposite side of a membrane from that at which phosphorylation occurs then yields two energy-coupled systems for rejecting Na^+ and Ca^{2+} and perhaps also for carrying K^+ and Mg^{2+} in.

Let us next assume that we understand the transport problem. How does the cell utilize the concentrations of cations which it has generated? We shall now elaborate somewhat on the early statements regarding cation function. We start with the outside of the cell.

Crystallization of Salts in Biology

The solubility product of many calcium salts is less than that of magnesium salts (radius-ratio effect) and precipitation of calcium carbonate, oxalate, phosphate, and even fluoride commonly occur in biological systems. This precipitation is assisted by the rejection of calcium from the interior of cells. It would appear that there are fibrous protein structures outside cells and these proteins act as initiators of crystallization, possibly using sulfate and carboxylate groups of outer membranes. Given a fine control of precipitation and solution, bone and shell material can be transferred in the blood stream to be deposited in a new region. The growth of the skeleton of animals, the deposition of shells of eggs, and the building of many other structures demand this type of activity.

Given that the system is in such perfect balance, very small changes can bring about catastrophic faults. Let us assume that proteins and polysaccharides slowly become more oxidized to more anionic polymers with age. These binding sites could lead to the initiation of crystal growth and thus the deposition of calcium salts. Is this why aging is associated with calcium deposition in cataracts, stones, cartilage, hardening of soft tissues and arteries?

We can now see that the laying down of hard structures is a consequence of the radius-ratio effect as much as is the binding of calcium to the outer saccharides and proteins of cells. The walls of bacteria and spores have been extended by evolution to the celluloses of plants and bones and shells of animals.

As we have described earlier, there are enzymes which are extracellular and are used in digestion. A considerable number of proteases, saccharrases, and nucleases are calcium-dependent. However, many other processes which are as complicated as digestion require calcium. Blood clotting, for example, is assisted by calcium-activated processes and may be controlled by altering the level of available calcium. The part played by sodium in all these processes is unclear but, in general, sodium accumulates with calcium.

Intracellular Ions

The inside of a cell depends very much upon the cell type. At one extreme, the nerve cell interior is largely a conduction path for cations. In the squid axon, the counter-anion is ethyl sulfate to a large degree, which makes certain that the cations will be free ions—mobility is then very high. In most other cells, the concentration of small anions of weak acids is very high—*e.g.*, phosphates and carboxylates—and the strong acid anions are usually associated with large structures—*e.g.*, the membranes and the polynucleic acids. Proteins only contain weak acid groups on the whole. The equilibria between this complicated system of anions, scattered in what are tantamount to various phases, and the cations with which this article is concerned are bound to be exceedingly difficult to unravel, and they have to be viewed against the ionic pumping by the membranes and sudden fluctuations in membrane permeability. In what follows, we make an effort to look at the over-all system. Table XI gives some gross concentrations of ions associated with biological systems.

Table XI. Ion Content of Living Systems, Mmoles/1000 Grams *(16)*

	K	Na	Mg	Ca
Human red cell (wet)	92.0	11.0	2.5	0.1
Squid nerve extract	5.0	1.0	0.5	0.1
Yeast cells (dry)	110.0	13.0	13.0	1.0
Euglena cells (wet)	103.0	5.0	4.8	0.3
E. Coli cells (wet)	250.	20.	20.	5.
Banana fruit (dry)	10^3	0.1	200.	0.2

Inside cells, where free (Ca^{2+}) is around $10^{-7}M$ and free (Mg^{2+}) is around $10^{-3}M$, there are clearly many substrates and proteins which will form complexes with either cation of stability constant around 10^3 to 10^4. Thus, only magnesium can be bound, *e.g.*, to ATP, ADP, pyrophosphate, and enzymes like enolase and phosphoglucomutase. However, in many cells there are also sites with binding constants of 10^6 for calcium and less than or about 10^3 for magnesium. On altering the membranes by an imposed force, calcium moves to the inside of the cell and its concentration rises to $10^{-5}M$ so that these sites become occupied by calcium and action is triggered. The distinction between the first group and the second group of sites is, we suggest, no greater than that between $NH(CH_2CO_2^-)_2$, NTA, and $(CH_2CO_2^-)_2N \cdot CH_2 \cdot O \cdot CH \cdot CH_2 \cdot O \cdot$

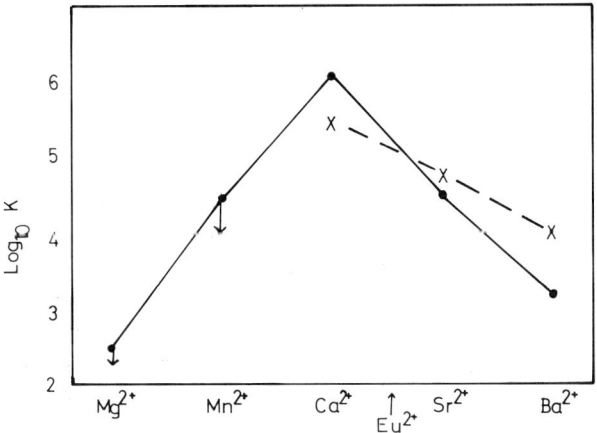

Figure 3. The stability sequence for the binding of cations to troponin from skeletal muscle (full curve) and for heart muscle (broken curve), after Ebashi and Endo (45)

$CH_2 \cdot N(CH_2CO_2^-)_2$, EGTA. However, the consequences are profound, for nerve pulses produce the necessary membrane changes, allowing calcium concentration to be pulsed, and the influx of calcium then generates activities of a wide variety including hormone release, transmitter release, and muscle contraction.

The best example of calcium binding is provided by troponin, the protein of muscle which is responsible for triggering muscle action. The binding constants are given in Figure 3. Compare the data on binding by EGTA in Table II. The high values for binding by EGTA to manganese contrasts with the binding to troponin, suggesting that troponin does not bind through nitrogen but does so through (probably) four carboxylate residues alone. This type of protein may be common to many contractile devices but it can also be common to the basic protein used in the initial membrane modification. Thus calcium releases transmitters from vesicles in cells. It does so by altering the vesicle and cell membranes. It is known from colloid and emulsion science that calcium is very effective in breaking disperse phases, and we would argue that this is because of its multi-chelating ability. By analogy, we suspect that this is the mechanism by which calcium can release chemicals trapped in vesicles. Could the calcium entering with sodium in the action phase of the nerve also be the tool for propagating the nerve depolarization wave? It will certainly alter the outer cell membrane as well as the vesicle membrane. All such ideas are open to study using the lanthanide cations, for they have similar properties to those of calcium. The relative magnesium concentration inside and outside cells is very dependent upon the living system. It may well be that magnesium flows into some cells with calcium and assists in the attack on vesicle systems—*e.g.*, in the squid axon. Alternatively, the magnesium cation can flow out of a system, making the attack by calcium more effective.

Inside cells, K^+ is > 100 mM while Na^+ is 10 mM so that a binding site which binds sodium ten times more strongly than potassium will be equally occupied by the two cations. The evidence of binding strengths of potassium-activated enzymes is that potassium is bound at least ten times more strongly than sodium so that sodium does not interfere. The sites must be comparable with the membrane sites for potassium. As yet there is only the suggestion of one function for sodium in cells, that is, as an activator of chromosome "puffing." Perhaps the sodium/potassium ratios in cells change during growth.

Once again, the level of the cations is controlled by membrane processes and metabolism from bacteria to animals and plants so that the same dependences on cations are found in all forms of life. For example, intracellular control of glycolysis is partly through the magnesium and potassium concentrations. Here we observe, much as has been seen in the chemistry of zinc, iron, cobalt, and copper, that biology has few variants on particular systems but these systems are chosen with remarkable ingenuity.

A further intriguing problem associated with ionic distribution is its evolution. The sea is a high-sodium, relatively low-potassium, high-magnesium, and relatively low-calcium medium. Fresh water contains virtually no cations. In tidal regions there are brackish waters of every kind of ionic composition. Life flourishes in all these environments and clearly it is important for life (as we know it) to produce an intracellular concentration which is roughly 100 mM K^+, 10 mM Na^+, 10 mM Mg^{2+}, and $10^{-7}M$ Ca^{2+}. The control of the ions lies in membranes, whether they be cell membranes of bacteria, the villi of the kidney, the skin of the frog, or the nasal salt gland of a bird. Species which have poor control mechanisms have restricted access to different environments and their evolution is limited. Species which carry their environment around with them have developed hugely. However, these very species are susceptible to damage at the membrane level. These systems have been evolved with great subtlety and we are ill-advised, as yet, to disturb their balance.

Though the ultimate refinement of this competition of the four cations in and outside cells cannot yet be appreciated, it must play some part in all complicated apparatus, including the brain. Looking at the inorganic elements of life, I have the constant feeling that life is as much inorganic as organic and is as limited by the cosmic abundances as by the possibility of organic synthesis. The absence of an attack on living systems by the inorganic chemist, on the scale that has been launched by the organic chemist, has left us with a totally false impression of the chemistry of living things. Biochemistry is as much inorganic as it is organic.

Literature Cited

(1) Phillips, C. S. G., Williams, R. J. P., "Inorganic Chemistry," Vol. 2, p. 48, Oxford University Press, Oxford, 1966.
(2) Williams, R. J. P., Tilden Lecture, *Quart. Rev. (London)* (1970) **24**, 331.
(3) Williams, R. J. P., *J. Chem. Soc.* (1952) 3770.
(4) Williams, R. J. P., *J. Phys. Chem.* (1954) **58**, 121.
(5) Williams, R. J. P., *Analyst* (1953) **78**, 586.
(6) Pauling, L., "The Nature of The Chemical Bond," p. 335, Cornell University Press, Ithaca, New York, 1945.
(7) "Stability Constants," A. Martell and L. G. Sillen, Eds., Spec. Publ. No. 17, Chemical Society, London, 1964.
(8) Dietrich, B., Lehn, J. M., Sauvage, J. P., *Tetrahedron Letters* (1969) **34**, 2889.
(9) Pedersen, C. J., *J. Am. Chem. Soc.* (1970) **92**, 391.
(10) Moore, C., Pressman, B. C., *Biochem. Biophys. Res. Commun.* (1964) **15**, 562.
(11) Pressman, B. C., *Red. Proc.* (1968) **27**, 1278.
(12) Izatt, R. M., Rytting, J. M., Nelson, D. P., Haymore, B. L., Christensen, J. J., *Science* (1968) **164**, 443.
(13) Rais, J., Krys, M., *J. Inorg. Nucl. Chem.* (1969) **31**, 2903.
(14) Williams, R. J. P., *in* "The Protides of the Biological Fluids," Vol. 14, p. 25, H. Peeters, Ed., Elsevier, Amsterdam, 1967.
(15) Eisenman, G., *Biophys. J.* (1962) **2** (Supplement 2), 259.
(16) Wacker, W. E. C., Williams, R. J. P., *J. Theoret. Biol.* (1968) **20**, 65.
(17) Williams, R. J. P., *Bioenergetics* (1970) **1**, 214.
(18) Cope, F. W., *Proc. Natl. Acad. Sci. (USA)* (1966) **54**, 225.
(19) Manners, J. P., Morallee, K. G., Williams, R. J. P., *Chem. Commun.*, (1970) 965.
(20) Foster, M. E., Williams, R. J. P., to be published.
(21) Radda, G. K., Waller, S., Williams, R. J. P., to be published.
(22) Inturrusi, C. E., *Biochim. Biophys. Acta* (1969) **173**, 567.
(23) *Ibid.*, (1969) **174**, 630.
(24) Britten, J. S., Blank, M., *Biochem. Biophys. Acta* (1968) **159**, 160.
(25) Gehring, P. J., Hammond, P. B., *J. Pharmacol. Exptl. Therap.* (1967) **155**, 187.
(26) *Ibid.*, (1964) **145**, 215.
(27) Mullins, L. J., Moore, R. D., *J. Gen. Physiol.* (1960) **43**, 759.
(28) Manners, J. P., Williams, R. J. P., to be published.
(29) Williams, R. J. P., *Proc. Intern. Congr. Pharmacology, 4th*, Berne, 1969, in press.
(30) Takata, M., Pickard, W. F., Lettvin, J. Y., Moore, J. W., *J. Gen. Physiol.* (1966) **50**, 1499.
(31) Peacocke, A. R., Williams, P. A., *Nature* (1968) **211**, 1140.
(32) Cautrecases, P., Fuchs, S., Anfinsen, C. B., *J. Biol. Chem.* (1966) **242**, 1541.
(33) Blanstein, M. P., Goldman, D. E., *J. Gen. Physiol.* (1968) **51**, 279.
(34) Lehninger, A. A., *Biochem. J.* (1970) **119**, 129.
(35) Mela, L., Chance, B., *Biochemistry* (1968) **7**, 4059.
(36) Hille, B., *J. Gen. Physiol.* (1968) **51**, 221.

(37) Eisinger, J., Shulman, R. G., Szymanski, B. M., *J. Chem. Phys.* (1962) **36,** 1721.
(38) Cohn, M., *Quart. Rev. Biophys.* (1970) **3,** 61.
(39) Peck, E. J., Ray, W. J., *J. Biol. Chem.* (1969) **244,** 3748.
(40) Williams, R. J. P., "The Enzymes," Vol. I, p. 391, P. D. Boyer, H. Landy, and K. Myrback, Eds., Academic Press, New York, 1959.
(41) Dwek, R. A., Morallee, K. G., Nieboer, E., Rossotti, F. J. C., Xavier, A., Williams, R. J. P., *Chem. Commun.* (1970) 1132.
(42) Nieboer, E., Johnson, C. E., Williams, R. J. P., Xavier, A., to be published.
(43) Kennard, O., Isaacs, N. W., Coppola, J. C., Kirby, A. J., Wamer, S., Motherwell, W. D. S., Watson, D. G., Wampler, D. L., Chenerg, D. H., Larson, A. C., Kern, K. A., di Sansevercino, L. R., *Nature* (1970) **225,** 333.
(44) Arnone, A., Bier, C. J., Cotton, F. A., Hazen, E. E., Richardson, D. C., Richardson, J. S., *Proc. Natl. Acad. Sci.* (1969) **64,** 420.
(45) Ebashi, S., Endo, M., *Progr. Biophys. Mol. Biol.* (1968) **18,** 123.

RECEIVED June 26, 1970.

9

Structure of the Second Coordination Sphere of Metal Complexes and its Role in Catalysis

D. R. EATON

McMaster University, Hamilton, Ontario, Canada

> *Catalysis by transition metal complexes and metalloenzymes involves a sequence of ligand exchange reactions. Such a reaction can be viewed as an interchange of an inner sphere and outer sphere ligand. It is argued therefore that the structure of the second coordination sphere is of direct relevance to catalysis. A method of studying second coordination sphere structure based on dipolar NMR shifts in paramagnetic complexes is discussed. Even in weakly bound complexes, there is a definite preferred structure for the complex which may or may not be favorable for a subsequent substitution reaction.*

Although the validity of specific models for specific enzyme-catalyzed reactions may be open to question, the most general physical chemical principles are likely to be common to both enzymatic and nonenzymatic catalysis. I would therefore like to make a plea for the consideration of the significance of such a general phenomenon to catalysis, both by metalloenzymes and in nonbiological systems. I refer to the structure of the so-called second coordination sphere of metal complexes and its relevance to catalysis. This is an aspect of the field which has not been too widely discussed previously, either at the present symposium or elsewhere.

Many of the characteristics of a typical homogeneously catalyzed reaction are, I believe, equally applicable to enzymatic reactions. A suitable example is provided by the hydroformylation reaction catalyzed by cobalt carbonyl hydride. The mechanism of this reaction is relatively well understood and may be represented by the series of steps:

$$HCo(CO)_4 \rightleftarrows HCo(CO)_3 + CO$$

$$R\,CH_2{=}CH + HCo(CO)_3 \rightleftarrows (RCH{=}CH_2)\,HCo(CO)_3 \rightleftarrows RCH_2{-}CH_2Co(CO)_3$$

$$RCH_2CH_2Co(CO)_3 + CO \rightleftarrows RCH_2CH_2Co(CO)_4$$

$$RCH_2CH_2Co(CO)_4 + CO \rightleftarrows RCH_2CH_2COCo(CO)_4$$

$$RCH_2CH_2COCo(CO)_4 + H_2 \rightleftarrows RCH_2CH_2CHO + HCo(CO)_4$$

Thus, initially there is loss of a carbon monoxide ligand to give a new complex which in turn adds olefin to produce a third complex, hydride transfer gives a fourth compound, etc., and eventually the original cobalt carbonyl hydride is reformed with the simultaneous generation of product aldehyde. The details of this mechanism are not important for our purposes but it is reasonable to suppose that all catalytic reactions can be broken down in this way. All the significant chemistry occurs around the metal ion. We can be more specific than this and say that each step is, in effect, a ligand exchange reaction. The definition of ligand exchange reaction may have to be stretched a little but, I think, remains meaningful. For example, there may be loss of a strongly bound carbon monoxide and temporary replacement by a weakly bound solvent molecule. There may be an intramolecular reaction changing the nature of a ligand (olefin to alkyl) and possibly involving the coordination of solvent. Given such an extended definition, it is apparent that each step is a ligand exchange or interchange reaction. Finally, if the over-all reaction rate is to be fast as demanded by the condition of efficient catalysis, each individual step must be fast. The basic problem in catalysis, enzymatic or otherwise, is therefore reduced to one of understanding the factors which will induce a metal ion to rapidly exchange one ligand environment for another.

Not all ligand exchange reactions proceed by the same mechanism. Langford and Gray (1), for example, have introduced a classification involving associative mechanisms in which there is an intermediate of higher coordination number, dissociative mechanisms involving an intermediate of lower coordination number, and concerted mechanisms (I_a and I_d) with no clear intermediate. Whatever the detailed mechanism, there will be no efficient ligand exchange unless the incoming ligand is present and suitably positioned to take its place in the coordination shell. Another description of ligand exchange, according to Tobe (2), which I think is equally valid is "a rearrangement process of the aggregate whereby an inner sphere ligand changes place with an outer sphere ligand (3)." The structure and the binding of peripheral outer sphere ligands in a metalloenzyme or catalyst system may be just as important as that of the ligands which are directly attached to the metal. Perhaps some indication that this is correct is provided by the observation that enzymatic catalysis often produces large differences in the entropy of activation rather than in the enthalpy. Such a result might be expected if there were particularly favorable modes of holding potential ligands in a second coordinative sphere.

What is known of second coordination sphere structure? The term is an old one, first introduced by Werner (3) in the nineteenth century. Most of the examples which have been studied involve ion pairing of one kind or another. Work in this area has involved spectroscopic methods, polarography, the use of ion exchange resins, and a variety of other techniques (4). In spite of this, many of the details of the subject are not well understood. Turning to nonionic outer sphere complexes, the situation is much worse. Further, nonionic complexes are certainly most important in homogeneous catalysis since this usually occurs in nonaqueous and nonpolar solvents. In all probability, nonionic ligands are also of considerable significance in enzymatic reactions.

If the second coordination sphere is to have importance to ligand exchange reactions, we require answers to the following questions. Do the outer sphere ligands occupy spatially well-defined positions as do inner sphere ligands? If they do, there are obvious possibilities of catalytic importance. Secondly, is there any preferred orientation for a second sphere ligand with respect to the complex? Favorable or unfavorable orientation may be of very great significance for a facile substitution mechanism. Thirdly, what are the binding energies for second sphere coordination? Specifically, it would be very interesting to know how these compare with first coordination sphere binding energies. It seems probable that there will be a whole range of binding energies from almost zero upwards. From the catalytic point of view, the interesting region is likely to be that of intermediate energies where the second sphere ligands are held sufficiently strongly and sufficiently well to give some kind of an intermediate but not so tightly that they are themselves nonlabile. Following from the above question, we would like to know what type of forces holds the second sphere ligands in place. Is it purely electrostatic interaction and Van der Waals forces or is covalent bonding a factor to take into account? We need answers to this last question before we can expect to develop any predictive ability in this area.

From the catalytic point of view, the most interesting type of complex for study would be one where second sphere coordination was relatively strong and there is facile exchange between first and second sphere ligands. Experimentally, this is a difficult situation to deal with; because of the multiplicity of exchange processes involving free ligands, second sphere ligands, and first sphere ligands, interpretation of experimental data presents problems. I would therefore like to start by describing some studies involving weakly bound outer sphere ligands which do not exchange with the primary ligands. This simplifies matters very much from the experimental point of view. Hopefully, this approach will at least serve to sort out some of the factors involved in the stability of the second coordination sphere. Our studies have been extended to more strongly held complexes but I will not discuss this area.

The method we have used (5) in our studies of weakly bound second sphere complexes involves some properties of the NMR spectra of paramagnetic materials. With paramagnetic complexes, there are dipole–dipole interactions between the electron spin on the metal and the nuclear spins on ligands. Normally, the form of the geometric dependence on dipole–dipole interactions ensures that they are averaged to zero by tumbling of the molecules in solution. McConnell (6) first showed that in certain cases the dipole–dipole interactions will be manifested by shifts of the NMR lines. The essential criterion is that the g tensor of the paramagnetic complex be anisotropic. In this case, there will be a shift of the NMR resonances of interacting nuclei even in solution. The shift is given by an expression in the form:

$$\frac{\Delta H_i}{H} = \frac{\Delta \nu_i}{\nu} = \frac{(3\cos^2\Theta - 1)}{r^3} \frac{\beta^2 S(S+1)}{27kT} (g_{11} + 2g_\perp)(g_{11} - g_\perp)$$

This formula gives the shift in field (ΔH_i) or frequency ($\Delta \nu_i$) of the ith nucleus interacting with the unpaired electrons. It reduces to zero unless $g_{11} - g_\perp \neq 0$; i.e., there must be magnetic anisotropy. The second important feature is the geometric term $(3\cos^2\Theta - 1)/r^3$, in which r is the distance of the interacting nucleus from the metal ion and Θ is the angle between the principle magnetic axis of the molecule and a line joining the nucleus and the metal atom. It is this geometric dependence of the shift which opens up the possibility of structural studies. There are a variety of other terms in the equation, the exact nature of which depends to some extent on the relationship between the electron spin relaxation time and the correlation times for molecular tumbling. Since our interests are qualitative, the exact form of this third term will not be important. In addition to the above dipolar interaction, a Fermi contact interaction is also possible in these paramagnetic complexes. This latter effect is operative only if there is a finite possibility of finding the unpaired electron at the nucleus in question, and this condition demands metal–ligand covalent bonding. The weight of the evidence indicates that such bonding does not occur in the second coordination sphere complexes we will be discussing. The strategy of these experiments is, therefore, to use a suitable paramagnetic complex—*i.e.*, one with a large *g* value anisotropy as a nucleus for second coordination sphere formation. The dipole interaction then serves as a convenient probe for locating the positions of the second coordination sphere ligands.

The choice of a suitable system is dictated by several considerations. To avoid problems in disentangling the anisotropy term from the geometric term, it is desirable to use a complex with known geometry and *g* values. Secondly, it should be a nonlabile complex for reasons indicated above. Thirdly, the second sphere ligands should be rigid to avoid problems associated with averaging the shifts over the internal motions of the molecule. With these considerations in mind, we chose Co(II)trispyrazolylborate (the structure in Figure 1 with X = B, Y = H) as the complexing molecule. The *g* values of this complex are well known from the detailed ESR work of Jesson (7). The anisotropy is very large (g_{11} = 8.46; $g_\perp$ = 0.98) so a large dipolar effect is expected. Otherwise it is a stable, almost octahedral, complex with tridentate ligands and would not be expected to exhibit any unusual properties with regard to second coordination sphere bonding. As second coordination sphere ligands, aniline and pyridine were chosen initially since these are fairly rigid molecules with comparable dipole moments (aniline, 1.53D, pyridine 2.23D). They differ in that the functional groups are at the opposite ends of the dipole. The nitrogen is the negative end in pyridine but the positive end in aniline. An inert solvent is also required, and carbon tetrachloride was chosen for this purpose. The effects will be much reduced if the solvent, which is present in large excess, can compete favorably for a position in the second sphere.

In Figure 2, a series of traces of the NMR spectra of the aromatic protons of aniline in carbon tetrachloride with increasing concentrations of Co(II)trispyrazolylborate is shown. All the lines are shifted to high field with increasing cobalt concentrations. That this effect is attributable to the paramagnetic dipolar interactions was demonstrated by carrying out similar experiments with the analogous zinc complex. In the latter case, the shifts are very much smaller (1 or 2 Hz compared with 40 or

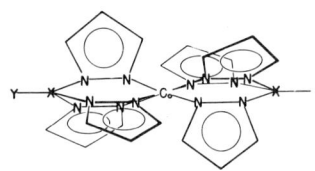

Figure 1. Geometry of Co-(II)trispyrazolylborate

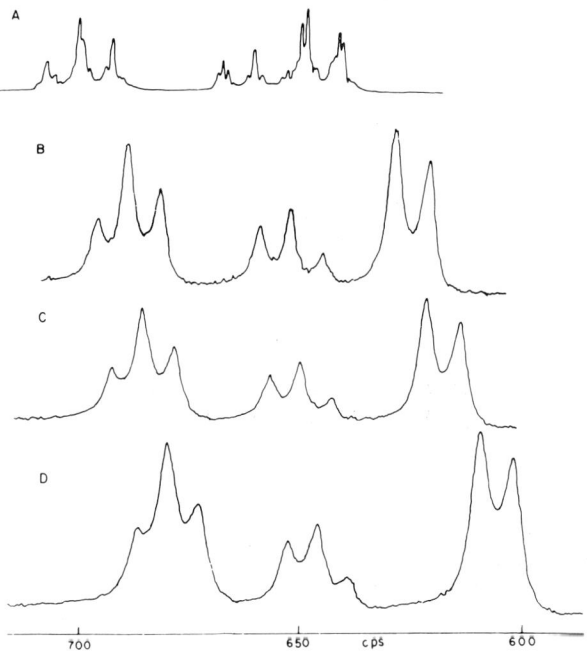

Figure 2. The NMR spectra of aniline. 100 MHz CCl_4 solution. A = pure aniline; B, C, and D = increasing amounts of $Co[HBpz_3]_2$. Shifts are in Hz to low field of TMS. The lines are assigned as meta, para, and ortho from left to right.

50 Hz) and served as reference frequencies. Assignment of the lines follows from the spin–spin splitting patterns and it is apparent from the traces that the shifts diminish in the order ortho > meta > para. The NH_2 protons (not shown in the figure) have the largest shift of all. One can immediately make several qualitative inferences. Firstly, the aniline occupies a preferred position in the second coordination sphere. If this were not so, the shifts would be averaged to zero. Secondly, the high field direction of the shifts shows that $(3 \cos^2\Theta - 1)$ is negative and hence that the preferred direction of approach is perpendicular to the symmetry axis of the molecule. Thirdly, there is a preferred orientation of the aniline in which the NH_2 group is closest to the cobalt. This result follows from the $1/r^3$ dependence of the shifts. These qualitative deductions regarding the geometry of the second sphere complex are illustrated in Figure 3.

Next we might look at some analogous experiments with pyridine as a ligand. Some NMR spectra are shown in Figure 4.

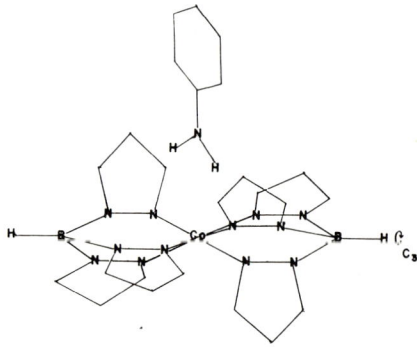

Figure 3. Preferred orientation of second sphere aniline in $Co[HBpz_3]_2$ complex

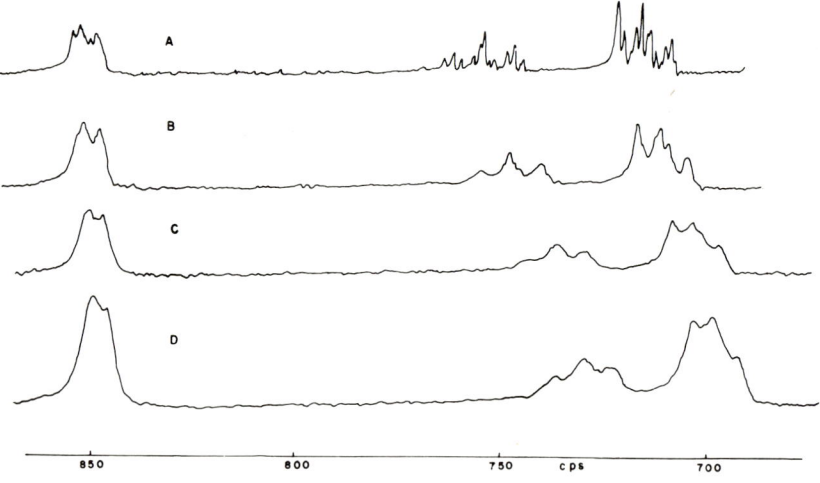

Figure 4. The NMR spectra of pyridine. 100 MHz CCl_4 solution. A = pure pyridine; B, C, and D = increasing amounts of $Co[HBpz_3]_2$. Shifts in Hz to low field of TMS. The lines are assigned as α, γ, and β from left to right.

Again the shifts are to high field but in this case the α proton is shifted very little, the β proton somewhat more, and the γ proton most of all. We can deduce therefore that in this case the preferential approach of the second coordination sphere ligand involves a structure with the proton closest to the metal atom. In both cases, the positive end of the dipole is closest to the metal. This is exactly the opposite of the usual electrostatic situation for first sphere complexes in which a negative dipole is directed toward a positive metal ion. One suitable model would involve hydrogen bonding between the amino protons and the pyrazoyl nitrogens but such a model is less attractive when applied to pyridine.

These are very labile complexes, as might have been anticipated, and the observed shifts are the average of free ligand and second coordination sphere ligand. If we wish to attempt a more qualitative treatment of the data, the simplest model would involve the equilibrium:

$$X + L \rightleftarrows XL \quad (L = \text{aniline or pyridine})$$

This model predicts that the shifts will be given by

$$\Delta \nu_{obs} = \frac{K[X]\Delta \nu_o}{1 + K[X]}$$

and hence will be independent of [L] and directly proportional to [X] if K is small. Experimentally, both of these conditions are verified. Figures 5 and 6 show plots for aniline and pyridine illustrating the second

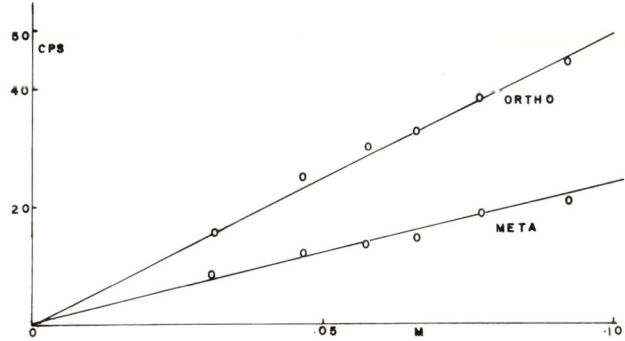

Figure 5. Plot of the dipolar shifts of the ortho and meta protons of aniline vs. concentration of $Co[HBpz_3]_2$

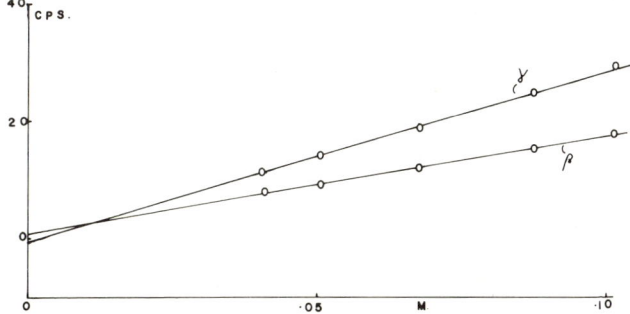

Figure 6. Plot of the dipolar shifts of the β and γ protons of pyridine vs. concentration of $Co[HBpz_3]_2$

point. If K is small, the slopes of these plots give the product $K\Delta\nu$ (K is the equilibrium constant and $\Delta\nu_o$ the shift of the undissociated second sphere complex). There is evidence (*vide infra*) that this model is not completely satisfactory but the slopes of these plots nevertheless give a measure of the extent of the participation of L in the second sphere bonding. The analysis can be carried further by investigating the temperature dependence of the shifts. After applying a correction which allows for the temperature dependence of the magnetic susceptibility, it may be shown that a plot of ln $\Delta\nu$ vs. $1/T$ has a slope corresponding to $\Delta H/R$ if the above simple model is assumed. Such a plot is shown in Figure 7 for the meta protons of aniline.

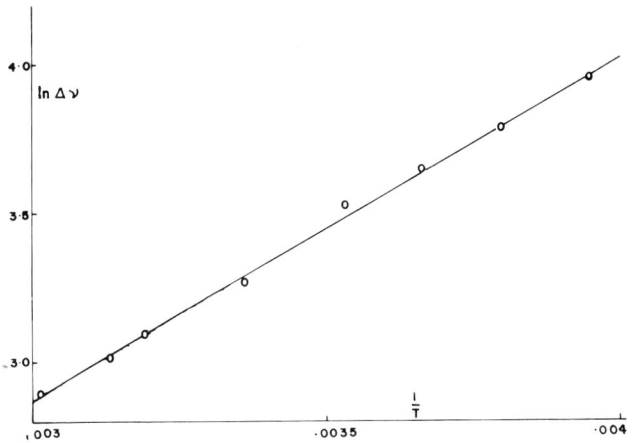

Figure 7. Plot of $\Delta\nu_{corr}$ vs. $1/T$ for the meta protons of aniline

In this way, values of ΔH of -2.4 ± 0.3 Kcal for aniline and -2.9 ± 0.3 Kcal for pyridine have been obtained.

This static model for second sphere complex formation is not fully valid and the above values of ΔH do not have full thermodynamic significance. Thus, if one assumes a "static" model for the second sphere complex, Equation 1 enables the relative shifts of the ortho, meta, para, and NH_2 protons to be calculated for any position of the aniline. Such calculations give qualitative but not quantitative agreement with experiment. The nature of the disagreement is illustrated by the data in Table I.

The dependence is closer to $1/r^4$ than $1/r^3$ and the closest fit is obtained with a Co–N separation of around 7 or 8Å, corresponding nicely to the Van der Waals radii of the molecules. It seems that the correct interpretation of these results is that we should not regard a second sphere complex as a rigid well-defined entity. Rather, there is a prefer-

Table I. Calculated Relative Shifts of Aniline Protons

Model	r	Ortho	Meta	Para
6Å	$\dfrac{1}{r^3}$	1.0	0.504	0.437
8Å		1.0	0.547	0.471
6Å	$\dfrac{1}{r^4}$	1.0	0.385	0.305
8Å		1.0	0.439	0.348
6Å	$\dfrac{1}{r^5}$	1.0	0.295	0.212
8Å		1.0	0.352	0.258
Experimental		1.0	0.470	0.320

Table II. Results in Carbon Tetrachloride

	pH[a]	$p\,CF_3$	$p\,N(CH_3)_2$	$p\,OCH_3$
$K\,\Delta\nu_o$ ortho[b]	500	740	360	410
$K\,\Delta\nu_o$ meta	240	343	214	189
$K\,\Delta\nu_o$ NH_2	532	889	–	–
Ratio meta/ortho	0.48	0.46	0.59	0.46
ΔH ortho	−2.5 Kcal	−3.5 Kcal	−2.7 Kcal	–
ΔH meta	−2.3 Kcal	−3.7 Kcal	−2.5 Kcal	−2.1 Kcal
Hammett σ	0.0	+0.55	−0.60	−0.27
Dipole moment	1.55 D	4.28 D	1.43 D	1.82 D

[a] Data from Reference 1.
[b] $K\,\Delta\nu_o$ values in Hz.

ential direction of approach, a preferential orientation of the complexing molecule, and an optimum separation, but other configurations are not much less probable and calculation of the dipolar shifts would require averaging over a large number of configurations, each weighted according to its probability. More extensive calculations indicate that given this flexibility, agreement can be obtained between observed and calculated shifts. The significance of these results is probably more in the qualitative demonstration of ordering of molecules in the second sphere than in the numerical results. Thus, the values of 2.4 and 2.9 kcal for the enthalpies of complexation are not really firm parameters in the thermodynamic sense because it is not possible to separate the geometric factors from the energy factors using this type of experiment. Nevertheless, they reflect in at least a qualitative way the energies involved in this type of second sphere complex.

The second sphere geometry may or may not be favorable for a substitution reaction. Thus, if the aniline prefers to approach with the NH_2 towards the metal, we might regard this as a favorable orientation. The pyridine results clearly indicate an unfavorable orientation for substitu-

Table III.

Substituent	K∆ ν_o ortho, Hz	K∆ ν_o meta, Hz
p NO_2	455	90
p CN	737	175
p CF_3	500	230
H	250	108
p CH_3	225	113
p OCH_3	218	106
p $N(CH_3)_2$	130	80

tion. If electrostatic interactions predominate in determining the mode of orientation, we should be able to predict the behavior of substituted anilines. Electron-withdrawing substituents should increase the dipole moment and enhance the binding in the second sphere. Electron-donating substituents should have the opposite effect. Table II shows some results with para-substituted anilines obtained in carbon tetrachloride.

The significant data are the slopes of the $\Delta \nu$ vs. [X] plots which give a measure of the relative preferences of the anilines for second coordination sphere interaction. Thus, the ortho value of 500 for aniline is increased by 740 by substitution of the electron-withdrawing CF_3 group and decreased to 410 and 360, respectively, by the donating substituents $N(CH_3)_2$ and OCH_3. Values for the other proton positions change similarly. Carbon tetrachloride is not a suitable solvent for many anilines but some additional results have been obtained in benzene. This allows a wider range of anilines to be used but benzene competes more favorably than carbon tetrachloride for a position in the second sphere and as a result the shifts are smaller by a factor of about two. The temperature range available is also more restricted. Some results are shown in Table III.

The general trend is maintained but there is substantial variation in the ortho-to-metal ratio, indicating that the flexibility of the second sphere complexes is varying.

A further question which might be asked is how the second sphere structure of these neutral molecules compares with that of ionic complexes. Some information pertinent to this question has been obtained by investigating compounds with the structure of Figure 1 with $X = C$ and $Y = H$. The resulting complex is a dipositive ion, and ion pairing with PF_6^- counter ions has been investigated. The ^{19}F shifts were measured relative to the analogous zinc compound. Qualitatively, the important result is that the shifts are to low field rather than to high field; since the sign of the ^{19}F nuclear moment is the same as that of ^1H, this indicates that the preferred direction of approach of the PF_6^- anion is along the

Results in Benzene

Ratio	ΔH, Kcal	Hammett σ	Dipole Moment
0.20	–	+0.78	6.29
0.24	3	+0.63	5.96
0.46	3	+0.55	4.28
0.43	–	0	1.55
0.50	3	−0.17	1.66
0.49	5	−0.27	1.82
0.62	5	−0.60	1.43

symmetry axis rather than perpendicular to it. The details of this situation are a little more difficult to disentangle than in the previous cases since there are definitely two equilibria to consider and the equilibrium constants are not small. Our estimation of the dipolar shift for the undissociated complex has been obtained by examining the shifts in mixed solvents with different dielectric constants and extrapolating to zero dielectric constant. The results indicate a separation between the Co and the P of about 8 Å, which again corresponds rather nicely to a contact ion pair rather than a solvent separated ion pair. An obvious rationalization of this result is that a neutral molecule (in this case, solvent acetonitrile) will take the closest position and the ion occupies the next most favorable position. This is perhaps not entirely surprising since ion–ion interactions fall off less rapidly with distance than ion–dipole interactions and it is possible that there can be an over-all energy advantage in placing the dipoles close to the complex and the ions further away.

All of the above results have been concerned with weak second sphere coordination. It seems likely that there will be a whole range of binding energies culminating in a situation in which the second sphere ligand approaches the stability of a first sphere ligand. An example of such a case is provided by the work of Corsini et al. (9) on U(VI) complexes of 8-hydroxyquinoline. Preparation of such a complex leads to the production of a compound which analysis shows contains an extra molecule of neutral ligand. This additional molecule is held partly by hydrogen bonding to the oxygen of one of the primary ligands and partly by formation of a seventh covalent bond to uranium. Further, radioisotope studies have demonstrated facile exchange with the primary ligands. This molecule is, however, held too tightly to exchange readily with free ligand. The optimum situation for promoting ligand exchange is one in which the second sphere binding is of intermediate strength.

To summarize, I would like to suggest that the acquisition of a ligand in the second coordination sphere is an essential prerequisite for ligand exchange. The stability and geometric arrangement of ligands in this second coordination sphere will have a very large influence on the rate of ligand exchange. Since homogeneous catalytic and enzymatic processes proceed by a series of ligand exchange processes, the structure of the outer sphere ligands is likely to be of great significance. Even very weak second sphere complexes have some preferred molecular structure which may or may not be favorable for subsequent ligand exchange. These initial results seem to indicate that electrostatic interactions are dominant in determining this structure. In enzymes, functional groups distant from the active site may play a decisive role in determining second sphere structure.

Acknowledgment

I should like to thank the National Research Council of Canada for financial support of this research and S. Trofimenko of the Central Research Department, E. I. Du Pont de Nemours Co., for gifts of Co(II)-trispyrazolylborate and Co(II)trispyrazolylmethane. The results with substituted anilines were obtained in collaboration with H. O. Ohordnyk and Linda Seville.

Literature Cited

(1) Langford, C. H., Gray, H. B., "Ligand Substitution Processes," Benjamin, New York, 1965.
(2) Tobe, M. L., ADVAN. CHEM. SER. (1965) **49**, 7.
(3) Werner, A., *Ann. Chem.* (1912) **386**, 1.
(4) Beck, P., *Coord. Chem. Rev.* (1968) **3**, 91.
(5) Eaton, D. R., *Can. J. Chem.* (1960) **47**, 2645.
(6) McConnell, H. M., Robertson, R. E., *J. Chem. Phys.* (1958) **29**, 1361.
(7) Jesson, J. P., *J. Chem. Phys.* (1966) **45**, 1049.
(8) Heck, R. F., Breslow, D. S., *J. Am. Chem. Soc.* (1961) **83**, 4023.
(9) Corsini, A., Abraham, J., Thompson, M., *Chem. Commun.* (1967) 1101.

RECEIVED June 26, 1970.

10

Structure and Function of Metalloenzymes

D. D. ULMER and B. L. VALLEE

Biophysics Research Laboratory, Peter Bent Brigham Hospital and Department of Biological Chemistry, Harvard Medical School, Boston, Mass.

> *In metalloenzymes, metals which affect function directly by participation in catalysis appear to have ligand sites which differ markedly from those of metals which influence function indirectly through modulation of protein structure. Thus, in both E. coli alkaline phosphatase and equine liver alcohol dehydrogenase, metals positioned at the active site interact selectively with chelating agents, undergo isotope exchange, and display distinctive physical chemical characteristics. Such active site metals may have an irregular geometry which facilitates their catalytic role. In contrast, nonactive site metals exhibit physical properties more like those of simple, bidentate model complexes; they frequently appear to stabilize structure or influence subunit interactions as shown by their effects on sedimentation or hydrogen exchange rates of proteins. Such "structural" metals may function importantly in control mechanisms for biochemical reactions.*

During the past two decades, considerable progress in understanding the role of metals in biological processes has resulted from the discovery and purification of a large number of metalloproteins and from efforts to ascertain their function—*e.g.*, in catalysis, electron transport, or gas exchange (*1*). Studies of the physical chemical properties of this group of proteins have been of particular interest because of the presence of the inorganic moiety. In principle, metals should be excellent labels of the protein sites to which they bind owing to the distinctive physical properties of metal complexes with common ligands—*e.g.*, color, in the case of copper, cobalt, or iron salts. Hence, one might have hoped that the basic characteristics of the metal, known from studies of inorganic complexes, would be preserved in biological systems and aid identification of the ligand groups and their properties at protein loci of special interest, *i.e.*, the active sites of metalloenzymes. Over many years, we have attempted to relate such characteristics, particularly the spectra of simple bidentate metal complexes, to those of metalloenzymes and have found surprisingly little correspondence; rather, the spectra of metalloenzymes are often quite unusual, a circumstance which of itself may be most significant.

It now seems possible that some of the unusual chemical features of metalloenzymes, not duplicated in any simple metal chelates, may well reflect their biological specificity. In this regard, metals in metalloenzymes may be likened to certain amino acid side chains in nonmetalloenzymes which display unique chemical reactivity toward specific organic reagents, while the underlying chemical reactions are not observed in simple peptides or denatured proteins. Thus, the inactivation of certain seryl proteases by diisopropylfluorophosphate (2) and that of sulfhydryl, histidyl, and lysyl enzymes by other specific organic reagents under mild conditions (3) arises from the unique chemical properties of these amino acid side chains when they are positioned critically at active enzyme sites. Presumably, the basis of such reactivity is related to an unusual microscopic environment about the residues involved and poses questions as to the relation of atypical chemical reactivity to biologic function.

Significantly, spectroscopic, optical rotatory, and magnetic properties of metalloenzymes (1) suggest that characteristics of their metal atoms are also affected uniquely by the protein environment at active enzymic sites. Moreover, while the physical properties of metalloenzymes are quite unusual when compared with those of simple metal complexes, substrates or inhibitors may simplify their absorption and rotatory dispersion spectra to resemble more nearly those of well defined model systems (4). Taken together, such observations have emphasized the singular nature of metals and their coordination in metalloenzymes, perhaps reflecting features of their participation in catalysis.

The most intensive investigative efforts in this field have for many years involved single chain enzymes, for example, carboxypeptidase and carbonic anhydrase. With such enzymes, it has been possible reversibly to remove and restore metal atoms at the active site, substitute different metals for those normally found there, and examine the resulting spectrochemical and functional consequences without major interference from structural alterations which might complicate interpretation of the data. More recently, similar lines of investigation have been successfully extended to multichain enzymes. However, such efforts have been complicated by problems arising from the presence of subunit structure, including variations in metal content, protein stability, ease of metal removal and reconstitution, and the existence of isoenzymes. Despite these problems, investigations of multichain metalloenzymes have extended our understanding of the biological functions of metals, particularly their apparent frequent participation in control of protein structure (5).

Indications of a structural role for metals in a multichain enzyme were actually noted nearly a decade ago; e.g., the removal of zinc from yeast alcohol dehydrogenase with chelating agents resulted in dissociation of the protein into subunits (6). During the intervening years, additional investigations have suggested that this structural role for metals may be rather common and, indeed, have important implications to enzymatic function.

Today, for the purpose of illustrating these points, we wish to review selected segments of work performed in our laboratory on two multichain enzymes—alkaline phosphatase from E. coli and equine liver alcohol dehydrogenase—in which intrinsic metal atoms appear to play both functional and structural roles. Thereafter, we hope to discuss data suggesting that metals may be important also to the structure of certain single-chain proteins and to the general phenomenon of stabilization of protein structure.

Alkaline Phosphatase from E. coli

The preparation of E. coli alkaline phosphatase by the method of Simpson et al. (7), employing osmotic shock to release the enzyme from the periplasmic space and subsequent purification by chromatography on DEAE cellulose, reproducibly results in an ultracentrifugally and electrophoretically homogeneous protein of high specific activity and characteristic metal composition. The zinc content of different preparations measured by emission spectrography, atomic absorption spectroscopy, and chemically, employing dithizone, varies from 2500 to 3040 μg/gm of protein, equivalent to 3.4 to 4.2 gram atoms of zinc per mole of phosphatase. The purified enzyme also contains small amounts of magnesium and iron for which no function has, as yet, been identified.

The chelating agent, 8-hydroxyquinoline-5-sulfonic acid, rapidly removes two of the four zinc atoms from alkaline phosphatase when the incubation is carried out as indicated in Figure 1. Catalytic activity concomitantly falls to nearly negligible values. Hence, two zinc atoms appear to be located at the active site of alkaline phosphatase and are critical to its function. More prolonged incubation with this metal-binding agent results in removal of the remaining more firmly bound zinc, resulting in a metal-free apoenzyme.

While only two of the zinc atoms of alkaline phosphatase are essential for enzymatic function, all four of the metal atoms appear to contribute to the structural stability of the protein, as demonstrated by ultracentrifugal experiments. The molecular weight of alkaline phos-

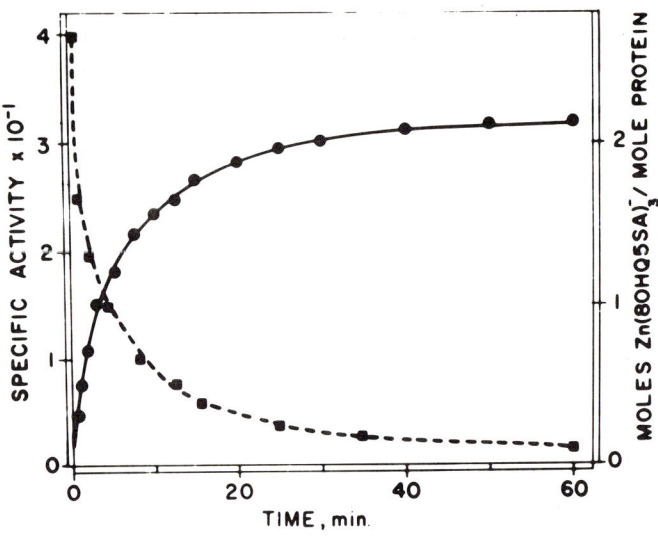

Figure 1. Kinetics of zinc removal from alkaline phosphatase
[Protein] = 1.33 × 10^{-5}M, [8-OHQ5SA] = 6.65 × 10^{-4}M, 0.1 tris-*chloride, pH 8, 25°*

phatase, obtained by low-speed equilibrium centrifugation, is approximately 89,000. This weight is little affected by removing completely the metal from the protein, as indicated by the sedimentation values in water for the native and apoenzyme (Table I). In solvents which fully denature alkaline phosphatase, permitting its dissociation into subunits—e.g., guanidine hydrochloride—both the native and apoenzymes again sediment as symmetrical boundaries with identical sedimentation values, $s_{20,w} \cong 1.2S$, under these conditions. However, in two solvent systems examined, 8 M urea and 40% dioxane, the apoenzyme dissociates into subunits while the native zinc enzyme is only slightly altered. In these instances, the zinc atoms help to maintain protein structure (Table I). The nature of the solvents utilized suggests that both metal ion stabilization and hydrophobic forces may affect the dimerization of alkaline phosphatase (8).

Table I. Sedimentation of Native and APO Phosphatase
20°, pH 7.0, 0.05 M *tris*-Cl, 0.1 M KCl

Solvent	$S_{20,w}$, Native	$S_{20,w}$, APO
Water	6.1	6.1
5 M Guanidine HCl	1.2	1.2
8 M Urea	5.8	1.4
40% Dioxane	5.9	1.8

The stoichiometry of stabilization of the quaternary structure by zinc, identified by the experimental procedure just discussed, is depicted in Figure 2. For this purpose, increments of zinc were added to the apoenzyme which was then diluted into 40% dioxane and sedimented. In the absence of added zinc, no protein with an s value of 5.9S is discerned; all of the enzyme is dissociated into subunits under these conditions. However, when a full complement of zinc—*i.e.*, 4 gram atoms per mole— is added to the enzyme, all of the protein sediments with an s value of 5.9S. At intermediate zinc contents, two boundaries are observed. These experiments indicate that all four zinc atoms are necessary to stabilize the quaternary structure of this enzyme fully (8).

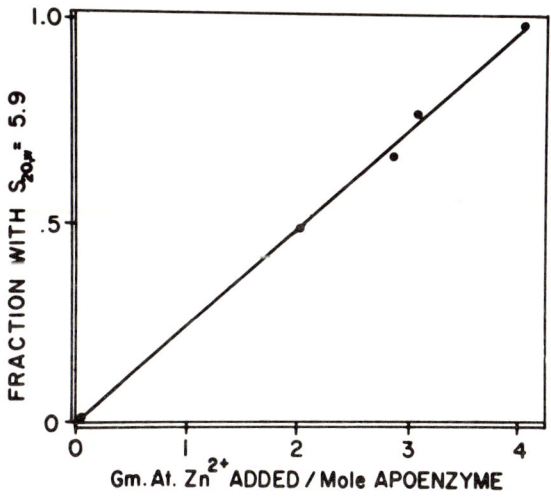

Annals of the New York
Academy of Sciences

Figure 2. Stabilization of quaternary structure of phosphatase by zinc
40% Dioxane, 0.05M tris-Cl, 0.1M KCl, pH 7.0, 20°

© *The New York Academy of Sciences;
1969; reprinted by permission.*

The substitution of cobalt for the native zinc ions of alkaline phosphatase results in an active enzyme with distinctive optical properties, generated by the interaction of cobalt with the ligands of the protein. These properties may be employed to investigate the modes of binding of cobalt to the enzyme and also serve in a remarkable fashion to distinguish the catalytically essential metal atoms from those which play only a structural role.

Table II. Competition of Zinc and Cobalt for the Metal Binding Sites of Alkaline Phosphatase

Zinc Added, Gram Atoms/Mole	Cobalt Bound, Gram Atoms/Mole	Zinc + Cobalt, Gram Atoms/Mole
0	3.9	3.9
1.0	2.8	3.8
2.0	2.0	4.0
3.0	1.2	4.2
4.0	0	4.0

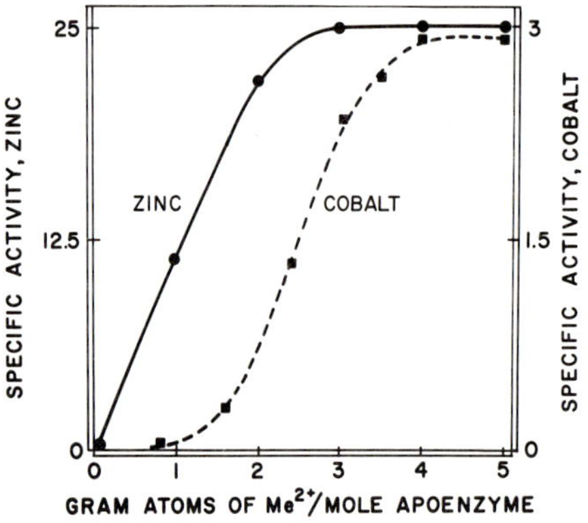

Biochemistry

Figure 3. Phosphatases
Assay: 10^{-3}M NPP, 1M NaCl, 0.01M tris-Cl, pH 8, 25°

Zinc and cobalt appear to occupy the same protein sites in alkaline phosphatase, as demonstrated by the results of competition between the two metals for these sites (Table II). Varying molar proportions of zinc are added to the apoprotein prior to addition of a five-fold molar excess of cobalt. Excess metal is removed by gel filtration and the cobalt incorporated is then measured. When cobalt alone is added to the apoprotein, nearly 4 gram atoms of the metal are incorporated (line 1, Table II). In the remaining experiments, the sum of zinc and cobalt bound is nearly 4 in all instances. Hence, the four protein sites can apparently bind zinc and cobalt interchangeably (9).

Two distinct types of binding sites in phosphatase can be distinguished on the basis of their differential affinity for cobalt and zinc. The restoration of activity upon addition of increments of zinc to the apoenzyme is compared with the activity resulting from addition of similar increments of cobalt in Figure 3. Notably, the first 2 gram atoms of zinc restore nearly all the activity of native phosphatase: the last two zincs bound have little effect on it.

In attempts to regenerate the catalytically active metalloenzyme from apophosphatase, the method chosen for the preparation of the apoenzyme has proven critical. For the present studies, apophosphatase was prepared by removing zinc with 8-hydroxyquinoline-5-sulfonic acid, followed by extensive dialysis to remove both the zinc complex and the free metal-binding agent. Addition of 2 gram atoms of zinc per mole of enzyme restores activity, and additional metal ions do not increase it further. In contrast, when apophosphatase is prepared by Sephadex chromatography in the presence of EDTA, the initial two zinc atoms added to the apoprotein fail to restore enzymatic function, while activity does reappear upon addition of the third and fourth gram atoms per mole [Petticlerc et al., Eur. J. Biochem. (1970) 14, 301.]

Employing ^{14}C-EDTA in a similar manner, it can be shown that this agent binds stubbornly to apophosphatase, thereby accounting for the altered sequence of restoration of the zinc atoms (Kaden, T., Vallee, B. L., unpublished observations). When zinc is added to the EDTA-treated enzyme which has been dialyzed for 24 hours only, the first two zinc atoms fail to restore activity and, indeed, the third and fourth atoms per mole will do so, as was observed in the Sephadex experiments (*loc. cit.*). At this juncture, several moles of ^{14}C-EDTA remain bound to the apoenzyme. However, when the apoenzyme is dialyzed for prolonged periods, further EDTA is removed; the addition of zinc then gives results which progressively resemble those obtained with the 8-hydroxyquinoline-5-sulfonic acid-treated enzyme, becoming identical with them after 5 days of dialysis when virtually no ^{14}C-EDTA remains bound to the enzyme.

Apparently, after preparation of the enzyme with EDTA and owing to incomplete removal of this agent, zinc atoms will first interact with EDTA which is bound to the enzyme, rather than with the enzyme itself. Metal is bound at the active enzymatic sites only when such EDTA sites are saturated. Thus, the preparative procedures employed seem to account for the difference in results observed.

Only two of the cobalt atoms restore activity but, in the case of cobalt, it is the last 2 gram atoms of metal bound which result in catalytic function (Figure 3). The first two cobalt atoms bound affect activity but little. Apparently, zinc first occupies the catalytic sites and the remaining structure-stabilizing sites only thereafter. In contrast, cobalt initially occupies only structural sites and thereafter the catalytic sites. Presumably, such a difference in site selection might arise from the known differences in preferred ligand configuration for each metal (9).

Cobalt, in contrast to zinc, is an excellent physical-chemical probe. The activity data just presented suggest two different modes of interaction of cobalt with phosphatase, and the spectra of the cobalt enzyme can be

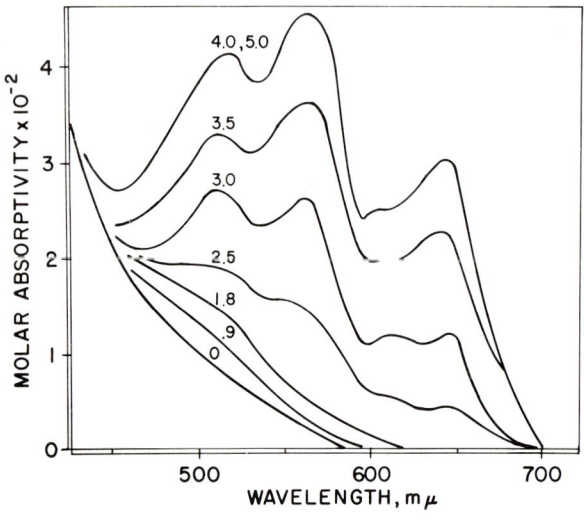

Figure 4. Absorption spectra of cobalt alkaline phosphatases
0.01M tris-Cl, pH 8.0, 25°

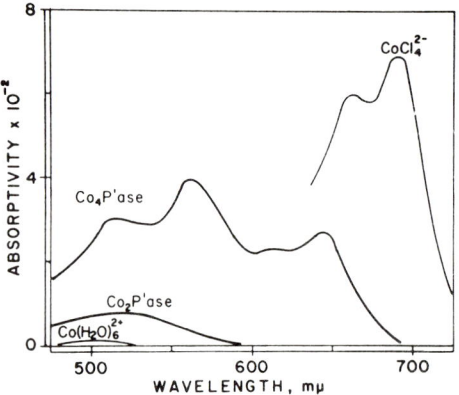

Figure 5. Cobalt absorption spectra

utilized to confirm this hypothesis. As shown in Figure 4, the addition to apophosphatase of the molar excesses of cobalt indicated above the curves result in spectra which differ significantly for the first two atoms of cobalt when compared with the last two cobalt atoms bound. The featureless curve at the lower left of the figure, labeled zero, represents the absorption spectrum of apoenzyme at the high concentrations of protein required for such experiments. The spectrum generated by the addition of the first 2 moles of cobalt is simple, of low absorptivity, and centered about 500 mμ. It indicates cobalt bound at the structural sites. In contrast, the spectrum generated by addition of the second pair of cobalt ions is complex and of significantly higher absorptivity with bands at 640, 610, 555, and 510 mμ. This unusual spectrum arises from cobalt bound at the catalytic sites (9).

The two different types of spectra of cobalt phosphatase are compared with those of known cobalt complex ions in Figure 5, which suggests that there may be differences between the geometry of the structural and catalytically active cobalt ions in the enzyme. The spectra of octahedral cobalt complexes, such as the hexaquo complex in the lower left portion of the figure, are simple, of low absorptivity, and generally centered at about 500 mμ. The resemblance of this spectrum to that of the structural cobalt spectrum in the enzyme, the curve designated "Co_2 P'tase," is apparent. In contrast, the complicated multibanded spectrum of the catalytic site cobalt, designated "Co_4 P'tase," is dissimilar to both the octahedral model and to the spectrum of tetrahedral cobalt complexes, illustrated by the tetrachloro complex in the upper right. Such unusually complex spectra suggest an irregular geometry of metal binding and have thus far only been observed in metalloenzymes with cobalt at their active sites—e.g., carboxypeptidase, carbonic anhydrase, and here, alkaline phosphatase.

For convenience, the absorptivity of the cobalt protein at 640 mμ, a spectral region free of the influence of extraneous bands at either shorter or longer wavelengths, may be used as a measure of the presence of the active site cobalt spectrum as shown in Figure 6. Thus, if increments

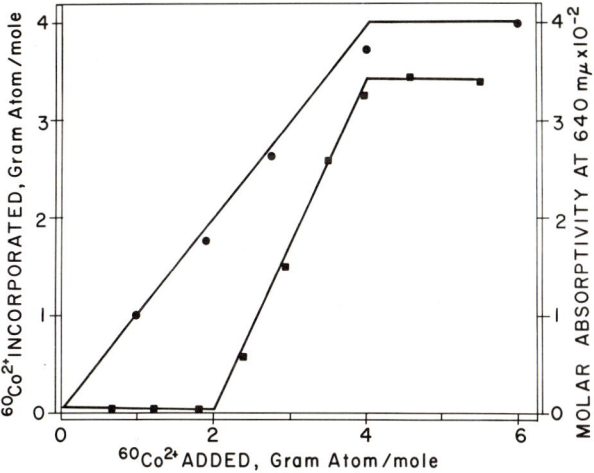

Figure 6. Cobalt incorporation into alkaline phosphatase

of isotopically labeled $^{60}Co^{2+}$ are added to the apoenzyme, a total of 4 gram atoms of metal is required to reconstitute the protein fully, the titration curve on the left. However, only the last two of these four cobalt atoms, those associated with restoration of catalytic activity, generate the spectrum at 640 mμ, the titration curve on the right.

If we are correct in assuming that the unusual active site cobalt spectrum reflects a geometry about the metal atom which in some way is advantageous for catalytic activity, conditions which alter activity of the enzyme should be paralleled by alterations of the spectra. This can be tested by comparison of the pH dependence of the cobalt spectra with that of enzymic activity (Figure 7). Like the zinc enzyme, the activity

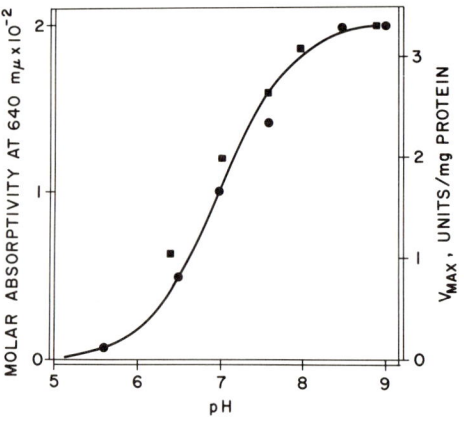

Biochemistry

Figure 7. Effect of pH on absorption spectrum and activity of cobalt alkaline phosphatase

of cobalt phosphatase reversibly decreases over the pH range from 8 to 6 with an apparent pK of 7.0, as shown by plotting V_{max} (the squares) vs. pH. Similarly, from pH 8 to 6, the distinctive active site cobalt spectrum, presented as absorptivity at 640 mμ (the circles), is reversibly abolished. The presence of the spectrum of the active site cobalt atoms exactly parallels the enzymatic activity (9).

It might also be expected that inhibitors, such as phosphate ion, which alter the activity of the enzyme would also alter the spectrum of the active site cobalt atoms. This is borne out by the data shown in Figure 8. The upper solid line again identifies the complex multibanded spectrum of the active site cobalt atoms. The addition of 2 moles of phosphate per mole of protein simplifies this spectrum to that shown by the

10. ULMER AND VALLEE *Metalloenzymes* 197

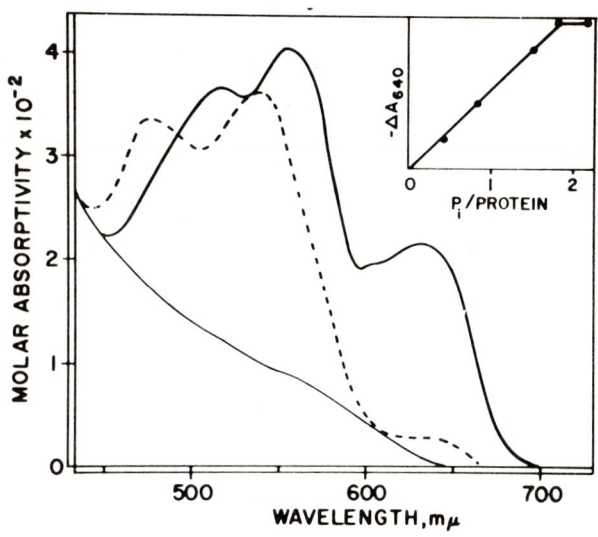

Figure 8. *Absorption spectra of cobalt phosphatase ± phosphate*
0.01M tris-Cl, pH 8.0, 25°

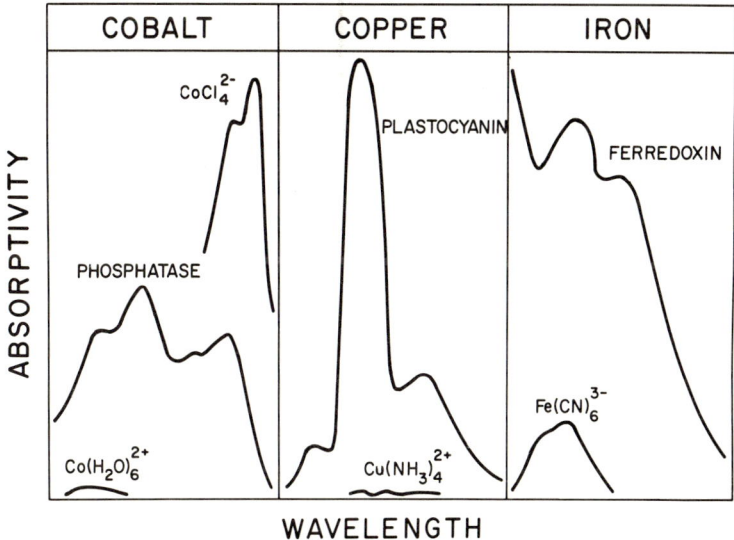

Figure 9. *Spectral comparison of metal complexes and metalloenzymes*

dashed line, having maxima only at 540 and 475 mμ. This spectrum of the enzyme-inhibitor complex is similar in band position to that of an octahedral cobalt complex but is of much higher absorptivity, perhaps indicating filling of the coordination sphere of the catalytically active cobalt ions. In an analogous manner, the complex circular dichroic spectrum of the cobalt enzyme is markedly simplified upon formation of the enzyme complex with phosphate (9).

The Entatic Site Hypothesis

We have mentioned earlier the dissimilarities between the spectral properties of chromophoric metal ions at the active sites of metalloenzymes and the properties of simple bidentate model complexes of the same metals. Cobalt phosphatase has served well to illustrate such a dissimilarity and, in Figure 9, the data for phosphatase, representative of a cobalt enzyme, are shown again along with those for plastocyanin, a copper enzyme, and ferredoxin, an iron enzyme. Each enzyme spectrum is unusual compared with the simple model complexes shown at the bottom of the figure. More detailed spectral data as well as comparison of other physical properties of metalloenzymes—e.g., electron paramagnetic resonance spectra—with those of model complexes have been summarized previously (10).

The unusual properties of metals in metalloenzymes, when compared with those of simple, bidentate metal complexes, are presumed to arise through conformationally imposed constraints upon amino acid side chains which comprise the multidentate ligands of the metal. Such constraints may well force irregular geometries on the metal atoms (10). The imposition of a new geometry on a chemical system is prone to induce reactivity. Therefore, irregular geometry, such as appears to exist about the metal atom in phosphatase and other metalloenzymes, might be looked upon as an intrinsic characteristic of such systems designed to achieve a condition energetically favorable to catalysis (10, 11). It would appear that in many enzymes the properties of the catalytically active metal atoms might be consistent with those to be expected if the metal had a distorted geometry approximating that of the plausible transition state for the very reaction in which it is involved. Such possible distortions and/or the existence of an open coordination position could then signal the existence of an activated energy state, pre-existing the entry of substrate, and lowered by the formation of enzyme–substrate or inhibitor complexes. We have termed such a postulated activated energy state, intrinsic to the enzyme, entatic, i.e., in the state of tension or stretch (10, 11). This view in no way precludes, of course, that the entry of substrate itself could lower the activation energy further as has been suggested. In fact, both of these could be characteristic of the catalytic process.

Mildvan ("The Enzymes," 3rd ed., P. D. Boyer, Ed., Vol. II, p. 493, Academic, New York, 1970), in evaluating the entatic site hypothesis, cites as his primary objection that "rates of ligand substitution, ligand deprotonation, and electron transfer which are found in simple model systems are often as high or higher than those found in metalloenzymes." Reasoning further on this basis, he concludes that since the metals "are inherently more reactive than the limiting rates of enzyme catalysis," the protein has not "conferred kinetic reactivity" upon them. One can not easily take exception to this argument; unfortunately, it greatly confuses the issue since it refutes what the entatic site hypothesis never stated. Hence, it is in the nature of a "non sequitur."

The entatic site hypothesis does note the existence of distorted geometries of metal coordination sites in metalloenzymes and explores their possible relationship to enzymatic function. It does not imply enhanced general chemical reactivity of the metal atoms in metalloenzymes; on the contrary, the hypothesis proposes that the distorted geometry of the coordination sphere of metals in metalloenzymes is responsible for increases in a critical step or steps in catalysis of an enzymic reaction. Indeed, it could be argued that decreased reactivity in all reactions other than the enzymatic reaction might be a general feature of metalloenzymes, although this had not been suggested.

The additional comment that the high affinity of metalloenzymes for their metals as "compared with the stability of chelates which use the same ligands, argues against a thermodynamically strained coordination" is similarly not relevant and based upon a misinterpretation of the entatic site hypothesis. Entasis implies that the difference in energy between the ground state and transition state for the enzymatic reaction is reduced, not that the metal–enzyme complex is thermodynamically less stable, as was inferred. Indeed, there is no reason to suppose that the distorted environment of a metal ion in an enzyme as opposed to a simple metal complex leads necessarily to an increase in free energy. The studies of alkaline phosphatase just presented certainly seem consistent with the entatic state hypothesis.

Alcohol Dehydrogenase from Equine Liver

Horse liver alcohol dehydrogenase, LADH, provides a somewhat different example of an enzyme in which the intrinsic metal atoms appear

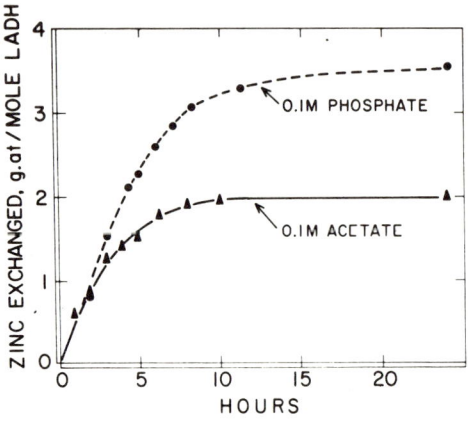

Proceedings of the National Academy of Sciences

Figure 10. Exchange of Zn in LADH
1.2×10^{-5}M LADH, 1.3×10^{-4}M $^{65}Zn^{2+}$, pH 5.5, 4°

to be heterogeneous in terms of their chemical bonding and, correspondingly, serve both functional and structural roles.

In 1955, Theorell and coworkers (*12*) first reported that LADH contained 2 gram atoms of zinc per mole, a value later confirmed in our laboratory (*13*) and consistent with the demonstration of two active sites on the basis of coenzyme binding (*14, 15*). However, during the past 15 years, methods for purification of the enzyme and estimation of its specific activity and molecular weight have been revised repeatedly, resulting in uncertainties concerning the molar stoichiometry of protein and metal. Recently, this problem has been further complicated by recognition of the subunit structure and the presence of isoenzymes of LADH. During the past few years, the zinc contents of uniformly treated samples of the enzyme prepared in our laboratory have varied from 3.1 to 4.3 gram atoms of zinc per mole, based on a molecular weight now thought to be 80,000 (*16*). Significantly, however, a variety of experiments performed in our laboratory still indicate that only two of the zinc atoms in this molecule are involved in enzymatic activity (*17*).

When dialyzed at pH 5.5 in the presence of isotopic zinc, $^{65}Zn^{2+}$, LADH remains perfectly stable for several days without loss of enzymatic activity and undergoes zinc exchange. However, the number of zinc atoms which exchange depends critically upon the nature of the buffer anion employed. In 0.1 M phosphate buffer, all of the 3.6 gram atoms of zinc in this particular preparation of LADH exchange with $^{65}Zn^{2+}$ within 24 hours (Figure 10). In contrast, if exchange is carried out under comparable conditions but in 0.1 M acetate buffer, only 2 gram atoms of zinc exchange. If both acetate and phosphate buffer are present, only 2 gram atoms exchange also. Hence, acetate appears to prevent the exchange of a portion of the zinc atoms in LADH (*18*).

In acetate buffer, all of the labile zinc undergoes exchange with a single rate constant, the absolute value of which depends upon enzyme concentration (*18*) (Figure 11A). In contrast, in phosphate buffer, het-

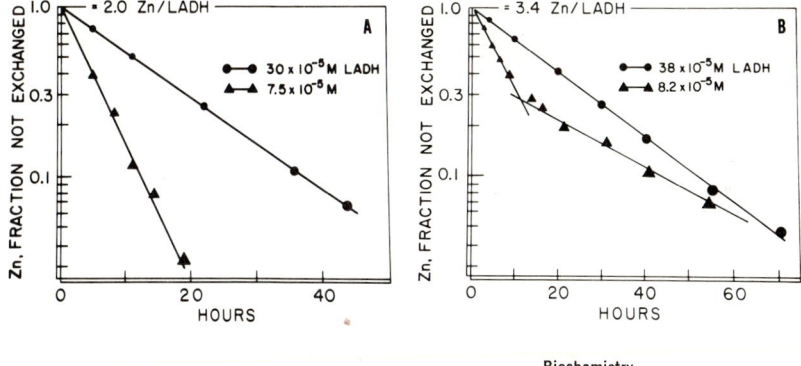

Figure 11. Zinc isotope exchange in LADH
A. 0.1M acetate buffer, pH 5.5
B. 0.1M phosphate buffer, pH 5.5

erogeneity in the exchange rate is apparent (*18*). At high enzyme concentrations, a single exchange reaction occurs with a rate comparable with that observed in acetate, while at lower enzyme concentrations two exchange rates are evident (Figure 11B). Thus, under proper conditions, exchange in phosphate reveals two groups of zinc atoms which appear to be bound to LADH with differing stability. Notably, enzymes in which the zinc atoms have been fully exchanged in either phosphate or acetate buffer are fully active catalytically and physicochemically indistinguishable both from each other and from the native enzyme.

A number of lines of evidence indicate that the two zinc atoms which exchange in acetate are related to catalytic activity while the additional 1 to 2 gram atoms which exchange in phosphate are not involved in enzymic activity but, rather, may serve a role in maintenance of the structure of LADH. The study of metalloenzyme–inhibitor complexes provides one approach to delineating these two categories of zinc.

As is well known, LADH is inhibited by chelating agents either reversibly, through their binding to zinc to form a mixed complex, or irreversibly, by removing the metal from the enzyme. OP, 1,10-phenanthroline, inhibits reversibly through formation of an enzyme–zinc–chelate mixed complex (*13*). As shown in Figure 12, this complex generates a

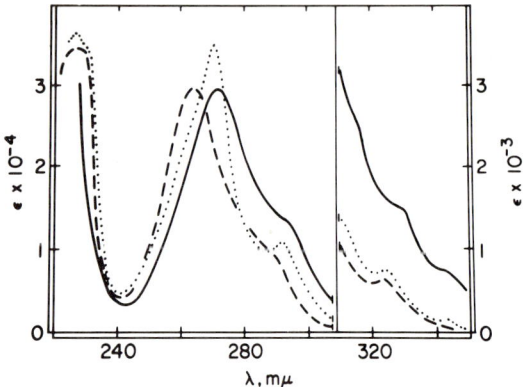

Figure 12. Absorption spectra of OP (– –), $Zn(OP)_1^{2+}$ (····), and LADH · Zn · OP (———) 0.1M tris-Cl, pH 7.5

characteristic absorption spectrum with peaks at 271, 297, 316, 329, and 345 mµ (19). For comparison, the absorption spectrum of OP, itself, and that of the complex of OP with inorganic zinc are also indicated.

The LADH–Zn–OP chromophore is optically active (15) and generates a positive extrinsic Cotton effect as shown by the rotatory dispersion spectrum in Figure 13. The corresponding ellipticity band, observed by circular dichroism, is shown in the same figure. Such Cotton effects reflect the asymmetry of the metal binding site of the enzyme, and may be usefully employed to signal the interactions of other molecules at this locus (15).

Based upon the chromophoric properties of the enzyme–metal–inhibitor mixed complex, either spectrophotometric or spectropolarimetric methods may be employed to study the stoichiometry of binding of OP to LADH. In Figure 14, the difference absorbance between the mixed complex and the component enzyme and OP at two separate wavelengths, 329 and 297 mµ, as well as the difference rotation between the mixed complex and the enzyme alone are plotted according to the method of molar proportions. It is evident that only 2 moles of chelating agent bind to each mole of enzyme when the titration is carried out either spectrophotometrically or spectropolarimetrically (20). In an identical manner, optical rotatory titration indicates that only 2 moles of another chelating agent, α,α'-bipyridyl, bind to each mole of LADH. Hence, these experiments show that only a portion of the total zinc in LADH is available for reaction with chelating inhibitors.

Since it appears that only 2 of the 3.5 to 4 gram atoms of zinc in LADH react with OP, it was of interest to examine further the stoichi-

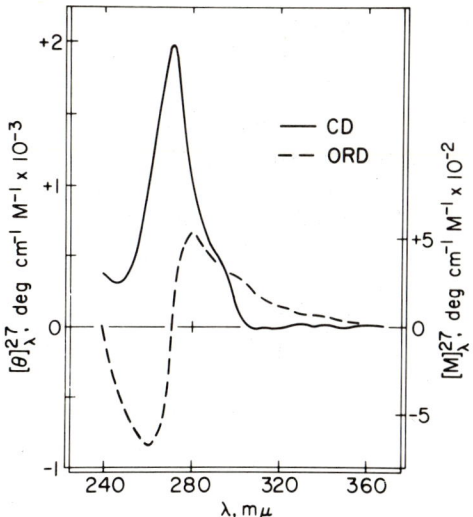

Figure 13. ORD and CD spectra of the LADH · Zn · OP complex
0.1M tris-Cl, pH 7.5

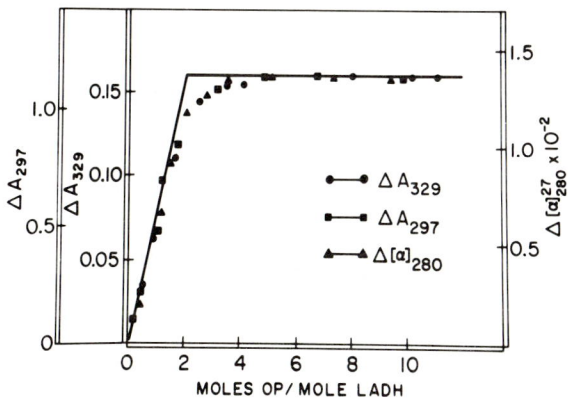

Figure 14. Spectrophotometric and ORD titration of LADH with OP

ometry of OP binding and the nature of the complexes formed, by means of Job's methods of continuous variations (21). In this procedure the first component, LADH · Zn, is varied from 0 to 100 mole % while the second component, OP, is simultaneously varied from 100 to 0 mole %, the final sum of the molar concentrations remaining constant. The spectra for the enzyme, OP, and their mixture are recorded separately and the difference absorbance, obtained at 329 mμ, is plotted to obtain the relationships shown in Figure 15.

Based upon the total zinc content of this preparation of enzyme, 3.5 gram atoms per mole, the lines drawn through the linear portions of the experimental data intersect at 64 mole % of zinc. The point of intersection would be expected to be 50 mole % of zinc if OP were to form a 1:1 complex with each zinc atom of LADH. However, if only 2 of the 3.5 zinc atoms of the enzyme react with OP under these conditions, maximal formation of the 1:1 Zn · OP complex occurs when the molar ratio of OP to total zinc bears the relationship 2/3.5—i.e., at the point of 3.5/(3.5 + 2.0) corresponding to 64 mole % of zinc, as here observed. Hence, it appears that the method of continuous variations is capable of signaling the presence of "nonreacting" zinc atoms, in accord with the titration data shown previously (20).

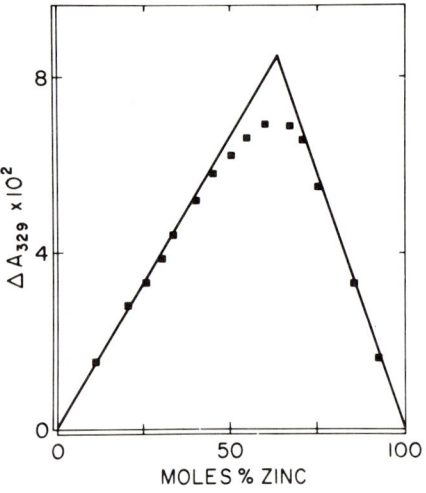

Biochemistry

Figure 15. Method of continuous variations: interaction of OP with zinc in LADH
0.1M phosphate, pH 7.5

In principle, of course, such data obtained by Job's method might be accounted for statistically by the formation of complexes other than the enzyme-bound 1:1 Zn · OP complex, hence with a different stoichiometry than that suggested. The experimental parameters pertinent to each of the various statistically possible complexes of OP with enzyme-bound and free zinc, e.g., dissociation constants and molar absorptivities,

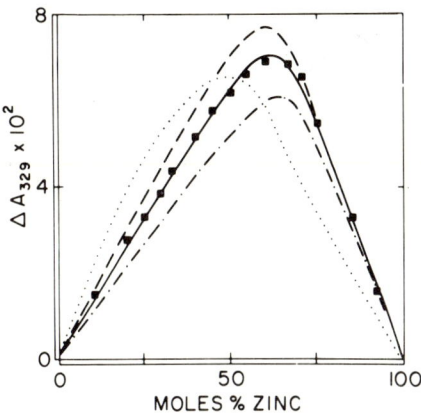

Biochemistry

Figure 16. Method of continuous variations: theoretical curves for interaction of OP with zinc in LADH

	Model	$Zn_{reactive}/Zn_{total}$
———	$E \cdot (Zn)_1 \cdot (OP)_1$	2.0/3.5
– – –	$E \cdot (Zn)_1 \cdot (OP)_2$	1.0/3.5
· · · · ·	$E \cdot (Zn)_2 \cdot (OP)_1$	3.5/3.5
–·–·–·	$E \cdot (Zn)_1 \cdot (OP)_1$	2.0/4.0

are known or can be reasonably approximated (20). On this basis, it is possible to design a computer program to permit systematic examination of these parameters and predict the form of Job's curve which should apply in each instance. The resulting theoretical curves for a 1:2 Zn · (OP)$_2$ complex, a 2:1 (Zn)$_2$ · OP complex (which, of course, is chemically unknown), and for two 1:1 Zn · OP complexes, based on either the measured 3.5 gram atoms of zinc present or upon an assumed value of 4 gram atoms, are shown in Figure 16. The experimental values actually determined, the solid squares, coincide only with the theoretical curve for formation of two separate 1:1 Zn · OP complexes in a system where the total enzyme-bound zinc present is 3.5 gram atoms (the amount determined analytically for this preparation) (20). Hence, these data confirm that only two of the zinc atoms of LADH react with OP, forming a 1:1 mixed complex, while the remaining 1.5 gram atoms are occluded from reaction with this chelating agent.

The differential reactivity of the zinc atoms in LADH has also been shown by a somewhat different type of experiment which is summarized in Table III. The native enzyme again contained a total of 3.5 gram atoms of zinc, of which only two are "exposed" and react with OP, as just shown. The remaining zinc is designated "buried." When the enzyme is exposed to the metal-binding agent, sodium diethyldithiocarbamate,

Table III. Differentiation of Zinc Atoms in Native and Modified LADH[a]

	Zinc (G. At./M.W. 80,000)		
	Total	Exposed[b]	Buried
Native	3.5	2.0	1.5
DDC[c]—modified	1.5	0	1.5
Carboxymethylated	2.0	2.0	0

[a] Drum et al., Proc. Natl. Acad. Sci., U.S., 57, 1434, 1967.
[b] Measured by ^{65}Zn exchange in acetate buffer, pH 5.5, and by rotatory dispersion titration.
[c] Diethyldithiocarbamate.

DDC, which has a much greater affinity for LADH zinc than does OP, the enzyme is irreversibly inhibited owing to removal of 2 gram atoms of zinc (17). This "DDC-treated" enzyme no longer generates a Cotton effect when exposed to OP, suggesting that OP and DDC interact with the same two zinc atoms of LADH. Moreover, when two of the zinc atoms of LADH are first labeled with ^{65}Zn^{2+} in acetate buffer, under the conditions described previously, and the enzyme is then exposed to DDC, only the labeled or "exposed sites" but none of the unlabeled zincs are removed. Hence, the two zinc atoms which exchange in acetate buffer appear to be the same two atoms which interact with OP to form a mixed complex and which are preferentially removed by DDC. Thus, by several criteria, these two zinc atoms are chemically more reactive than the remaining zinc and seem to be the metal atoms involved directly in catalytic function of the enzyme.

We have shown previously that carboxymethylation of the active center cysteinyl residues of LADH also inactivates the enzyme and renders the zinc atoms less stable (17). If enzyme labeled with ^{65}Zn^{2+} in acetate buffer is carboxymethylated and then dialyzed, only the unlabeled zinc atoms are lost. In this instance, the two ^{65}Zn^{2+} labeled atoms remaining do interact with OP and, therefore, are "exposed" (Table III). Thus, although carboxymethylation completely inactivates the enzyme, this is not because of the loss of its catalytically essential zinc atoms. Rather, the zincs which are lost upon carboxymethylation appear to be those most critical in maintenance of the quaternary structure as shown earlier in ultracentrifugal experiments (17).

To summarize, two of the zinc atoms of LADH appear to be located at the active site of the enzyme; they exchange readily in acetate buffer and are either removed by or form mixed complexes with chelating agents, resulting in a loss of catalytic activity. An additional 1.5 to 2.0 gram atoms of zinc seem to be buried in the molecule elsewhere than at the active site; they do not exchange with isotopic zinc in acetate, but do exchange in phosphate buffer. These metal atoms do not interact with chelating agents but are labilized by carboxymethylation of LADH; they appear to influence catalysis only indirectly by maintaining the structural integrity of the protein.

The intrinsic zinc atoms of LADH not only can be exchanged for extraneous zinc, but they can also be fully exchanged with cobalt or cadmium. Two new enzymes result, both of which are catalytically active and should prove useful in delineating characteristics of the active site region of this enzyme (*22*).

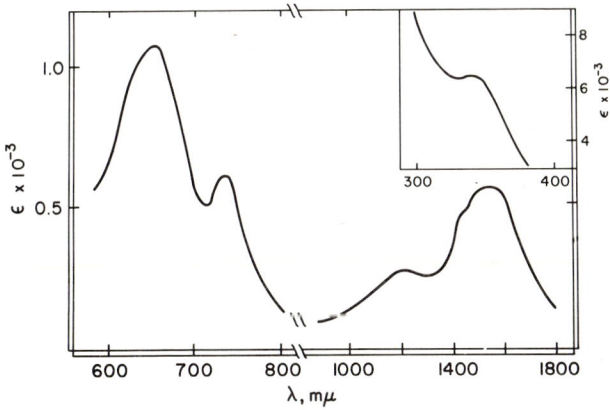

Figure 17. Absorption spectrum of Co-LADH
0.1M tris-Ac, pH 7.0

As shown in Figure 17, cobalt LADH has unusual visible, infrared, and near ultraviolet absorption spectra. In the visible region, maxima occur at 655 and 730 mμ, with molar absorptivities of 370 and 220 per cobalt atom, respectively. In the near infrared, broad bands are identified between 1000 and 1800 mμ with absorptivities of 75 to 150 per cobalt atom. The locations and intensities of this series of visible and infrared bands differ markedly from those of alkaline phosphatase and other cobalt metalloenzymes studied previously and should help ultimately to serve for more definitive evaluation of the geometry about the metal atoms, which again appears to be quite unusual. The insert at the upper right in Figure 17 shows the near ultraviolet spectrum of cobalt LADH which exhibits a maximum at 340 mμ with a molar absorptivity of almost 2000 per cobalt atom, perhaps consistent with a charge transfer band (*22*). Not surprisingly, the complex absorption bands of cobalt LADH generate remarkable circular dichroic spectra (Figure 18). Two positive and three negative dichroic bands occur between 300 and 450 mμ, with

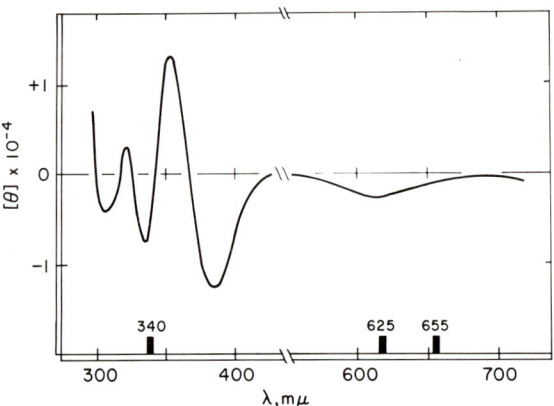

Figure 18. Circular dichroic spectrum of Co-LADH
0.1M tris-Ac, pH 7.0

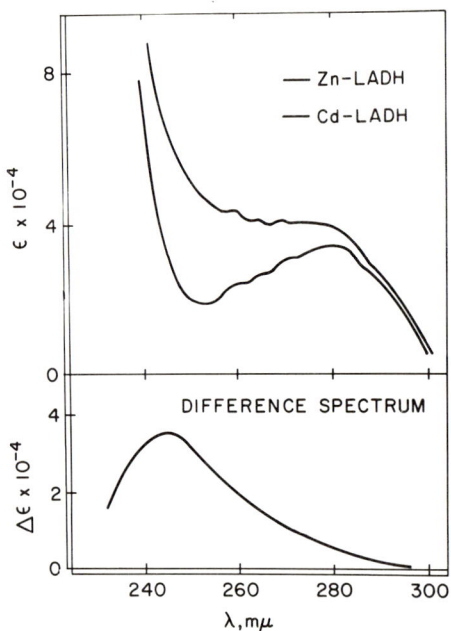

Figure 19. Absorption spectra of Zn-LADH and Cd-LADH
0.1M phosphate, pH 7.5

molecular ellipticities varying from 3000° to 13,000° (22). In addition to these five intense transitions, a broad negative band of lower amplitude is centered at 625 mμ. These data indicate that the ligand environment of cobalt in LADH is asymmetric and the rather remarkable absorption and CD spectra again suggest an unusual geometry which may reflect the entatic state.

The absorption spectrum of cadmium LADH differs markedly from that of the zinc or cobalt enzyme and perhaps bears upon the nature of the metal-binding ligands (Figure 19). An intense band centered at 245 mμ with a molar absorptivity per cadmium atom of 10,200 is shown by the difference spectrum at the bottom of the figure: zinc LADH is employed as the reference. Notably, the molar absorptivity of this band is nearly 14,000, close to that reported for the cadmium mercaptide chromophores of metallothionein (23). This is consistent with the hypothesis that sulfhydryl groups may serve as metal ligands in LADH.

Both cadmium and cobalt enzymes modulate in a distinctive fashion, the optical properties of asymmetrically bound extrinsic chromophores such as the coenzyme, DPNH, and the chelating inhibitor, 1,10-phenanthroline, providing a useful means to study the interaction of these molecules with LADH (22).

The data cited thus far indicate that both alkaline phosphatase and liver alcohol dehydrogenase contain heterogeneous populations of metal atoms of the same species. In both instances, only two of the zinc atoms native to the enzyme appear to be involved in enzymatic activity. The remaining metal atoms do not have a catalytic role but appear to influence the quaternary structure of the protein, although the details of the manner in which this is accomplished are as yet uncertain. These observations have induced us to reexamine the possible effects of metals on structure in other proteins, including those having only single chains.

Metal Stabilization of Protein Structure

Recently, hydrogen isotope exchange techniques have proven to be a particularly promising experimental approach to delineation of the structural roles of metals in proteins (24, 25, 26) and to the general problem of stabilization of protein structure (5). Hydrogen exchange studies of metallothionein serve to illustrate the power of this technique. This small molecular weight protein first isolated from mammalian kidney contains nearly 6% cadmium, 2–3% zinc, 8.5% sulfur, and smaller amounts of other metals including copper and iron (23, 27). Cadmium and zinc compete for the same binding sites and are likely isomorphic in metallothionein. Most of the sulfur is accounted for by cysteine and it appears that one atom of cadmium or zinc is bound to three sulfhydryl

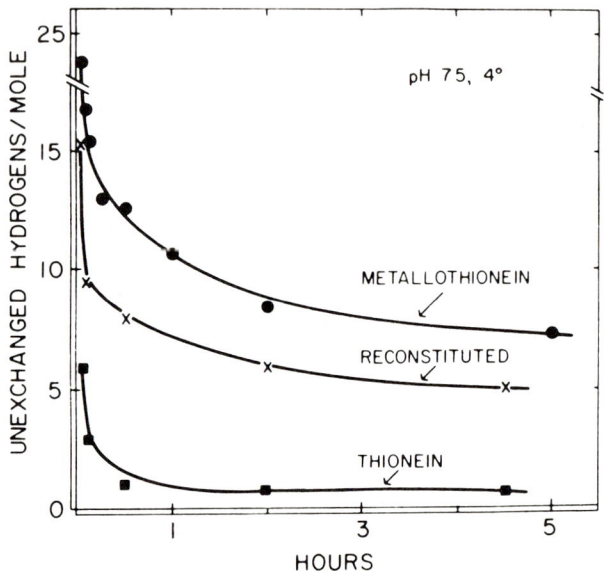

Figure 20. Hydrogen–tritium exchange of thionein and metallothionein

groups. The unique metal–sulfur composition of metallothionein provides more than a dozen metal mercaptide crosslinks per mole which might be expected to have a pronounced stabilizing effect on structure.

The hydrogen–tritium exchange of this protein was determined by the two-column gel filtration technique of Englander (28) (Figure 20). Metallothionein, the native protein, retains 26 of its 94 potentially exchangeable hydrogen atoms after initial gel filtration at pH 8.0, 4°C; approximately 10 of these exchange slowly over the next five hours. Such retardation of exchange indicates that the metalloprotein has an ordered structure. In contrast, thionein, the metal-free protein, exchanges nearly all of its hydrogens virtually instantaneously, typical of random or disordered structure. Addition of either cadmium or zinc partially reconstitutes the metalloprotein, again retarding exchange and suggesting restoration of native conformation to a significant degree.

Thus, metals critically affect the hydrogen exchange of metallothionein and likely play a dominant role in determining its structure. However, a single metal atom can influence conformation in other metalloproteins, as illustrated by comparable studies of trypsin.

Delaage and Lazdunski (29) have shown that the binding of a single calcium atom per mole of trypsin accounts for its well known stabilization by this metal. Because trypsin undergoes rapid autocatalytic degradation at alkaline pH, the diisopropylfluorophosphate-treated enzyme, also stabilized by calcium, was employed for hydrogen exchange measurements. Figure 21 illustrates the exchange of DFP–trypsin in the presence and absence of calcium. Calcium retards the exchange of approximately 10 hydrogens per mole indicating that, in the presence of the metal, protein conformation is significantly more compact.

The effect of metals on hydrogen exchange in a number of other proteins determined by the gel-filtration technique is summarized in Table IV. The difference after one hour in the number of hydrogens exchanged in each protein, in the presence and absence of the appropriate metal, is expressed both as total hydrogens per mole and as percent of peptide

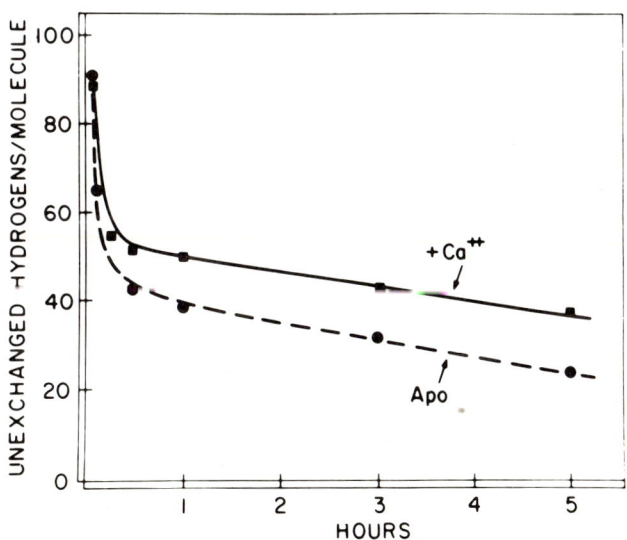

Figure 21. Effect of calcium on H–T exchange of trypsin DFP-trypsin in 0.1M tris-Cl, pH 7.0, 4° ± 0.1M Ca^{2+}

Table IV. Effect of Metals on H–T Exchange

Proteins	Metal	$\Delta H/Mole$ ($\pm Metal$)	
		Total H	% Peptide H
CPK [a]	Mg^{2+}	0	0
CPD A (α) [b]	Zn^{2+}	0	0
CPD A (γ)	Zn^{2+}	20	7
Alk P'Tase	Zn^{2+}	50	6
Conalbumin	Fe^{3+}	50	8
Transferrin	Fe^{3+}	70	11
Liver ADH [c]	Zn^{2+}	70	10

[a] Creatine phosphokinase.
[b] Carboxypeptidase.
[c] Alcohol dehydrogenase.

hydrogens—most likely the slowly exchanging groups being measured (*31*). Metals do not appear to affect the exchange of creatine phosphokinase and bovine pancreatic carboxypeptidase A-α. However, metals do retard the exchange of from 20 to 70 hydrogens per mole in carboxypeptidase A-γ, alkaline phosphatase, conalbumin, transferrin, and liver alcohol dehydrogenase (*30*). Hence, hydrogen isotope exchange appears to be a sensitive experimental technique for investigation of metal–protein interactions.

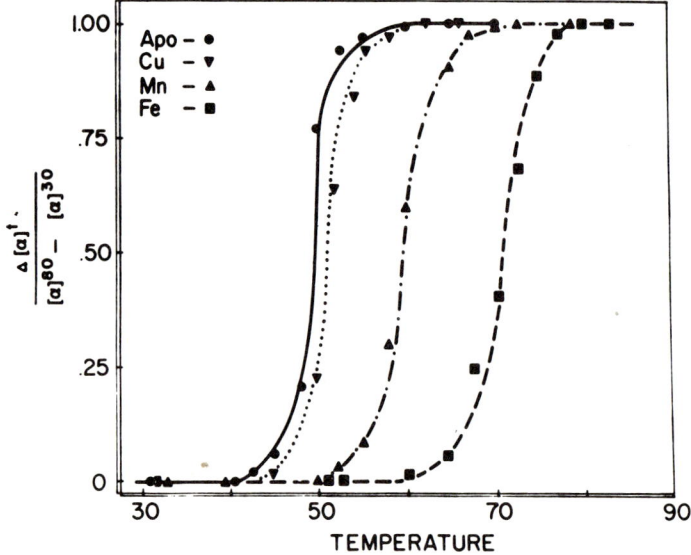

Figure 22. Thermal transitions of APO and metalloconalbumins

We have postulated that metals retard the hydrogen exchange of proteins owing to structural alterations. If this hypothesis is valid, then the degree of retardation of exchange should be proportional to the influence of metals on structure as measured by other means such as thermal perturbation. Experiments designed to examine this hypothesis are shown in Figures 22 and 23. Figure 22 shows the effect of several different metals on the optical rotation of conalbumin when this protein is heated, at pH 8, in 2 M urea. As the temperature is increased, the apoprotein undergoes a highly cooperative transition with a midpoint of about 48°. At this pH, conalbumin tightly binds 2 atoms per mole of copper, manganese, or iron, shifting the midpoint of the thermal transition to 51°, 59°, and 72°, respectively. Hence, all these metals stabilize conalbumin against heating, but there is a marked difference in their effectiveness, in the order, iron > manganese > copper. Precisely the same sequence

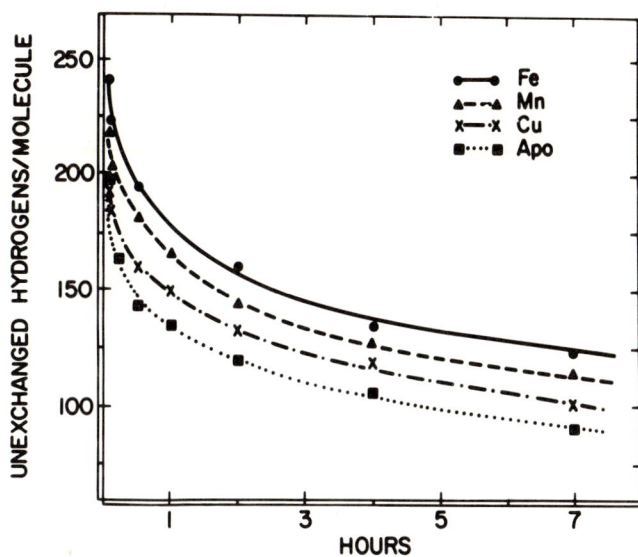

Figure 23. *Effect of metals on H–T exchange of conalbumin*

of metals is evident when comparing their effectiveness in retarding the hydrogen–tritium exchange of conalbumin, measured at the same pH, but at 4°C (Figure 23). Throughout the course of the exchange, the iron protein retains the most hydrogens, the apoprotein the fewest number of hydrogens, and the manganese and copper proteins are again intermediate. Hence, the same metal-induced structural alterations which increase the resistance of conalbumin to heat denaturation appear to regulate its hydrogen exchange.

As noted earlier, the hydrogen–tritium exchange of transferrin is markedly altered by iron. At pH 8, 4°C, a nearly constant difference of at least 70 hydrogens per mole differentiates the iron and apoprotein throughout exchange (Figure 24). Such a displacement of the two curves could indicate that the metal affects a localized area of structure, thereby occluding completely a discrete cluster of hydrogens. Alternatively, the metal, through modulation of macromolecular conformation, might perturb exchangeability of all hydrogens in the protein approximately to an equivalent degree. In principle, these possibilities may be distinguished experimentally by analysis of the exchange kinetics.

The rates of hydrogen exchange in proteins are both pH and temperature dependent. Figure 25 displays schematically the analysis of exchange data as a function of either pH or temperature. Examples of typical experimental data are shown on the left; extent of exchange is plotted *vs.* time. It is evident that increases in either pH or temperature

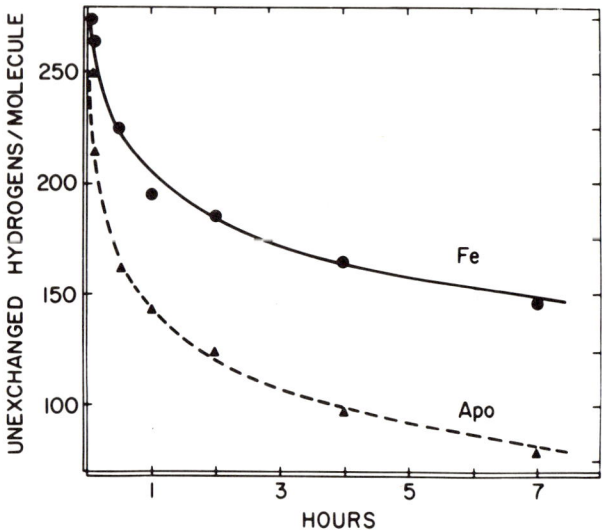

Figure 24. Hydrogen–tritium exchange of APO- and iron transferrin
0.1M tris-Cl, pH 8.0, 4°

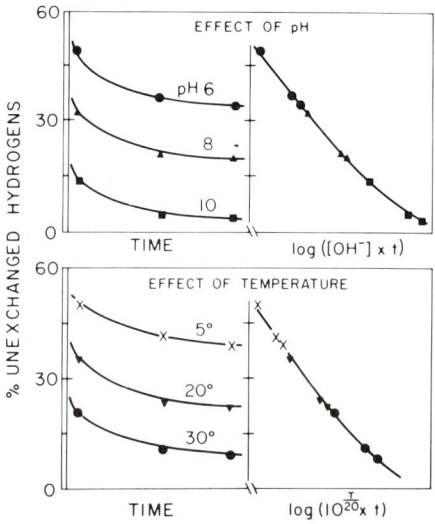

Figure 25. H–T exchange: graphic analysis

permit stepwise sampling of progressively more inaccessible groups until virtually all measurable hydrogens in the molecule have exchanged. On the right in Figure 25, both pH and temperature dependence of exchange are integrated in terms of the model proposed by Hvidt (*31*) and plotted as suggested by Willumsen (*32*). Each set of symbols represents one value for either pH or temperature, but at differing times. When such a plot describes a simple monotonically decreasing function, as shown, it indicates that the rate-limiting step in exchange is the temperature-dependent bimolecular reaction between exposed peptide groups and catalytic hydroxyl ions. Moreover, according to the model, a displacement of the decay curve to the left, along the abscissa, signifies a less stable macromolecular conformation of the protein (*25*). The extent of such a displacement is a measure of the difference in conformational stability of the two forms.

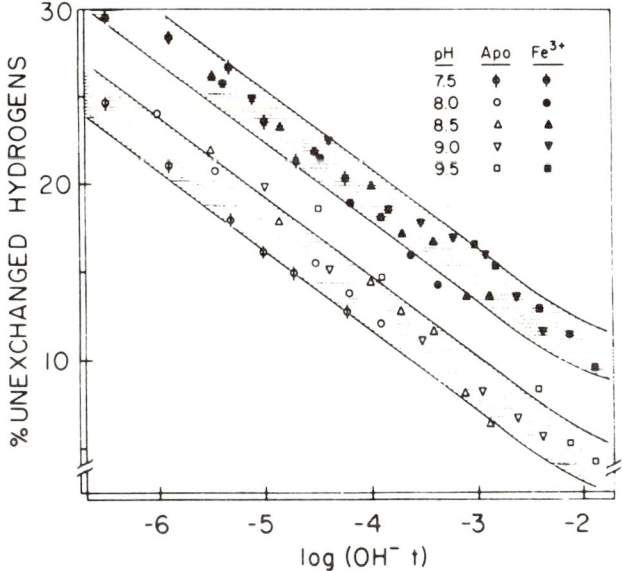

Figure 26. Effect of pH on H–T exchange of transferrin

Iron transferrin and apotransferrin conform to such a model when examined at discrete intervals over a pH range of 7.5 to 9.5 (Figure 26). The hatched areas represent two standard deviations about the mean but the data are not corrected for electrostatic effects which should further decrease the deviation. The difference between the two forms of transferrin is nearly constant over the pH range examined and the exchange of both decreases in a monotonic fashion, indicating their like response to hydroxyl ion catalysis. Most importantly, the exchange of the apoprotein is displaced more than 1.5 log units to the left, along the abscissa, as expected for a substantially less stable conformation.

The results are remarkably similar on varying temperature from 4° to 54°C (Figure 27). Increasing temperature accelerates the exchange of both iron and apoproteins, and their difference again is virtually constant over the whole range of hydrogens exposed. Therefore, in transferrin, iron appears to influence all classes of exchangeable hydrogens uniformly whether one examines the most labile groups, those of intermediate stability, or those rendered exchangeable only at extremes of temperature or pH. Hence, iron seems to affect the exchange of hydrogens distributed throughout the molecule, not a discrete group of them, and the metal appears to influence total macromolecular conformation rather than inducing a localized structural change.

Such data seem entirely consistent with Linderstrom-Lang's motility model (33) which suggests that proteins fluctuate continuously between closely related conformations. In this sense, it can be visualized that the metal operates to retard hydrogen exchange by shifting the equilibrium between open, exchanging and closed, nonexchanging forms of transferrin. We have shown previously (24, 25) that under these circumstances, it is possible to calculate a minimal conformational stabilization energy because of metal binding from the extent of displacement of the two curves as shown in both Figures 26 and 27. Apparently, the binding

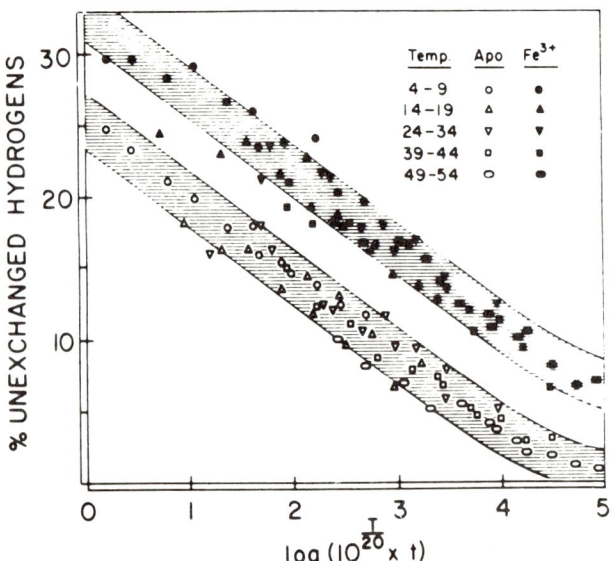

Figure 27. Effect of temperature on H–T exchange of transferrin

of iron to the ligands of transferrin supplies a minimal stabilization energy of approximately 1.5 to 2 kcal per mole.

From studies of this type, it is apparent that hydrogen–tritium exchange provides a sensitive kinetic means to assess quantitatively the effects of metal on stabilization of protein structure. Moreover, such data suggest that metals, by modulating equilibria between closely related states of proteins which differ in conformational energy, may well influence enzymic activities, protein–protein interactions, or turnover rates of proteins in biologic fluids, thereby affecting important regulatory processes and homeostatic mechanisms.

In conclusion, it is evident that metal atoms play diverse roles in the function and structure of metalloproteins. In both alkaline phosphatase and liver alcohol dehydrogenase, metals which are positioned at the active enzymic sites are essential for catalytic activity and appear to be bound differently and display distinctive physical chemical characteristics as compared with metals located elsewhere in the molecule. The unusual properties of metals at the active sites of enzymes suggest an irregular geometry and may well relate to their participation in the catalytic process. In contrast, metals at other sites in these enzymes appear to play primarily a structural role; however, they may influence enzymic function indirectly by modulation of subunit interactions. In many other proteins, metals appear to be involved in subunit interaction or in stabilization of structure, as evidenced by their effects upon hydrogen–tritium exchange; in this manner, they may exert important regulatory constraints upon biochemical reactions. It seems likely that further delineation of the effects of metals on protein structure may add substantially to our understanding of the functional role of metals in biologic systems.

Acknowledgment

D. D. Ulmer is a Research Career Development awardee of the National Institutes of Health. Portions of the work cited were supported by Grants-in-Aid GM-11639 and GM-15003 from the National Institutes of Health of the Department of Health, Education and Welfare.

Literature Cited

(1) Vallee, B. L., Wacker, W. E. C., "Proteins," H. Neurath, Ed., Vol. 5, Academic Press, New York, 1970.
(2) Oosterbaan, R. A., Jansz, H. S., "Comprehensive Biochemistry," M. Florkin and E. H. Stotz, Eds., Vol. 16, p. 1, Elsevier, Amsterdam, 1965.
(3) Vallee, B. L., Riordan, J. F., *Ann. Rev. Biochem.* (1969) **38**, 733.
(4) Vallee, B. L., Latt, S. A., "Structure–Function Relationships of Proteolytic Enzymes," P. Desnuelle, H. Neurath, and M. Ottesen, Eds., Munksgaard, Copenhagen, 1969.

(5) Ulmer, D. D., *Proc. Rochester Conf. Toxicity, 2nd, 1969* (in press).
(6) Kagi, J. H. R., Vallee, B. L., *J. Biol. Chem.* (1960) **235**, 3188.
(7) Simpson, R. T., Tait, G. H., Vallee, B. L., *Biochemistry* (1968) **7**, 4336.
(8) Simpson, R. T., Vallee, B. L., *Ann. N. Y. Acad. Sci.* (1969) **166**, 670.
(9) Simpson, R. T., Vallee, B. L., *Biochemistry* (1968) **7**, 4343.
(10) Vallee, B. L., Williams, R. J. P., *Proc. Natl. Acad. Sci. U.S.* (1968) **59**, 498.
(11) Vallee, B. L., Williams, R. J. P., *Chem. Brit.* (1968) **4**, 397.
(12) Theorell, H., Nygaard, A. P., Bonnichsen, R. K., *Acta. Chem. Scand.* (1955) **9**, 1148.
(13) Vallee, B. L., Hoch, F. L., *J. Biol. Chem.* (1957) **225**, 185.
(14) Theorell, H., Bonnichsen, R. K., *Acta. Chem. Scand.* (1951) **5**, 1105.
(15) Ulmer, D. D., Li, T.-K., Vallee, B. L., *Proc. Natl. Acad. Sci. U.S.* (1961) **47**, 1155.
(16) Drum, D. E., Li, T.-K., Vallee, B. L., *Biochemistry* (1969) **8**, 3783.
(17) Drum, D. E., Harrison, J. H., IV, Li, T.-K., Bethune, J. L., Vallee, B. L., *Proc. Natl. Acad. Sci. U.S.* (1967) **57**, 1434.
(18) Drum, D. E., Vallee, B. L., *Biochemistry* (1969) **8**, 7392.
(19) Vallee, B. L., Coombs, T. L., Williams, R. J. P., *J. Am. Chem. Soc.* (1958) **80**, 397.
(20) Drum, D. E., Vallee, B. L., *Biochemistry* (1970) **9**, 4078.
(21) Job, P., *Ann. Chim. Rome* (1928) **9**, 113.
(22) Drum, D. E., Vallee, B. L., *Biochem. Biophys. Res. Commun.* (1970) **41**, 33.
(23) Kagi, J. H. R., Vallee, B. L., *J. Biol. Chem.* (1961) **236**, 2435.
(24) Ulmer, D. D., Kagi, J. H. R., *Biochemistry* (1968) **7**, 2710.
(25) Kagi, J. H. R., Ulmer, D. D., *Biochemistry* (1968) **7**, 2718.
(26) Ulmer, D. D., *Biochim. Biophys. Acta* (1969) **181**, 305.
(27) Kagi, J. H. R., Vallee, B. L., *J. Biol. Chem.* (1960) **235**, 3460.
(28) Englander, W. S., *Biochemistry* (1963) **2**, 798.
(29) Delaage, M., Lazdunski, M., *Biochem. Biophys. Res. Commun.* (1967) **28**, 390.
(30) Ulmer, D. D., *Federation Proc.* (1970) **29:2**, 1227.
(31) Hvidt, A., Nielsen, N. O., *Advan. Protein Chem.* (1966) **21**, 287.
(32) Willumsen, L., *Biochim. Biophys. Acta* (1966) **126**, 382.
(33) Linderstrom-Lang, K., *Symp. Peptide Chem. Spec. Publ. No. 2*, p. 1, Chemical Society, London, 1955.

RECEIVED June 26, 1970.

11

The Biochemistry of N_2 Fixation

R. W. F. HARDY, R. C. BURNS, and G. W. PARSHALL

Central Research Dept., E. I. du Pont de Nemours and Co., Wilmington, Del. 19898

> *The nature of the N_2ase reaction is described with respect to electron donation, energy requirement, and reduction characteristics, with particular analysis of the seven classes of substrates reducible by N_2ase, a complex of a Mo–Fe and Fe protein. Chemical and physical characteristics of Fe protein and crystalline Mo–Fe protein are summarized. The two-site mechanism of electron activation and substrate complexation is further developed. Reduction may occur at a biological dinuclear site of Mo and Fe in which N_2 is reduced to NH_3 via enzyme-bound diimide and hydrazine. Unsolved problems of electron donors, ATP function, H_2 evolution and electron donation, substrate reduction, N_2ase characteristics and mechanism, and metal roles are tabulated. Potential utilities of N_2 fixation research include increased protein production and new chemistry of nitrogen.*

The facile reduction of N_2 to NH_3 by the natural catalyst, nitrogenase, is not understood and has never been equalled with synthetic catalysts. Developments of the past decade may change this. A renaissance in N_2 fixation research in both biochemistry and inorganic chemistry is providing the first opportunity for bioinorganic interaction between these previously independent approaches to catalytic N_2 reduction. Remarkable progress—from crude extracts to crystalline protein and from a single N_2ase substrate to recognition of seven classes of substrates—in the elucidation of the complex biochemistry of N_2 fixation has been made during the past decade (Figure 1) (*1, 2, 3, 4, 5, 6, 7, 8, 9, 10, 11*). Moreover, advances in the inorganic chemistry of N_2-fixation—synthesis of transition metal complexes of N_2 and reduction of N_2 under ambient conditions—are yielding the first examples of bioinorganic models.

We will summarize the status of the biochemistry of N_2 fixation, including nitrogenase (N_2ase) reactions and characteristics, with exten-

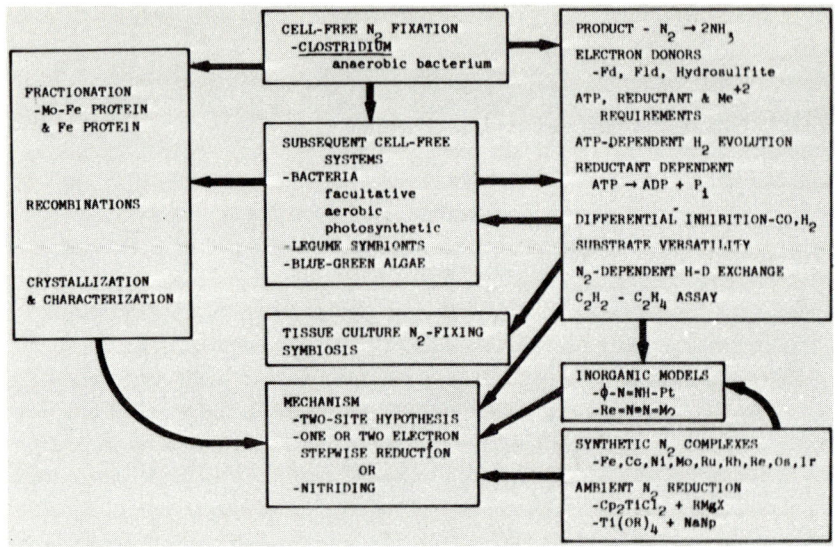

Figure 1.   Advances in the biochemistry of N_2 fixation, 1960–1970

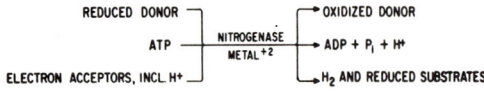

Annual Review of Biochemistry

Figure 2.   Reactants and products of interdependent reaction catalyzed by N_2ase

sive analyses of substrate reductions and characteristics of the recently crystallized Mo–Fe protein, the major protein of N_2ase. A mechanism compatible with the known biochemistry of N_2ase and supported by model studies will be presented. Current problems in the biochemistry of N_2 fixation will be summarized, and important potential utilities of N_2 fixation research from this most productive decade will be suggested.

Nitrogenase Reactions

The integrated N_2-fixing reaction catalyzed by N_2ase is represented in its simplest form in Figure 2 (6). Reactants are (1) an electron donor, (2) adenosine triphosphate (ATP) and divalent cation (M^{2+}), and (3) an electron acceptor. Protons of the aqueous media as well as any one of six additional classes of reducible substrates can function in this capacity. Products are (1) oxidized electron donor, (2) adenosine diphosphate (ADP), inorganic phosphate (P_i), and M^{2+}, (3) H_2 and the reduced product characteristic of any added reducible substrates, e.g., NH_3 from

N_2. In contrast to inorganic reactions, activity is restricted to aqueous media and a pH between 6 and 8.

Electron Donors. The known natural and artificial electron donors which react directly with N_2ase are listed in Table I. Clostridial ferredoxin (Fd) and flavodoxin (Fld) are the natural electron donors for clostridial N_2ase, and Fd and Fld may also be natural donors for other N_2ases. All donors are characterized by an oxidation–reduction potential of $-0.28V$ or less, but several different redox groups are represented. The iron–sulfur protein, clostridial ferredoxin, the first ferredoxin to be isolated, was a by-product of research directed towards the purification of N_2ase. The characteristics of clostridial and other ferredoxins will be discussed in a later paper (10). Clostridial Fd functions as an electron-transferring protein with a variety of enzymes, including clostridial and other N_2ases (15, 16, 17). Based on turnover number, it is the most active known natural electron donor for N_2ase. In the N_2-fixing anaerobe, *Clostridium pasteurianum*, it transfers electrons to N_2ase from pyruvate in the sequence: pyruvate $\rightarrow$ pyruvate dehydrogenase $\rightarrow$ Fd $\rightarrow$ N_2ase; it also fills an essential role in H_2 metabolism, mediating electron transfer to and from hydrogenase (H_2ase): Fd $\rightleftharpoons$ H_2ase $\rightleftharpoons$ H_2. Attempts to obtain apoferredoxin from *C. pasteurianum* grown on limiting-iron media led to the discovery, isolation, and crystallization of flavodoxin, a novel electron-transferring protein containing flavin mononucleotide, but no iron–sulfur group (18, 19, 20). Fld is interchangeable with Fd in all reactions tested. On an equimolar basis, Fld has about one-third of the activity of Fd for electron transfer to N_2ase. Recently, Fds from the N_2-fixing aerobe *Azotobacter vinelandii* (12) and from soybean nodules (13) and an Fld called azotoflavin from *A. vinelandii* (21) have been

Table I. Natural and Artificial Electron Donors for N_2ase

Donor	Turnover Number, Moles of $2e/Min/Mole$
Natural	
Clostridial ferredoxin	32
Clostridial flavodoxin	10
Azotobacter ferredoxin [a]	2.5
Azotobacter flavodoxin [a,b]	0.4
Soybean nodule ferredoxin [c]	–
Artificial	
Methyl viologen	2
Benzyl viologen [d]	–
$Na_2S_2O_4$	–

[a] Ref. *12*.
[b] Azotoflavin.
[c] Ref. *13*.
[d] Ref. *14*.

isolated. These will couple electrons from an illuminated modified chloroplast system to their respective N$_2$ases. Their low turnover numbers relative to clostridial Fd and Fld may be attributed to poor coupling between the added chloroplast system and the donors of these nonphotosynthetic organisms; alternatively, a poor interaction between these carriers and N$_2$ase and the implication that these carriers may not serve N$_2$ase physiologically cannot be ruled out.

Artificial donors include the viologen dyes and hydrosulfite. Methyl viologen can replace clostridial Fd or Fld. Oxidation of the monomer of poly-β-hydroxybutyrate, a major component of soybean nodule bacteroids, has been coupled to N$_2$ase to give very low activity *via* a complex system composed of a dehydrogenase, NAD$^+$, a diaphorase, and a viologen dye (*14*). Hydrosulfite is the most useful donor for *in vitro* N$_2$ase assays since it can be used in substrate amounts in contrast to all other natural and artificial donors (*22*).

One- and two-electron reduction forms of Fd (*23*) and Fld (*20, 24*) have been observed, but it is not known whether they react with N$_2$ase as one- or as two-electron donors. Fld and Fd are acidic proteins; N$_2$ase is itself composed of acidic proteins. The type of interaction involved in complex formation between these acidic proteins is unknown.

Table II. Specificity of ATP Requirement for N$_2$ase (*26*)

Phosphagen	C_2H_4, mμMoles/Incubation
ATP	1440
CTP	0.80
UTP	<0.05
GTP	<0.05

Energy. The reaction $N_2 + H_2 \rightarrow 2\, NH_3$ is exothermic with $\Delta F°_{25} = -8$ Kcal; however, energy is required for N$_2$ase activity. Of the common nucleoside triphosphates, only ATP has any detectable activity (Table II) (*25, 26*); furthermore, no substitutes for ATP have been found. The K_m for ATP is $1-3 \times 10^{-4} M$ (*27, 28*).

Nitrogenase catalyzes the hydrolysis of ATP $\rightarrow$ ADP + P$_i$ in a reaction that is reductant-dependent (*15, 29, 30*) and results in electron transfer and activation to provide a low-potential species capable of reducing a unique range of substrates (*see* Reducible Substrate section). The rate-determining step of ATP utilization may be bimolecular (*25, 31*). The product ADP is an inhibitor of ATP utilization (*32*), and thus an ATP-generating system is used *in vitro* to reconvert ADP to ATP (*15, 33*).

Exchange, reversal, or intermediates of the reaction ATP $\xrightarrow[\text{e's}]{\text{N}_2\text{ase}}$ P$_i$ + ADP
have not been detected. A divalent metal cation is required; Mg^{2+}, Mn^{2+}, Co^{2+}, Fe^{2+}, and Ni^{2+} function in that order of effectiveness (27, 30). Inhibitors of electron transfer or uncouplers of oxidative and photosynthetic phosphorylation inhibit neither ATP utilization nor electron transfer by N$_2$ase, but chelating agents such as o-phenanthroline and Tiron are inhibitory (30, 32). The rate of ATP hydrolysis is independent of the presence of added reducible substrates or inhibitors of N$_2$ fixation, such as CO and H$_2$. The activation energy for the interdependent electron-transfer–ATP hydrolysis reaction is about 14 Kcal/mole above 20°C and 35–50 Kcal/mole below 20°C (26, 27); surprisingly, this reaction, not reducible substrate activation, is the limiting step in all N$_2$ase catalysis.

The reported stoichiometry of ATP consumed per 2e transferred has shown considerable variation; values as high as 4–5 have been obtained with most N$_2$ase preparations (34, 35, 36), although recently values of two or less have been reported (37, 38). Calculated values as low as one have been obtained in vivo (39, 40), suggesting that higher values for the isolated N$_2$ase, tested in vitro, may be physiologically atypical.

Reducible Substrates. The proton and the compounds shown in Figure 3 represent the seven distinct classes of substrates reducible by N$_2$ase. Characteristics of the reductions of the primary member of each class will be described, followed by information on reduction of analogs and a summary of the general characteristics of N$_2$ase-catalyzed reductions.

SUBSTRATES		PRODUCTS	ELECTRONS	K$_m$ mM	RELATIVE RATE	RATE X ELECTRONS
N≡N		2NH$_3$	6	0.1	1	6
[N=N≡N]$^-$		NH$_3^{(1)}$, N$_2^{(1)}$	2	1.0	3	6
O=N≡N		N$_2^{(1)}$, H$_2$O$^{(1)}$	2	1.0	3	6
HC≡CH		C$_2$H$_4^{(1)}$ (no C$_2$H$_6$ or CH$_4$)	2	0.1–0.3	4	8
CH$_3$N≡C		CH$_4^{(1)}$, C$_2$H$_6^{(V)}$, C$_2$H$_4^{(V)}$, C$_3$H$_6^{(V)}$, C$_3$H$_8^{(V)}$, CH$_3$NH$_2$	6, 8, 10, 12, 14	0.2–1.0	0.8	4.8
HC≡N		CH$_4^{(1)}$, C$_2$H$_6^{(V)}$, C$_2$H$_4^{(V)}$, NH$_3^{(1)}$, CH$_3$NH$_2^{(0.1)}$	6, 4	0.4–1.0	0.6	3.6

Figure 3. *Classes of substrates reduced by N$_2$ase*

CLASSES. Molecular nitrogen, the physiological substrate of N_2ase, consumes six electrons and is reduced to 2 NH_3 (*41*). The lack of specific inhibition of N_2 reduction by NH_4^+ indicates that ammonia does not complex strongly with N_2ase. No enzyme-free intermediates have been detected (*42, 43, 44*) and, on the basis of 2-, 4-, 6-, 8-, 10-, 12-, and 14-electron addition products found in other N_2ase-catalyzed reactions (Figure 3), enzyme-bound diimide and hydrazine are the obvious, but not yet detected, intermediates. The N_2 concentration for half maximal rate is about 0.1 mM (*15, 29, 45, 46, 47, 48, 49, 50*).

Azide is reduced by two electrons to NH_3 and N_2 (*46, 51*); nitrous oxide is reduced by two electrons to H_2O and N_2 (*52*). Surprisingly, N_2 produced from either N_3^- or N_2O is not further reduced to ammonia unless conditions are imposed to favor the complexation of product N_2. The affinity of N_3^- or N_2O for N_2ase is about one-tenth that of N_2 for N_2ase. The rate of reduction of N_3^- or N_2O is about three times that of nitrogen; the rate of electron consumption is similar to that for N_2.

Acetylene is reduced by two electrons to C_2H_4; further reduction to ethane or methane, if any, is <0.01% (*26, 53, 54*). The extreme specificity of this reaction is striking; however, recent reports suggest that a thiol–molybdenum system and a platinum complex will reduce C_2H_2 specifically to C_2H_4 (*55, 56*). Ethylene, like ammonia, does not complex strongly with N_2ase, being neither inhibitory to N_2 fixation nor reduced by N_2ase (*26, 53, 54*). The affinity of C_2H_2 for N_2ase is similar to that of N_2 for N_2ase (*26, 50, 53, 54, 57*). The rate of reduction is four times that of N_2 and the rate of electron consumption is about 1.3 times that of N_2 (*58*). The reduction of acetylene is stereospecific without loss of hydrogens; the major product of reduction in D_2O is *cis*-1,2-dideuteroethylene with a trace of *trans*-1,2-dideuteroethylene (*26, 53, 59*).

Hydrogen cyanide is reduced by six electrons to equivalent amounts of NH_3 and CH_4 (*46, 60*), a reaction observed with no other homogeneous catalyst and not reported for the molybdenum–thiol–borohydride system proposed as a model of N_2ase (*55*). Methylamine, a 4-electron product, may also be formed in amounts equivalent to 10% of the major products (*46*). It is not known whether HCN or CN^- is the actual substrate, since both species are present at reaction pH. The affinity for cyanide based on equilibrium concentration of HCN is intermediate between that for N_2 or C_2H_2 and N_2O or N_3^-, while the affinity based on CN^- concentration is substantially greater (*28, 46, 50*). The rate of electron consumption is less than that seen in N_2 reduction, which may indicate partial inhibition of N_2ase by cyanide and/or the generation of undetermined products.

Methyl isonitrile, a nitrile isomer, is, as expected, reduced by six electrons to CH_4 and CH_3NH_2 as the major products (*50, 60, 61, 62*). Methane is produced from the isonitrile carbon, as shown by the formation of perdeuteromethane in D_2O (60). Variable amounts of the 8-, 10-, 12-, and 14-electron addition products—ethylene, ethane, propylene, and propane—are also formed (50, 61, 62). The rate of electron consumption is again less than that in N_2 reduction, probably for the same reasons as suggested in the cyanide case.

Protons are reduced by N_2ase to H_2 (32, 63, 64, 65). The reaction has been referred to as ATP-dependent H_2 evolution in order to distinguish it from conventional hydrogenase activities. Protons, the ultimate source of the H_2 evolved based on ratios of H_2:HD:D_2 evolved from H_2O:D_2O mixtures, are non-rate-limiting (66). Attempts to reverse H_2 evolution have been unsuccessful. The reaction is not inhibited by 1 atm of H_2 or CO. In the absence of an added reducible substrate, all electrons go into H_2 production; on addition of a reducible substrate, the amount of H_2 evolution is diminished as some electrons are diverted to reduction of the added acceptor, but total electron consumption and rate of transfer are unaffected (Table III). Saturating concentrations of C_2H_2

Table III. Partitioning of Electrons Between Proton Reduction and C_2H_2 Reduction by *A. vinelandii* N_2ase

	μMoles per Mg Protein per 30 Min, 30° [b]		
pC_2H_2 [a]	H_2	C_2H_4	$H_2 + C_2H_4$
0	8.6	0.0	8.6
0.004	6.4	2.3	8.7
0.007	4.8	3.4	8.2
0.010	4.2	4.2	8.4
0.015	3.5	5.0	8.5
0.025	2.6	6.4	9.0
0.100	0.4	8.2	8.6

[a] Plus argon to 1.0 atm.
[b] Reaction conditions as described in Ref. 26.

consume almost all the electrons transferred by N_2ase and practically eliminate H_2 evolution; saturating concentrations of N_2 consume only 75% of the electrons transferred by N_2ase, and the remaining 25% are evolved as H_2, explaining the apparently high rate of electron consumption previously indicated for C_2H_2 reduction relative to N_2. Thus, the ratios of N_2ase activity measured as H_2 evolution, C_2H_2 reduction, and N_2 fixation are 4:4:1.

Additional classes of substrates that are reduced by N_2ase may well remain to be found. Certain classes—e.g., NO, CO, CH_2O, CH_3I, CH_3NH_2, CNO^-, dimide, and hydrazine—have been tested, but no evidence was obtained for reduction products (42, 43, 46, 51, 60).

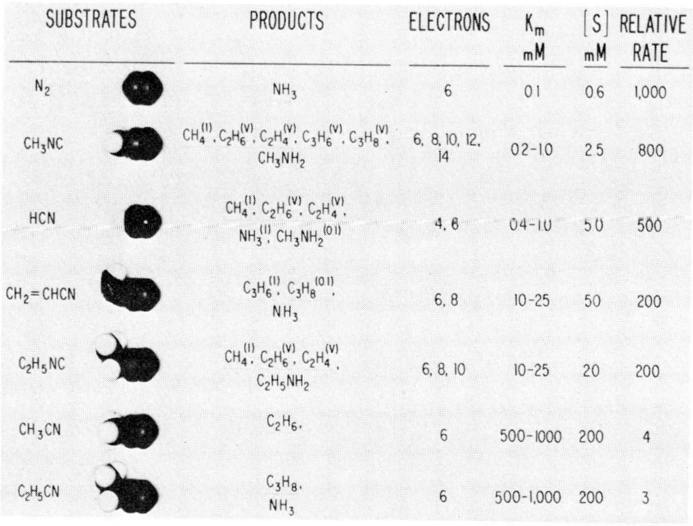

SUBSTRATES	PRODUCTS	ELECTRONS	K_m mM	[S] mM	RELATIVE RATE
N_2	NH_3	6	0.1	0.6	1,000
CH_3NC	$CH_4^{(I)}, C_2H_6^{(V)}, C_2H_4^{(V)}, C_3H_6^{(V)}, C_3H_8^{(V)}$, CH_3NH_2	6, 8, 10, 12, 14	0.2–10	2.5	800
HCN	$CH_4^{(I)}, C_2H_6^{(V)}, C_2H_4^{(V)}$, $NH_3^{(I)}, CH_3NH_2^{(OI)}$	4, 6	0.4–10	5.0	500
$CH_2=CHCN$	$C_3H_6^{(I)}, C_3H_8^{(OI)}$, NH_3	6, 8	10–25	50	200
C_2H_5NC	$CH_4^{(I)}, C_2H_6^{(V)}, C_2H_4^{(V)}$, $C_2H_5NH_2$	6, 8, 10	10–25	20	200
CH_3CN	C_2H_6,	6	500–1000	200	4
C_2H_5CN	C_3H_8, NH_3	6	500–1,000	200	3

Figure 4. Comparison of the reduction of nitriles and isonitriles by N_2ase

Table IV. Rate Constants (M^{-1}) for Isonitrile Reduction by Azotobacter N_2ase

Rate Constant	CH_3NC		C_2H_5NC	
	H_2O	D_2O	H_2O	D_2O
$\dfrac{d\ C_2H_6/dt}{d\ CH_4/dt \times [RNC]}$	6.2	33	0.3	0.9
$\dfrac{d\ C_2H_4/dt}{d\ CH_4/dt \times [RNC]}$	1.1	3–54	0.4	1.0
$\dfrac{d\ C_2H_6/dt}{d\ C_2H_4/dt}$	5.6		0.7	0.9
$\dfrac{d\ C_3H_8/dt}{d\ C_2H_6/dt \times [RNC]}$	1.1	2.3		
$\dfrac{d\ C_3H_6/dt}{d\ C_2H_6/dt \times [RNC]}$	2.0	22		
$\dfrac{d\ C_3H_8/dt}{d\ C_3H_6/dt}$	0.55	0.10		

NITRILE, ISONITRILE, AND ALKYNE ANALOGS. Exploration of N_2ase-catalyzed reductions of selected nitrile, isonitrile, and alkyne analogs provides examples of novel chemical reactions and yields information about the topography of the active sites of N_2ase.

Isonitriles are more effective substrates of N_2ase than the corresponding nitriles (Figure 4) (60, 61). Methyl isonitrile has about 500× the affinity of acetonitrile for N_2ase; ethyl isonitrile is about 50× more reactive than propionitrile. The greater reactivity of isonitriles compared with nitriles for the presumed active site metal is attributed to the superior end-on binding capability of the isonitrile carbon vs. the nitrile nitrogen. This may also explain the trend of reactivity among the isonitriles, though steric effects are undoubtedly important.

The production of variable amounts of C_2- and C_3-hydrocarbons from CH_3NC and C_2-hydrocarbons from C_2H_5NC indicates unusual catalysis, but can be accommodated by an alternating insertion–reduction mechanism (61, 62). This insertion reaction of N_2ase-catalyzed isonitrile reduction was the first proposed isonitrile insertion reaction in a biochemical or chemical system. Subsequently, an isonitrile insertion reaction has been demonstrated with a chemical system (67). The rate of C_2H_6 or C_2H_4 formation by N_2ase is directly related to the rate of CH_4 formation and isonitrile concentration for a given isonitrile and N_2ase at constant proton activity (Table IV) (62). An alternate mechanism also appears to yield C_2H_4 from CH_3NC (59), and its activity is favored by high isonitrile concentrations and low proton activity. Similarly, the rate of C_3H_8 or C_3H_6 formation is directly related to the rate of C_2H_6 formation and CH_3NC concentration at constant proton activity (Table IV). Decreasing proton activity by replacing H_2O with D_2O or increasing pH increases the rate constants; CO and CH_2O alter the rate constants. The

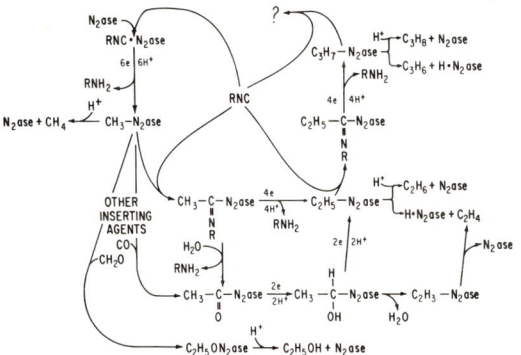

Figure 5. Reduction and insertion mechanism for isonitrile reduction by N_2ase

products, kinetics, effect of proton activity, and effects of other inserting agents suggest a major mechanism of alternating reduction and isonitrile insertion to form R · N$_2$ase (R = CH$_3$, C$_2$H$_5$, and C$_3$H$_7$ with all carbons arising from the isonitrile carbon), which yields the observed alkanes by proton attack and alkenes by hydride elimination as shown in Figure 5.

The isonitrile reductions provide considerable indirect information regarding N$_2$ase. A somewhat open N$_2$ase site(s) that is able to accommodate substantially larger molecules than N$_2$ is required for the isonitrile reactions. However, steric limitations appear to exist, since the rate constants and C$_2$H$_6$:C$_2$H$_4$ ratio for C$_2$H$_5$NC are different from those for CH$_3$NC. Minor variations in the environment of the active site of N$_2$ases from different sources is suggested by the different but characteristic C$_2$H$_6$:C$_2$H$_4$ product ratios for different N$_2$ases. Cross recombination of Mo–Fe and Fe proteins from various sources indicates that the C$_2$H$_6$:C$_2$H$_4$ ratio is specified by the Mo–Fe protein, not the Fe protein (59).

Saturated unbranched nitriles up to n-butyronitrile are reduced by N$_2$ase (Figures 4 and 6) (61), further demonstrating that substantially larger substrates than N$_2$ can be accommodated by the active site(s). The failure to reduce the branched nitrile, isobutyronitrile, indicates steric limitations and the need for an exposed carbon of the nitrile group. The affinity of cyanide is about 500× that of either acetonitrile or propionitrile, which may suggest that CN$^-$, not HCN, is the substrate with

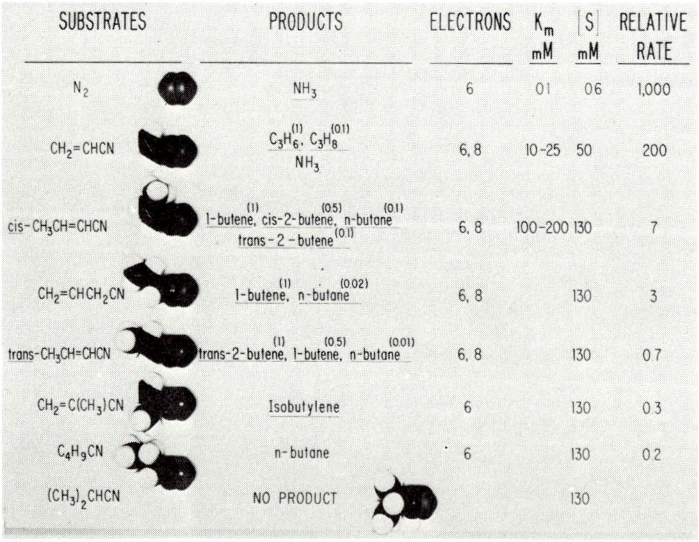

SUBSTRATES	PRODUCTS	ELECTRONS	K$_m$ mM	[S] mM	RELATIVE RATE
N$_2$	NH$_3$	6	0.1	0.6	1,000
CH$_2$=CHCN	C$_3$H$_6^{(1)}$, C$_3$H$_8^{(0.1)}$ NH$_3$	6, 8	10–25	50	200
cis-CH$_3$CH=CHCN	1-butene$^{(1)}$, cis-2-butene$^{(0.5)}$, n-butane$^{(0.1)}$ trans-2-butene$^{(0.1)}$	6, 8	100–200	130	7
CH$_2$=CHCH$_2$CN	1-butene$^{(1)}$, n-butane$^{(0.02)}$	6, 8		130	3
trans-CH$_3$CH=CHCN	trans-2-butene$^{(1)}$, 1-butene$^{(0.5)}$, n-butane$^{(0.01)}$	6, 8		130	0.7
CH$_2$=C(CH$_3$)CN	Isobutylene	6		130	0.3
C$_4$H$_9$CN	n-butane	6		130	0.2
(CH$_3$)$_2$CHCN	NO PRODUCT			130	

Figure 6. *Reduction of three and four carbon nitriles by N$_2$ase*

C, not N, involved in complexation for the unsubstituted nitrile, while N is involved in complexation of all of the substituted nitriles.

Unsaturated nitriles (61) and isonitriles (50) such as acrylonitrile, methyl substituted acrylonitriles, 3-butene nitrile, or vinylisocyanide have much greater affinity for N_2ase than the corresponding saturated nitriles or isonitriles (Figures 4 and 6). These differences may be attributed to the increased electron density on nitrogen in the case of an olefinic group in conjugation with the nitrile group or to coordination through the olefin as well as the nitrile nitrogen.

Olefinic bonds are only reduced by nitrogenase when they and —C≡N are in conjugation. Ethylene and 1,3-butadiene are not reduced. Acrylonitrile is reduced by six electrons to propylene and ammonia and by eight electrons to propane and ammonia (68). Double bond migration occurs in propylene formation. No 2- or 4-electron products have been found.

A two- to five-fold increase in reduction rate of all nitriles except HCN is produced by replacement of H_2O with D_2O (68); in contrast, no marked effect of D_2O vs. H_2O on rate of reduction of any other substrate is noted. The increased rate of reduction of added reducible nitriles in D_2O is caused by an altered partitioning of electrons in favor of RCN reduction vs. proton reduction, at least in the case of acrylonitrile. Thus, the reduction of RCN by N_2ase appears to be less facile than the reduction of RNC, N_2, etc. This difference between RCN and RNC suggests nonequivalence of the C≡N bond of RCN and RNC toward electron addition from N_2ase.

Single substitution of each of the hydrogens of acrylonitrile with methyl groups produces an analog series of substrates whose variation in rate of reduction can be largely attributed to steric effects (Figure 6) (61). Cis-crotononitrile is a more effective substrate than trans-crotononitrile, which in turn is more effective than methacrylonitrile. These results, in addition to those already noted for isobutyronitrile, suggest that effective catalysis requires that the nitrile carbon be exposed, possibly for attack by a metal involved in electron donation.

In the alkyne series (50, 61) methyl- and ethylacetylene are reduced, but not dimethylacetylene. Allene is also reduced by N_2ase, possibly via isomerization to methylacetylene (Figure 7).

SUMMARIZED CHARACTERISTICS OF N_2ASE-CATALYZED REDUCTIONS. Substrates possess either NN, NC, or CC but not CO triple or potential triple bonds. This range of substrates makes N_2ase the most versatile homogeneous reducing catalyst yet described.

Reduction cleaves NN, NC, or NO (so far only in N_2O but not in NO) bonds, but not CC bonds.

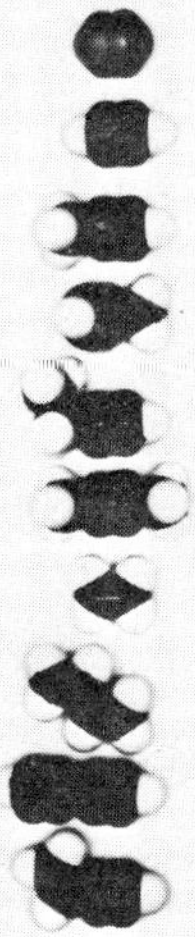

SUBSTRATES		PRODUCTS	ELECTRONS	K_m mM	[S] mM	RELATIVE RATE
N_2		NH_3	6	0.1	0.6	1
$HC{\equiv}CH$		C_2H_4	2	0.1–0.3	3	4
$CH_3C{\equiv}CH$		C_3H_6	2			
$CH_2{=}C{=}CH_2$		C_3H_6	2			
$CH_3CH_2C{\equiv}CH$		1-BUTENE (?)	2			
$CH_3C{\equiv}CCH_3$		NO PRODUCT				
$CH_2{=}CH_2$		NO PRODUCT				
$CH_2{=}CHCH{=}CH_2$		NO PRODUCT				
$HC{\equiv}CC{\equiv}CH$		(?)				
$CH_2{=}CHC{\equiv}CH$		(?)				

Figure 7. Reduction of alkynes by N_2ase

Alkynes are reduced specifically to alkenes.

Only alkynes possessing an acetylenic H are reduced.

The C=C bond is not reduced unless it is conjugated with a C≡N bond.

Products incorporating 2, 4, 6, 8, 10, 12, and 14 electrons are detected, indicating that N_2ase catalyzes stepwise reduction; however, electron transfer may occur in 1- or 2-electron increments.

Unbranched substrates two to three times as large as N_2 are reduced; branching adjacent to the CN group inhibits reduction of nitriles, suggesting that an interaction between N_2ase and the nitrile C is required for reduction. A dinuclear pocket site, as discussed under mechanism, will accommodate all substrates and explain observed steric limitations.

Vinyl groups increase the affinity of nitriles or isonitriles for N_2ase.

Nature of all substrates, affinity for N_2ase, and reactions implicate transition metals in catalysis with end-on complexation for all substrates except C_2H_2.

The reduction of the isonitrile carbon to CH_4 and to C_2- and C_3-hydrocarbons is readily accommodated by an alternating reduction and insertion mechanism involving a metal.

HD Formation. The following exchange, $D_2 + H_2O \rightarrow HD + HDO$, is catalyzed by N_2ase while N_2 is being reduced (*66, 69*). Carbon monoxide inhibits both this exchange and N_2 reduction. The ratio of HD to NH_3 formed is directly related to pD_2 and is 4.5 at 0.6 atm of D_2. The above characteristics suggest that the exchange involves hydrogens of intermediates of N_2 reduction, and the $HD:NH_3$ ratios of >3 indicate that the rate of exchange exceeds the rate of reduction. A similar exchange occurs between D_2 and the hydrogens of an aryldiimide– and arylhydrazine–platinum complex, and this exchange is catalytic in a protonic solvent (66). An N_2-independent HD formation has also been reported for N_2ase (70).

Inhibitors. Carbon monoxide is a potent inhibitor of all N_2ase-catalyzed reductions with the exception of H_2 evolution (32, 64, 65). Inhibitor constants (K_i) of about $4 \times 10^{-7}M$ have been reported (48, 49, 50, 71); however, the type of inhibition is in dispute and may not be totally competitive (72, 73). Nitric oxide is also reported to be a competitive inhibitor of N_2ase-catalyzed reductions with a K_i similar to CO (71). The various substrates are mutual inhibitors of each other. H_2 is a competitive inhibitor of only N_2 reduction (68, 72, 73, 74) with a K_i of about 1×10^{-4} mM (47, 48, 49, 50, 66, 71).

Nitrogenase

Soluble N_2ases have been extracted from anaerobic [*Clostridium pasteurianum* (41)], aerobic [*Azotobacter vinelandii* (22, 75), A. chroococcum (76), *Mycobacterium flavii* (77)], facultative [*Bacillus polymyxa* (78), *Klebsiella rubiacearum* (79), *K. pneumoniae* (80)], and photosynthetic [*Rhodospirillum rubrum* (81, 82), *Chromatium* (83)] bacteria, from legume nodules [soybean (2, 48, 49, 84, 85) and seradella (48)], and possibly from blue-green algae [*Anabaena cylindrica* (86)], but not yet from nonlegume nodulated plants (87). Most pioneering studies have used Clostridium or Azotobacter N_2ase with complementary studies on N_2ases from the other bacteria and soybean nodules.

The various N_2ases are similar but not identical, suggesting a somewhat common active site and mechanism for all N_2ases with variations in the amino acid composition in nonactive site areas. All are a complex of (26, 34, 59, 88, 89, 90, 91, 92) a Mo–Fe protein and an Fe protein for which we propose the names azofermo and azofer (93).

The N_2ase enzyme is in association–dissociation equilibrium with its component proteins (26) and is probably composed of one Mo–Fe protein and two Fe proteins (Figure 8) (94, 95). All N_2ases have identical reaction requirements and catalyze identical reactions. However, the effectiveness of the interchange of the Mo–Fe protein and the Fe protein from various species parallels, in general, the degree of physiological similarities of the donors (26, 59, 90). The nature of the interaction involved in complex formation from the acidic protein components is unknown but probably involves specific rather than general charge effects.

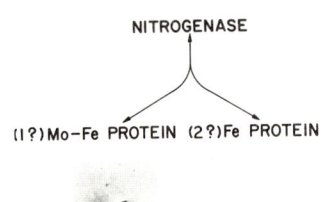

PROTEIN CRYSTALS

Figure 8. Nitrogenase and its component proteins with a phase contrast micrograph of crystals of the Mo–Fe protein

Table V. Summarized Characteristics of Azotobacter and Clostridia N_2ase

	Azotobacter	Clostridium
Homogeneity	–	–
Molecular weight	–	–
Mo:Fe:S^{2-}:CySH	–	–
Other metals	–	–
Activities		
Reactions	$N_2 \rightarrow 2NH_3$; $N_3^- \rightarrow N_2 + NH_3$	
	$N_2O \rightarrow N_2 + H_2O$; $C_2H_2 \rightarrow C_2H_4$	
	$RCN \rightarrow RCH_3 + NH_3$	
	$RNC \rightarrow CH_4, C_2H_6, C_2H_4, C_3H_8, C_3H_6 + RNH_2$	
	$2H^+ \rightarrow H_2$; ATP $\rightarrow$ ADP + Pi	
Specific activity	~80	~80
(nmoles N_2 reduced/min/mg)		
ATP/2e	4–5	4–5
Activation energy >20°	14 Kcal/mole	–
<20°	35–50 Kcal/mole	–
EPR spectrum		
Native protein	g = 4.30, 3.67, 2.01	–
+ $Na_2S_2O_4$	g = 4.30, 3.67, 2.01, 1.94	–

No homogeneous N_2ase preparations have been obtained, and the purest preparations are those reconstituted from Mo–Fe protein and Fe protein preparations. Complete correspondence between native N_2ase and reconstituted N_2ase has not been established; unusual stoichiometries associated with activities of reconstituted N_2ase (*37, 91, 92, 96*) suggest that native N_2ase is preferred for investigations of reaction parameters and was used in all of the work described here.

N_2ase. Characteristics of N_2ases from Azotobacter and Clostridia are summarized in Table V. A molecular weight of about 220,000 and about 1 Mo, 20 Fe, and 20 S^{2-} are suggested for Azotobacter N_2ase based on analyses of its components (*see* Mo–Fe and Fe protein sections) but definitive physical and chemical analyses of N_2ase must await the availability of high-activity homogeneous preparations which hopefully might be crystallized in a form suitable for x-ray analysis. Purified N_2ase is rapidly inactivated by O_2 and is cold-sensitive (*34, 81, 96, 97*); the Mo–Fe protein and Fe protein are both O_2-sensitive, but only the Fe protein is cold-sensitive (*34, 97*).

Useful procedures for the purification of Azotobacter N_2ase and its fractionation into the Mo–Fe and Fe proteins are shown in Figure 9

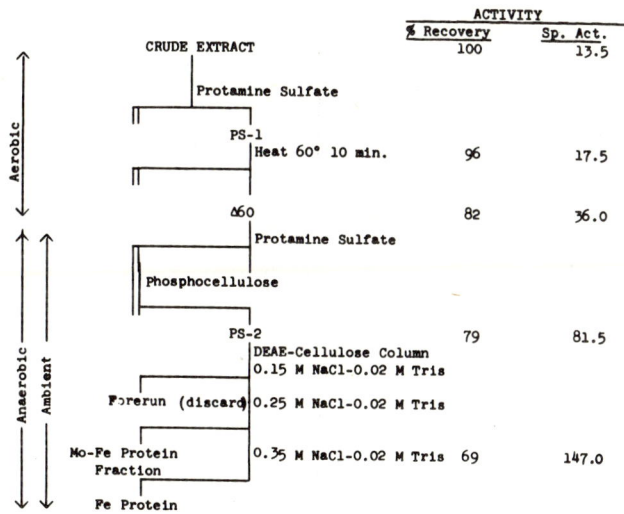

Figure 9. Purification and fractionation of Azotobacter N_2ase; specific activity is nmoles N_2 reduced/min · mg protein of indicated fraction

(*58, 98*). Activity of the purified unfractionated N_2ase is about 80 nmoles N_2 reduced/min · mg protein or about 40% of the theoretical activity of homogeneous N_2ase calculated on the basis of activity of the purified individual components.

Mo–Fe Protein. The chemical and physical characteristics of purified Mo–Fe protein from Azotobacter (*58, 98, 99*) and Clostridium (*94, 100*) are summarized in Table VI. This protein will be described in detail since our current experimentation implicates the Mo and Fe of this protein at the complexation and reduction site(s) of N_2 and other substrates. Furthermore, it is the only protein for which definitive data on homogeneous preparations are available.

Homogeneous Azotobacter Mo–Fe protein has been isolated recently by crystallization (*98*). A phase contrast micrograph of the crystalline

Table VI. Characteristics of Mo–Fe Protein

	Azotobacter	Clostridium[a]
Homogeneity	3× Crystallized	Purified–95+%
Molecular weight		
Native	270,000	160,000–200,000
Subunits	40,000 (2 types)	Multiple types
S_W^{20}	10.1	
$Mo:Fe:S^{2-}.CySH$	2:32:25:37	1–2:15:15:–
Other metals	None	–
Activities		
Uncombined	None	None
Combined with Fe protein	All N_2ase activities tested	
Specific activity	362	345
(nmoles N_2 reduced/min/mg)		
UV–visible absorption spectrum		
Native protein ε_{278}	$470 \times 10^3 M^{-1} cm^{-1}$	–
ε_{412}/Fe	$2.6 \times 10^3 M^{-1} cm^{-1}$	–
Mossbauer spectrum		
Native protein δ_o	0.39 mm/sec	–
ΔEQ	0.78 mm/sec	–
$+Na_2S_2O_4$ δ_o	–	–
ΔEQ	–	–
EPR spectrum		
Native protein	$g = 4.30, 3.67, 2.01$	–
$+Na_2S_2O_4$	$g = 4.30, 3.67, 2.01, 1.94$	–
Oxidized	$g = 4.30, 2.01$	–
Magnetic susceptibility	3.0 ± 0.3 B.M./Fe	–
Heat of solution	–5.4 Kcal/mole	–

[a] Ref. 94 and 100.

protein is shown in Figure 8. Crystal dimensions are $1-4 \times 30-60\mu$. Larger crystals will provide the opportunity for x-ray analyses. The isolation and crystallization procedure are somewhat unusual for a protein. Isolation requires only a 30-fold purification from crude extract to homogeneous protein with a recovery of over 50% of the initial activity (Figures 9 and 10). The simplicity of the procedure makes it possible for a single technician to convert whole cells to 500 mg of crystalline Mo–Fe protein in four days. Crystallization is based on the insolubility of the purified protein in a solution $\leq 0.08M$ NaCl or other mono- or multivalent salt of equivalent ionic strength. The crystalline protein can be redissolved and recrystallized to constant specific activity and amino acid composition. Solutions of the recrystallized protein are dark brown while crystals are essentially white, suggesting that the environment of the chromophore is altered on crystallization. Similar color changes are known for many transition metals that undergo ligand dissociation during solubilization (103). The heat of solution is about –5.4 Kcal/mole. It is suggested that minor modifications of this procedure will be useful for the crystallization of purified Mo–Fe proteins from other N_2-fixing organisms, especially those that show effective recombination with Azotobacter Fe protein.

Molecular weight of the crystallized Mo–Fe protein is 270,000 based on ultracentrifugal analyses in $0.25M$ NaCl–$0.01M$ tris-HCl, pH 7.2 (Table VI). The protein tends to associate, and this molecular weight may represent a dimer. Various dissociating agents (104) produce two types of similar subunits of molecular weight of about 40,000 each (99). The corresponding clostridial protein has a reported molecular weight of 160,000–200,000 (94, 100). Six types of subunits have been suggested for it (100), but may simply represent various recombinations of the sulfur-rich monomers.

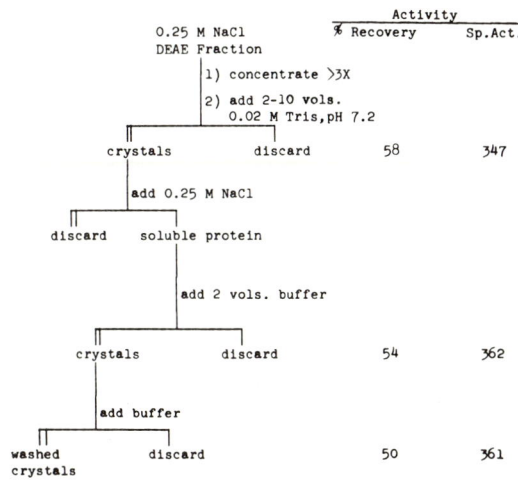

Figure 10. Crystallization of Azotobacter Mo–Fe min · mg protein; recovery is percent of crude extract activity.

Analyses of the Azotobacter protein (98, 99) show 2 Mo, 32 Fe, and 25 S^{2-} per 270,000 M.W. Sulfide analyses are frequently low with iron–sulfur proteins so that actual S^{2-} may be more equivalent to metal content. Cysteine content of about 40 indicates equivalency of iron and cysteine and very little cysteine not involved in iron ligation, both features common to typical iron–sulfur proteins. No other metals are found in amounts equivalent to one atom per 270,000. The protein contains all the common amino acids; most iron–sulfur proteins are acidic, and this one is no exception, with almost a two-fold excess of acidic amino acid residues over basic.

The purified Mo–Fe proteins show no enzymatic activities or even partial reactions of N_2ase unless combined with the Fe protein. Attempts to elucidate specific functions of either the Fe or Mo–Fe protein by binding studies with ATP or $^{14}CN^-$ (105) have not shown the specificity required for assignment of roles (28). The specific activities of purified protein from Azotobacter or Clostridia (when recombined with saturating amounts of their respective Fe proteins) are similar—about 350 nmoles N_2 reduced/min · mg protein.

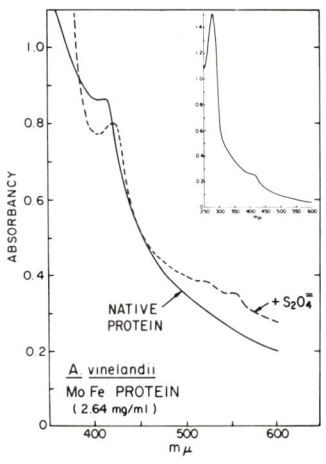

Biochemical and Biophysical Research Communications

Figure 11. Visible–UV spectra of native and reduced crystallized Azotobacter Mo–Fe protein; cuvettes contained 2.64 mg or (inset) 0.87 mg protein in 2 ml

The UV and visible absorption spectra of the native protein prepared under the usual anaerobic conditions are shown in Figure 11 (98). A maximum occurs at 280 nm with a shoulder at 412 nm. This spectrum is typical of iron–sulfur proteins containing four or more irons, such as clostridial ferredoxin, and the extinction coefficient per iron atom at 412 nm is similar to that for clostridial ferredoxin. Reduction with hydrosulfite shifts the shoulder at 412 nm to about 420 nm with the appearance of weak maxima at 525 and 557 nm.

The ESR spectra of the native and reduced protein were obtained at 4°K (99). The native protein exhibits weak resonances at g values of 2.01, 3.67, and 4.30. Reduction with hydrosulfite enhances the amplitude of the resonances with g values of 3.67 and 4.30 and introduces a resonance at $g = 1.94$. The latter is typical of all iron–sulfur proteins of the ferredoxin type and thus may reflect a conventional iron–sulfur protein character (*i.e.*, electron transfer function) in N_2ase, which is presumably unrelated to the unique N_2-fixing aspect of N_2ase. The resonances at $g = 3.67$ and 4.30 have not been observed in iron–sulfur proteins. A resonance at 4.30 is observed in rubredoxin, but reduction decreases instead of increases its amplitude. The resonance at a g value of 3.67 is lost on exposure to air and is only partially restored by subsequent treatment with hydrosulfite. The resonances at g values of 4.30 and 3.67 have been variously attributed to high-spin Fe^{2+} and intermediate-spin Fe^{3+}. No resonances attributable to Mo have been observed with Mo–Fe protein in contrast with earlier reports with crude material (*106*). The absence of resonance may be owing to the absence of Mo^{5+}, masking by the nearby iron resonance, or spin coupling. ESR spectra of native and reduced N_2ase show only resonances similar to those of the Mo–Fe protein.

The Mossbauer spectra were obtained for crystallized Mo–Fe protein isolated from Azotobacter grown on ^{57}Fe-enriched media (*99*). Between 20° and 200°K, the spectrum of the native protein is a doublet with an isomer shift of 0.39 mm/sec (relative to Fe metal) and a quadrupole split of 0.84 mm/sec indicative of high-spin Fe^{3+}. A small shoulder on the high-field side may indicate high-spin Fe^{2+} or Fe^{1+} (*107*) and may represent the unique portion of the iron involved in N_2 complexation. Reduction with hydrosulfite converts at least 25% of the iron to a form representative of high-spin Fe^{2+} or Fe^{1+}.

The magnetic susceptibility of the native protein is 3 B.M./Fe by either NMR or Faraday cage technique (*99*). The large amount of iron makes assignment of specific values to various forms of iron impossible.

Fe Protein. Characteristics of the Fe protein from Azotobacter (*101*) and Clostridia (*94, 108*) are summarized in Table VII. Clostridial Fe protein is stated to be 90–95% pure and the absence of tryptophan (*102*) suggests that almost all of the contaminating proteins have been removed. Recent preparations of Azotobacter Fe protein show specific activities even higher than those of clostridial Fe protein (*101*). The clostridial protein has a molecular weight of 39,000 and contains 2–3 Fe and 2 S^{2-} atoms (*94, 108*). Molybdenum is absent, no ESR is detectable in the native protein (*102*) and, surprisingly, a resonance at g value of 1.94 has not been observed on reduction. All other purified iron–sulfur proteins with the exception of the atypical HIPIP and rubredoxin exhibit resonance in this area on reduction. One may suggest that the Fe protein is an atypical iron–sulfur protein or requires additional examination at 4°K for a resonance at $g = 1.94$. The individual protein has no activity alone but has all N_2ase activities in combination with the Mo–Fe protein.

Table VII. Characteristics of Fe Protein

	Azotobacter[a]	Clostridium[b]
Homogeneity	–	Purified (tryptophan-free)
Molecular weight	–	39,000
Fe:S^{2-}:CySH	–:–:–	2–3:2:8(free CySH)
Other metals	Mo absent	Mo absent
Activities		
Uncombined	None	None
Combined with Mo–Fe protein	All N_2ase activities tested	
Specific activity	532	460
(nmoles N_2 reduced/min/mg)		
Mossbauer spectrum		
Native protein δ_o	–	–
ΔEQ	–	–
EPR spectrum		
Native protein	–	None
+$Na_2S_2O_4$	–	–
Oxidized	–	$g \cong 2$

[a] Ref. *101*.
[b] Ref. *94*, *102*, and *108*.

Metals. Neither the structure nor the specific function of the Fe or Mo prosthetic groups of N_2ase are known. Mossbauer and ESR spectra are proving useful in characterizing the iron. Spectral data and S^{2-} analyses suggest that much of the iron in the Mo–Fe protein is of the Fe–S type, similar to that of clostridial ferredoxin, and that it may, therefore, function in electron transport and reduction. The development of an appropriate model of these Fe–S proteins or definition of the structure of bacterial ferredoxins may well have direct relevance to our understanding of iron in N_2ase. Other types of iron may also be present, and not all of the iron may be functional or necessary. Particularly informative data on this point may be obtained with N_2ase isolated from bacteria grown on low Fe levels—*e.g.*, from Clostridia grown on the Fe level which is insufficient for Fd synthesis, but sufficient for N_2ase development (*18*, *19*). Definitive involvement of Mo has not been reported.

Mechanism

Electron Activation. A scheme based on the two-site hypothesis (*6*, *7*, *30*, *46*, *65*, *94*)—electron-activation site and substrate complexation site—for all types of reactions catalyzed by N_2ase is shown in Figure 12 (*7*). The reactions can be clearly divided into two types. One type is the limiting reaction of N_2 fixation and involves the interdependent reaction of electrons from a suitable donor and of ATP with N_2ase to produce ADP, P_i, and in the absence of added reducible substrate to evolve H_2.

We designate this reaction electron activation and the site(s) involved as the electron-activating site. The other type involves reduction of added reducible substrates and the additional site involved is referred to as the substrate-complexing site. The two sites are distinguished by inhibition experiments. Carbon monoxide inhibits reactions at the substrate-complexing site, but not at the electron-activating site.

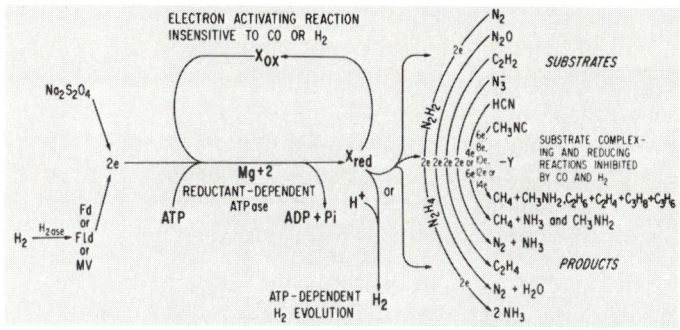

Progress in Phytochemistry

Figure 12. Nitrogenase reactions based on electron-activation substrate-complexation two-site hypothesis

We suggest that electron activation is the major, if not only, function of ATP in N_2 fixation. Other proposed functions of ATP include dehydration and conformation change (32). Inability to detect exchange, reversal, or intermediates of the N_2ase-catalyzed ATP hydrolysis has limited the opportunity to explore the electron-activating mechanism, and the designation of transient reaction complexes—e.g., $P_i \cdot N_2$ase or ADP $\cdot N_2$ase—remains speculative. Inhibitor experiments suggest that ATP utilization involves metals of N_2ase in a mechanism which increases reducing potential and is uniquely different from other biochemical reactions of ATP. The apparent irreversibility of H_2 evolution is compatible with the formation of a strong nucleophile, or possibly a hydride, by electron activation. This nucleophile may then evolve H_2 or, as described below, transfer electrons to an added reducible substrate.

Substrate Reduction. Two major proposals exist for the mechanism of the reduction of N_2 and other added substrates. One proposal suggests stepwise reduction with enzyme-bound diimide and hydrazine as intermediates; the other suggests nitriding. A multiplicity of mechanisms intermediate between these two extremes is possible.

We believe that all of the known information on reactions and characteristics of N_2ase can be accommodated by the following mechanistic scheme involving stepwise reduction, with enzyme-bound diimide and

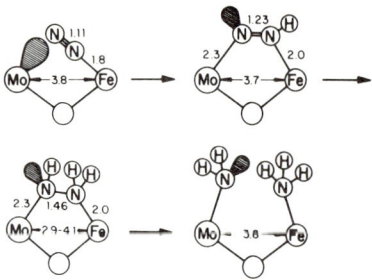

Figure 13. Proposed model for active site and reduction of N_2 with enzyme-bound diimide and hydrazine as intermediates

hydrazine as intermediates in N_2 reduction (Figure 13). The proposed active site is composed of Mo and Fe bridged by S, which has the proper size and electronic characteristics to give a Mo–Fe distance of about 3.8 Å. The site is located in a pocket in the protein sufficiently large to accommodate substrates like n-butyronitrile, but also possessing steric limitations compatible with the substituted acrylonitriles and branched saturated nitriles. The initial reaction in N_2 reduction involves the formation of a linear complex of N_2 with a unique Fe of N_2ase in much the same way as in the transition metal complexes (*see* below). Attack by a nucleophilic filled orbital of Mo (possibly a hydride) produces a dinuclear diimide species. This nucleophilic Mo is the terminus of the electron-activating system and may be the site of H_2 evolution, or, alternatively, is coupled to the site of H_2 evolution. The unfavorable thermodynamics of this step are overcome by coordination of the diimide and possible H-bonding with adjacent amine or hydroxyl groups of the protein. Addition of two more electrons and protons produces a dinuclear hydrazine, and addition of the final two electrons and protons cleaves the remaining N–N bond. Dissociation of NH_3 from Fe reopens the site for coordination of N_2 and hydrolysis of Mo–NH_2 gives the second molecule of NH_3. Increase in the N–N bond length from 1.09 to 1.46 Å is accompanied by changes in bond angle so that the Mo–Fe distance can remain constant. Inhibition by H_2 can be explained by interaction (oxidative addition?) with the Fe site. In contrast to N_2 reduction, C_2H_2 reduction may only require the Mo site, explaining H_2 inhibition of N_2 but not C_2H_2 reduction.

Various inorganic models provide support for this mechanism. These include the simple inorganic model consisting of the interaction of a platinum hydride with a benzenediazonium salt (the earliest example of an N_2 complex) (*109*). The intermediates in this reaction resemble those postulated in Figure 13:

$$[\text{Ar}-\text{N}\equiv\text{N}]^+ + \text{H}-\underset{\underset{\text{PEt}_3}{|}}{\overset{\overset{\text{PEt}_3}{|}}{\text{Pt}}}-\text{Cl} \rightarrow \left[\text{Ar}-\text{N}=\overset{\overset{\text{H}}{|}}{\text{N}}-\underset{\underset{\text{PEt}_3}{|}}{\overset{\overset{\text{PEt}_3}{|}}{\text{Pt}}}-\text{Cl}\right]^+ \xrightarrow{\text{H}_2}$$

$$\left[\text{Ar}-\text{NH}-\text{NH}_2-\underset{\underset{\text{PEt}_3}{|}}{\overset{\overset{\text{PEt}_3}{|}}{\text{Pt}}}-\text{Cl}\right]^+ \xrightarrow{\text{H}_2} [\text{ArNHNH}_3]^+ + \text{H}-\underset{\underset{\text{PEt}_3}{|}}{\overset{\overset{\text{PEt}_3}{|}}{\text{Pt}}}-\text{Cl}$$

The similarity between the hydrogen–deuterium exchange reaction of the aryldiimide– and arylhydrazine–platinum intermediates of this model and of N_2ase is the best evidence for enzyme-bound diimide and hydrazine as intermediates (*66*). The correlation between the relative affinity of various ligands for N_2ase and for some of the synthetic complexes of Fe, Co, Ni, Mo, Ru, Rh, Re, Os, and Ir that bind, but have not yet been shown to reduce, N_2 (Table VIII) indicate that these metal complexes

Table VIII. Comparison of the Affinity of Ligands for N_2ase and Inorganic Complexes

Nitrogenase	$CO\sim NO >> N_2 \sim C_2H_2 \sim H_2 > N_3^-$, N_2O, RNC, RCN $>>$ NH_3, H_2O, C_2H_4
$(Ph_3P)_3Co(N_2)H$	$CO >> N_2 \sim H_2 > C_2H_4$, NH_3
$(Ph_3P)_3Ru(N_2)H_2$	$CO >> N_2 \sim H_2 > NH_3$
$[(NH_3)_5RuN_2]^{2+}$	$CO \sim N_2 \sim NH_3 > H_2O >> H_2$

may be valid models of the initial step of N_2ase-catalyzed N_2 fixation—*i.e.*, the formation of a N_2ase · N_2 complex. The chemistry of these complexes (*3, 8*) has been discussed in an earlier lecture (*1*). The characteristics of these transition metal complexes support linear complexation of N_2 to a metal of N_2ase—*e.g.*, Mo or Fe in a low valence state with donation of the nonbonding electrons of N_2 to the metal and back donation of *d*-electrons of the metal to N_2.

Further support for diimide formation in the N_2ase system is provided by results on the interaction of a nitrogen complex of Re with Mo^{4+} (*110*). A postulated Re≡N≡N≡Mo is formed with a corresponding large decrease in N–N stretching frequency towards that characteristic of diimide.

The specific reduction of C_2H_2 to C_2H_4 by a molybdothiol–borohydride system, but not by an iron–thiol system, is in agreement with C_2H_2 reduction at the Mo site of N_2ase (55).

The titanium systems for N_2 reduction under ambient conditions were discussed in an earlier paper (11). They are assumed to involve nitride formation and require anhydrous conditions which are not characteristic of the preceding models or of N_2ase.

Problems

Biochemical Problems of N_2 Fixation. An extensive and liberal tabulation of biochemical problems of N_2 fixation is summarized in Table IX. The major problem for the bioinorganic chemist is mechanism: What is the nature of the active site and process of reduction? Are metals involved in complexation and reduction, as indirect evidence suggests?

Table IX. Biochemical Problems in N_2 Fixation

Natural electron donors	Identification 1- or 2-Electron transfer Complex with N_2ase
ATP	Electron activation Conformation change Dehydration ATP and reductant reaction sequence—ordered or random Apparent irreversibility—ATP→ADP + P_i
H_2 evolution and electron donation	Similar or different sites Redox potential Hydride Role of H_2 evolution 1- or 2-Electron additions
Reducible substrates	Other classes Similar or different sites Mono- or multi-nuclear sites Versatility Reaction specificity Nonreduction of CO and product N_2 from N_2O or N_3^- Superior coupling of electrons to C_2H_2 than to N_2 Double bond migration in CH_2=C—C≡N Easier reduction of RNC and N_2 than RCN Noninhibition of C_2H_2 and RCN reduction by H_2 Reduction and/or activation for complexation

Table IX. Continued

N_2ase complex	Cellular localization
	Crystallization and x-ray analysis
	Fe protein
	Reconstituted *vs.* native
	Cross combinants, active *vs.* inactive
	Subunits
	Function of each protein or subunit
	Natural and synthetic active fragments
Mechanism	Active site, metal, nonmetal, or combination
	Stepwise reduction *vs.* nitriding
Metals	Structure of Fe, Mo, and S^{2-} prosthetic groups
	Equivalent or different Fe's
	Functional or structural role
	Relationship to other Fe–S proteins
	Model for Fe and Mo groups
Other	Leghemoglobin function
	Cold inactivation of Fe protein
	Activation energy change at $\sim 20°$
	Nomenclature

Is the process nitriding, diimide–hydrazine, or intermediate between these two extremes? X-ray crystallographic studies will certainly provide answers to structure, but the dynamics of the system will only be revealed through collaborative studies by biochemists and inorganic chemists correlating N_2ase and its reactions with those of various models; this latter phase is gaining momentum and should include heterogeneous, as well as homogeneous, models in the future. The ATP function in electron activation, the limiting step in biological N_2 fixation, must not be ignored, and understanding of the mechanism of N_2 reduction probably will not be possible until the ATP reaction is defined. Other problems such as the nature of natural electron donors, localization of N_2ase, and function of the much-investigated leghemoglobin, although important to the overall biochemistry of N_2 fixation, are ancillary to the understanding of the facile reduction of N_2 by N_2ase.

Utilities

Potential Utilities of N_2 Fixation Research. Increased protein production and new insights into the chemistry of N_2 are potential utilities of the current biological and chemical research on N_2 fixation (Table X). We believe that the provision of additional protein is the most important objective. Furthermore, we favor enhanced symbiotic N_2 fixation and extension of the plant–microbe symbiosis to other agriculturally impor-

Table X. Potential Utilities of N_2 Fixation Research

Increased N_2 fixation by legumes—*e.g.*, soy beans, clover

Extension of N_2 fixing activity—*e.g.*, to corn, wheat, etc.

New chemistry of N_2

tant plants *vs.* new catalysts based on N_2ase and/or new N_2 complexes as the preferred route for increasing food protein. Fertilizer nitrogen will continue to play a major role in food protein production, but new catalysts probably will not produce a dramatic effect on fertilizer nitrogen production since the process is now quite efficient. New catalysts for N_2 based on N_2ase and/or new N_2 complexes and reducing systems may exert their greatest impact through the direct utilization of aerial N_2 to form important organo–nitrogen monomers.

Two major technique advances—the $C_2H_2 \rightarrow C_2H_4$ assay (*26, 111*) and *in vitro* callus-Rhizobium symbiosis (*112*)—provide the opportunity to enhance and eventually extend biological N_2 fixation.

Development of the $C_2H_2 \rightarrow C_2H_4$ reduction reaction of N_2ase into a universal, quantitative, high-volume, and sensitive assay for N_2-fixing activity provides the first opportunity to assess field parameters on a scale essential for the optimization of N_2 fixation, especially in important protein crops such as soybeans (*113*). This assay is the most important practical application to develop from the fundamental biochemical studies of N_2 fixation during the past decade.

The reduction of the complex symbiotic N_2-fixing system to the relative simplicity of an *in vitro* callus-Rhizobium symbiosis opens the door for understanding of the process of bacterial infection and induction of N_2ase activity in symbionts. Previous studies have been restricted to whole plants or roots, while the *in vitro* system permits studies in a defined liquid medium with callus. Knowledge of the factors involved in the development of the symbiosis may permit its extension to major crops, such as wheat, which must now be provided with fertilizer nitrogen.

Literature Cited

(1) Allen, A. D., ADVAN. CHEM. SER. (1971) **100**, 79.
(2) Bergersen, F. J., *Proc. Roy. Soc.* (1969) **B, 172**, 401.
(3) Borodko, Yu. G., Shilov, A. E., *Usp. Khim.* (1969) **38**, 761.
(4) Burris, R. H., *Proc. Roy. Soc.* (1969) **B, 172**, 339.
(5) Chatt, J., *Proc. Roy. Soc.* (1969) **B, 172**, 327.
(6) Hardy, R. W. F., Burns, R. C., *Ann. Rev. Biochem.* (1968) **37**, 331.
(7) Hardy, R. W. F., Knight, E., Jr., *in* "Progress in Phytochemistry," p. 407, L. Reinhold and Y. Liwschitz, Eds., Wiley, London, 1968.
(8) Murray, R., Smith, D. C., *Coordination Chem. Rev.* (1968) **3**, 429.

(9) Postgate, J. R., *Nature* (1970) **226**, 25.
(10) Rabinowitz, J. C., Advan. Chem. Ser. (1971) **100**, 322.
(11) Van Tamelen, E. E., Advan. Chem. Ser. (1971) **100**, 95.
(12) Yoch, D. C., Benemann, J. R., Valentine, R. C., Arnon, D. I., *Proc. Natl. Acad. Sci. U.S.* (1969) **64**, 1404.
(13) Yoch, D. C., Benemann, J. R., Arnon, D. I., Valentine, R. C., Russell, S. A., *Biochem. Biophys. Res. Commun.* (1970) **38**, 838.
(14) Klucas, R. V., Evans, H. J., *Plant Physiol.* (1968) **43**, 1458.
(15) Mortenson, L. E., *Proc. Natl. Acad. Sci. U.S.* (1964) **52**, 272.
(16) D'Eustachio, A. J., Hardy, R. W. F., *Biochem. Biophys. Res. Commun.* (1964) **15**, 319.
(17) Bulen, W. A., Burns, R. C., LeComte, J. R., *Biochem. Biophys. Res. Commun.* (1964) **17**, 265.
(18) Knight, E., Jr., D'Eustachio, A. J., Hardy, R. W. F., *Biochim. Biophys. Acta* (1966) **113**, 626.
(19) Knight, E., Jr., Hardy, R. W. F., *J. Biol. Chem.* (1966) **241**, 2752.
(20) Knight, E., Jr., Hardy, R. W. F., *J. Biol. Chem.* (1967) **242**, 1370.
(21) Benemann, J. R., Yoch, D. C., Valentine, R. C., Arnon, D. I., *Proc. Natl. Acad. Sci. U.S.* (1969) **64**, 1079.
(22) Bulen, W. A., Burns, R. C., LeComte, J. R., *Proc. Natl. Acad. Sci. U.S.* (1965) **53**, 532.
(23) Orme-Johnson, W. H., Beinert, H., *Biochem. Biophys. Res. Commun.* (1969) **36**, 337.
(24) Mayhew, S. G., Foust, G. P., Massey, V., *J. Biol. Chem.* (1969) **244**, 803.
(25) Moustafa, E., Mortenson, L. E., *Nature* (1967) **216**, 1241.
(26) Hardy, R. W. F., Holsten, R. D., Jackson, E. K., Burns, R. C., *Plant Physiol.* (1968) **43**, 1185.
(27) Burns, R. C., *Biochim. Biophys. Acta* (1969) **171**, 253.
(28) Biggins, D. R., Kelly, M., *Biochim. Biophys. Acta* (1970) **205**, 288.
(29) Dilworth, M. J., Subramanian, D., Munson, T. O., Burris, R. H., *Biochim. Biophys. Acta* (1965) **99**, 486.
(30) Hardy, R. W. F., Knight, E., Jr., *Biochim. Biophys. Acta* (1966) **132**, 520.
(31) Silverstein, R., Bulen, W. A., *Natl. Meeting, ACS, 158th, New York, Sept. 1969*, Abstr. 225.
(32) Bulen, W. A., LeComte, J. R., Burns, R. C., Hinkson, J., in "Non-Heme Iron Proteins: Role in Energy Conversion," p. 261, A. San Pietro, Ed., Antioch Press, Yellow Springs, Ohio, 1965.
(33) Hardy, R. W. F., D'Eustachio, A. J., *Biochem. Biophys. Res. Commun.* (1964) **15**, 314.
(34) Bulen, W. A., LeComte, J. R., *Proc. Natl. Acad. Sci. U.S.* (1966) **56**, 979.
(35) Winter, H. C., Burris, R. H., *J. Biol. Chem.* (1968) **243**, 940.
(36) Hadfield, K. L., Bulen, W. A., *Biochemistry* (1969) **8**, 5103.
(37) Jeng, D. Y., Morris, J. A., Mortenson, L. E., *J. Biol. Chem.* (1970) **245**, 2809.
(38) Kelly, M., *Biochim. Biophys. Acta* (1969) **171**, 9.
(39) Dalton, H., Postgate, J. R., *J. Gen. Microbiol.* (1969) **56**, 307.
(40) Daesch, G., Mortenson, L. E., *J. Bacteriol.* (1967) **96**, 346.
(41) Carnahan, J. E., Mortenson, L. E., Mower, H. F., Castle, J. E., *Biochim. Biophys. Acta* (1960) **44**, 520.
(42) Burris, R. H., Winter, H. C., Munson, T. O., Garcia-Rivera, J., in "Non-Heme Iron Proteins: Role in Energy Conversion," p. 315, A. San Pietro, Ed., Antioch Press, Yellow Springs, Ohio, 1965.
(43) Schollhorn, R., Burris, R. H., *Federation Proc.* (1966) **25**, 710.

(44) Hardy, R. W. F., D'Eustachio, A. J., Knight, E., Jr., *in* "1964 Colloq. Biol. Nitrogen Fixation," *Science* (1965) **147**, 310.
(45) Schneider, K. C., Bradbeer, C., Singh, R. N., Wang, L. C., Wilson, P. W., Burris, R. H., *Proc. Natl. Acad. Sci. U.S.* (1960) **46**, 726.
(46) Hardy, R. W. F., Knight, E., Jr., *Biochim. Biophys. Acta* (1967) **139**, 69.
(47) Strandberg, G. W., Wilson, P. W., *Proc. Natl. Acad. Sci. U.S.* (1967) **58**, 1404.
(48) Koch, B., Evans, H. J., Russell, S. A., *Proc. Natl. Acad. Sci. U.S.* (1967) **58**, 1343.
(49) Bergersen, F. J., Turner, G. L., *J. Gen. Microbiol.* (1968) **53**, 205.
(50) Kelly, M., *Biochem. J.* (1968) **107**, 1.
(51) Schollhorn, R., Burris, R. H., *Proc. Natl. Acad. Sci. U.S.* (1967) **57**, 1317.
(52) Hardy, R. W. F., Knight, E., Jr., *Biochem. Biophys. Res. Commun.* (1966) **23**, 409.
(53) Dilworth, M. J., *Biochim. Biophys. Acta* (1966) **127**, 285.
(54) Schollhorn, R., Burris, R. H., *Proc. Natl. Acad. Sci. U.S.* (1967) **58**, 213.
(55) Schrauzer, G. N., ADVAN. CHEM. SER. (1971) **100**, 1.
(56) Tripathy, P. B., Roundhill, D. M., *J. Am. Chem. Soc.* (1970) **92**, 3825.
(57) Koch, B., Evans, H. J., *Plant Physiol.* (1966) **41**, 1748.
(58) Burns, R. C., Hardy, R. W. F., *in* "Photosynthesis and Nitrogen Fixation," A. San Pietro, Ed., *in* "Methods in Enzymology," S. P. Colowick and N. O. Kaplan, Eds., Academic, New York, 1970, in press.
(59) Kelly, M., *Biochim. Biophys. Acta* (1969) **191**, 527.
(60) Kelly, M., Postgate, J. R., Richards, R., *Biochem. J.* (1967) **102**, 1c.
(61) Hardy, R. W. F., Jackson, E. K., *Federation Proc.* (1967) **26**, 725.
(62) Hardy, R. W. F., Parshall, G. W., *Natl. Meeting, ACS, 158th, New York, Sept. 1969*, Abstr. 226.
(63) Burns, R. C., Bulen, W. A., *Biochim. Biophys. Acta* (1965) **105**, 437.
(64) Burns, R. C., *in* "Non-Heme Iron Proteins: Role in Energy Conversion," p. 289, A. San Pietro, Ed., Antioch Press, Yellow Springs, Ohio, 1965.
(65) Hardy, R. W. F., Knight, E., Jr., D'Eustachio, A. J., *Biochem. Biophys. Res. Commun.* (1965) **20**, 539.
(66) Jackson, E. K., Parshall, G. W., Hardy, R. W. F., *J. Biol. Chem.* (1968) **243**, 4952.
(67) Otsuka, S., Nakamura, A., Yoshida, T., *J. Am. Chem. Soc.* (1969) **91**, 7196.
(68) Fuchsman, W. H., Hardy, R. W. F., *Bacteriol. Proc.* (1970) 148.
(69) Turner, G. L., Bergersen, F. J., *Biochem. J.* (1969) **115**, 529.
(70) Kelly, M., *Biochem. J.* (1968) **109**, 322.
(71) Lockshin, A., Burris, R. H., *Biochim. Biophys. Acta* (1965) **111**, 1.
(72) Hwang, J. C., Burris, R. H., *Federation Proc.* (1968) **27**, 639.
(73) Hwang, J. C., Ph.D. Thesis, University of Wisconsin, Madison, 1968.
(74) Burns, R. C., Fuchsman, W. H., Hardy, R. W. F., unpublished results.
(75) Oppenheim, J., Fisher, R. J., Wilson, P. W., Marcus, L., *J. Bacteriol.* (1970) **101**, 292.
(76) Kelly, M., *Intern. Congr. Microbiol., 9th, Moscow, 1966*, 277.
(77) Biggins, D. R., Postgate, J. R., *J. Gen. Microbiol.* (1969) **56**, 181.
(78) Witz, D. F., Detroy, R. W., Wilson, P. W., *Arch. Microbiol.* (1967) **55**, 369.
(79) Silver, W. S., *in* "1964 Colloq. Biol. Nitrogen Fixation," *Science* (1965) **147**, 310.
(80) Mahl, M. C., Wilson, P. W., *Can. J. Microbiol.* (1968) **14**, 33.
(81) Burns, R. C., Bulen, W. A., *Arch. Biochem. Biophys.* (1966) **113**, 461.
(82) Munson, T. O., Burris, R. H., *J. Bacteriol.* (1969) **97**, 1093.
(83) Winter, H. C., Arnon, D. I., *Biochim. Biophys. Acta* (1970) **197**, 170.

(84) Koch, B., Evans, H. J., Russell, S. A., *Plant Physiol.* (1967) **42,** 466.
(85) Bergersen, F. J., *Biochim. Biophys. Acta* (1966) **130,** 304.
(86) Smith, R. V., Evans, M. C. W., *Nature* (1970) **225,** 1254.
(87) Sloger, C., Silver, W. S., in "Non-Heme Iron Proteins: Role in Energy Conversion," p. 299, A. San Pietro, Ed., Antioch Press, Yellow Springs, Ohio, 1965.
(88) Mortenson, L. E., Morris, J. A., Jeng, D. Y., *Biochim. Biophys. Acta* (1967) **141,** 516.
(89) Klucas, R. V., Koch, B., Evans, H. J., Russell, S. A., *Federation Proc.* (1968) **27,** 593.
(90) Detroy, R. W., Witz, D. F., Parejko, R. A., Wilson, P. W., *Proc. Natl. Acad. Sci. U.S.* (1968) **61,** 537.
(91) Kelly, M., Klucas, R. V., Burris, R. H., *Biochem. J.* (1967) **105,** 3c.
(92) Kelly, M., *Biochim. Biophys. Acta* (1969) **171,** 9.
(93) Hardy, R. W. F., Burns, R. C., Hebert, R. R., Holsten, R. D., Jackson, E. K., *Intern. Biol. Program Meeting on Nitrogen Fixation, Wageningen, The Netherlands, Sept. 1970; Plant and Soil,* in press.
(94) Vandecasteele, J. P., Burris, R. H., *J. Bacteriol.* (1970) **101,** 794.
(95) Jeng, D., Mortenson, L. E., *Natl. Meeting, ACS, 158th, New York, Sept. 1969,* Abstr. 227.
(96) Dua, R. D., Burris, R. H., *Proc. Natl. Acad. Sci. U.S.* (1963) **50,** 169.
(97) Moustafa, E., Mortenson, L. E., *Anal. Biochem.* (1969) **24,** 226.
(98) Burns, R. C., Holsten, R. D., Hardy, R. W. F., *Biochem. Biophys. Res. Commun.* (1970) **39,** 90.
(99) Burns, R. C., Fry, K. T., Weiher, J. F., Hardy, R. W. F., *J. Biol. Chem.,* in preparation.
(100) Dalton, H., Mortenson, L. E., *Bacteriol. Proc.* (1970) 148.
(101) Moustafa, E., *Biochim. Biophys. Acta* (1970) **206,** 178.
(102) Jeng, D. Y., Devanathan, T., Moustafa, E., Mortenson, L. E., *Bacteriol. Proc.* (1969) 119.
(103) Gosser, L. W., Tolman, C. A., *Inorg. Chem.* (1970) **9,** 2350.
(104) Gvozdev, R. I., Yakovlev, V. A., Linde, Y. R., Vorob'ev, L. V., Alfimova, E. Ya., *Izv. Akad. Nauk SSSR, Ser. Biol.* (1969) (No. 2), 215.
(105) Bui, P. T., Mortenson, L. E., *Proc. Natl. Acad. Sci. U.S.* (1968) **61,** 1021.
(106) Nicholas, D. J. D., Wilson, P. W., Heinen, W., Palmer, G., Beinert, H., *Nature* (1962) **196,** 433.
(107) Novikov, G. V., Syrtsova, L. A., Likhtenshtein, G. I., Trukhtanov, V. A., Rachek, V. F., Gol'danskii, V. I., *Dokl. Akad. Nauk SSSR* (1969) **181,** 1170.
(108) Moustafa, E., Mortenson, L. E., *Biochim. Biophys. Acta* (1969) **172,** 106.
(109) Parshall, G. W., *J. Am. Chem. Soc.* (1967) **89,** 1822.
(110) Chatt, J., Dilworth, J. R., Richards, R. L., Sanders, J. R., *Nature* (1969) **224,** 1201.
(111) Stewart, W. D. P., Fitzgerald, G. P., Burris, R. H., *Proc. Natl. Acad. Sci. U.S.* (1967) **58,** 2071.
(112) Holsten, R. D., Hebert, R. R., Burns, R. C., Hardy, R. W. F., *Bacteriol. Proc.* (1970) 149.
(113) Hardy, R. W. F., Holsten, R. D., Jackson, E. K., Burns, R. C., *Agron. Abstr.* (1969) 95.

RECEIVED June 26, 1970.

12

Structure–Function Relationships in Cytochrome *c* Oxidase and Other Hemeproteins

WINSLOW S. CAUGHEY

Arizona State University, Tempe, Ariz. 85281

> *Changes in porphyrin structure, axial ligand, oxidation state, and medium result in significant effects upon a number of independently observable physical parameters which may serve to establish in much greater detail those aspects of structure which determine hemeprotein function. Infrared spectroscopy has become a useful probe of ligand binding in hemoglobins, myoglobins, and cytochrome c oxidase. The highly characteristic properties (e.g., NMR, EPR, IR, and electronic spectra) of μ-oxohemin dimers made possible recognition that hemin a of the oxidase can form a μ-oxo-dimer despite the presence of the bulky 1-hydroxy-2(trans, trans-farnesyl)ethyl group—a group which can assume a conformation in which the external double bonds could couple with iron porphyrin in electron-transfer or phosphorylation functions.*

Our recent studies on the chemistry of cytochrome oxidase function are discussed after briefly considering the value of certain physical observables for the study of hemeproteins.

The porphyrin proteins and enzymes have an intriguing bioinorganic chemistry, not only because of their critical biological roles but also owing to the challenge of being able to correlate a large number of independent observable physical parameters with those aspects of structure which determine biological functions. It is particularly significant that these physical observables frequently can be applied to intact tissues as well as to isolated protein and also that pure hemes and hemins can be used in the refinement of interpretations of the effect of structure upon these parameters. It is frequently most important that the native natural hemes be used if such refinements are to be significantly relevant to the *in vivo* environment. This point will become particularly clear when we consider unique aspects of heme A structure.

The significant variables found among hemeproteins obviously include porphyrin structure, axial ligand, oxidation state, and medium provided by protein and/or solvent at the heme site (1). Changes in axial ligand affect the strength of binding interactions between iron and porphyrin nitrogens; conversely, changes in groups at the periphery of the porphyrin ring alter in turn the basicity of porphyrin nitrogens, the bonding between porphyrin nitrogens and iron, and the bonding between iron and the axial ligands. Such cis and trans effects have been well documented for iron(II) porphyrins and to a lesser extent for hemeproteins (1, 2, 3, 4). Recently, more detailed interpretations of such effects in iron(III) compounds have been obtained (2).

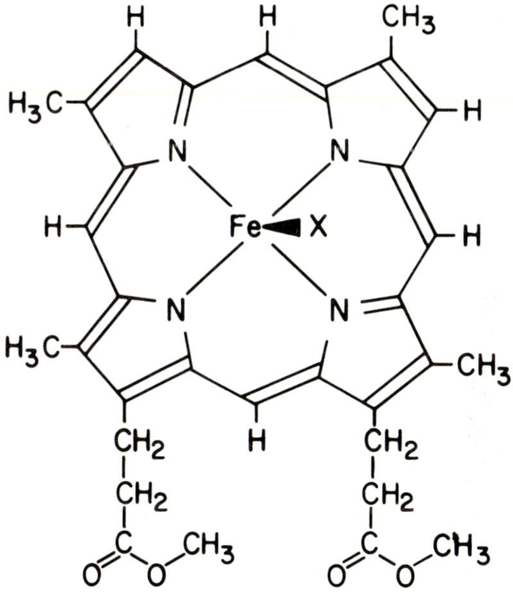

Figure 1. Deuteroporphyrin IX dimethyl ester iron(III) (deuterohemin dimethyl ester)

The magnitude of changes in several physical properties as the axial ligand is varied has been shown for a series of deuterohemins (Figure 1). "X" represents the axial ligand varied in these high-spin iron(III) compounds. Rather marked effects of ligands upon the near infrared band, the quadrupole splitting of Mossbauer spectra, and zero-field splitting measured directly from far infrared spectra are illustrated by data in

Table I. Spectroscopic Properties of Deuteroporphyrin IX Dimethyl Ester Iron (III) Compounds

Ligand	NIR Band max. ($CHCl_3$), mµ	ΔE, 298°K, Mm/Sec	2_D, 4.2°K, Cm^{-1}
F^-	800	0.59	11.1^-
Cl^-	910	0.91	17.9
Br^-	934	1.12	23.6
I^-	958	1.30	33.0
CH_3O^-	761	0.54	9.3
ϕO^-	769		10.9
CH_3COO^-	836	0.6	13.8
N_3^-	868	0.7	14.8

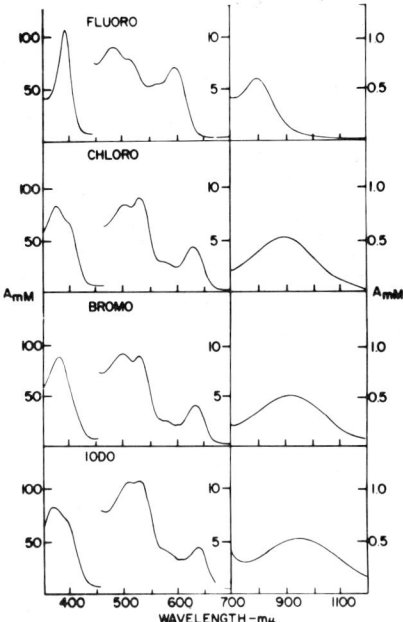

Figure 2. Absorption spectra for deuteroporphyrin IX dimethyl ester iron-(III) derivatives in chloroform; ligands correspond to "X" of Figure 1

Table I (2, 5, 6, 7). Figure 2 further illustrates changes in electronic spectra for chloroform solutions. In benzene, electronic spectra, though slightly different, are generally similar (Figure 3). Such differences caused by solvent effects may not be unlike subtle differences found between certain hemeproteins. The shifts in electronic spectra attributed to changes in axial ligand follow the same ligand order as that for the other physical

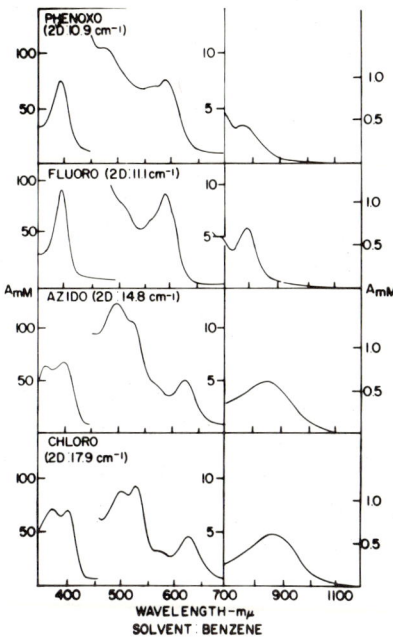

Figure 3. Absorption spectra for deuteroporphyrin IX dimethyl ester iron-(III) derivatives in benzene. 2D values refer to zero-field splitting (7). Ligands correspond to "X" of Figure 1.

properties including ESR spectra which show significant differences in the width of the $g = 6$ band and NMR spectra where magnitudes of paramagnetic shifts vary (5). Proton NMR spectra for azido and phenoxo derivatives in $CDCl_3$ are shown in Figure 4. Particularly noteworthy are the large paramagnetic shifts not only for porphyrin protons but also for protons on the phenoxo ligand. Assignments are not difficult to make with such hemins and provide a basis for the application of NMR to the study of paramagnetic hemeproteins. However, assignments in the case of the proteins are more difficult; separation of contact, pseudo-contact, and ring current field effects represent a major problem. The magnitude of ring current field effects is illustrated by the influence of concentration upon the pattern of ring methyl and ester methyl resonances for the diamagnetic 2,4-dipropionyldeuteroporphyrin IX dimethyl ester in $CDCl_3$ (Figure 5). As the concentration is increased, an equilibrium between monomeric and dimeric species is shifted in favor of more dimer (8). Plots of chemical shift *vs.* concentration (Figure 6) reveal the 5-methyl group resonance as less sensitive to dimer formation than are 1, 3, and 8 methyls—a reflection of the manner in which the dimer is formed.

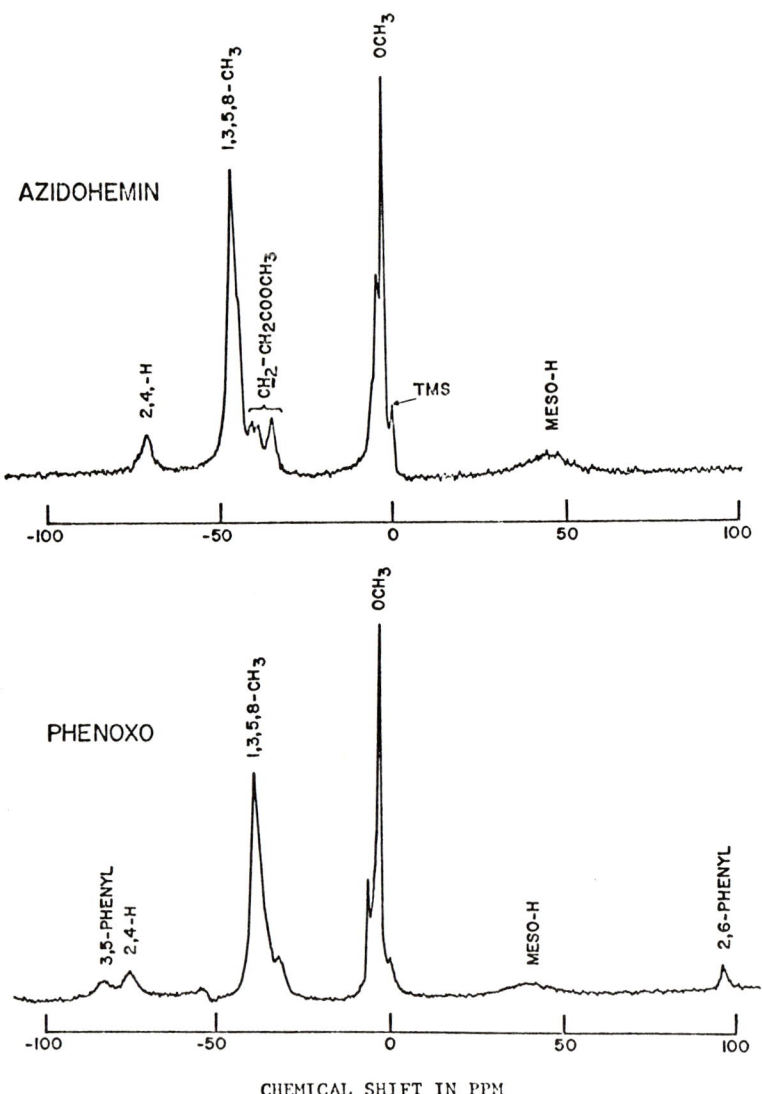

Figure 4. Proton magnetic resonance spectra at 100 MHz of azido (upper) and phenoxo (lower) derivatives of deuteroporphyrin IX dimethyl ester iron(III) in $CDCl_3$ at 35°C

Shifts in frequency can occur as a result of intermolecular ring current field effects present in the dimer but not in the monomer (8, 9, 10, 11).

The other spectroscopic techniques mentioned above have also been applied to hemeproteins. The far infrared determination of zero field splitting has been extended to myoglobin fluoride (11.9 cm^{-1}) and hemoglobin fluoride (12.5 cm⁻¹) and compared with protohemin fluoride (10.0 cm⁻¹) and deuterohemin fluoride (11.1 cm⁻¹) (*12*). Also, a nonzero E value (E/D ca. 0.085) was found for protohemin azide in support of a contribution of the azide ligand to rhombic distortion in the absence of either the protein or a *trans*-histidine, as is the case for myoglobin and hemoglobin azides.

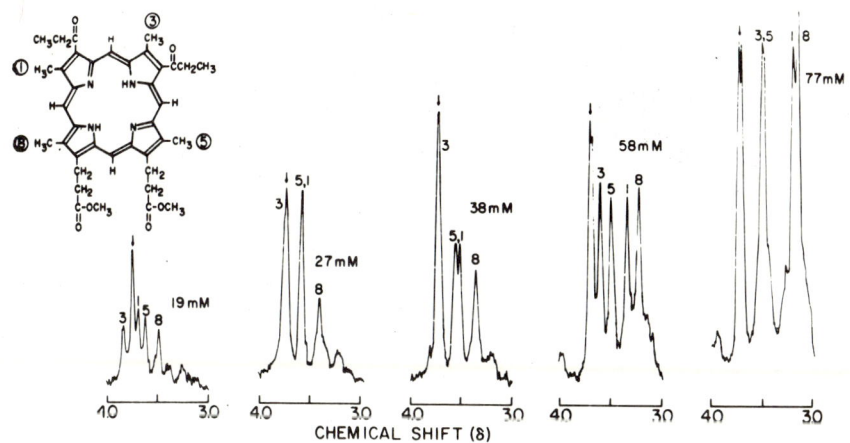

Figure 5. Proton magnetic resonance spectra at 60 MHz of 2,4-dipropionyl-deuteroporphyrin IX dimethyl ester at different concentrations in $CDCl_3$ at 35°C. Arrows indicate ester methyls.

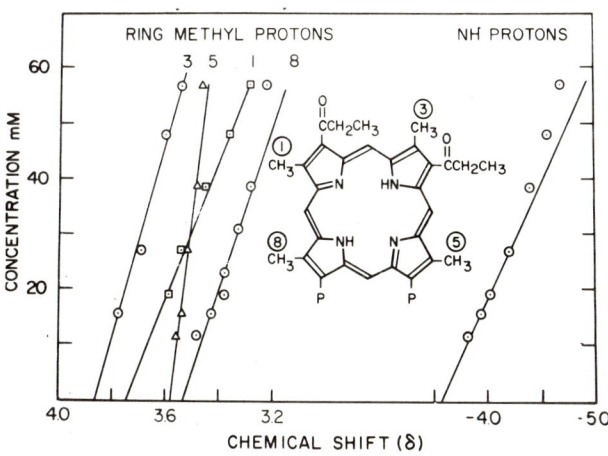

Figure 6. Plots of chemical shifts vs. concentration for ring methyl and NH protons from 60 MHz proton magnetic resonance spectra of 2,4-dipropionyldeuteroporphyrin IX dimethyl ester in $CDCl_3$ at 35°C

While it is clear that these approaches along with x-ray structure determinations represent highly promising probes of hemeprotein structure, one can nevertheless be severely limited in his interpretation of data in terms necessary to explain enzyme function. The identity of the axial ligands bound to metal in a given hemeprotein under a given set of conditions is frequently unclear. Furthermore, even if ligands are known, one cannot necessarily assume knowledge of the stereochemistry, strength of binding, and protein or solvent environment of the ligands—each a factor of importance to the properties exhibited by the hemeprotein. If the ligand contains protons, NMR spectra can be useful. More recently, we have explored the use of infrared techniques in this regard.

Infrared spectra of aqueous solutions of hemoglobins and myoglobins exhibit a "window" of relatively low absorption from about 1750 to 2800 cm^{-1}. Bands for carbon monoxide, azide, and cyanide ligands bound to these hemeproteins have been studied by an infrared difference technique developed by Alben and McCoy in our laboratory. The CO band was readily observed bound to hemoglobin within red blood cells (13). The azide derivatives for each protein exhibited two bands (Figure 7) assigned to low- and high-spin forms (14). With myoglobin, two CO bands are observed (Figure 8) (15), whereas only one CO band was found for normal hemoglobins (13). Thus, with both CO and N_3 derivatives, myoglobin shows a greater tendency to bind ligands in two

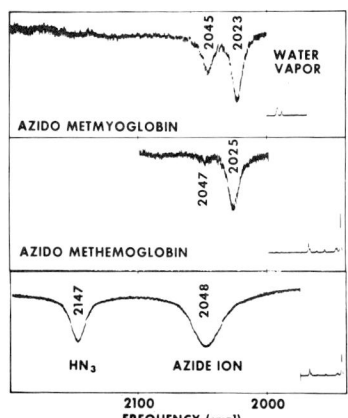

Figure 7. Infrared difference spectra. Top: of azidometmyoglobin vs. metmyoglobin (0.02M). Middle: of azidomethemoglobin vs. bovine plasma albumin (0.01M). Bottom: of sodium azide in 0.05M citrate buffer (pH 3) vs. buffer.

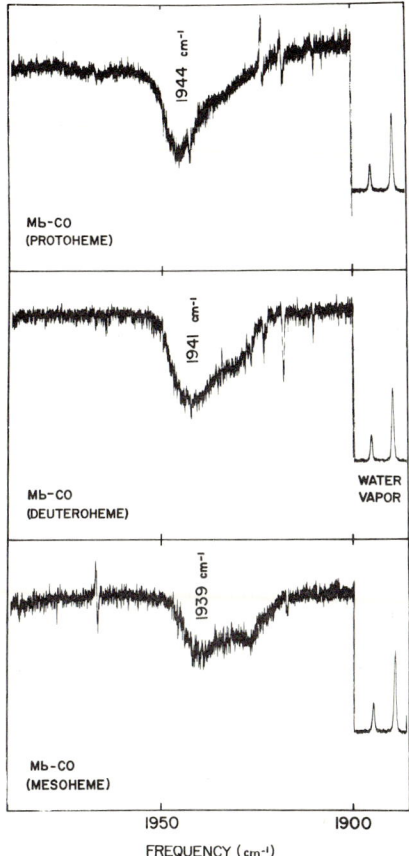

Figure 8. Infrared difference spectra of carbonyl sperm whale myoglobins reconstituted from different hemes vs. metmyoglobin in phosphate buffer, pH 6.4. Top: protoheme. Middle: deuteroheme. Bottom: mesoheme.

ways than does hemoglobin. We conclude from infrared data that both CO and N_3 form significantly bent Fe–ligand bonds in hemoglobins and myoglobins (Figure 9) (*16*). Bending appears greater in the case of myoglobins since ν_{CO} is <1944 cm^{-1} for COMbs compared with 1951 cm^{-1} for COHbA; a lower frequency is expected to accompany a greater deviation from linearity. Lack of sensitivity of ligand frequencies to pH changes in normal proteins (*14, 15*) as well as effects of amino acid substitutions for the distal histidine in abnormal hemoglobins (*17*) suggest a bonding interaction between the distal histidine and ligand—not hydrogen bonding as suggested in an x-ray study of N_3Mb but rather an

n–σ donor–acceptor interaction with histidine as donor and the ligand atom bound to iron as acceptor. Thus, the bond labeled (?) in Figure 9 represents a bond between a partially positive carbon of CO and a partially negative nitrogen of the distal histidine. Although infrared bands for bound O_2 are not available nor have high-resolution x-ray data been obtained for O_2Hb or O_2Mb, undoubtedly O_2 occupies essentially the same "pocket" as does CO and experiences similar bonding interactions with Fe(II) and distal histidine. A linear Fe–O_2 bond is excluded on steric and theoretical grounds. However, a bent FeO_2 bond is entirely consistent with all available data. And the greater affinity of O_2 for myoglobin than for hemoglobin is in accord with greater bending and a "tighter pocket" in the case of oxymyoglobin. On the other hand, available x-ray and infrared data do not demand bending in the cyanides but rather are compatible with the presence of linear FeCN bonds in the cyano myoglobins and hemoglobins (*14*).

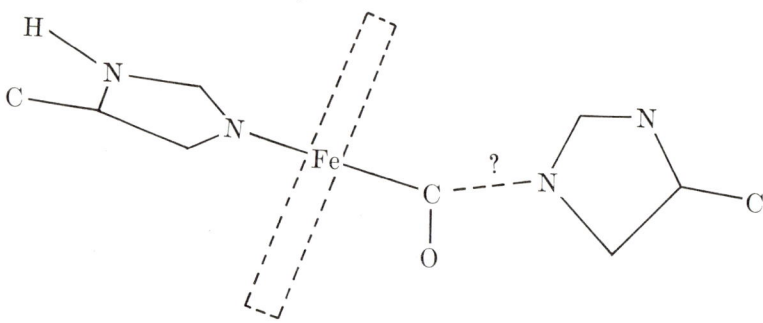

Figure 9. Schematic representation of CO binding to myoglobins and hemoglobins

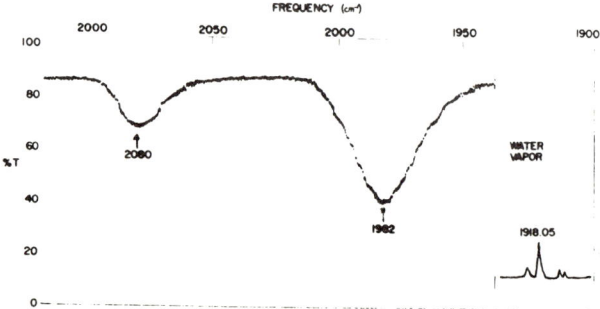

Figure 10. Infrared spectrum of carbonyl heme A (10mM) in bromoform with 0.13M pyridine

Bayne, McCoy, and I have recently obtained evidence that this technique can be extended to carbonyl cytochrome c oxidase wherein one CO ligand is bound per protein molecule of weight in excess of 200,000. Here, the CO-to-protein ratio is an order of magnitude smaller than in hemoglobins and myoglobins. One band, apparently a CO stretch, was found with a frequency and band width consistent with a terminal CO ligand bound to heme A iron within a nonpolar environment well isolated from external solvent (18).

The reactions of cytochrome c oxidase, the site of oxygen utilization for cellular energy production *via* oxidative phosphorylation, have received much study but the nature of the oxygen binding site *per se* has remained unclear. Two hemes (heme A) and two copper atoms are considered to be at the active site. Oxygen binding at copper, at heme iron, or at more than one metal at the same time are possibilities that have been mentioned.

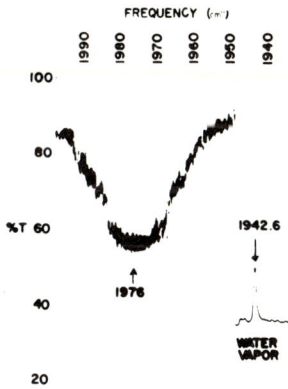

Figure 11. Infrared spectrum of carbonyl heme A (4mM) in 1% Tween 80–0.1M potassium phosphate pH 7.4

The spectrum for bovine heart carbonyl heme A in pyridine–bromoform exhibited a band at 1982 cm^{-1}; a second as yet unexplained weaker band was found at 2080 cm^{-1} (Figure 10). In Tween 80–phosphate, the buffer used for the oxidase, a broad band at 1976 cm^{-1} was obtained for CO heme A (Figure 11). A difference spectrum for CO oxidase *vs.* oxidized oxidase in 1% Tween 80–0.1M potassium phosphate buffer pH 7.4 at a concentration 0.1mM in heme A was obtained (Figure 12). Calcium fluoride cells, path length of 0.025 mm, maintained at 10° ± 2°C were

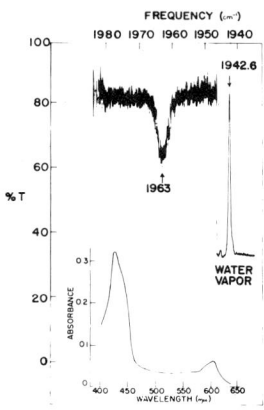

Figure 12. Infrared difference spectrum of carbonyl cytochrome c oxidase vs. oxidized oxidase. Water vapor band at right used for wavelength calibration. Insert at bottom: Soret and visible spectra of the CO derivative in the CaF_2 cell used to obtain the infrared spectrum.

used. The oxidase was isolated from bovine heart by the method of Yonetani (19).

The single sharp band at 1963 cm^{-1} for the oxidase is near the frequency found for CO heme A and thus is consistent with CO binding to the oxidase at heme iron as a terminal CO ligand but is inconsistent with CO serving as a bridging ligand (*e.g.*, where carbon is bound to two metal atoms) in which case a significantly lower frequency is expected. The extremely narrow half-band width of *ca.* 3 cm^{-1} shows the environment about the CO ligand in the oxidase to be quite different from that of the heme in solution where a half-band width greater than 20 cm^{-1} was found. Thus the CO ligands, though terminal, are not exposed to the external solvent environment but rather experience a stable environment in terms of solvation interactions. The narrowness of the band is consistent with, but does not absolutely require, a very nonpolar ligand environment—even less polar than the environment of CO when bound in myoglobins and hemoglobins. Presumably, O_2 and CO bind at the same sites. The infrared evidence suggests that the O_2 binding site in the oxidase where oxygen is reduced is even less polar than the site in hemoglobin or myoglobin where oxygen is bound reversibly. These infrared data thus provide still further evidence that the frequently stated reason for the reversible oxygen binding to hemoglobin and myoglobin, namely, the nonpolar environment of the heme site, is inadequate.

In view of the considerable discussion of nitrogen complexes in earlier papers, I should mention in passing that infrared spectra have also been most useful in our studies of heme–nitrogen complexes in continuation of a recent observation that azido protohemin may be converted to an Fe–N_2 complex in pyridine solution (*14*).

Another aspect of cytochrome *c* oxidase structure we have considered is the possibility of FeOFe bridging between the heme A components of the fully oxidized oxidase. This possibility was suggested by our discovery of μ-oxo hemin dimers as the products from heme autoxidation reactions (*20*) and was further supported by properties of the oxidase, notably the evidence for magnetic coupling between paramagnetic centers of the oxidized oxidase obtained by Beinert and others (*21*). On the other hand, we recognized that the bulkiness of certain substituents on the porphyrin ring of heme A might sterically preclude such dimer formation.

The nature of the substituents on heme A was recently clarified by structural studies by George Smythe and others in our laboratory (*22*). Although it is not appropriate to discuss here the relevant organic chemistry which provide the requisite support for the structure (Figure 13), the elucidation of the structure did require due consideration to the inorganic chemistry involved. A prominent feature of the structure is the C_{17} side chain—a 1-hydroxy-2(*trans, trans* farnesyl) ethyl group. "X" of Figure 13 may be represented as "H" for, although preparations have not been obtained for which an OH group was measured directly

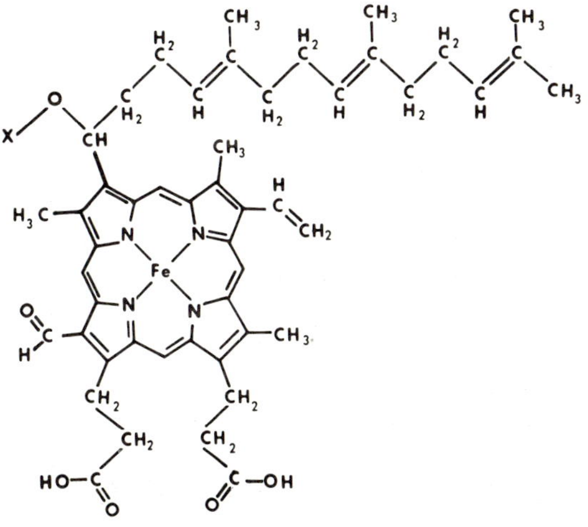

Figure 13. Heme A

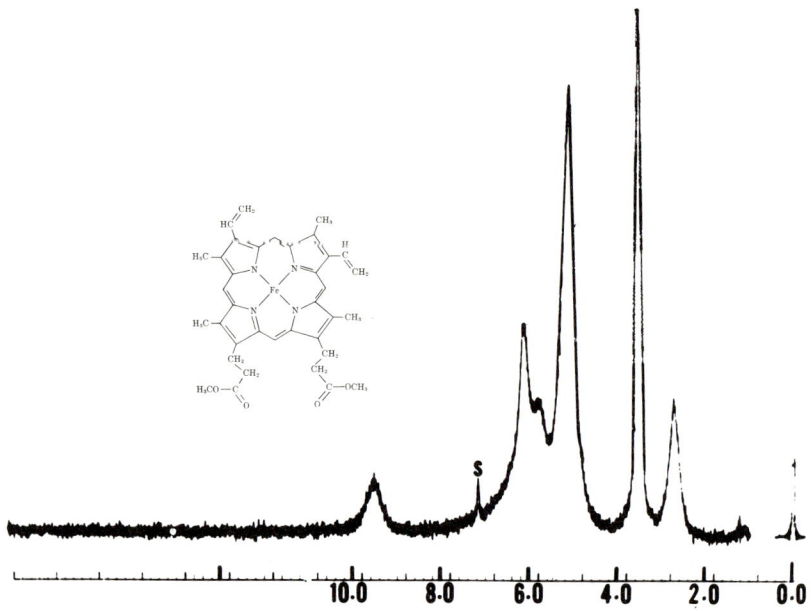

Figure 14. Proton magnetic resonance spectrum at 100 MHz of μ-oxo-bis(protoporphyrin IX dimethyl ester iron(III)) in $CDCl_3$ at 35°C

(*e.g.*, by IR or NMR spectra), much indirect evidence suggests a free OH group is readily obtained, at least as an intermediate. In the native environment we suspect the OH is involved in interactions with acid labile groups.

To evaluate the effect of FeOFe bridging on properties of hemins, we prepared variously substituted μ-oxo hemin dimers related in structure to protohemin. The single oxygen bridge endowed these compounds with highly characteristic properties of great value for identification of such bridging in proteins as well as in iron porphyrins.

The NMR spectra are unusual, as shown in the spectrum for μ-oxo-bis(protohemin dimethyl ester) (Figure 14). Although the chemical shifts fall within the usual range for diamagnetic metal porphyrins, most resonances are much broader and shifted somewhat from the case of low-spin iron(II) porphyrins. In both low-spin (23) and high-spin (5) iron(III) porphyrins, many protons experience marked paramagnetic shifts to high and low fields well outside the usual diamagnetic region. That sufficiently strong coupling between iron(III) atoms occurs to make this compound essentially diamagnetic is shown by the absence of changes in NMR spectrum as temperature is varied (paramagnetic shifts exhibit temperature dependence), by the absence of an EPR spectrum, and by determination of magnetic susceptibilities over the range 4° to 77°K by Moss of IBM-Watson Laboratories. These compounds also exhibit characteristic infrared, electronic, and Mossbauer spectra.

The NMR spectra, which differ markedly from spectra for low- and high-spin iron(III) and low-spin iron(II) porphyrins in chemical shifts and in line widths, and infrared bands between 800 and 900 cm^{-1} have been particularly useful in establishing that FeOFe compounds can in fact be prepared from hemin A derivatives. For example, the diethyl ester, monoethyl ether derivative of heme A (Figure 15) when reduced

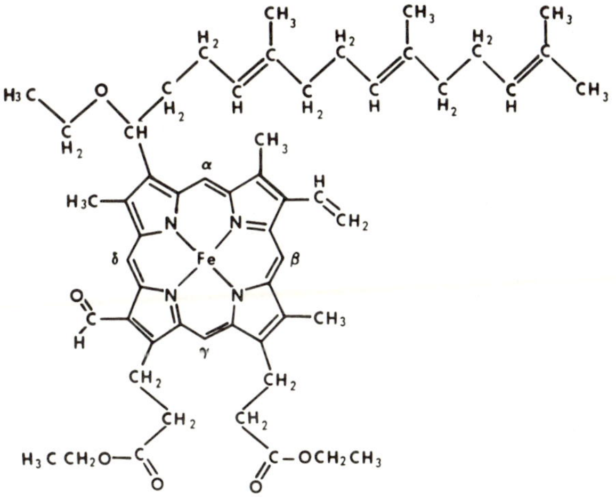

Figure 15. *Heme A monoethyl ether, diethyl ester*

in pyridine d_5 exhibits well-resolved sharp bands typical of a low-spin iron(II) spectrum (Figure 16), in sharp contrast to the spectrum of the μ-oxo dimer in CDCl$_3$ (Figure 17) wherein all protons except for those associated with the end of the C$_{17}$ group are much broader than in the iron(II) compound. The close correspondence of chemical shifts for protons common to both protohemin and hemin A derivatives as FeOFe and dipyridine iron(II) species can be noted in Tables II and III. It is clear from such NMR data as well as infrared and other evidence that we have prepared FeOFe species for a number of hemin A derivatives and thereby have demonstrated that the presence of the bulky C$_{17}$ group does not preclude the formation of an FeOFe bridge in these compounds.

It has also been of interest that a copper–heme A complex—which Smythe and Bayne obtained from bovine heart by shortening our isolation procedure (24)—exhibits similarities in NMR and visible spectra to those for pure FeOFe compounds. Other experiments suggest copper ions may either promote formation of FeOFe linkages or form CuOFe

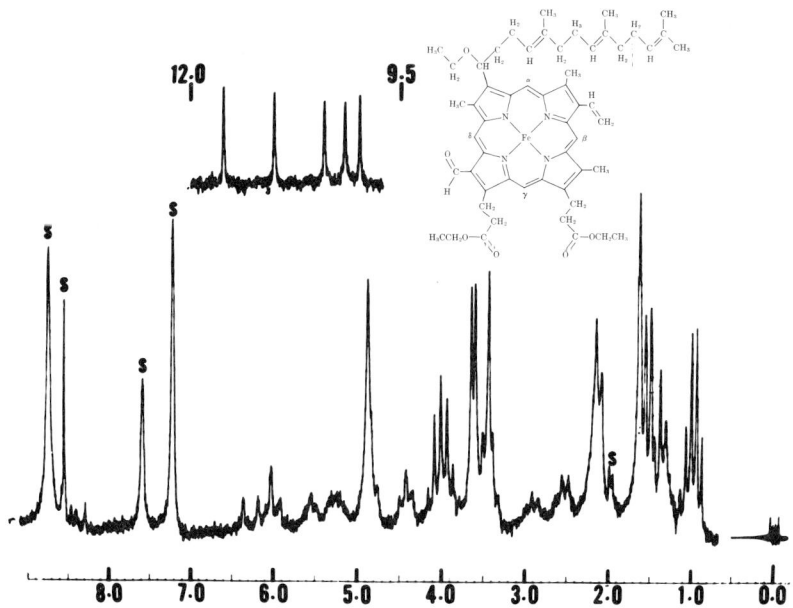

Figure 16. Proton magnetic resonance spectrum at 100 MHz of the low-spin iron(II) species of heme A monoethyl ether, diethyl ester in pyridine-d_5. Solvent resonances are designated by "S."

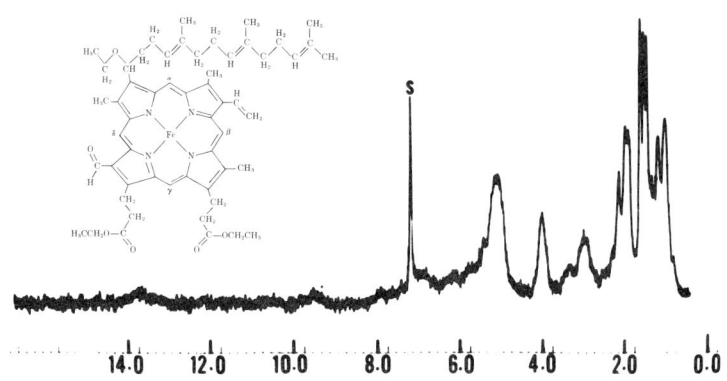

Figure 17. Proton magnetic resonance spectrum at 100 MHz of the μ-oxo dimer of hemin A monoethyl ether, diethyl ester in $CDCl_3$

Table II. Chemical Shifts for FeOFe Dimers

Proton	Chemical Shift[a]	
	Hemin A Derivatives	Protohemin IX
$-C\underline{H}=CH_2$	9.7	9.6
$-CH=C\underline{H}_2$	5.8	5.8
Ring Methyls	5.2	5.2
$\alpha-CH_2$	7.0, 6.2	6.2
$\beta-CH_2$	3.4, 3.0	2.8
Ester $-OCH_3$	3.6	3.6

[a] In ppm from TMS; spectra taken in CDCl$_3$ solution at 35°C.

Table III. Chemical Shift Comparison for Dipyridine Heme Monomers of Hemin A Derivatives and Protohemin IX

Proton	Chemical Shift[a]	
	Hemin A Derivatives	Protohemin IX
$-C\underline{H}=CH_2$	8.4	8.5
$-CH=C\underline{H}_2$	6.2	6.1
Ring Methyls	3.65, 3.61, 3.44	3.74, 3.60
$\alpha-CH_2$	4.7, 4.4	4.46
$\beta-CH_2$	3.44, 2.90	3.40
Ester $-OCH_3$	3.49, 3.39	3.50

[a] In ppm from T—S, spectra taken in pyridine-d_5.

linkages—possibilities of obvious interest in regard to the oxidase. Clearly, the consequences of these and other types of bridging must be examined in an attempt to provide a unique interpretation of the structural arrangements in the intact oxidase.

Now that the structure for heme A is available, we considered why it has such a structure (Figure 13). Nature evidently chooses to modify protoheme at two places to prepare heme A: the 8-methyl of protoheme is converted to a formyl and the 2-vinyl to a 1-hydroxy-2-(*trans,trans*-farnesyl) ethyl group. An explanation for the first transformation may simply be related to the greater electron-withdrawing effect of the formyl shown to be significant in several studies (*1*). Several speculations on the role(s) of the C_{17} group have arisen. Earlier we mentioned the possibility of "X" being other than "H" in the intact oxidase. At least in the heme A as isolated from beef heart, lipids tenaciously accompany the heme through isolation procedures; possibly lipids interact at OX. Also, in the Cu–heme A preparation isolated (*24*), effects of the presence of paramagnetic Cu(II) ion on the NMR spectrum of the porphyrins were greatest for the protons near OX—an observation consistent with a Cu(II) location near OX. The nature of "X" may of course be different in the two hemes present in the oxidase. The fact that two different heme A fractions may be isolated from bovine heart (25) now appears best explained in terms of differences in associated lipid. Other evidence suggests that the double bonds may participate in cytochrome c oxidase function.

Once the nature of the unsaturation in the C_{17} chain was established, we considered if this structural feature had an active role in function or could we say nature has just been too lazy to reduce these double bonds to give a saturated polyisoprenoid group. Possibly related in function are the similarly unsaturated, though frequently much longer, isoprenoid chains of ubiquinones, vitamin K's, and related compounds, substances also implicated in electron-transfer and/or oxidative phosphorylation processes. Thus, the C_{17} group in heme A may be directly related to either electron transfer or coupling of phosphorylation in the oxidase.

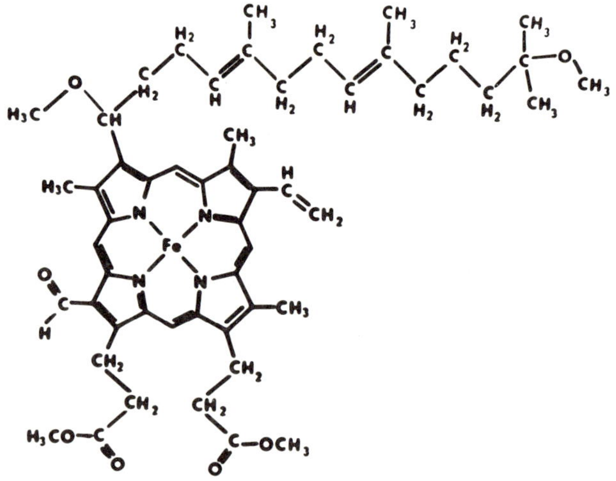

Figure 18. Heme A dimethyl ether dimethyl ester

An obvious question is how could such a group participate in electron-transfer or coupling reactions? A few experimental observations for the terminal (*i.e.*, C_{12}–C_{13}) double bonds may be relevant to this question. This terminal bond undergoes certain addition reactions much more readily than do either of the two inner double bonds. For example, in dilute solutions of sulfuric acid in methanol, a dimethyl ester, dimethyl ether derivative (Figure 18) can be prepared without evidence of other products with methoxy groups at C_5 or C_9. We were surprised to find that as a consequence of this addition reaction at C_{13}, a shift of 3 nm

Figure 19. Model of heme A with the trans,trans-*farnesyl group in an extended conformation*

Figure 20. Model of heme A with the farnesyl ethyl group in a conformation that allows pi-orbital overlap between adjacent double bonds and between the porphyrin and the C_4–C_5 bond

in the alpha band of the hemochromogen spectrum occurred. This shift is the more unexpected because several differences in "X" (*i.e.*, at OX) resulted in little, if any, shift in wavelength in the hemochromogen spectrum.

An examination of models suggests at least two ways in which the unsaturation of the C_{17} chain may be involved in oxidase function. Figure 19 illustrates the case with the *trans, trans*-farnesyl ethyl group fully extended. However, as shown in Figures 20 and 21, it is readily possible

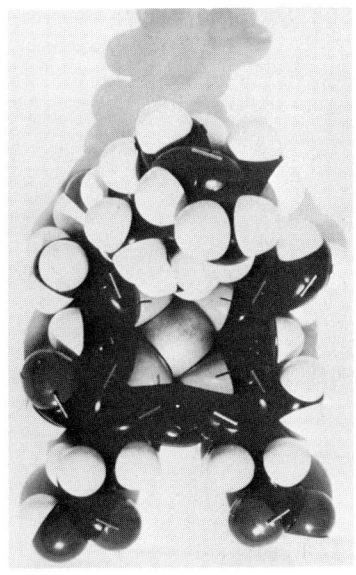

Figure 21. Another view of the conformation in Figure 20

to attain a conformation in which the first double bond (C_4–C_5) lies parallel to the heme plane over the pyrrole moiety containing the 1 and 2 ring positions, with the second (C_8–C_9) and third (C_{12}–C_{13}) double bonds positioned directly above the first to permit pi-orbital overlap between each double bond and between the first double bond and porphyrin. The arrangement is sketched in Figure 22. It is reasoned that in such a conformation, electron transfer between the terminal double bond and porphyrin ring is possible. Indeed, similar conformations may explain an electron transfer role for similarly unsaturated isoprenoid chains, frequently of much longer length, found in compounds such as the ubiquinones. Other possibilities include the use of overlap of the three double bonds as described to transfer electrons to or from a copper ion. Of course, the reactions of the terminal double bond could also involve addition reactions in a manner which in effect couples the terminal double bond with the porphyrin or copper systems.

Models reveal still another way in which the terminal double bond may participate, namely, by reactions near the iron atom. As shown in

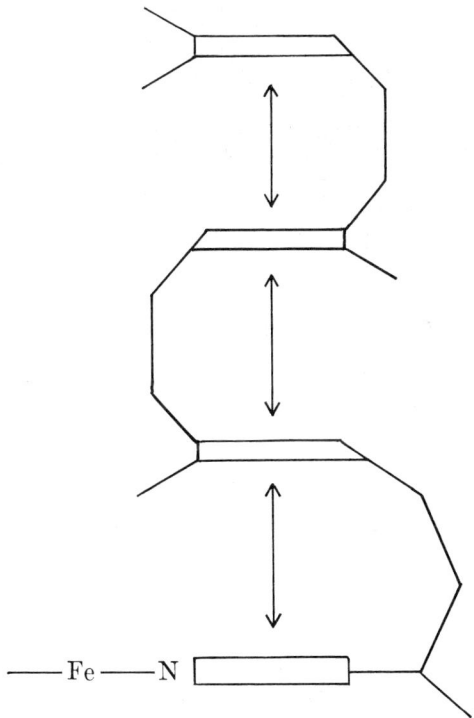

Figure 22. A schematic representation of the conformation illustrated in Figures 20 and 21

Figures 23 and 24, there is no obvious steric restraint in achieving a conformation of the C_{17} chain in which the C_{12}–C_{13} bond is adjacent to the iron and parallel to the porphyrin plane. Clearly, the C_{12}–C_{13} bond could readily take part in reactions with oxygen or other ligands at iron or with iron itself. Such reactions could obviously function in processes of electron transfer or coupling with phosphorylation. Since there are

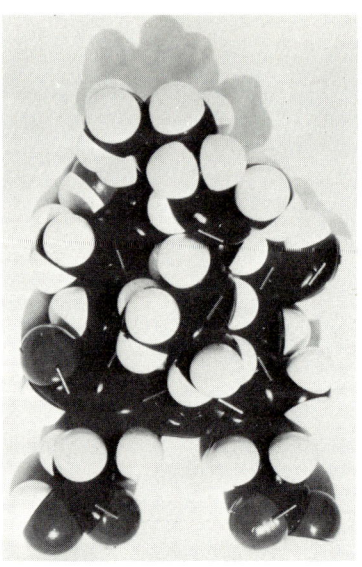

Figure 23. Model of heme A with the farnesyl ethyl group in a conformation where the terminal (C_{12}–C_{13}) double bond is adjacent to the iron atom

Figure 24. Another view of the conformation in Figure 23

two hemes in the oxidase, it is possible that the C_{17} groups do not function in the same way for both hemes.

Though speculative, these suggestions are now amenable to experimental evaluation. I am sufficiently optimistic to believe that truly significant progress will soon be made in the evaluation of structural features and reactions of the very complex enzyme cytochrome c oxidase as well as for other hemeproteins. This progress will result in no small part from the applications of independent physical methods at several levels—the hemin, the protein, and the tissue levels.

Acknowledgment

The support given this work by the United States Public Health Service (Grant #HE 13190) is gratefully acknowledged.

Literature Cited

(1) Caughey, W. S., *Ann. Rev. Biochem.* (1967) **36**, 611.
(2) Caughey, W. S., Eberspaecher, H., Fuchsman, W. H., McCoy, S., Alben, J. O., *Ann. N. Y. Acad. Sci.* (1969) **153**, 722.
(3) Alben, J. O., Caughey, W. S., *Federation Proc.* (1962) **21**, 46.
(4) Caughey, W. S., Alben, J. O., McLees, B. D., *Proc. Intern. Conf. Coordination Chem.,* 7th (1962) 136.
(5) Caughey, W. S., Johnson, L. F., *Chem. Commun.* (1969) 1362.
(6) Moss, T. H., Bearden, A. J., Caughey, W. S., *J. Chem. Phys.* (1969) **51**, 2624.
(7) Richards, P. L., Caughey, W. S., Eberspaecher, H., Feher, G., Malley, M., *J. Chem. Phys.* (1967) **47**, 1187.
(8) Caughey, W. S., York, J. L., Iber, P. K., in "Magnetic Resonance in Biological Systems," A. Ehrenberg, B. G. Malmstrom, and T. Vanngard, Eds., p. 25, Pergamon, Oxford, 1967.
(9) York, J. L., Caughey, W. S., *Abstr. Papers, 143rd Meeting, ACS, Cincinnati* (1963) 31A.
(10) Abraham, R. J., Burbidge, P. A., Jackson, A. H., Kenner, G. W., *Proc. Chem. Soc.* (1963) 134.
(11) Closs, G. L., Katz, J. J., Pennington, F. C., Thomas, M. R., Strain, H. H., *J. Am. Chem. Soc.* (1963) **85**, 3809.
(12) Bracket, G. C., Richards, P. L., Caughey, W. S., Hickman, H. H., *J. Chem. Phys.,* in press.
(13) Alben, J. O., Caughey, W. S., *Biochemistry* (1968) **7**, 175.
(14) McCoy, S., Caughey, W. S., *Biochemistry* (1970) **9**, 2387.
(15) McCoy, S., Caughey, W. S., in "Probes for Membrane Structure and Function," B. Chance, M. Cohn, C. P. Lee, T. Yonetani, Eds., Academic, New York, in press.
(16) Caughey, W. S., *Ann. N. Y. Acad. Sci.,* in press.
(17) Caughey, W. S., Alben, J. O., McCoy, S., Charache, S., Hathaway, P., Boyer, S., *Biochemistry* (1969) **8**, 59.
(18) Caughey, W. S., Bayne, R. A., McCoy, S., *J. Chem. Soc. D* (1970) 950.
(19) Yonetani, T., in "Methods in Enzymology," Vol. X, R. W. Estabrook and M. E. Pullman, Eds., p. 332, Academic, New York, 1967.

(20) Alben, J. O., Fuchsman, W. H., Beaudreau, C. A., Caughey, W. S., *Biochemistry* (1968) **7**, 624.
(21) Van Gelder, B. F., Beinert, H., *Biochim. Biophys. Acta* (1969) **189**, 1.
(22) Smythe, G. A., Caughey, W. S., *J. Chem. Soc. D* (1970) 809.
(23) Wuthrich, K., Shulman, R. G., Wyluda, B. J., Caughey, W. S., *Proc. Natl. Acad. Sci. U.S.* (1969) **62**, 636.
(24) Bayne, R. A., Smythe, G. A., Caughey, W. S., *in* "Probes for Membrane Structure and Function," B. Chance, M. Cohn, C. P. Lee, and T. Yonetani, Eds., Academic, New York, in press.
(25) Caughey, W. S., Davies, J. L., Fuchsman, W. H., McCoy, S., *in* "Structure and Function of Cytochromes," K. Okunuki, M. D. Kamen, and I. Sekuzu, Eds., p. 20, University of Tokyo Press, Tokyo, 1968.

RECEIVED June 26, 1970.

13

Low–Spin Compounds of Heme Proteins

W. E. BLUMBERG and J. PEISACH

Bell Telephone Laboratories, Inc., Murray Hill, N. J. 07974 and the Departments of Pharmacology and Molecular Biology, Albert Einstein College of Medicine, Yeshiva University, Bronx, N. Y. 10461

> *Oxidation of diamagnetic oxyhemoglobin yields the paramagnetic derivative high-spin ferrihemoglobin and five low-spin derivatives which can be studied with electron paramagnetic resonance. For most low-spin heme proteins, only five low-spin z ligand combinations, each with its own range of g values, can occur naturally or through chemical modifications, utilizing ligands endogenous to the heme. Four of these ligand combinations necessarily have an N atom of histidine. The other z ligands possible are ^-OH, histidyl N, methionyl S, and a nitrogenous ligand of as yet undetermined chemical composition. The fifth z ligand combination has a cysteine S as one z ligand and one of various nitrogenous bases as the other. One can separately quantitate each low-spin species in a mixture and assay for all paramagnetic cytochromes in microsomes or mitochondria or for paramagnetic forms of hemoglobin in intact red cells.*

This paper is concerned with recent research on certain compounds which can be made from heme proteins. Of course, all heme proteins are themselves compounds but some of them have very interesting derivatives. The conversion between heme proteins and their various derivatives gives some insight as to what kind of chemistry is going on in these proteins. The technique that we have used to study these compounds is electron paramagnetic resonance (EPR), and thereby hangs a definition of the importance of the ferric heme protein compounds. According to R. J. P. Williams (this symposium), compounds are important if they can be studied with the equipment at hand.

This presentation contains two sections: what is being called bioinorganic chemistry at this symposium as it relates to ferric heme compounds and something of the biology of low-spin ferric heme compounds, indicating why we have been studying them.

Quite a number of the interesting heme proteins occur in the ferrous state, and we cannot study those by EPR. We would not want to give the impression they are not important just because we cannot study them using this technique. However, we can study the ferric ones. Ferric compounds exist in two spin states: high-spin, having a spin of 5/2, and low-spin, having a spin of 1/2. The EPR of each of these is very distinctive when examined at low temperatures on samples which are frozen solutions of porphyrins or heme proteins. Figure 1 is a drawing of typical

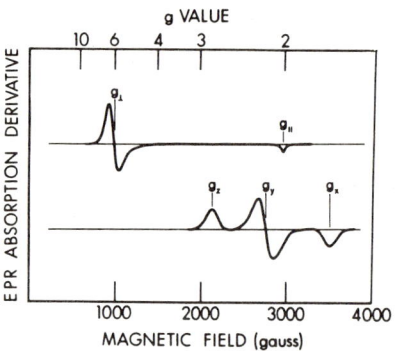

Figure 1. Typical X-band EPR spectra of high-spin (upper) and low-spin (lower) ferric heme compounds as examined in frozen solutions

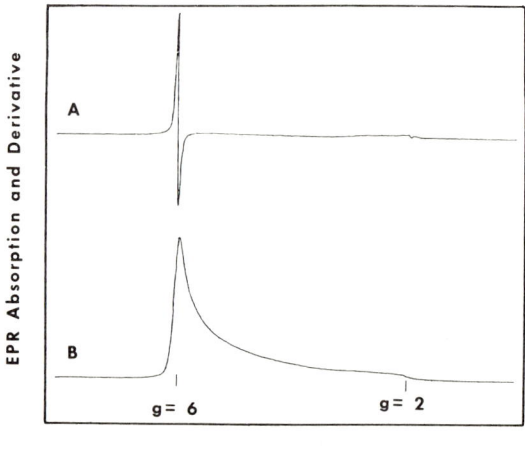

Figure 2. EPR spectra of a solution of hemin chloride in N,N-dimethylformamide

EPR absorption derivative spectra as seen at low temperatures in the X-band apparatus (9000 Mc/sec). The high-spin state is the upper one. There is an absorption extending from g values of about 6 to g values of about 2 (1). The g value is a scale factor which is inverse in the magnetic field, and people usually quote that rather than a magnetic field because it already has the frequency of the spectrometer divided out of it. The spectra are taken in the derivative form, and so this derivative represents an absorption which extends over the range $g = 6–2$. This is an axial spectrum, which one might expect from a molecule such as a porphyrin which seems to have four-fold symmetry if one neglects the difference between the substituent groups on the periphery. The lower curve is for a typical low-spin ferric compound and has three absorption derivative features. With sufficient accuracy for our purposes, the three g values can be read off at the three places indicated (2).

Let us now examine some real EPR spectra taken from porphyrin samples. Hemin chloride dissolved in N,N-dimethylformamide gives the EPR spectra shown in Figure 2. The lower curve (absorption) extends from $g = 6–2$ and the upper curve (absorption derivative) excursion is very large near $g = 6$ but barely discernable at $g = 2$. As far as one can tell from such a spectrum, the system is axial; that is, g_x and g_y are both equal to 6, while g_z is equal to 2. One can easily convert the hemin to a low-spin compound by adding ligands which will replace chloride. One such ligand is mercaptoethanol. By adding mercaptoethanol to this

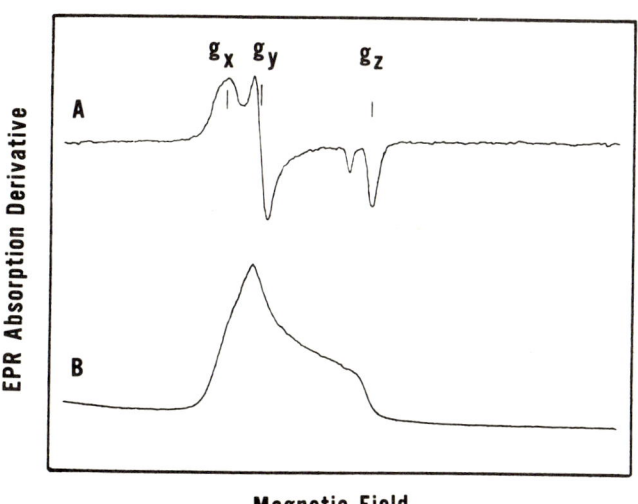

Figure 3. EPR spectra of the same sample of hemin chloride as was used previously (Figure 2) to which had been added mercaptoethanol

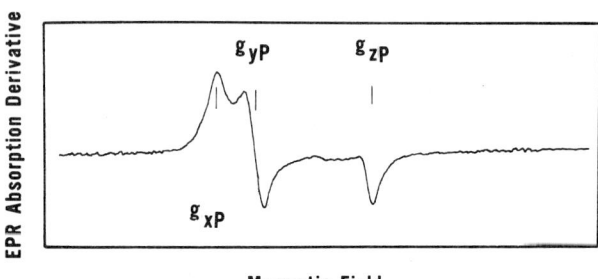

Figure 4. EPR spectrum of a sample of myoglobin to which mercaptoethanol has been added

same sample and freezing it again, we obtain the spectra shown in Figure 3. This is the EPR spectrum (absorption, lower; absorption derivative, upper) of the protoporphyrin IX mercaptoethanol compound. The three g values can be read from it in the manner illustrated in Figure 1. The absorption extends between the two extreme g values, in this case 2.37 and 1.93. The sharp derivative notch (second feature from the right) is owing to the remaining high-spin hemin chloride and is at $g =$ 2. This compound is not biologically very interesting, but an exactly analogous compound can be made simply by adding mercaptoethanol to a heme protein, such as myoglobin (3). Figure 4 shows the EPR spectrum obtained by doing just that, and it is essentially identical. Thus we have made a heme mercaptoethanol compound inside the heme protein itself, and it is just such a series of compounds that we have studied.

Myoglobin is a heme protein which is found in red muscle and which binds oxygen. It is normally in the ferrous state but can readily be converted to the ferric state. It consists of a single polypeptide chain and a single heme.

The studies we are discussing mainly involve hemoglobin. Hemoglobin is a more complicated molecule than is myoglobin although evolutionally they are very closely connected. Hemoglobin consists of four polypeptide chains, each with its own heme. The molecule consists of a pair of like chains (designated alpha) and another pair of like chains (designated beta). To the naked eye, neither the alpha chains nor the beta chains can be distinguished from the chain of myoglobin. Certainly, they differ in quite a number of details but they have the same number of helical and nonhelical regions, and these are arranged in almost the same tertiary structure (4, 5).

First let us look at an interesting experiment which will only have a phenomenological interpretation until we examine a model of this molecule. We start with ferric hemoglobin A, the oxidized form of the normal hemoglobin that probably all of us have, consisting of two alpha chains and two beta chains. It is oxidized to the ferric high-spin form with ferricyanide. It is stable for a long time (6), and its EPR spectrum is shown in Figure 5 (upper curve). If one now adds histidine to this sample, one finds that there is a change to the extent that the histidine is added. The high-spin material decreases, and a low-spin compound is formed (Figure 5, middle curve). This low-spin compound is called a hemichrome, that being a collective name for certain ferric low-spin compounds. An interesting thing happens if instead we start with hemoglobin H (7), which is an abnormal hemoglobin consisting of four beta chains. It is not very stable (8), unlike hemoglobin A, and given sufficient time (around three hours) it will make the same hemichrome without the addition of histidine. That is, when two experiments are run side by side, the oxidation of hemoglobin A produces no further product

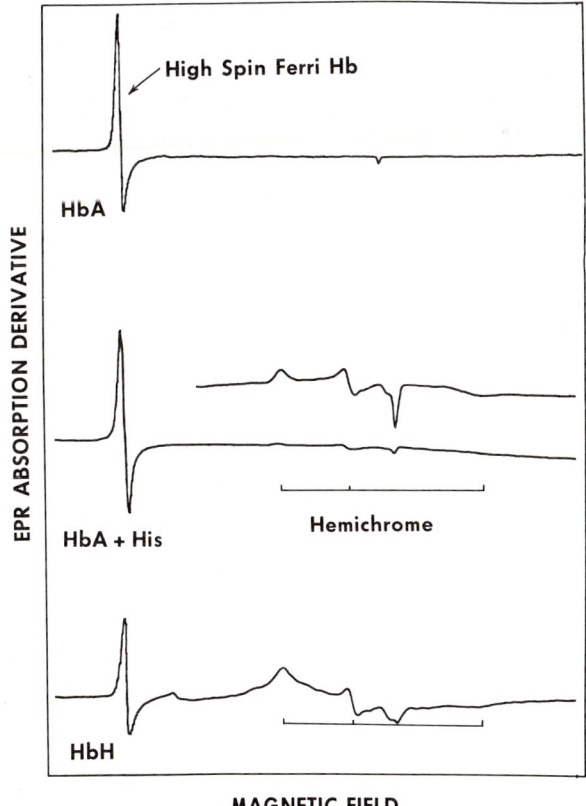

Figure 5. EPR spectra of a sample of high-spin ferric hemoglobin A (upper) and the same sample to which histidine has been added (middle), and hemoglobin H (lower)

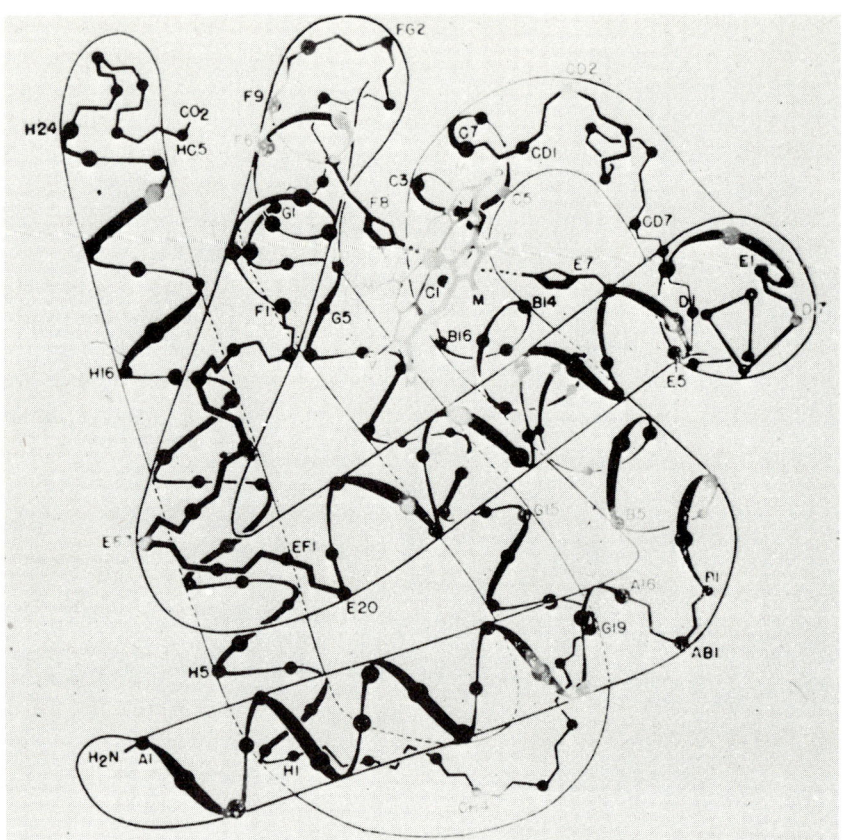

R. E. Dickerson, "The Proteins," Academic Press

Figure 6. α-Carbon diagram of myoglobin molecule obtained from 2-Å analysis

Stretches of α-helix are represented by smooth helix with exaggerated perspective. The nonhelical segments are represented by three-segment zigzag lines between α-carbon atoms. Fainter parallel lines outline high-density region as revealed by 6-Å analysis. Heme group framework is sketched in forced perspective, with side groups identified as follows: M = methyl, V = vinyl, P = propionic acid. Five-membered rings at F8 and E7 represent histidines associated with heme group on the proximal and distal sides, respectively.

while the oxidation of hemoglobin H is followed by the same reaction which takes place when one adds histidine to hemoglobin A.

A look at the model of a chain will reveal what is happening. Figure 6 is the Dickerson model (9) of myoglobin, but one can pretend that it is either of the chains of hemoglobin. The closest nonporphyrin ligand of the iron atom is the nitrogen atom from the imidazole ring which belongs to histidine and is called the proximal ligand. On the other or distal side of the porphyrin, there is a relatively empty space which carries oxygen in these molecules when they are in the ferrous state. Farther away, and too far away to make a bond, there is another imidazole ring with its nitrogen atom directed toward the heme.

In this empty pocket one can add very small molecules, and that is exactly what happens when one adds mercaptoethanol or histidine to the molecule. On standing, the unstable hemoglobin will readjust the tertiary structure so that the endogenous nitrogen atom can now coordinate with the iron atom. Under these conditions, the heme exists as a dihistidine compound (*10*). That produces exactly the same ligand atoms as leaving the tertiary structure alone and adding another histidine in the pocket. As far as the iron is concerned, it is coordinated to the porphyrin and two nitrogen atoms from imidazole. It does not seem to care about the other details, and thus those two compounds appear identical by EPR.

One can make quite a number of other interesting compounds this way, with ligand atoms which are exogenous to the molecule and with ligand atoms which are endogenous to the molecule. For example, oxidized alpha chains from hemoglobin A which have been separated from the beta chains show a high-spin EPR spectrum similar to those we have already seen (*11*). When the pH is raised, there appears a compound which is in pH equilibrium with the high-spin compound. This is just a hydroxide compound of the normal ferric alpha chain. The EPR spectrum of a sample almost completely shifted to this low-spin form is shown in Figure 7. (A) Hydroxy form: oxy alpha chains were oxidized with five molecular proportions of ferricyanide in 0.02 M *tris*-hydrochloride

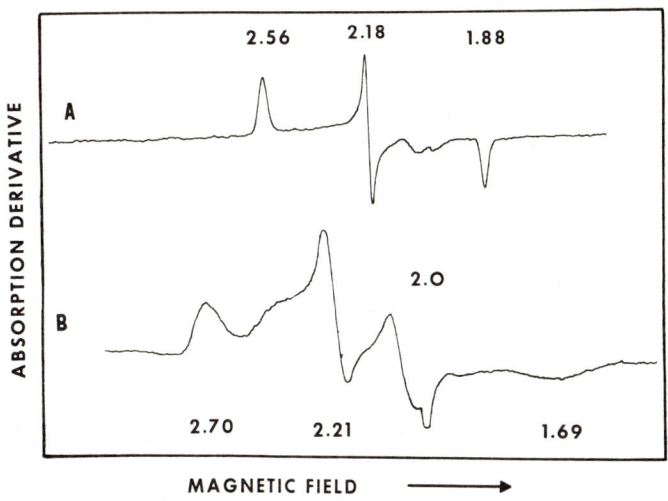

Figure 7. *EPR spectra of low-spin forms of isolated ferric alpha chains of hemoglobin A*

buffer, pH 8.0. Immediately the ferricyanide was removed and the buffer exchanged for 0.15 M *tris*-sulfate buffer, pH 8.7, by passage over a small column of Biogel P-2. (B) Dihistidyl form: the high-spin ferric alpha chains were allowed to stand for one hour in 0.05 M phosphate buffer, pH 5.6. For a short time this is very freely reversible back to the high-spin compound upon changing the pH. However, upon standing, this compound, too, will spontaneously readjust so that it transforms into the same hemichrome with the same g values that we saw before from the beta chains of hemoglobin H (*11*). The same thing is happening as in the beta chains; that is, the tertiary structure is relaxing so the distal histidine can come in and touch the iron, thus producing a dihistidine iron compound. Several other hemichromes can be made under selected conditions (*10*) and will be discussed later.

What can a theoretical chemist do with such a low-spin EPR spectrum in order to elucidate structural information? Let us look at the quantum mechanical Hamiltonians for high-spin and low-spin ferric systems.

$$H_{h.s.} = g\beta H \cdot S + D(S_z^2 - \tfrac{1}{3} S(S+1)) + E(S_x^2 - S_y^2)$$

$$H_{l.s.} = g\beta H \cdot (S + L) + \lambda[L \cdot S + (\Delta/\lambda)Y_2^0 + (V/\lambda)(Y_2^2 + Y_2^{-2})]$$

The two terms involving Y_2^k represent symmetries of the electrostatic crystal field (*12*) and are proportional to $z^2 - r^2/3$ and $x^2 - y^2$, respectively. The terms involving H are Zeeman interaction terms and give no structural information. The term $L \cdot S$ (the spin orbit coupling) in the low-spin Hamiltonian also gives none. It is the remaining terms which indicate the geometry of the structure. The coefficients D and Δ/λ represent the departure of the ligand arrangement from octahedral toward tetragonal; *e.g.*, the proximal and distal ligands becoming inequivalent to the four porphyrin ligands. The coefficients E and V/λ represent the departure of this lowered symmetry from tetragonal toward rhombic. Under the conditions of $E/D = 1/3$ or $V/\Delta = 2/3$, the symmetry has been termed completely rhombic (*13*). These coefficients completely determine the three g values in either spin case and are all the information one can extract from an EPR spectrum of a frozen solution. The method of analysis has been well summarized by Griffith (*14*) and by Weissbluth (*15*).

If we have the analysis of a large number of low-spin ferric compounds, they may be conveniently summarized on a crystal field diagram such as Figure 8. Here we have plotted the tetragonal field (Δ/λ) as abscissa and V/Δ, which we have termed the rhombicity, as ordinate. All of these compounds, with the exception of the ones labelled with

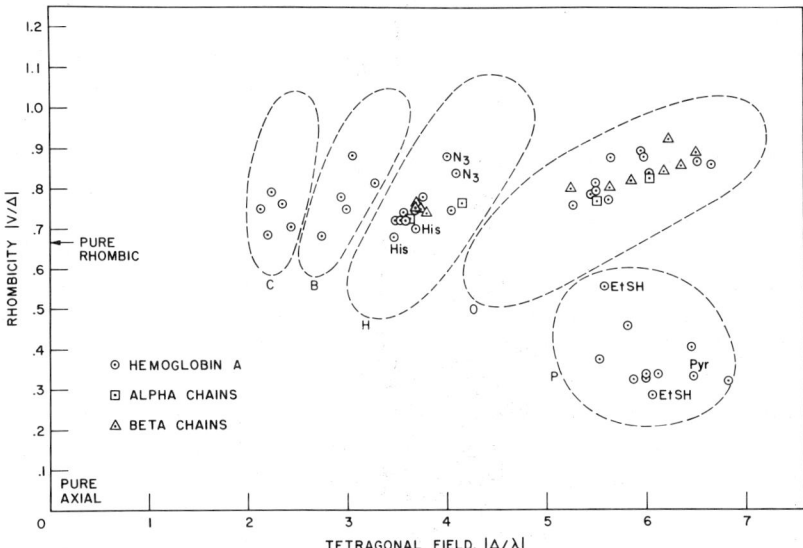

Figure 8. Crystal field parameters for ferric low-spin compounds of hemoglobin A and its isolated constituent chains

some exogenous materials, are formed with atoms which are endogenous to hemoglobin and thus represent the various classes of endogenous low-spin compounds. Hemoglobin A was isolated from human red cell hemolysates without the use of toluene. Isolated alpha chains were prepared from oxyhemoglobin by the method of Bucci and Fronticelli (4) as modified by Parkhurst, Gibson, and Geracci (5). Beta chains were obtained as oxyhemoglobin H. Oxidation to the ferric form was performed with ferricyanide in a Biogel P-2 column at pH 7.

Low-spin compounds can be formed using the following reagents referred to in the figure: EtSH, mercaptoethanol; Pyr, pyridine; His, histidine; N_3, azide. The five areas enclosed by the dashed lines (drawn with some artistic license) define the regions where parameters for the five different compounds may be expected to lie. The groups labelled O and H contain the hydroxide and dihistidine compounds, respectively, which have been discussed above. The group labelled P contains the mercaptoethanol and other sulfhydryl compounds. The group labelled C contains compounds which have histidine on one side and a thioether from methionine on the other, by analogy with cytochrome *c*, where this structure is known to exist. The remaining group, B, contains a histidine on one side and an unknown ligand on the other. This unknown ligand is the same as is found in the cytochromes *b*, whatever that may be.

The hemichromes which fall into this B group can be made from hemoglobin in essentially 100% yield, even though we do not know what the distal ligand is. The recipe is as follows: almost any denaturing agent which disrupts hydrophobic bonding between different parts of the tertiary structure will lead to this compound. Most water-soluble aromatic molecules meet this requirement. Thus, this type of hemichrome was discovered after salicylate treatment (8), but imidazole will also produce it.

We have digressed a long way from the study of native hemoglobin. Which of these hemichromes can be renatured under the appropriate conditions to functioning hemoglobin? Our experiments have shown (6) that the hydroxide and dihistidine (O and H) types can be renatured while the C, B, and P types seem to have crossed the point of no return. Referring to the myoglobin chain model again (Figure 6), one can see that of the two reversible compounds, the one with the hydroxide as the distal ligand needs no tertiary structure change at all, and it is understandable why that is freely reversible. The one in which the distal imidazole nitrogen comes over to bond to the iron requires only a very small tertiary structure change to make that possible, and it is understandable why that might be reversible. The others require much more of a change in the tertiary structure in order that the required distal ligand atoms can approach the iron atom.

There are several theoretical things that can be pointed out on this diagram (Figure 8). First of all, the way we have chosen to plot the theoretical constants here is that the abscissa is a function of the total electron donation to the iron, but perhaps not a linear function. The farther out one goes to the right, the farther one departs from a cubic situation. The ordinate, being just the ratio of two symmetry parameters, is a pure number of geometrical significance. Therefore, compounds lying on the same horizontal line would have the same geometry while compounds lying on the same vertical line would have the same total electron density at the iron atom. This electron density is determined by a property of the six ligand atoms which may loosely be called electronegativity. We were advised during the discussion following the oral presentation of this paper both that this is an improper use of the term electronegativity and that the use is perfectly in order, requiring no apology.

It is interesting to note that the center of gravity of the four compounds labelled C, B, H, and O lie along more or less the same horizontal line, and indeed they share one thing in common. They all have a nitrogen atom of the proximal histidine which is presently in the native heme protein, so that all of those compounds are in the same geometry. Furthermore, that ligand almost completely determines the geometry of the compound. Changing the sixth ligand only changes the electron density and does not disturb the geometry. On the other hand, making mercaptide compounds destroys the geometry. That is, the mercaptide does not impose the same geometry on heme as does the nitrogen atom of the proximal histidine. Azide and imidazole, although they look very different to an inorganic chemist, look very similar as far as an iron porphyrin is concerned. All the iron "knows" is that there is a double bonded nitrogen there, and it "cares" very little about the geometry of that particular ligand. The geometry is determined by the other ones.

Whether we start with compounds from hemoglobin A, which is an equal proportion of alpha and beta chains, or alpha chains, or beta chains, one ends up with indistinguishable compounds in these various groups. That is particularly noticeable in the hydroxide group, where we have made the most extensive study of this finding. One could not possibly tell from these low-spin compounds which of these chains is present or whether there was a mixture of the two, and this just points out that the iron in ferric low-spin compounds does not care about the second coordination sphere. It only cares about the first coordination sphere, and the first coordination sphere is identical in all of these compounds belonging to a single group Of course, such important properties as oxidation–reduction potential and reaction kinetics are indeed determined by the combined effects of the nearest coordination sphere and those further out.

The meaning of the arrows marking the symmetry "pure rhombic" (13) requires some explanation. If one has a more or less octahedral ligand situation like structure I, where the iron atom is coordinated say to four A's and to two B's, there is clearly perfect tetragonal symmetry as is usually observed (16) in the case of high-spin ferric compounds. One can draw a square connecting the A's and say, "That's the heme" because there were four atoms in the parent molecule which were essentially equivalent. Now I pose a riddle to you: in these low-spin compounds near the line indicating a completely rhombic field (structure II), where the A's are as different from the B's as the B's are from the C's, where is the square?

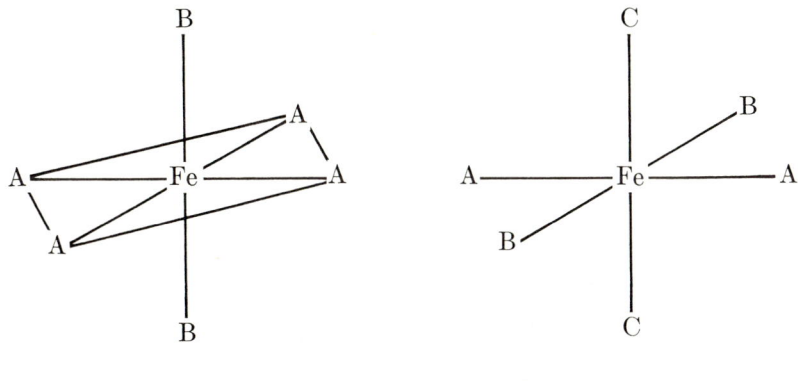

I   II

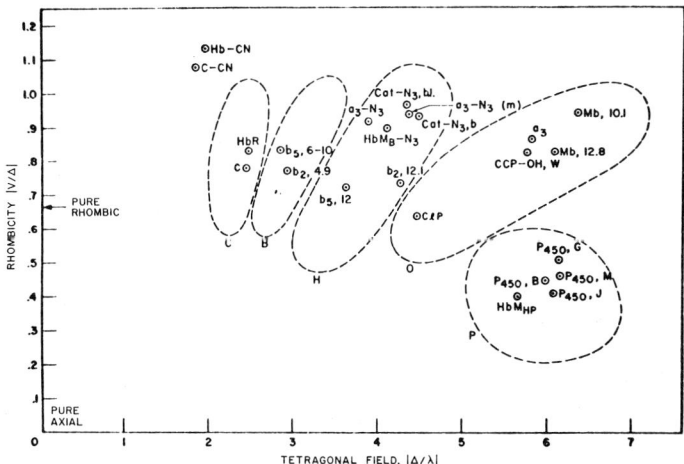

Figure 9. Crystal parameters for ferric low-spin forms of various heme proteins

The abbreviations used are: P-450, J, P-450, M, rabbit liver microsomal cytochrome P-450 (17, 18); P-450, B, rat liver microsomal P-450 (19); P-450, G, bacterial cytochrome P-450 (20); HbM_{HP}, Hemoglobin $M_{Hyde\ Park}$ (21); ClP, chloroperoxidase (22); CCP-OH, W, cytochrome c peroxidase (23); Mb, 12.8, sperm whale myoglobin, pH 12.8 (24); Mb, 10.1, sperm whale myoglobin, pH 10.1 (24); a_3, cytochrome a_3 (25); b_5, 12, cytochrome b_5, pH 12.1 (26); b_2, 12.1, cytochrome b_2, pH 12.1 (27); HbM_B–N_3, Hemoglobin M_{Boston} azide (21); a_3–N_3, cytochrome a_3 azide (28); Cat–N_3, b, horse erythrocyte catalase azide (29); a_3–N_3(m), cytochrome a_3 azide, minority component (28); Cat–N_3, b.l., beef liver catalase azide (30); b_2, 4.9, cytochrome b_2, pH 4.9 (27); b_5, 6–10, cytochrome b_5, pH 6 to 10 (28); c, cytochrome c (31); Hb_R Hemoglobin$_{Riverdale}$ (32); C–CN, cytochrome c cyanide (30); Hb–CN, ferrihemoglobin cyanide (30). The analysis for ClP, C–CN, and Hb–CN are based on two g values, while all other points are based on three g values.

Let us compare the low-spin compounds of hemoglobin to those of some other heme proteins. Figure 9 is a similar crystal field diagram for some other heme proteins where the contours (drawn with some artistic license) are the same ones as on the hemoglobin figure. What other low-spin compounds of heme proteins are like the hemichromes of hemoglobin? Some compounds which are unlike hemichromes are hemoglobin and cytochrome *c* cyanide (and also myoglobin cyanide), which is not surprising as cyanide ion is certainly unlike any endogenous ligand. All the cytochromes P-450, which are now known to be mercaptide heme compounds, fall in the same region as the hemoglobin mercaptide heme compounds.

Some hydroxide compounds of myoglobin (24) give an opportunity to indicate what effects outer sphere interaction, using the phrase loosely, have on these compounds. Myoglobin at pH 10.1 and 12.8 are shown, and there are also intermediate points. The crystal field very smoothly travels between the two points shown as the pH is raised by almost three log units.

Two native cytochromes b have been studied (26, 27), cytochromes b_2 and b_5, and they fall in the region labelled B. Cytochrome c falls in the region labelled C. Cytochrome c contains histidine and methionine as the two nonporphyrin ligands of iron (33).

Catalase azide, for example, falls exactly where hemoglobin azide does; the same geometry, same electron density. The structure is probably the same. Nobody knows what the proximal ligand to the heme iron is in catalase, but we are willing to speculate that it is imidazole, exactly as in hemoglobin.

In the previous paper (34), there were speculations about the configuration of heme a. Two compounds of heme a_3 are observed in half-reduced cytochrome c oxidase. One is a normal hydroxide, and, in the case when azide is added, there is a normal a_3 azide. This tells us that under these conditions, the heme a_3 is behaving as a normal isolated heme, and does not have a peculiar configuration in cytochrome c oxidase under these conditions.

The question has been asked as regards the great difference in reactivity between the various heme proteins: what structural differences are responsible for this? Here are three examples of cytochromes (a, b, c) which have exactly the same inner sphere coordination as do certain compounds of hemoglobin. Thus the great differences in reactions of the cytochromes are brought about by differences in structure further out than that. Clearly, the cytochromes and hemoglobin differ greatly as one proceeds further out from the iron atom.

There are some exceptions which do not fit on this diagram (Figure 8), and in general they are the peroxidases. The low-spin compounds of the peroxidases have different electron densities, and so I would venture to say that the peroxidases do not have the same proximal ligand as do the proteins listed in Figure 8.

Let us return to hemoglobin and discuss some of the biological applications of the study of these low-spin compounds. Hemoglobin, of course, is intended to carry oxygen. In human beings it carries oxygen in the erythrocyte. The lifetime of the human erythrocyte is about 120 days (35). The erythrocyte does not make any more hemoglobin once it is transformed from a reticulocyte into an erythrocyte. The initial charge of hemoglobin has to last for the lifetime of the red cell. Hemoglobin binds and unbinds oxygen several times a second; thus, a given heme will have to turn over an oxygen molecule about 10^8 times during its lifetime, and clearly it cannot afford to make many mistakes in that process. But mistakes are made indeed, both accidentally and otherwise. There can be accidental mistakes in that the dissociation can somehow go awry, and the heme can be left oxidized to the ferric state during the course of it, or oxidizing agents can pass into the erythrocyte which then oxidizes the heme and destroys its function. These processes must be reversed, and there are repair mechanisms in the erythrocyte which do this reversal.

First let us look at the rates of oxidation and denaturation of hemoglobin and its various derivatives. Hemoglobin A can be split into its constituent chains by binding p-mercuribenzoate to the chains, which causes them to dissociate (*36, 37*). Figure 10 shows the rate of oxidation by an oxidant (ferricyanide) at 6°, with a 4:1 molar excess of oxidant. The first order kinetic constants $k_{1,2}$ and $k_{2,3}$ are the reciprocal of the speed of the reactions observed under these conditions, not a thermodynamic constant. Whether the hemoglobin A is in its very stable tetrameric form or separated into beta chains or alpha chains, the rate of attack by this oxidant is about the same. On the other hand, the rate of the formation of the first hemichrome, that is the dihistidine hemichrome, from this ferric state is very different for these compounds (*6*). It is fastest for alpha chains, considerably slower for beta chains, and essentially unobservable for hemoglobin A. We kept a sample of ferrihemoglobin A for seven months, and in that time it was about half converted to the first hemichrome. The first hemichrome, as was mentioned, is a reversible one. The next hemichromes are irreversible, and we do not as yet have

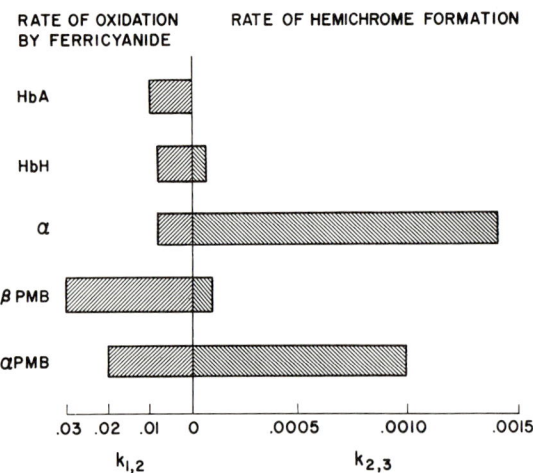

Figure 10. Rates of ferricyanide oxidation of oxyhemoglobin A, oxyhemoglobin H, and oxy alpha chains and their oxy PMB derivatives and the rates of spontaneous conversion of the ferric forms of these proteins to hemichromes

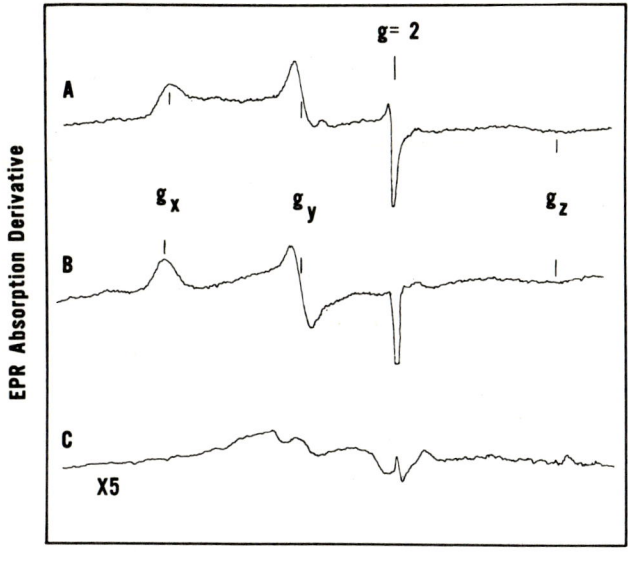

Figure 11. The reduction of dihistidine hemichrome of hemoglobin H as studied by EPR
The features of the EPR absorption derivative, labelled g_x, g_y, and g_z, arise from the hemichrome. The feature at $g = 2$ is the high field end of the EPR spectrum of the high-spin ferric protein not yet converted to hemichrome.

any quantitative data on the times in which this hemichrome goes to the irreversible ones but they are of the order of hours.

That this first hemichrome is reversible can be illustrated by a simple experiment. Figure 11 shows the EPR spectrum of the dihistidine hemichrome, and part of the spectrum of the remaining high-spin material (upper curve). Three hours after adding a 50-fold excess of ascorbate to this sample, nothing has changed (middle curve). But then after about a day, here observed at five times the gain (lower curve), dihistidine hemichromes and the high-spin material have disappeared. All one sees by EPR is a trace of ascorbate radical and some absorptions which are owing to the irreversible hemichromes. The optical spectrum of the sample indicates that it has been converted almost entirely to oxyhemoglobin.

The red cell has a system of reducing enzymes (38). It gets reducing equivalents from glucose and eventually these are converted to DPNH. There is a reductase which runs on DPNH and recognizes the ferric hemoglobin molecules which do not have their tertiary structure disrupted. There are cases where hemichromes will proceed to the irreversible form in large quantities; something has to be done about them as they cannot be reduced. There is a back-up mechanism for repairing a red blood cell where the reductive process has been saturated and irreversible hemichromes have been formed. This mechanism is the formation of the Heinz bodies, which are clusters of denatured hemoglobin. The red blood cell concentrates on its membrane molecules which have been disrupted irreversibly from their normal configuration (*39*). Figure 12 shows erythrocyte ghosts (red cells lysed and washed of soluble hemoglobin) from a normal and two pathological situations. In a normal case (A) there are no Heinz bodies, that is, there has been no need for this back-up process. In B, C, and D, Heinz bodies are visible by phase contrast. The samples of blood were taken from patients who had alpha thalassemia (*40*), a genetically determined disease in which there is an unbalance in the production of alpha and beta chains; the beta chains are in excess. The beta chains are not as stable as hemoglobin A (*cf*. Figure 10), and so when they are in excess, they will proceed to hemichrome and in certain cases lead to the production of Heinz bodies. These Heinz bodies are ultimately removed from the erythrocyte membrane by the

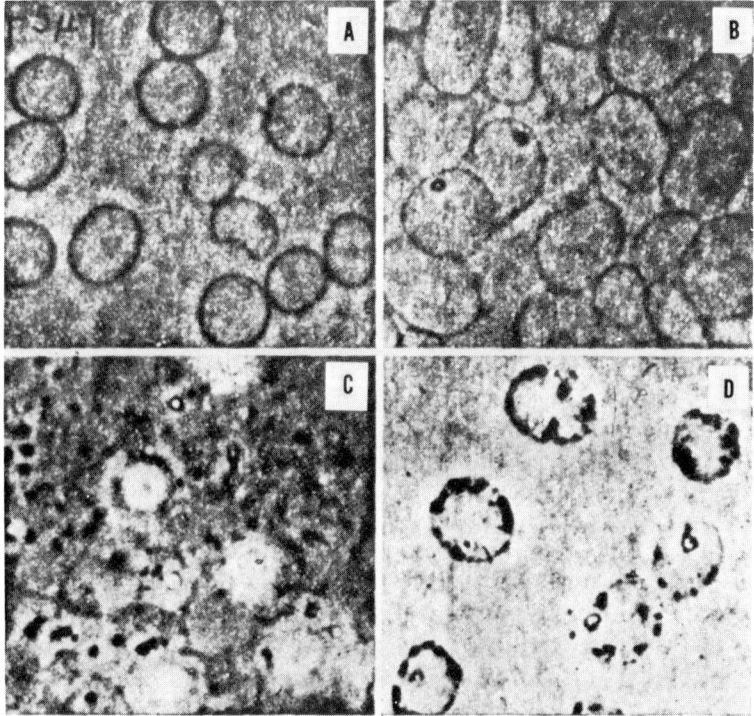

Figure 12.   Phase contrast micrograph of red cell ghosts

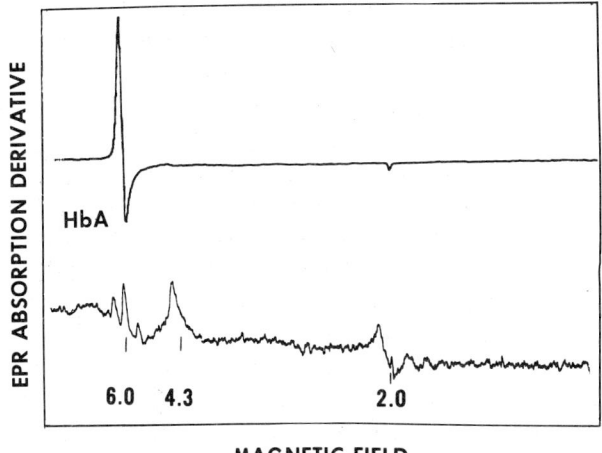

Figure 13. EPR spectrum of purified ferrihemoglobin A (upper) and of 0.5 ml of packed red cells obtained from a normal adult (lower)

spleen. After incubation for 12 hours outside the patient, there are many more Heinz bodies visible in C. In D there are many Heinz bodies in a sample from a patient who had no spleen, and thus no mechanism to clean up the Heinz bodies. Small inclusion bodies appear dark while the larger more dense inclusion bodies are bright, being outside the focal plane.

There are several other ways in which Heinz body formation can be promoted. The red blood cell may lack its reductive capacity. There are such diseases—*e.g.*, glucose-6-phosphate reductase deficiency (*38*)—and they are transmitted genetically. In other cases, there are unstable variant hemoglobins (*41*). An important case, which may apply more to the general public, is the accidental or intentional ingestion of oxidants. A loading dose of oxidants will make a large quantity of ferric hemoglobin and will saturate the reductive capacity of the cell. The excess may then proceed to the irreversible hemichromes, and one will need this back-up mechanism of Heinz body formation.

What are the normal levels of high-spin ferric hemoglobin and the low-spin compounds which we have been studying? Figure 13 shows the EPR spectrum of a sample of blood from a girl who had a very low level of high-spin ferric hemoglobin (probably from having led a very sheltered life at home). The absorption derivative spectrum of ferrihemoglobin extends from $g = 6$–2. In the sample of red cells, the absorption at $g = 6$ arises from the high-spin ferrihemoglobin in the sample while the signals at slightly higher and lower fields ($g = 6.7$ and 5.4)

arise from the erythrocyte catalase which is naturally ferric in its resting state. No low-spin derivatives of ferric hemoglobin are visible. The absorptions at $g = 4.3$ and near $g = 2$ arise from iron impurities not associated with the heme. A quantitative comparison between the intensity of the two high-spin absorptions (A and B) shows that the hemoglobin in this red cell sample is only one part in 10^4 ferric. There are no observable hemichromes. This low level is a result of the fact that this girl must have had a very low intake of oxidants. The more customary level of ferric hemoglobin in the blood is about 1% or so of the total. One can easily demonstrate the effect of oxidants. Blood examined before and after taking oxidants will show that the proportion of ferric hemoglobin has gone up. Nitrite is a good oxidant for hemoglobin and amyl nitrite is the quickest oxidant for getting into the blood, as it is volatile. One can just breathe it, and it will pass into the blood stream, making the ferric hemoglobin level go up immediately. Amyl nitrite is used as an antidote for cyanide poisoning (42) as the ferric hemoglobin thus produced will scavenge the cyanide ion in solution and prevent it from binding to the mitochondrial respiratory enzymes. The cyanide may be removed from the ferric hemoglobin spontaneously (a very slow process) or by the administration of thiosulfate (42). The remaining ferric hemoglobin must then either be renatured or disposed of *via* Heinz bodies. We incubated the sample used for Figure 13 for several minutes in isotonic nitrite, and about 5% of the hemoglobin was oxidized. In Figure 14, one can immediately recognize the high-spin ferric hemoglobin ($g = 6$) and quite a number of absorptions (denoted by vertical lines) from the various hemichromes which may be identified with the compounds described by the points in Figure 8. The total amount of paramagnetic

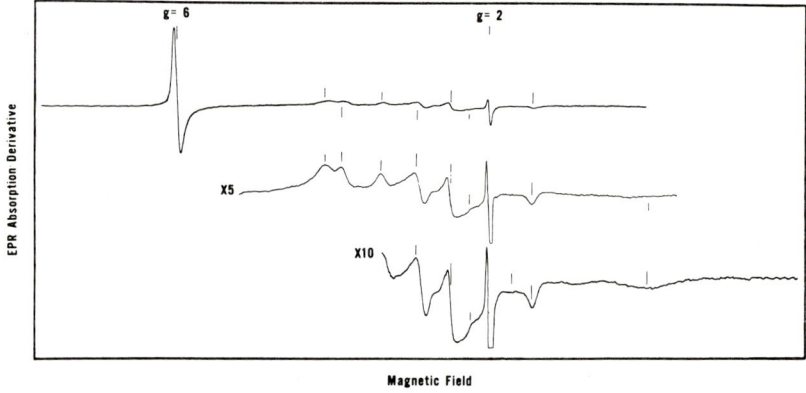

Figure 14. EPR of packed red cells which were incubated with isotonic nitrite

material in this sample consists of less than 5% of the total heme. The nitrite incubation has produced both reversible and irreversible hemichromes. All of these compounds, regardless of the state, bind cyanides. However, it would be desirable to have an oxidant to use as a cyanide antidote which did not produce irreversible hemichromes.

As pointed out in a previous talk (43), nitrates were being used extensively as fertilizer, and it is a widely held opinion that they are not yet a pollution problem. However, if one drinks water which has run off a nitrate-fertilized field, one will ingest the nitrate. Nitrate does not oxidize hemoglobin, but the intestinal flora convert it to nitrite, and then the nitrite is instantaneously taken into the blood and oxidizes hemoglobin. Is that a present danger? There are very few places in the world where the nitrate concentration in the drinking water is high enough to be a health hazard to adults, but there are certain places (e.g., Long Island, New York) where water systems have been ordered closed as the nitrate concentration was deemed high enough to be of danger to infants. Infants are far more susceptible to oxidants than are adults, as the enzyme systems for handling the products of hemoglobin oxidation are not as well developed.

Our research has elucidated the mechanism of an interesting finding that hemotologists have known a long time. In order to study the morphology of Heinz bodies in the red cells, they incubate erythrocytes from normal people with phenylhydrazine, because phenomenologically it is found that phenylhydrazine will promote the formation of Heinz bodies in several hours from normal erythrocytes (44). Phenylhydrazine is a bifunctional reagent as regards hemoglobin. It passes the erythrocyte membrane and the hydrazine part oxidizes the iron. Then the phenyl part makes the B-type hemichrome, which is an irreversible one, as the molecule falls into the category of a soluble aromatic. Finding a large excess of irreversible hemichrome, the erythrocyte proceeds to make Heinz bodies rapidly. These Heinz bodies are similar, if not identical, to those formed spontaneously in patients with unstable types of hemoglobin.

In summary, we may say that by studying ferric low-spin compounds of hemoglobin and other heme proteins, we have been able to demonstrate that information concerning the structure of these compounds may be obtained by EPR experiments and that a knowledge of the structure enables one to understand more fully the role of these compounds in physiological processes.

Acknowledgment

The portion of this investigation carried out at the Albert Einstein College of Medicine was supported in part by a Public Health Service

Research grant to J. Peisach (HE-13399) from the Heart and Lung Institute. This is Communication No. 214 from the Joan and Lester Avnet Institute of Molecular Biology. J. Peisach is a recipient of a Public Health Service Research Career Development Award (1-K3-GM-31,156) from the National Institute of General Medical Sciences.

Literature Cited

(1) Griffith, J. S., *Proc. Roy. Soc.* (1956) **A235**, 23.
(2) Kneubuhl, F. K., *J. Chem. Phys.* (1960) **33**, 1074.
(3) Bayer, E., Hill, H. A. O., Roder, A., Williams, R. J. P., *Chem. Commun.* (1969) 109.
(4) Kendrew, J. C., *Sci. Am.* (1961) **205** (6), 96.
(5) Perutz, M. F., Muirhead, H., Cox, J. M., Goamann, C. G., *Nature* (1968) **219**, 139.
(6) Rachmilewitz, E. A., Peisach, J., Blumberg, W. E., *J. Biol. Chem.*, in press.
(7) Ranney, H., Jacobs, A. S., Udem, L., Zalusky, R., *Biochem. Biophys. Res. Commun.* (1968) **33**, 1004.
(8) Rachmilewitz, E. A., "Second Symposium on Cooley's Anemia," *Ann. N.Y. Acad. Sci.* (1969) **165**, 171.
(9) Dickerson, R. E., *in* "The Proteins," H. Neurath, Ed., 2nd ed., Vol. II, p. 603, Academic, New York, 1964.
(10) Blumberg, W. E., Peisach, J., *in* "Structure and Function of Macromolecules and Membranes," B. Chance, C. P. Lee, T. Yonetani, Eds., Academic, New York, 1970, in press.
(11) Peisach, J., Blumberg, W. E., Wittenberg, B. A., Wittenberg, J. B., Kampa, L., *Proc. Natl. Acad. Sci. (U.S.)* (1969) **63**, 934.
(12) Bleaney, B., Stevens, K. W. H., *Rept. Progr. Phys.* (1953) **16**, 108.
(13) Blumberg, W. E., *in* "Magnetic Resonance in Biological Systems," A. Ehrenberg, B. G. Malmstrom, T. Vanngard, Eds., p. 119, Pergamon, Oxford, 1967.
(14) Griffith, J. S., "The Theory of Transition Metal Ions," p. 363, Cambridge University Press, 1961.
(15) Weissbluth, M., *Struct. Bonding* (1967) **2**, 1.
(16) Blumberg, W. E., Peisach, J., Ogawa, S., Rachmilewitz, E. A., Oltzik, R., *J. Biol. Chem.*, in press.
(17) Ichikawa, Y., Yamano, T., *in* "Recent Developments of Magnetic Resonance in Biological System," S. Fujiwara and L. H. Piette, Eds., p. 108, Hirokawa Publishing Co., Tokyo, 1968.
(18) Miyake, Y., Gaylor, J. L., Mason, H. S., *J. Biol. Chem.* (1968) **243**, 5788.
(19) Klein, M., Blumberg, W. E., Peisach, J., unpublished observations.
(20) Gunsalus, I. C., Katagiri, M., Ganguli, B., Yu, C. A., *Symp. Membrane Function Electron Transfer to Oxygen, Miami, Jan. 20–24, 1969.*
(21) Watari, H., Hayashi, A., Morimoto, H., Kotani, M., *in* "Recent Developments of Magnetic Resonance in Biological System," S. Fujiwara and L. H. Piette, Eds., p. 128, Hirokawa Publishing Co., Tokyo, 1968.
(22) Palmer, G., Hager, L. P., private communication.
(23) Wittenberg, B. A., Kampa, L., Wittenberg, J. B., Blumberg, W. E., Peisach, J., *J. Biol. Chem.* (1968) **243**, 1863.
(24) Gurd, F. R. N., Falk, K.-E., Malmstrom, B. G., Vanngard, T., *J. Biol. Chem.* (1967) **242**, 5724.
(25) van Gelder, B. F., Orme-Johnson, W. H., Hansen, R. E., Beinert, H., *Proc. Natl. Acad. Sci. (U.S.)* (1967) **58**, 1073.
(26) Bois-Poltoratsky, R., Ehrenberg, A., *European J. Biochem.* (1967) **2**, 361.

(27) Watari, H., Groudinsky, O., Labeyrie, F., *Biochim. Biophys. Acta* (1967) **131,** 592.
(28) Beinert, H., personal communication.
(29) Toril, I., Ogura, Y., *in* "Recent Developments of Magnetic Resonance in Biological System," S. Fujiwara and L. H. Piette, Eds., p. 101, Hirokawa Publishing Co., Tokyo, 1968.
(30) Peisach, J., Blumberg, W. E., unpublished observations.
(31) Salmeen, I., Palmer, G., *J. Chem. Phys.* (1968) **48,** 2049.
(32) Ranney, H., Peisach, J., Blumberg, W. E., unpublished observations.
(33) Dickerson, R. E., Kopka, M. L., Weinzierl, J. E., Vernum, J. C., Eisenberg, D., Margoliash, E., *in* "Structure and Function of Cytochromes," K. Okunuki, M. D. Kamen, and I. Sekuzu, Eds., p. 225, University Park Press, Baltimore, 1968.
(34) Caughey, W. S., ADVAN. CHEM. SER. (1971) **100,** 248.
(35) Wintrobe, M. M., "Clinical Hematology," 6th ed., p. 165, Lea & Febiger, Philadelphia, 1967.
(36) Bucci, E., Fronticelli, C., *J. Biol. Chem.* (1965) **240,** PC 551.
(37) Parkhurst, L. J., Gibson, Q. H., Geracci, G., personal communication.
(38) Rapoport, S., *in* "Essays in Biochemistry," Vol. 4, P. N. Campbell and G. D. Grenville, Eds., p. 234, Academic, New York, 1968.
(39) Rachmilewitz, E. A., Peisach, J., Bradley, T. B., Blumberg, W. E., *Nature* (1969) **222,** 248.
(40) Nathan, D. G., Gunn, R. B., *Am. J. Med.* (1966) **41,** 815.
(41) Perutz, M. F., Lehman, H., *Nature* (1968) **219,** 903.
(42) Goodman, L. S., Gilman, A., "The Pharmacological Basis of Therapeutics," p. 920, MacMillan, New York, 1965.
(43) Hardy, R. W. F., ADVAN. CHEM. SER. (1971) **100,** 219.
(44) Wintrobe, M. M., "Clinical Hematology," 6th ed., p. 216, Lea & Febiger, Philadelphia, 1967.

RECEIVED June 26, 1970.

14

Ceruloplasmin, A Link Between Copper and Iron Metabolism

EARL FRIEDEN

Florida State University, Tallahassee, Fla. 32306

The evidence that ceruloplasmin (Cp) (E.C. 1.12.3.1) is a direct molecular link between copper and iron metabolism is summarized. Copper deficiency results in low plasma Cp and iron, reduced iron mobilization, and eventually anemia, even with high iron storage in the liver. Cp controls the rate of iron uptake by transferrin. Transferrin plays a key role in the availability of iron for the biosynthesis of hemoglobin in the reticulocytes. The ferroxidase activity of Cp results in the reduction of free iron ion generating a concentration gradient from the iron stores to the capillary system, thus promoting a rapid iron efflux in the reticulo-endothelial system. It has been confirmed both in vivo *and in the perfused liver that 10^{-7} M Cp specifically induces a rapid rise in plasma iron.*

Our understanding of the biological role of copper has escalated during the past decade. As with most of the essential transition metals, the importance of copper arises from the fact that it is a required constituent of numerous essential proteins and enzymes. Those copper enzymes whose catalytic functions are well understood are listed in Table I. They range from cytochrome oxidase, the terminal oxidase in all animals and most plants, to hemocyanin, the oxygen carrier for many invertebrates.

Interest in the biological role of copper has greatly increased as the recognition of its role in a number of key physiological processes has developed. These include its importance in elastin and collagen formation which prevent aneurisms, soft bones, and other defects (1, 2), the requirement for copper in the taste response (3), and its requirement for cytochrome oxidase and related systems (4). Finally, there is perhaps the best known biological role of copper—its involvement in hemoglobin formation (5, 6, 7). I propose to deal exclusively with the latter, the role of copper in Fe metabolism, and to advance the thesis that ceruloplasmin (Cp), the copper protein of serum, is at least one of the direct molecular links between copper and iron metabolism. In this discussion I will summarize the data relating copper to iron metabolism, describe the relevant enzymic activity of Cp, present the growing evidence relating Cp to iron mobilization in the serum, and discuss the mechanism of the ferroxidase activity of Cp with special emphasis on the role of other metal ions.

Copper and Hemoglobin Biosynthesis

The recognition of anemia in the copper-deficient animal was first reported in 1928 (5). This defect in iron metabolism in the copper-deficient animal has undergone searching analysis for the past 20 years by Cartwright, Wintrobe, and their associates at the University of Utah School of Medicine and other groups. In copper-deficient pigs, it was eventually revealed as a severe anemia, both hypochromic and microcytic (6, 8). The progression of these events is illustrated in Figures 1 and 2. Lahey et al. (6) compared key blood analyses of five-day-old pigs fed a milk diet supplemented with iron and litter-mate controls on the same diet supplemented with both iron and copper. As shown in Figure 1, serum copper falls rapidly, followed by serum iron, erythrocyte copper, and eventually a dramatic reduction in red cell volume. The sequence of events in copper-deficient rats is shown in some data (Figure 2) from a recent paper by Owen and Hazelrig (7). The fall in plasma Cp (PPD-oxidase activity) and copper is followed by a slower but steady decline in liver copper and in hemoglobin. In certain anemias, the plasma Cp level is increased up to three-fold.

This led the Utah group to renew the idea of a role for copper in the biosynthesis of hemoglobin. This role could be effected at one of the three main lines of hemoglobin biosynthesis: the biosynthesis of protoporphyrin or heme, the utilization of iron, or the biosynthesis of globin. Early hopes to find a copper-dependent step in heme biosynthesis were abandoned by Lee et al. (9). They found that as anemia developed in copper-deficient swine, there was a 2–3 fold increase in the activity of heme biosynthetic enzymes, δ-aminolevulinic acid synthetase, α-ketoglutaric acid-dependent glycine decarboxylase, and heme synthetase. They concluded that the anemia of copper deficiency is not the result of defective heme biosynthesis and copper was not a co-factor in any of these reactions. There is also no evidence for any impairment in globin biosynthesis. Thus, it seemed most reasonable to assume that copper is essential for the proper utilization of iron (9). While a connection between iron and copper metabolism has been appreciated for over 40 years, no authentic molecular mechanisms were in sight until our recent pro-

Table I. Important Copper

Name	Mol Wt, Kg	Where Found
Cytochrome oxidase	70	Most plants and animals
Hemocyanin	450–6680	Mollusc and arthropod plasma
Amine oxidase(s)	225	Most animals, etc.
Tyrosinase	35, 100	Animal skin, melanoma, insects, plants
Dopamine-β-hydroxylase	290	Adrenals
Erthrocuprein	34	Red blood cells of most animals
Ceruloplasmin	160	Plasma of most animals
Ascorbic acid oxidase	150	Many plants
Plastocyanin	21	Algae, green leaves, and other plants
Galactose oxidase	75	Molds
Ribulose diphosphate carboxylase	560	Many plants

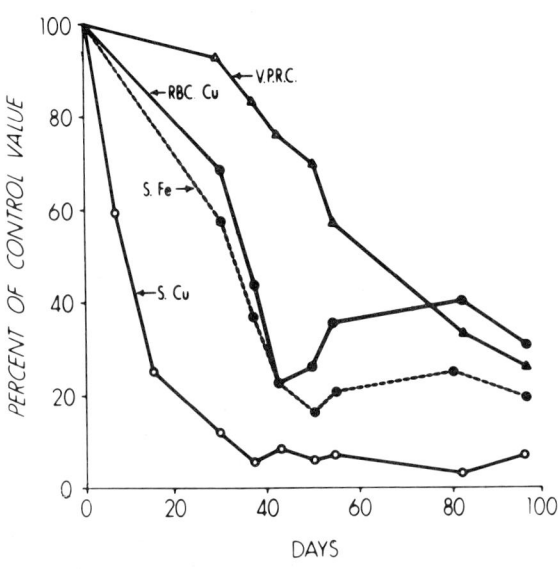

Figure 1. The rate of decline in serum copper (SCu), serum iron (SFe), red blood cell copper (RBC Cu), and volume of packed red cells (VPRC) with time. Adapted from Lahey et al., Ref. 6.

Proteins and Enzymes

Biochemical Function

Terminal oxidase
Oxygen carrier

Elastin, collagen formation
Tyrosinase oxidation, skin pigment (melanin) formation
Epinephrine biosynthesis

Superoxide dismutase

Fe(III)–transferrin formation
Copper transport
Terminal oxidase
Photosynthesis

Galactose oxidation
CO_2 fixation

posal. In 1966, we reported on the possible significance of the ferroxidase activity of Cp and proposed that this catalytic activity was involved in promoting the rate of iron saturation of transferrin in the plasma and in iron utilization (10). We now wish to present the background and information to support this idea and to summarize what is known about this interesting and, we think, useful catalytic activity of Cp.

Ceruloplasmin and its Biological Role

Ceruloplasmin (Cp) is the blue copper protein first reported in the plasma of most vertebrates by Holmberg and Laurell in 1948 (11). In normal circumstances, it accounts for over 98% of the copper in the plasma. Cp has a molecular weight of 160,000 with 7 Cu/mole. Some of the other vital statistics of this serum protein are included in Table II. A recent extensive summary of the chemical properties of Cp has been published by Jamison (12).

Four possible functions of Cp have been proposed in one form or another. These are not necessarily mutually exclusive, since there is no reason why a single prominent protein may not have several important functions—e.g., hemoglobin's oxygen-binding properties and its participation in the pH-homeostatic mechanism of the blood.

Cp transports utilizable copper from the liver to other tissues as needed. Cp is not believed to be involved in copper absorption, as em-

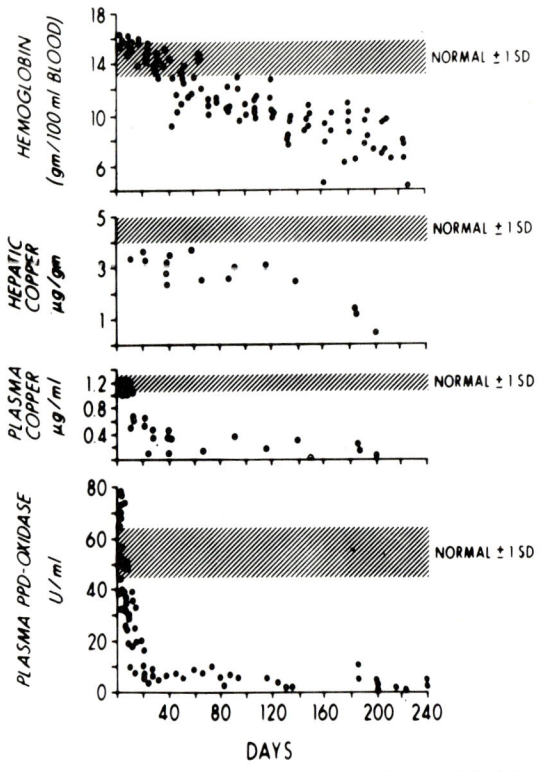

Figure 2. Blood hemoglobin, hepatic copper, plasma copper, and plasma oxidase (PPD–oxidase) values in copper-deficient rats after various periods on a copper-free diet. From Owen and Hazelrig, Ref. 7.

Table II. Molecular Parameters of Human Ceruloplasmin

Mol wt	160,000 (132,000)	Carbohydrates, %	7–8
$S_{20, \omega}$	7.1	Hexose, moles/mole	25–36
Cu, %	0.32	Sialic acid, m/m	9–12
per mole	7	Glucosamine, m/m	12–26
D, cm^2 sec^{-1} × 10^7	3.8	Fucose, m/m	1–2
Axial ratio	11	Gal/man	1/2–2
E_{280} nm, (1%)	14.9	Plasma	Mg %
E_{610} nm, (1%)	0.68	Normal	34 (± 5)
		Wilson's disease	9 (± 7)
Iso elec. pt., pI	4.4	Pregnancy	84 (± 20)
N-terminal a. a.	Lys. Val.	Infectious diseases	68 (± 10)

phasized in a recent review by Scheinberg (*13*). After copper is absorbed and carried into the bloodstream *via* albumin or albumin–amino acid complexes, it is taken up by the liver, incorporated into Cp, and then released into the bloodstream as a potential prosthetic form of Cu.

A corollary proposal related to this, discussed by Broman (*14*), is that Cp provides copper in a "prosthetic" group form for cytochrome oxidases and other Cu enzymes. The close correlation found between the level of cytochrome oxidase activity in the leucocytes and Cp in the serum of Wilson's disease patients, heterozygous carriers, and normal humans is in accord with this idea (*15*).

An idea which has been advanced in a recent discussion (*16, 17*) is that Cp is a "stress" enzyme of the circulation, possibly controlling levels of serum serotonin, epinephrine, melatonin, and other biogenic amines which are substrates for this enzyme.

It has also been suggested that Cp may have some relation to erythropoietin but attempts to confirm this have failed (*18*).

Finally, the proposal from our laboratory (*10, 19*), suggesting a role for Cp in the rate of Fe(III)–transferrin formation and, in turn, the biosynthesis of hemoglobin and other Fe-utilizing systems. We believe there is now convincing evidence to support the idea that Cp is a key link between iron and copper metabolism.

The Ferroxidase Activity of Cp

Our appreciation of the ferroxidase activity of Cp has a tortuous history. The catalytic activity of this circulating copper protein was first recognized by Holmberg and Laurell (*11*), who discovered the protein in 1948. Their preparations of Cp were capable of catalyzing the oxidation of aryldiamines, diphenols, and a group of oxidizable substances such as ascorbate, hydroxylamine, and thioglycolate. In addition to oxidation, another criterion for substrates of Cp is their ability to reversibly decolorize this blue protein. Later, Curzon (*20*) found that the enzymic oxidation of an aryldiamine could be either activated or inhibited by varying the concentrations of certain transition metal ions. Curzon and O'Reilly (*21*) found that Fe(II) could reduce Cp and suggested a coupled iron–Cp oxidation system (*22*).

Our interest in this copper protein was originally stimulated in 1957 by the possibility that Cp was an animal ascorbate oxidase, an enzyme that had been clearly identified in plants but had eluded detection in animal tissues. Initially there was considerable confusion as to whether the ascorbate oxidate activity of Cp was owing to traces of free Cu(II). We found that the oxidation of ascorbate by Cp and free copper differed in so many significant aspects that Cp had an ascorbate oxidase activity

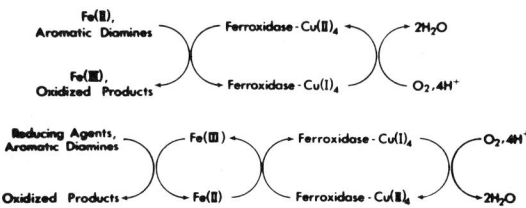

Figure 3. Diagram illustrating the various substrate groups of ceruloplasmin and how they react

independent of free Cu(I) (23). This activity was later attributed to traces of iron in most Cp preparations (24). Using ^{59}Fe, it was found that the concentration of iron eluted with Cp even from a Chelex-100 column was as high as 10^{-8} M. The molecular activity with respect to ascorbate was greatly reduced by dialyzing Cp against apotransferrin. Many additional supporting experiments were reported by McDermott et al. (24).

These data prompted a reinvestigation of the proposed range of substrates of Cp; these compounds fit into three groups as follows:

1. Fe(II), whose rate of oxidation is accelerated by Cp.

2. Certain aryldiamines and polyphenols—e.g., p-phenylenediamine and its methyl derivatives, epinephrine, norepinephrine, dopamine, and serotonin, the oxidation of which is not completely inhibited by iron chelators. These compounds are also directly oxidizable in the presence of the enzyme. The rates of oxidation of many of these substrates can be increased by the presence of iron via a Fe(II)–Cp coupled reaction. Curzon and Young (25) have recent evidence that Cp catalyzes the oxidation of ascorbate at very high substrate concentrations (5–100 mM) under conditions which appear to preclude the intervention of iron ions.

3. Numerous compounds which reduce Fe(III)—e.g., ascorbate, hydroquinone, catechol, hydroxylamine, thioglycolate, cysteine, ferrocyanide, and DOPA, whose oxidation was completely inhibited by iron chelators. Thus, they appear not to be directly oxidized by the enzyme, but function in an Fe–Cp coupled reaction and are iron-dependent substrates.

These catalytic reactions in which Cp is involved are summarized in Figure 3. Cp has significant ferroxidase activity to explain its broad capacity to catalyze the oxidation of a variety of reducing agents, all of which react with ferric ion. It was proposed (10) that when describing the enzymic activity of Cp, it be designated as a ferro-O_2-oxido-reductase (ferroxidase) and numbered E.C. 1.12.3.1 to conform with the recommendations of the Enzyme Commission of the International Union of Biochemistry. In referring to this protein as a serum copper carrier, the name ceruloplasmin may be preferred because of its historical significance; however, ferroxidase is a more descriptive and useful name in identifying its catalytic activity and its specific role in iron mobilization.

In comparing the ferroxidase activity with the p-phenylenediamine oxidase activity in serum, we found it desirable to develop a highly sensitive micro method for the determination of ferroxidase activity (26). This is based upon capturing the ferric ion formed as the transferrin complex according to the reaction scheme shown in Figure 4. Correlation over a wide range of Cp levels in a variety of human serum was excellent, as shown in Figure 5, except for the points representing Wilson's disease

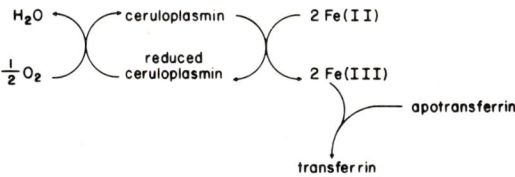

Clinical Chemistry

Figure 4. Reaction sequence used to determine ceruloplasmin activity using apotransferrin as an acceptor for Fe(III)

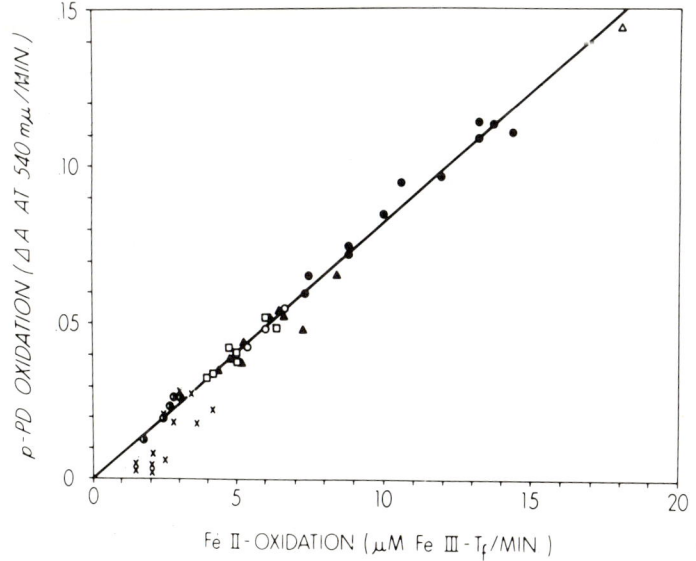

Clinical Chemistry

Figure 5. Correlation between p-phenylenediamine (PPD) and Fe(II) oxidation by human serum

Open square indicates male serum; open circle, female; solid circle, pregnant; half closed circle, cord blood; X, lyophilized and reconstituted from patients with Wilson's disease; solid triangle, serum of "disturbed" patient; open triangle, anemia (probably caused by extensive blood loss). From Johnson et al., Ref. 26.

Table III. Comparison of Some Properties of a New
Serum Ferroxidase with Ceruloplasmin[a]

	Cp	New Fox
Ferroxidase activity, K_m, μM	0.6, 50	40 μM
Wt, kg/ Cu atom	23	@ 80
Serum fraction	Cohn IV–I (α–2 glob.)	Cohn IV–I
PPD oxidase	Yes	No
Inhibited by azide, 1 mM	>98%	No
M.W. in kg	159 (132)	@ 800
Blue color	Yes	No
Chromatographic pattern	————Different————	
Immunoelectrophoretic mobility	————Different————	

[a] Topham and Frieden (Ref. 27).

serum. This latter fact may be of some importance in explaining the lack of extensive disturbance of iron metabolism in this pathology. In Figure 5, Fe(II) oxidation by human serum is plotted against the corresponding PPD oxidase activity. These data give correlation coefficient $r = 0.725$, indicating strongly positive correlation of the two activities by human serums with at least 98% probability of significance except for the Wilson's disease sera. Our test conditions were as follows: (1) PPD oxidation—200 μl serum, 9.2 mM PPD, and 0.2 mM neocuproine in 0.2 M acetate of pH 5.2 at 30° in total volume of 0.600 ml; (2) Fe(II) oxidation—10 μl serum, 30 μM ferrous ammonium sulfate, and 55 μM apotransferrin in 0.2 M acetate of pH 6.0 at 30° in total volume of 1.00 ml.

A New Serum Ferroxidase

Numerous observations in our laboratory suggested the possible presence of another non-Cp ferroxidase activity in normal human serum. The key experimental tool for this study was the fact that there always appeared to be a residual nonazide-sensitive ferroxidase activity in human serum and even in Cohn IV–I fractions. Based on this, a new ferroxidase activity has been identified in our laboratory by Topham et al. (27). As shown in Table III, this new enzyme differs from Cp in a number of key properties. The K_m's are in the μM range but the new ferroxidase has only one. Copper appears to be present but in a different ratio, and the estimated molecular weight appears to be much larger. The new enzyme seems to reside in the same fraction as Cohn IV–I but can be separated by appropriate DEAE–Sephadex chromatography. We have not fully exhausted the possible substrate range of the new enzyme except that we know that it has no PPD oxidase, mono-, or di-amine oxidase activity. There is no blue color. The distribution pattern in serum appears to be quite different, and we are in the process of studying this more completely. While Cp seems to drop very low in Wilson's disease serum, the new enzyme remains relatively unaffected. This point perhaps accounts for the fact that PPD oxidase activity did not correlate with ferroxidase activity in Wilson's sera and in sera from other animals. Naturally we're excited about this new enzyme and will have a further announcement about it in the literature (28).

Significance of the Ferroxidase Activity in Plasma

Despite the fact that the possible significance of the ferroxidase activity of Cp had been rejected by earlier workers, we reexamined this idea. In sera in which Cp activity is very low or inhibited by azide or cyanide, the rate of Fe(III)–transferrin formation starting with ferrous ion is greatly reduced (10). The only protein in human serum which contributes appreciably to the rate of Fe(III)–transferrin formation was a constituent of Cohn fraction-IV–I Cp. A careful comparison of the rates of enzymic and nonenzymic oxidation of Fe(II) under conditions approximating the physiological state was made using oxygen and iron ion as the variable parameters. The nonenzymic rate of Fe(II) oxidation is first order with respect to both Fe(II) and oxygen throughout a wide range of concentrations. In contrast, the Fe(II) oxidation catalyzed by Cp is zero order at oxygen concentrations greater than 10 μM (Figure 6)

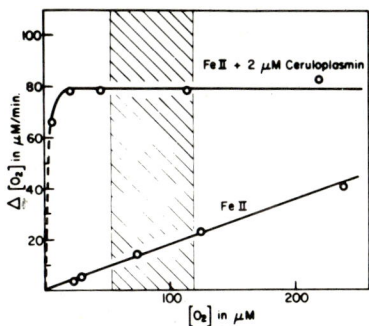

Journal of Biological Chemistry

Figure 6. The effect of oxygen concentration on the rate of enzymic and nonenzymic oxidation of Fe(II) at 30°. The reaction mixture contained 70 μM ferrous ammonium sulfate in 0.0133 M phosphate buffer (pH 7.35), with or without ceruloplasmin. From Osaki et al., Ref. 10.

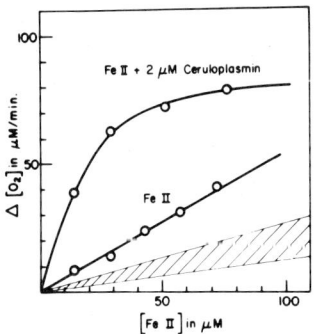

Journal of Biological Chemistry

Figure 7. The effect of Fe(II) concentration on the rate of enzymic and nonenzymic oxidation of Fe(II) at 30°. From Osaki et al., Ref. 10.

and at Fe(II) concentrations larger than 50 μM (Figure 7). Thus maximum enzymic activity is achieved below the typical range of oxygen concentrations found in human serum. The shaded area in Figure 6 indicates the oxygen concentration range in human vein (minimum) and artery (maximum). The Cp-catalyzed rate approaches first order when the ferrous concentration is reduced to less than 1.0 μM, with a K_m value of 10^{-6} M (29). The nonenzymic rate of Fe(II) oxidation is further reduced when plasma oxygen tension ranges are used (*see* shaded area in Figure 7). The oxygen concentration change per minute was measured, with or without 2 μM ceruloplasmin. The reaction mixture contained 211 μM oxygen and variable amounts of ferrous ammonium sulfate in 0.0133 M phosphate buffer (pH 7.35). The estimated nonenzymic oxidation rate at a lower oxygen concentration (55 to 120 μM) is indicated in the shaded part of the figure. These calculations show that the Cp-catalyzed oxidation of Fe(II) is 10 to 100 times faster than the nonenzymic oxidation under conditions that might be expected in plasma. The presence of reducing agents, such as ascorbate at modest concentrations of 40 μM, was sufficient to reduce greatly the net rate of the nonenzymic oxidation of Fe(II). In contrast, the Cp-catalyzed oxidation of Fe(II) was not affected by an excess of ascorbate, one of the principal reducing substances in plasma.

The free iron of plasma is essentially nondetectable. From the normal urinary excretion (30, 31), it is estimated that the free iron concentration in plasma could be about 0.1 $\mu g/100$ ml or 0.02 μM. It would be reasonable to consider that Fe(II) is oxidized to Fe(III) in plasma under these conditions. With 0.02 μM as the free Fe(II) concentration in dynamic equilibrium in plasma, 3.3 liters as the plasma volume, and the experimentally obtained maximum first order rate constant for nonenzymic oxidation of Fe(II) as 1.12 min^{-1}, the maximum amount of Fe(II) oxidized during 24 hours, without Cp, is estimated to be 95 μmoles/24 hours. This is uncorrected for the possible effect of reducing agents—e.g., ascorbate. It has been estimated that the average daily turnover of iron in the plasma is 30 to 40 mg per day (ca. 10 g hemoglobin/day) or about 540 to 720 μmoles per day, including 1 to 3 mg of iron absorbed from the intestine (31, 32). Thus, the rate of nonenzymic oxidation of iron is below the estimated requirement for the conversion of Fe(II) to Fe(III). The minimal enzymic oxidation of Cp, on the other hand, is sufficient: 1025 μmoles per day, many times more active than the nonenzymic oxidation.

Outline of the Ferroxidase Hypothesis

Our hypothesis (19, 33) that Cp could be rate-determining in the formation of Fe(III)–transferrin was based on several essential points as follows.

Iron must be converted to Fe(III) prior to binding by apotransferrin. The preferential binding of Fe(III) by apotransferrin is well known.

Iron must go through a Fe(II) to Fe(III) oxidation step in the plasma. This may be owing to the entry of iron into plasma as Fe(II) or to the fact that plasma has sufficient reducing power (e.g., ascorbate, glutathione) to convert spontaneously Fe(III) to Fe(II). Cp itself exists in plasma in a reduced (colorless) state. The oxidation state and precise molecular species of the emerging iron might be significantly affected by the presence of iron chelators which may play a role in the release of iron from ferritin as suggested from studies on model systems by Pape et al. (34). However, the precise form of iron which enters the plasma from the reticulo–endothelial cells or other iron storage organs is not yet known.

The rate of iron oxidation in the plasma or iron entry in the absence of Cp is not adequate to meet the Fe(III)–transferrin demands of the erythroid bone marrow cells or other tissues for iron. The impetus for the movement of iron into the plasma may arise from the generation of significant concentration gradients by Cp between intracellular iron in the iron storage organs and the iron concentration of the capillary systems, thus promoting a rapid efflux of iron from the reticuloendothelial system.

The Role of Transferrin

Under normal conditions, almost all of the iron bound to transferrin is rapidly taken up by the marrow (35, 36, 37). The observed half-life of ^{59}Fe(III)–transferrin *in vivo* was reported to be 1.3 hours, whereas the half-life for the protein moiety was much slower, 7 days (38). Any erythroid stimulation leads to a faster turnover of iron *in vivo*, and erythroid suppression has the opposite effect (38). Only the reticulocytes are capable of utilizing the Fe(III) bound to transferrin although both the reticulocytes and the mature red cell can take up free Fe(III). Katz and Jandl (38) concluded that the transferrin directs the entry of iron into those cells which are still actively making hemoglobin and prevents any accumulation of iron by the mature cell.

The extreme importance of transferrin in iron transport and in RBC formation is also suggested from a study of iron metabolism in congenital atransferrinemia (39). This appears to be a fatal disease in which the patient can only be kept alive for a period with blood transfusions. Injected ^{59}Fe disappears very rapidly from the serum of such patients (half-time of 5 minutes), and the ^{59}Fe in the RBCs is only a small fraction of the normal range (Figure 8). Autopsy revealed severe accumulation of

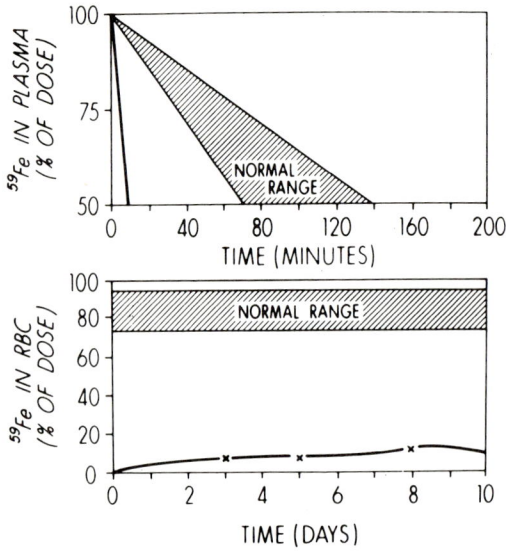

F. Heilmeyer, in "Iron Metabolism," Springer-Verlag

Figure 8. Plasma iron clearance and ^{59}Fe incorporation into erythrocytes in a juvenile patient with congenital atransferrinemia. From L. Heilmeyer, Ref. 39.

iron deposits in numerous tissues, particularly the liver, kidneys, heart, etc., but less iron in the spleen and none in the bone marrow. The explanation offered for these findings is that since iron could not be retained in the bloodstream, the iron diffused in all the organs of the body, leaving too little for utilization by the bone marrow. However, our appreciation of the vital role of transferrin is still seriously limited by the lack of key information about the biological activity of this protein.

The Effect of Copper on Iron Metabolism

We now turn to the question of how copper may be directly involved in iron metabolism, relying on the comprehensive paper of Lee *et al.* (*40*) which reports an *in vivo* study on iron metabolism in copper-deficient swine. They found evidence for abnormalities of iron metabolism at four major sites: the duodenal mucosa, the reticuloendothelial system (R–E), the hepatic parenchymal cells, and the normoblasts. For the first three tissues, each of which can concentrate, store, and release iron, copper deficiency leads to an impairment in the release of iron.

Iron Absorption

Let us first consider the question of the rate of iron absorption. *In vitro* studies on rat duodenum led Manis and Schachter (*41*) to propose that iron absorption occurs in two distinct steps: the transfer of iron from the intestinal lumen in mucosa and the release of mucosal iron into the bloodstream. The importance of chelating mechanisms in iron absorption and transport has been emphasized by Helbock and Saltman (*42*). Lee *et al.* (*40*) demonstrated conclusively that copper deficiency does not interfere with iron absorption. Iron stains of mucosa from copper-deficient pigs showed large amounts of iron in the lamina propria and within the epithelial cells of the mucosa. What is affected is a slower release of mucosal iron into the plasma in copper deficiency, resulting in excessive mucosal iron storage (probably as ferritin), despite a total body deficiency of iron. Brittin and Chee (*43*) have recently reported that there was no relation between serum ferroxidase activity and the changes in iron absorption known to occur after acute bleeding, nutritional iron deficiency, or iron loading.

The Reticuloendothelial System

The reticuloendothelial (R–E) system maintains the principal storage pool of iron and is the major supplier of iron to the plasma. The R–E cells acquire iron by phagocytosis of senescent RBCs. The normal

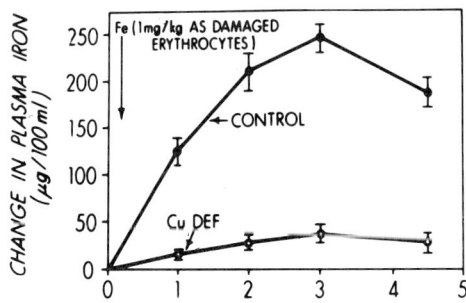

Journal of Clinical Investigation

Figure 9. The effect of an infusion of damaged red cells on plasma iron. Values for the mean ± SEM are depicted at each point. From Lee et al., Ref. 40.

plasma iron level reflects the balance between the availability of iron from the R–E cells and the rate at which iron is removed, mostly by the erythroid cells of the bone marrow. In copper deficiency, iron removal is normal and Lee *et al.* (*40*) were able to show that the release of iron from the R–E cells is impaired. This is illustrated in Figure 9 in which an intravenous infusion of 1 mg/kg of iron as damaged RBC increased the plasma iron in control pigs but had little effect on the plasma of copper-deficient pigs. Since it was shown independently that iron is taken up by the R–E system even in the copper-deficient animal, the defect is assigned to the iron release mechanism. The injection of inorganic copper —sufficient to restore the serum copper level—to a severely copper-deficient pig raises the serum iron to the saturation level in several hours (Figure 10). But most significant was the dramatic and rapid rise in plasma iron when Cp was injected (Figure 11). Thus Cp induces a rapid mobilization of iron, presumably from the R–E system. Marston and Allen (*44*) had also suggested that copper was involved in the release of iron stores from the liver.

Effect of Ceruloplasmin on Iron Mobilization **In Vivo**

The best evidence that Cp plays a direct role in plasma iron levels comes from a recent paper of the Utah group. Ragan *et al.* (*45*) have studied the effect of injecting copper and homologous Cp into copper-deficient swine with adequate stores of iron. In addition to the typical copper- and iron-deficient milk diet, each pig was given intramuscularly a total of 2.0 g iron as iron dextrin (Pigdex) from 5–30 days of age. When a profound state of copper deficiency was evident (80 days), the

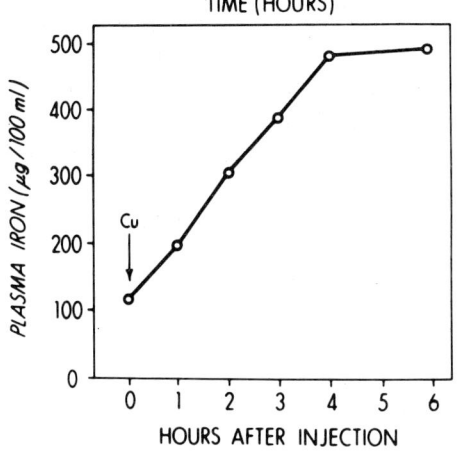

Figure 10. The effect of the administration of copper on the plasma iron in a copper-deficient pig; 100 μg/Kg body weight is sufficient to restore the plasma Cu to normal levels. From Lee et al., Ref. 40.

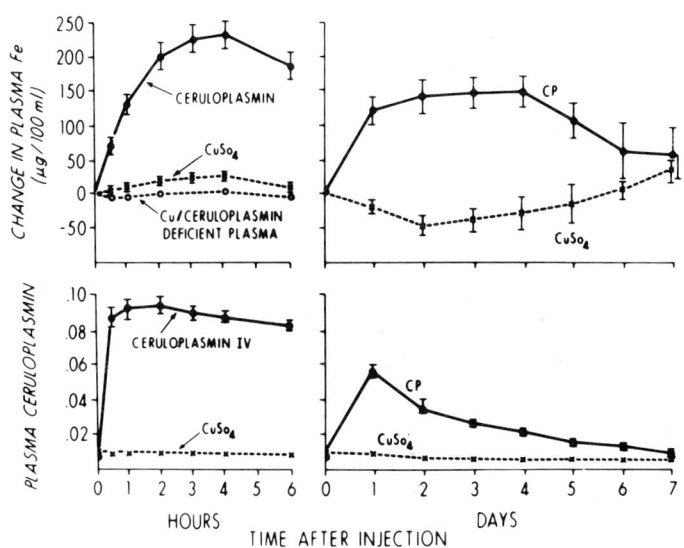

Figure 11. The effect of intravenous administration of ceruloplasmin (e———e), copper sulfate (x-----x), or copper-deficient plasma (o-----o) on the plasma iron and plasma ceruloplasmin (expressed as PPD oxidase activity) in copper-deficient swine. Brackets indicate mean standard error. If deviation is within the symbol it is not shown. From Ragan et al., Ref. 45.

effects of injected pig Cp, $CuSO_4$, or Cu-deficient pig plasma were determined (Figure 11). The amount of Cp injected was only enough to increase the plasma concentration to 15 $\mu g\%$, or 10% of the normal level. A remarkable rapid rise in plasma iron accompanies the Cp injection with a peak in 3–4 hours. The increase in plasma iron is significant after five minutes (Table IV). The maximum increase in plasma iron is about twice the normal level and persists for six days. Neither $CuSO_4$ nor other pig plasma factors could produce this increase in iron. In fact, $CuSO_4$ actually reduced the iron levels after two days, presumably by stimulating RBC formation.

Table IV. Increase in Plasma Iron During the First Hour after Injection of Ceruloplasmin or Copper Sulfate[a]

	Increase in Plasma Iron ($\mu g/100ml$)		
Time, Minutes	Ceruloplasmin-Injected pigs [b]	$CuSO$-Injected Pigs [b]	p[c]
5	18	3	0.01
15	26	3	0.005
30	84	2	0.005
60	136	14	0.001

[a] Ragan et al. (Ref. 53).
[b] Four pigs in each group.
[c] Paired t-test.

To determine whether the increase in plasma iron concentration after Cp administration was caused by an increased flow of iron into plasma or by decreased iron outflow, Ragan et al. (45) performed ferro-kinetic studies. Autologous plasma tagged with ^{59}Fe was injected intravenously, followed in 25 minutes by a Cp injection. Frequent blood samples were obtained over a two-hour period to determine plasma clearance and specific activity of the injected ^{59}Fe in relation to the plasma iron concentration. Attempts were then made to stimulate the observations by means of a computer program designed to calculate changes in specific activity and concentration in a pool as a function of altered inflow and outflow rates. The results of a representative study in a single copper-deficient pig given Cp are depicted in Figure 12. The best match between observed and computer-simulated data was achieved by holding plasma iron outflow constant and increasing inflow five-fold. In contrast, a less successful fit was obtained when inflow was held constant and outflow was decreased, or when a significant change in outflow was combined with an increased inflow. Therefore it appears that the primary effect of Cp is to increase plasma iron inflow.

During the past year, Cartwright, Lee, and coworkers have reported many additional *in vivo* experiments which further support the role of

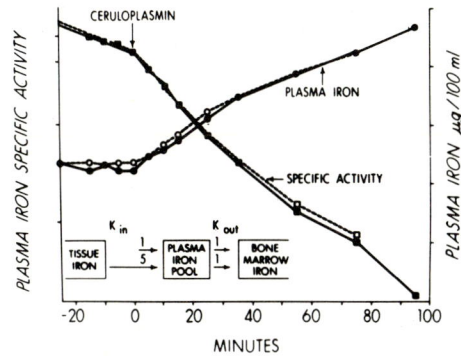

Figure 12. The effect of intravenous ceruloplasmin on plasma iron concentration (o——o) and specific activity (□——□) during a ferrokinetic study in a single copper-deficient pig. Adapted from Ragan et al., Ref. 45.

ferroxidase in iron metabolism (46, 47). Here are some of the results that have been obtained.

In copper-deficient pigs, the Cp level falls to less than 1% of the normal to usually about 0.5% ferroxidase activity. This deficiency of serum ferroxidase precedes the development of hypoferremia in the copper-deficient pig.

They have demonstrated that a rise in serum ferroxidase activity precedes a rise in serum iron following copper injection. Serum iron does not rise until the serum Cp reaches about 1% of normal. However, any hypoferremia can be corrected immediately *in vivo* by the administration of Cp. An empirical equation relating serum iron to Cp was derived; $\Delta Fe\ (\mu g/min) = 4 + 1.2 \log Cp$.

Injected I.V. ferrous iron disappears rapidly from the circulation in the absence of ferroxidase activity and does not bind as readily to apotransferrin as iron injected in the ferric form. In other words, in the absence of adequate ferroxidase, while there was no difference in the levels maintained when Fe(III) is injected, there was a 50% reduction in serum iron levels when Fe(II) was injected. This is interpreted as a demonstration of the direct physiological role of Cp in the control of serum Fe.

Finally, they prepared asialo–CP (48) which is rapidly removed by the liver and contains little or no iron-mobilizing activity when injected although it does show ferroxidase activity in *in vitro* tests. This supports the idea that Cp must function in the circulatory system.

These penetrating experiments by the Utah group lead to the conclusion that the defect in the release of iron in the copper-deficient animal can be reversed promptly by intravenous Cp, an effect that could not be attributed to copper alone. The effect of hypoceruloplasminemia and its correction were reflected on the R–E and the hepatic cells, since the mucosa was not involved. The Utah group (45) proposed that the data strongly support the role of Cp in serum iron utilization as proposed by the Florida State group. Moreover, the rapid success of Cp in mobilizing iron within five minutes after Cp injection makes it unlikely that an intracellular copper enzyme could be involved. In no case (45, 46, 47) have any of their *in vivo* observations been in conflict with the hypothesis that the ferroxidase activity of Cp is directly involved in iron mobilization.

Mobilization of Iron by Ceruloplasmin in the Perfused Liver

Prior to our knowledge of the success of the Utah group in demonstrating the *in vivo* mobilization of iron into the circulatory system by Cp, we had initiated a study of iron mobilization by perfusing livers from normal dogs and normal or Cu-deficient pigs (49, 50). The excised liver was thoroughly flushed with an appropriate perfusion medium, the perfusate was then allowed to recycle, and apotransferrin and other supplements were added. The iron appearing in the perfusate was detected in a flow cell at 460 nm as the Fe(III)–transferrin complex or at the 530 nm peak of the α,α-dipyridyl complex using an aliquot of the perfusate.

Table V. Demonstration of Iron Mobilization by Two Iron-Testing Methods[a]

Experiment	Absorbance Change	ΔFe in Perfusate, μM	Total Iron [b] Released, μG
1. a	A_{460},[c] 0.016	7.2	120
1. b	A_{530},[d] 0.020	7.0	115
2. a	0.020	9.1	150
2. b	0.027	9.6	158

[a] Data from Osaki and Johnson, Ref. 49.
[b] Cp concentration was 0.6 to 0.9 μM.
[c] A spectrophotometric method was used for Fe(III)–transferrin formation (26).
[d] Colorimetric determination of iron using α,α–dipyridyl.

A comparison between the two independent methods for detecting iron production is shown in Table V, in which good agreement between the two approaches was obtained.

A recording of the time course of the absorbance change at 460 nm is shown in Figure 13.

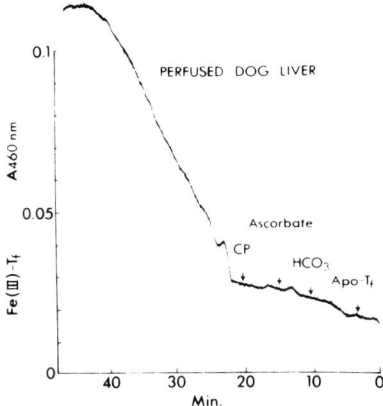

Figure 13. Time course of the absorbance change at 460 mµ [Fe-(III)–transferrin formation]. Data from Ref. 49.

The absorbance change of the perfusate was recorded with a Cary Model 15 spectrophotometer. The total volume of the perfusing solution at the end of all infusions was spectrophotometrically determined to be 120 ml using a dilution factor obtained by blue dextran (mol wt = 2,000,000) infusion. The perfusion was started with 0.01 M tris-HCl buffer, pH 7.4, + 0.9% NaCl solution. Each arrow indicates the time when additional infusions were made to the reservoir. It took approximately 100 sec before the effect of each infusion could be observed by the spectrophotometer. The infusions made at each point are: A, 17 ml of 220 μM apotransferrin (31 μM); B, 2 ml of 0.21 N HCO_3^- (0.0035 N); C, 1 ml of 0.1 M ascorbate (800 μM); and D, 8 ml of 73 μM human ceruloplasmin (4.9 μM). The concentrations in parentheses above were the final concentrations of each compound based on the total volume of 12 ml. The rate of Fe(III)–transferrin formation was estimated to be 3 μM per 100 sec or about 220 μg iron released from the liver in 20 minutes. If man were to mobilize Fe at the same rate as observed in normal dog liver, up to 192 mg Fe/day could be produced or several times the estimated daily need for iron.

These pilot experiments have been extended to find out how sensitive iron efflux from perfused livers was to Cp concentration (50). The rate of iron efflux was determined from the slopes of the recorded responses as in Figure 13. These were plotted as relative rates vs. Cp concentration. The effect of ferroxidase was observed at concentrations as low as 4×10^{-9} M with a maximum effect at 2×10^{-7} M, which is 10% of the normal human serum level. This data corresponds closely to the *in vivo* results of Ragan *et al.* (45) mentioned earlier in which only 1/10 of the normal level of Cp was required to produce maximum iron requirement for iron mobilization. However, an appreciable response is noted at 1% of the normal Cp level.

The specificity of the iron mobilization response in the perfusate system was also studied. Only Cp among the compounds tested proved to have any activity in the perfused livers. No activity was shown by 30 μM apo-transferrin, HCO_3^-, 10 μM $CuSO_4$, 5 mM glucose, 0.6 mM fructose, 120 μM citrate, or 36 μM bovine serum albumin, $\pm$ 21 μM $CuSO_4$. Further experiments designed to test apo–Cp, other copper oxidases, and other iron or copper binding compounds are in progress.

Exploratory studies on the mobilizable iron pool in the copper-deficient animal have provided some unexpected results (50). First of all, even in the copper-deficient animal with Cp-anemia, there remains an appreciable pool of mobilizable iron. Second, the intravenous injection of Cp will rapidly release this iron. The depletion of the pool is then followed by a reduction in the rate of mobilization. This supports the point of view by Ragan *et al.* (45) that the defect corrected by Cp *in vivo* is related to the inflow of iron into the plasma. The presence of an exhaustible pool of iron in the perfused liver system was observed. In this experiment, the liver was flushed for two hours before apotransferrin was added to the perfusion and solution; after 25 min, 0.2 μmole Cp, giving a 0.6 μM final Cp concentration in the perfusate, was added. The rapid appearance of iron in the perfusate was noted. The extent of iron mobilization depended on the length of flushing time prior to the enzyme infusion. In addition, ferroxidase did not stimulate further release of iron once the maximum level of iron mobilization was reached. There was no evidence for the presence of an inhibitor or the denaturation of the enzyme. These findings suggest that a pool of mobilizable iron was formed during the flushing period. The extent of the mobilization may depend on the size of the pool. This is supported by an additional experiment. A normal male dog liver was infused with 0.29 μM enzyme 50 min after the start of the perfusion, which resulted in the release of 80 μg of iron. Flushing was continued for two hours. Then the same amount of enzyme was infused again. Almost twice as much iron was rapidly mobilized by Cp.

Summary of the Iron Metabolism

The relevant economy of iron as affected by copper deficiency and Cp is summarized in the scheme depicted in Figure 14. Iron absorption by the mucosal cells appears to be unaffected in copper deficiency but release of iron into the plasma is impaired. The release of iron from

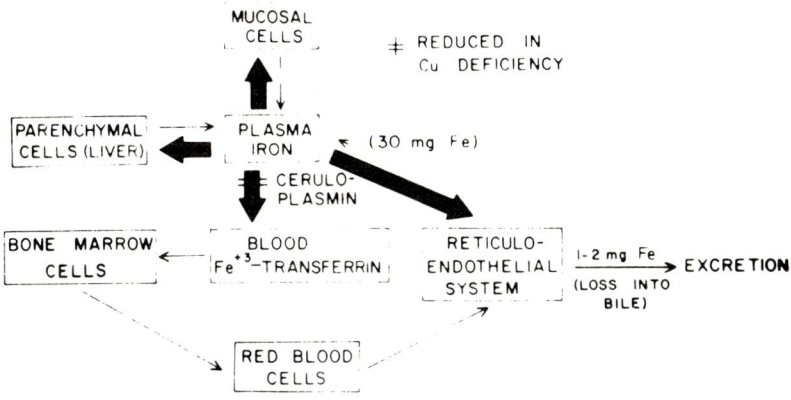

Figure 14. Schematic summary of iron metabolism and how iron release into the plasma is affected by copper deficiency and low plasma ceruloplasmin levels

the liver parenchymal cells is also reduced in copper deficiency. The most significant pathway, quantitatively, is the mobilization of iron from the R–E system into the plasma. This is lowered in copper deficiency and can be restored rapidly by injection of Cp or, more slowly, by copper administration. The remainder of the iron cycle is also depicted. The formation of Fe(III)–transferrin is dependent on Cp. The iron in the Fe(III)–transferrin is contributed directly to the developing reticulocyte in the bone marrow. When the red cell has run its course, it is scavenged by the R–E cells, principally in the spleen and liver, and the iron cycle is renewed. Cp can affect the mobilization of iron, particularly from the R–E and/or liver systems.

Iron Metabolism in Wilson's Disease

If iron mobilization is dependent on Cp, shouldn't there be a disturbance in iron metabolism, perhaps resembling copper deficiency, in Wilson's disease? This is a disorder characterized by low plasma Cp and the accumulation of copper in the liver and brain. It is treated by eliminating copper from the diet and/or removing copper by administering penicillamine. However, the evidence to support a concomitant upset in iron utilization is tenuous. There is one recent paper by O'Reilly et al. (51) in which eight patients with Wilson's disease were reported to have iron deficiency or low plasma iron or both, sometimes associated with anemia. Most of these subjects had low or low–normal levels of transferrin. One mitigating factor is that the size of the spleen in Wilson's disease is almost doubled and this may permit a more rapid turnover of plasma iron despite the low plasma ferroxidase activity.

Our data now provide a reasonable explanation why there may be no obvious defect in iron metabolism in Wilson's disease. Despite the reduction in Cp, sufficient ferroxidase activity remains to prevent anemia. Remember that less than one-tenth of the normal level of ferroxidase is sufficient to cause a maximum mobilization of iron in both *in vivo* and *in vitro* experimental systems (*46*, *50*). Thus, in Wilson's disease the level of serum ferroxidase may never become sufficiently low to be rate-limiting in iron utilization. It was also mentioned earlier that the ferroxidase activity in Wilson's disease serum was higher than expected from the aryldiamine oxidase activity. Richard W. Topham in our laboratory has been able to isolate a ferroxidase quite distinct from Cp which appears to remain relatively intact in Wilson's disease serum (*27*). Another possible explanation was offered by Hansen *et al.* (*52*) who proposed citrate as an alternative ferroxidase in low Cp serum.

Other Support for the Ferroxidase Proposal

I now wish to cite a number of relevant experiments which seem to lend further credibility to the role of ferroxidase. The first concerns

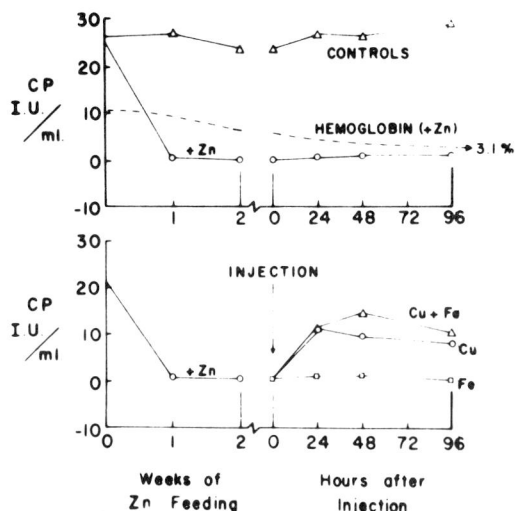

Proceedings of the Society for Experimental Biology and Medicine

Figure 15. Effect of feeding 0.75% zinc in the diet on the ceruloplasmin levels in otherwise normal rats. In the lower figure the injection of 100 μg Cu(II) and 400 μg iron or both are shown, indicating no restoration by iron alone and partial restoration by copper. Data adapted from Lee and Matrone, Ref. 53.

zinc-induced anemia in rats. It has been known for some time that feeding zinc to rats produces toxicity and a severe anemia. Following our suggestion that Cp may be a rate-limiting factor in hemoglobin biosynthesis, Lee and Matrone (53) studied changes in Cp as shown in Figure 15. They observed a dramatic fall in Cp to zero within seven days preceding the development of anemia in these rats. The Cp level could be partially restored by injecting 100 μg Cu. These experiments raise several fascinating questions about the effect of zinc. Part of the zinc effect may be owing to an interference with copper absorption as shown by Starcher (54), who found that both zinc and cadmium act as copper antagonists by displacing copper from a duodenal protein (M.S. 10,000) found normally in the duodenal mucosa. We suspect that zinc may interfere with copper utilization in the biosynthesis of Cp.

An interesting correlation was recently noted between serum iron and serum copper in a variety of experimental animals (55). In cattle, rabbits, chickens, etc., a high correlation was found between serum copper, serum iron, and total iron-binding capacity. The authors consider this to be a logical consequence of the role of ferroxidase in the transport of iron in the serum.

Ceruloplasmin Changes in Developmental Systems

The variation in plasma levels of Cp during pregnancy and other physiological states has been widely documented (13, 17, 56). More recent experiments indicate variation in the kind of Cp molecule that is produced in response to modified hormonal or other states. For instance, a recent paper of Milne and Matrone (57) identifies two ceruloplasmins in the growing neonatal pig. These two forms differ in their enzymic activity and even in their copper content, but not in their blue color or molecular weight. The sera from two-day-old pigs contained only one form of Cp with a low copper content. As the pigs grew older, the amount of the new Cp increased while the earlier form remained relatively constant. These changes may be related to switches in hemoglobin biosynthesis occurring during the parturition period.

Another interesting find in the comparative biochemistry of Cp is the observation that a significant change in the plasma oxidase activity occurs during the metamorphosis of the bullfrog tadpole. Despite earlier reports raising questions about the presence of plasma oxidase activity in amphibians, Inaba and Frieden (58) were able to show the existence of plasma oxidase activity and associate it with a blue copper protein similar to human Cp. The aryldiamine oxidase activity of bullfrog tadpoles was determined at various stages at the optimum pH 5.2 using dialyzed plasma. The oxidase activity increased gradually during the early stages of metamorphosis, reached a maximum level at metamorphic climax, and declined immediately after, although the frog level is somewhat higher. The frog enzyme was purified almost 900-fold and compared with human Cp. The properties of frog Cp are remarkably similar to crystalline human Cp in copper content, specific activity, etc., except for the sensitivity to inhibitors such as azide and HCO_3^-. This rapid change in Cp activity may be related to the remarkable switch in hemoglobin biosynthesis which takes place just before the peak of morphogenetic changes in the tadpole. Thus, this developmental increase in Cp may facilitate the mobilization of iron for rapid biosynthesis of the new hemoglobins which accompany metamorphosis.

In conclusion, we believe there is now ample evidence that the copper protein of serum, ceruloplasmin, is a molecular link between copper and iron metabolism. It has been shown both *in vitro* and *in vivo* to be directly involved in iron mobilization, the rate of formation of Fe(III) transferrin, and perhaps ultimately of hemoglobin biosynthesis. While the precise mechanism of this effect is still under investigation, we believe that it is directly related to the ferroxidase activity of this serum enzyme.

Kinetics of Ferroxidase Activity; Substrate Activation

Previously we have described the effect of Fe(II) and oxygen on the Cp-catalyzed oxidation of Fe(II) and compared these rates with the nonenzymic oxidation of Fe(II) under physiological conditions. The following is some of our recent work on the kinetics of ferroxidase. One of the fundamental questions that arose was the nonlinearity of the usual kinetic plots (29) leading to the assignment of two apparent K_m values for Fe(II), 0.6 μM, 50 μM. Curzon (59) has also reported nonlinear kinetics for an aromatic diamine substrate. Huber and Frieden (60) have excluded the possibility of two or more enzymes with experiments indicating the chemical and kinetic homogeneity of crystalline ferroxidase. The kinetics of Fe(II) oxidation could be accounted for by a mechanism represented in outline form in Figure 16 based on substrate activation resulting in a rate expression of the form

$$v = 240 \ E \ \frac{k_{xiii} \ (K_a[Fe(II)] + k_{13}/K_{xiii})}{K_a[Fe(II)] + 1}$$

Curves calculated from this expression fit the experimental points within allowed error at several temperatures as shown in Figure 17. The rate constant, k_{xiii}, for the proposed rate-determining step was 7–10 times greater for the Fe(II)-activated reaction than for the rate constant, k_{13}, for the nonactivated route. The reaction mixture contains 0.083 M che-

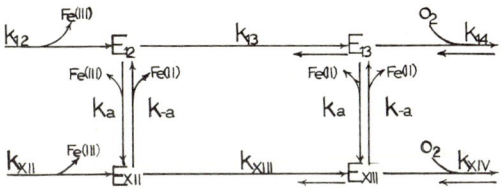

Figure 16. *The essential steps in the proposed activation mechanism. The enzyme forms in the unactivated pathway (arabic numbers) are converted to the corresponding activated forms (Roman numerals) by the binding of Fe(II) to the site of activation with the equilibrium constant* K_a. *Data from Huber and Frieden, Ref. 60.*

lexed acetate buffer, pH 6.0, 100 μM chelexed ascorbate, and 0.120 μM chelexed ferroxidase. Every point is the average of at least three experiments. The drawn curve is calculated from the activation model. The parameters used for calculation at each temperature are as follows:

Temp.	K_a, M^{-1}	k_{13}, Sec^{-1}	k_{xiii}, Sec^{-1}
36.9 ± 0.1	1.9 × 10^{+5}	0.43	3.0
30.0 ± 0.1	1.5 × 10^{+5}	0.29	2.1
25.0 ± 0.1	1.8 × 10^{+5}	0.17	1.3
20.0 ± 0.1	1.6 × 10^{+5}	0.11	1.1
15.3 ± 0.1	1.2 × 10^{+5}	0.091	0.91

The loss of absorption at 610 nm in the presence of low concentrations of Fe(II) and the activation observed with other divalent metal ions (Table VI), which are not substrates of ferroxidase, are also consistent with this mechanism (60).

Inhibition of Ferroxidase by Trivalent and Other Metal Ions

The inhibitory effects of trivalent and other metal ions on ferroxidase (ceruloplasmin) activity were investigated by Huber and Frieden (61) and the results are summarized in Tables VI and VII. All trivalent cations tested inhibited ferroxidase activity but the strong trivalent inhibitors had an ionic radius of 0.81 Å or less. The inhibition by Al(III) was mixed competitive and uncompetitive with respect to the substrate, Fe(II). The uncompetitive portion of the inhibition was not the result of competition by Al(III) with the other substrate, oxygen. A mechanism for the mixed inhibition by Al(III) was proposed consistent with these

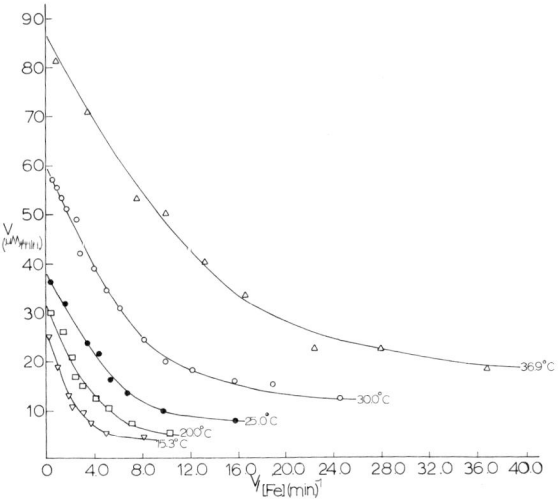

Journal of Biological Chemistry

Figure 17. Plot of velocity of Fe(II) oxidation vs. velocity/Fe(II) at various temperatures. Data from Huber and Frieden, Ref. 60.

Table VI. Effect of Cations on Ferroxidase Activity[a]

Inhibitory:
VO(II), ZrO(II) — Strong
In(III), Al(III), Sc(III), Ga(III)
Y(III), La(III), Ce(III)
Rh(III), Cr(III)
Ni(II), Zn(II), Pb(II), (high conc.) — Weak

Activating:
Fe(II) ≫ Co(II) > Mn(II) > Ni(II); Zn(II), Cd(II), Mg(II)

No Effect (at < 1 mM):
Li(I), Na(I), K(I), Sn(II), Ba(II), Ca(II)

[a] Data summarized from Huber and Frieden, Ref. *60, 61*.

Table VII. Kinetic Characteristics of the Inhibition of Ferroxidase by the Strongest Metal Ion Inhibitors[a]

Cation	Linear, Nonlinear	I/E	$I_{50}\%$, μM	K_I, M
Al(III)	L	1.9	2.1	6.2×10^{-12}
In(III)	L	1.0	0.45	4.7×10^{-7}
Sc(III)	L	1.2	13	—
Ga(III)	NL	—	~50	—
ZrO^{2+}	NL	—	0.5–2.0	—

[a] All reversible inhibition. Data from Huber and Frieden, Ref. *61*.

results. A comparison of the strong cationic inhibitors provided the following series in order of decreasing effectiveness of inhibition: In(III) > ZrO(II) > Al(III) > Sc(III) > Ga(III). Insofar as they overlap, these results are consistent with some earlier studies on the inhibition of aryldiamine oxidase activity of human Cp reported by Curzon (20) and McDermott et al. (24).

Effect of Anions on Rat Ceruloplasmin

In contrast to human and most other Cps, the oxidase activity of purified rat Cp and dialyzed rat serum was found by Lovstad and Frieden (62) to follow typical Michaelis–Menten kinetics. Therefore, the kinetics of the inhibition of the oxidase activity by several anions were investigated (62). Halide ions, NO_3^-, OCN^-, SCN^-, and N_3^- gave parallel lines in a double reciprocal plot (v^{-1} vs. S^{-1}), indicating uncompetitive inhibition kinetics. The order of inhibitory activity was: $N_3^- >$ $SCN^- >$ $OCN^- >$ $F^- >$ $I^- >$ $NO_3^- >$ $Br^- >$ Cl^-. The degree of inhibition of partially purified rat Cp by the weaker inhibitors (Cl^-, Br^-, NO_3^-, and I^-) was inversely proportional to the hydrated ionic radius of these anions, indicating a possible steric hindrance during binding. The binding of the anions to rat Cp was characterized by negative enthalpies and entropies. When dialyzed rat serum was used as the source of oxidase activity, the halide ions, NO_3^-, and SCN^- gave noncompetitive inhibition kinetics, whereas OCN^- and N_3^- inhibited uncompetitively as with the purified enzyme. Despite the kinetic differences, the effect of anions on rat Cp were comparable to their effect on human Cp as reported by Curzon et al. (63, 64).

Acknowledgment

This work was supported by grant No. 5 ROL He 08344 from the National Heart Institute, National Institutes of Health. Some of the material presented here was taken from a report by E. Frieden and S. Osaki to the Second Rochester Conference on Toxicity, "Heavy Metals and Cells," in June 1969. Appreciation is expressed to G. E. Cartwright for making available information in advance of publication. This is paper No. 37 from this laboratory in a series on copper biosystems.

Literature Cited

(1) Hill, C. H., Starcher, B., Kim, C., *Federation Proc.* (1967) **26**, 129.
(2) O'Dell, B. L., *Federation Proc.* (1968) **27**, 202.
(3) Henkin, R. I., Keiser, H. R., Jaffe, I. A., Sternlieb, T., Scheinberg, I. H., *Lancet* (1967) **2**, 1268.

(4) Cohen, E., Elvehjem, C. A., *J. Biol. Chem.* (1939) **107**, 97.
(5) Hart, E. B., Steenbock, H., Waddell, J., Elvehjem, C. A., *J. Biol. Chem.* (1928) **77**, 797.
(6) Lahey, M. E., Gubler, C. J., Chase, M. S., Cartwright, G. E., Wintrobe, M. M., *Blood* (1952) **7**, 1053.
(7) Owen, C. A., Jr., Hazelrig, J. B., *Am. J. Physiol.* (1968) **215**, 334.
(8) Cartwright, G. E., Gubler, C. J., Bush, J. A., Wintrobe, M. M., *Blood* (1956) **11**, 143.
(9) Lee, G. R., Cartwright, G. E., Wintrobe, M. M., *Proc. Soc. Exptl. Biol. Med.* (1968) **127**, 977.
(10) Osaki, S., Johnson, D. A., Frieden, E., *J. Biol. Chem.* (1966) **241**, 2746.
(11) Holmberg, C. G., Laurell, C. B., *Acta Chem. Scand.* (1948) **2**, 550.
(12) Jamison, G. A., in press.
(13) Scheinberg, I. H., *in* "The Biochemistry of Copper," p. 513, J. Peisach, P. Aisen, W. E. Blumberg, Eds., Academic, New York, 1966.
(14) Broman, I., *in* "Molecular Basis of Some Aspects of Mental Activity," O. Walaas, Ed., Academic, New York, 1967.
(15) Shokeir, M. H. K., Shreffler, D. C., *Proc. Natl. Acad. Sci.* (1969) **62**, 867.
(16) Frieden, E., Osaki, S., Kobayashi, H., *J. Gen. Physiol.* (1965) **49**, 213.
(17) Osaki, S., McDermott, J. A., Johnson, D. A., Frieden, E., *in* "The Biochemistry of Copper," p. 559, J. Peisach, P. Aisen, W. E. Blumberg, Eds., Academic, New York, 1966.
(18) Rambach, W. A., Shaw, R. A., Alt, H. L., *in* "Erythropoiesis," L. O. Jacobsen and M. Doyle, Eds., Grune and Stratton, New York, 1962.
(19) Frieden, E., *Nutr. Rev.* (1970) **28**, 87.
(20) Curzon, G., *Biochem. J.* (1960) **77**, 66.
(21) Curzon, G., O'Reilly, S., *Biochem. Biophys. Res. Commun.* (1960) **2**, 284.
(22) Curzon, G., *Biochem. J.* (1961) **79**, 656.
(23) Osaki, S., McDermott, J. A., Frieden, E., *J. Biol. Chem.* (1964) **239**, 3570.
(24) McDermott, J. A., Huber, C. T., Osaki, S., Frieden, E., *Biochem. Biophys. Acta* (1968) **151**, 541.
(25) Curzon, G., Young, S., personal communication.
(26) Johnson, D. A., Osaki, S., Frieden, E., *J. Clin. Chem.* (1967) **13**, 142.
(27) Topham, R., Frieden, E., unpublished data.
(28) Frieden, E., *J. Biol. Chem.* (1971) **246**.
(29) Osaki, S., *J. Biol. Chem.* (1966) **241**, 5053.
(30) Laurell, C. B., *in* "The Plasma Proteins," Vol. I, F. W. Putnam, Ed., Academic, New York, 1960.
(31) Moore, C. V., *in* "Iron Metabolism," F. Gross, Ed., Springer-Verlag, Berlin, 1964.
(32) Moore, C. V., Arrowsmith, W. R., Welch, J., Minnick, V., *J. Clin. Invest.* (1939) **18**, 553.
(33) Frieden, E., Osaki, S., *in* "Heavy Metals and Cells," Second Rochester Conference on Toxicity, Wiley, Somerset, N. J., 1970, in press.
(34) Pape, L., Multani, J. S., Stitt, C., Saltman, P., *Biochemistry* (1968) **7**, 613.
(35) Jandl, J. H., Katz, J. H., *J. Clin. Invest.* (1963) **42**, 314.
(36) Morgan, E. H., Laurell, C. B., *Brit. J. Haematol.* (1963) **9**, 471.
(37) Pollycove, M., Mortimer, R., *J. Clin. Invest.* (1961) **40**, 753.
(38) Katz, J. H., Jandl, J. H., *in* "Iron Metabolism," p. 118, F. Gross, Ed., Springer-Verlag, Berlin, 1964.
(39) Heilmeyer, F., *in* "Iron Metabolism," p. 201, F. Gross, Ed., Springer-Verlag, Berlin, 1964.
(40) Lee, G. R., Nacht, S., Lukens, J. N., Cartwright, G. E., *J. Clin. Invest.* (1968) **47**, 2058.
(41) Manis, J. G., Schachter, D., *Am. J. Physiol.* (1962) **203**, 73.
(42) Helbock, H., Saltman, P., *Biochem. Biophys. Acta* (1967) **135**, 979.

(43) Britten, G. M., Chee, Q. T., *J. Lab. Clin. Med.* (1969) **74,** 53.
(44) Marston, H. R., Allen, S. H., *Nature* (1967) **215,** 645.
(45) Ragan, H. A., Nacht, S., Lee, G. R., Bishop, C. R., Cartwright, G. E., *Am. J. Physiol.* (1969) **217,** 1320.
(46) Lee, G. R., Roeser, H. P., Nacht, S., Cartwright, G. E., *J. Clin. Invest.* (1970) Abstr. 177.
(47) Cartwright, G. E., Lee, G. R., personal communication.
(48) Morell, A. G., Van Den Hamer, C. J. A., Scheinberg, I. H., Ashwell, G., *J. Biol. Chem.* (1966) **241,** 3745.
(49) Osaki, S., Johnson, D. A., *J. Biol. Chem.* (1969) **244,** 5757.
(50) Osaki, S., Johnson, D. A., Topham, R., Frieden, E., *Federation Proc.* (1970) **29,** 695.
(51) O'Reilly, S., Pollycove, M., Bank, W. J., *Neurology* (1968) **18,** 634.
(52) Hansen, S. P., Nacht, S., Lee, G. R., Cartwright, G. E., *Proc. Soc. Exptl. Biol. Med.* (1969) **131,** 918.
(53) Lee, D., Jr., Matrone, G., *Proc. Soc. Exptl. Biol. Med.* (1969) **130,** 1190.
(54) Starcher, B., *J. Nutr.* (1969) **97,** 321.
(55) Planas, J., Balasch, J., *Rev. Espan. Fisiol.* (1970) **26,** 91.
(56) Evans, G. W., Cornatzer, N. F., Cornatzer, W. E., *Am. J. Physiol.* (1970) **218,** 613.
(57) Milne, D. B., Matrone, G., *Federation Proc.* (1969) **28,** 556.
(58) Inaba, T., Frieden, E., *J. Biol. Chem.* (1967) **242,** 4789.
(59) Curzon, G., *Biochem. J.* (1967) **103,** 289.
(60) Huber, C. T., Frieden, E., *J. Biol. Chem.* (1970) **245,** 3973.
(61) Huber, C. T., Frieden, E., *J. Biol. Chem.* (1970) **245,** 3979.
(62) Lovstad, R., Frieden, E., in press.
(63) Curzon, G., Speyer, B. E., *Biochem. J.* (1967) **105,** 243.
(04) Curzon, G., Cummings, J. N., *in* "The Biochemistry of Copper," p. 545, J. Piesach, P. Aisen, W. E. Blumberg, Eds., Academic, New York, 1966.

RECEIVED June 26, 1970.

15

Clostridial Ferredoxin: An Iron–Sulfur Protein

JESSE C. RABINOWITZ
University of California, Berkeley, Calif. 94720

> *Iron–sulfur proteins occur in animal, plant, and bacterial cells. The proteins are characterized by the presence of 1–0, 2–2, 4–4, 6–6, or 8–8 atoms of iron–sulfide. Only the structure of clostridial rubredoxin, a 1–0 protein, is known. It contains iron ligated to four sulfur atoms of cysteine residues of the polypeptide. With the exception of the "high potential iron protein," all the proteins show unexpectedly low redox potentials and function in biological oxidation–reduction reactions.*

A list of the known iron–sulfur proteins representative of the various classes now recognized and some of their properties is shown in Table I. One of the unusual characteristics of these proteins is the presence of "inorganic" sulfide. This is indicated by an S* since we are not sure of its exact chemical nature. It is released as H_2S when these proteins are acidified. The amount can be determined in a colorimetric procedure with dimethylphenylenediamine (40). There are equal amounts of iron and sulfur: 2, 4, 6, or 8 atoms of each in these proteins. Also listed, for reasons which will become apparent, is an example of an iron protein, rubredoxin, which does not contain the sulfide.

Most of the proteins listed have been isolated from bacteria but a very important example, the chloroplast ferredoxin, is found in green plants, and adrenodoxin is found in mammalian tissue.

The molecular weights of these proteins are quite small, and numbers of amino acids range from 55 to about 100. These proteins are not enzymes; they function as electron carriers and usually have to function in conjunction with some other high-molecular-weight protein that has enzymic activity.

Another noteworthy characteristic is the very low potential of the 2-iron–2-sulfur proteins and the 8-iron–8-sulfur proteins. However, these proteins exhibit a wide range of redox values. At one extreme is the "high potential iron protein" (HIPIP), isolated from chromatium, that has a redox potential of +350 mv. Rubredoxin has a redox potential of −57 mv, a value considerably less negative than the ferredoxins.

Most of these compounds function as one-electron carriers. However, it is very clear from more recent experiments that the 8-iron–8-sulfur protein from clostridium functions as a two-electron carrier (34, 35, 36).

The oxidation state of the iron in these proteins is really a difficult thing to determine. If one simply releases the iron by treatment with acid, one finds all the iron in the ferrous state, even in the so-called oxidized form of clostridial ferredoxin (28). This is because the iron gets reduced by the sulfur and sulfhydryl groups present in these proteins. If one treats the protein with a sulfhydryl reagent and prevents this kind of reduction by sulfhydryl groups, one finds 4 Fe^{2+} and 4 Fe^{3+} (34). But even this is misleading, and we think the concept of the oxidation state of the iron in these proteins is not a meaningful one; the electron is probably shared by the iron and sulfur atoms in the protein, and they function as electron carriers with the uptake of either one or two electrons by the iron–sulfur chromophore.

Another very important characteristic of the protein—at least, it was important for the isolation of several of these proteins (23)—is the EPR signal. The proteins exhibit varied properties with respect to this parameter. An important similarity between those proteins that are called ferredoxins, the 2-iron–2-sulfur and 8-iron–8-sulfur proteins, is the fact that they show a 1.94 EPR signal in the reduced form, usually achieved chemically. This signal is not observed with the oxidized form. Rubredoxin and "high potential iron protein" behave differently. This signal is very temperature-dependent, and for a long time the clostridial protein was not believed to show the EPR signal. My own feeling is that a little drop of HCl can tell you as much about the classification of these proteins as the EPR signal; you can easily smell H_2S if you just drop a little acid on any of these proteins.

The absorption spectra of the main types of the iron–sulfur proteins is shown in Figure 1. There are very significant differences that have been overlooked with regard to the nomenclature of these proteins. All the proteins examined—the chloroplast ferredoxin, the clostridial ferredoxin, and rubredoxin—show absorption at 280 nm in the oxidized form shown here, but the other absorption peaks differ. The clostridial ferredoxin has an absorption peak at 390 nm but that peak is missing in the plant type ferredoxin. Rubredoxin shows an absorption maximum at 390 nm, but it shows other absorption peaks at 500 and 580 nm which are absent from the clostridial type protein.

Table I. Iron–

Fe–S*	Source	M.W.	# Amino Acids	Absorption (Ox.), Nm
1–0 (Rubredoxin)				
	Clostridium	5,600 (*1*)	52–54 (*2,3*)	380 490
	Pseudomonas	20,000 (*5*)		377 495 (*6*)
2–2 (Ferredoxin)				
	Chloroplast	10,600	96–97 (*7–11*)	331 420 465 (*12*)
	Putidaredoxin	12,000 (*14*)		328 412 455 (*14*)
	Adrenodoxin	13,094 (*18*)	118 (*18*)	320 414 455 (*19*)
	Azotobacter-I (*23*)	21,000		331 419 460
	Azotobacter-II (*23*)	24,000		344 418 460
4–4 (High potential iron protein; HIPIP)				
	Chromatium (*24, 25*)	10,074		375
6–6 (Azotobacter ferredoxin)				
	Azotobacter (*26*)	20,000		320–335 375–415
	Azotobacter-III (*27*)	3,000		320–335 375–415
8–8 (Ferredoxin)				
	Clostridium	6,200 (*28*)	54–55 (*29–32*)	390 (*33*)
	Chromatium	9,600 (*38*)	81 (*38*)	390 (*39*)

It is interesting to note that the UV spectrum for the ferredoxin from chromatium, a photosynthesizing bacteria, suggests that it is very similar, if not identical, to that of the ferredoxin from clostridia, which are nonphotosynthetic organisms. It was originally suggested that this is a form of ferredoxin intermediate between the plant type and the clostridial type, because chemical analysis suggested that the iron content was intermediate between the chloroplast and clostridial type ferredoxin.

On either enzymatic or chemical reduction, these peaks disappear but the absorption does not go to zero. However, absorption at 280 nm remains unchanged in all cases. This absorption peak at 280 nm must be a function of the chromophore since removal of the iron and sulfide in a way to be described shortly causes this absorption to fall very markedly, whereas reduction has little effect on this absorption at 280 nm. I would like to emphasize the difference that exists between the spectra of the chloroplast ferredoxin and the clostridial ferredoxin. Even though they are both called ferredoxins, I think it remains to be determined how closely these compounds are related.

The biological function of these proteins is summarized in Table II. The biological function of rubredoxin isolated from the clostridia is not known. Lovenberg and Sobel (*1*) first isolated the beautiful crystals as

Sulfur Proteins

E'_o, mV	n	Electron Paramagnetic Resonance	
		Ox.	Red.
− 57 (1)	1 (1)	4.3, 9.4 (4)	None
− 37 (6)		4.3, 9.4 (6)	None
−432 (12)	1 (13)	None	1.94 (4)
−235 (15)	1 (16)	None	1.94 (17)
−367 (20)	1 (21)	None	1.94 (22)
	1	None	1.94
	1	None	1.94
+350	1	2.04	None
−410 (12)	2 (34–36)	None	1.94 (37)

a by-product in the purification of ferredoxin from the clostridia, but its biochemical role is still unknown. There is another compound isolated from pseudomonas that has been classified as a rubredoxin and functions as an electron carrier in ω-hydroxylation of the fatty acids (41). Oxygen is used in the system; there is an electron transfer through a chain to the pyridine nucleotides, and the rubredoxin undergoes oxidation–reduction in the process.

In a sense, this is very similar to the reactions catalyzed by the adrenodoxin and putidaredoxin, the 2-iron–2-sulfur proteins which catalyze oxidation of methylene groups. Adrenodoxin is required in the conversion of the steroid 11-β-methylene group to a hydroxy group, and again this occurs *via* an electron-carrying chain, ultimately coupled to pyridine nucleotides (19). The putidaredoxin catalyzes a hydroxylation of a camphor methylene group through a similar electron transfer chain (14). Adrenodoxin and putidaredoxin appear to be specific in these enzymic reactions.

The reaction catalyzed by the chloroplast ferredoxin is shown in Table II. It has been assumed that chloroplast ferredoxin acts as the primary electron acceptor in the photoreduction catalyzed by chloroplasts (42). However, there is some question as to whether ferredoxin is the

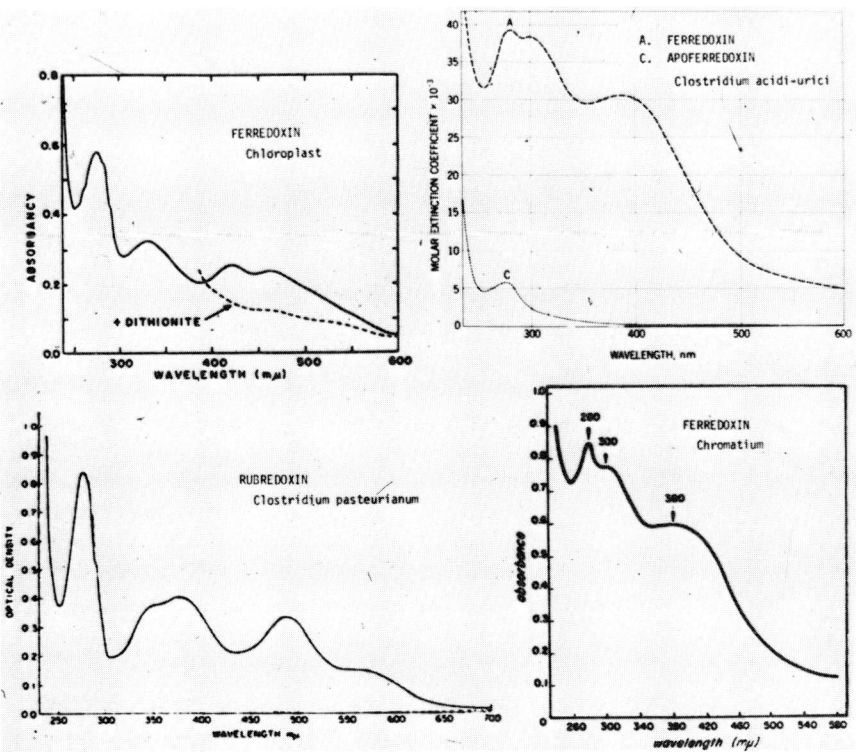

Figure 1.   Absorption spectra of iron–sulfur proteins

primary acceptor, or whether another factor, FRS (ferredoxin-reducing substance) (43), is the primary acceptor and ferredoxin acts as a component of the electron transfer chain.

Clostridial ferredoxin was discovered by virtue of its requirement in the oxidation of pyruvate by the clostridial species specifically involved in a nitrogen-fixing system (44). We can now interpret the requirement for pyruvate oxidation as a reaction leading to the formation of reduced ferredoxin, which in turn acted in the reduction of the various substrates, including N_2 (45). But there are many other substrates which can be reduced by the reduced ferredoxin—i.e., in the presence of the appropriate enzyme. For example, nitrite may be reduced to ammonia. Reduced ferredoxin can be formed by nonphotosynthetic bacteria by reactions other than puruvate oxidation; it can also be formed by the hydrogenase, or by the oxidation of a purine.

Clostridial ferredoxin is also photo-reduced by chloroplast preparations, and it is one of the chief reasons why the name ferredoxin has been used interchangeably for both the 8-iron–8 sulfur protein and the 2-iron–

Table II. Physiological Role of Iron–Sulfur Proteins

1 Fe–0 S*
   Rubredoxin (Fatty Acid ω-hydroxylation)

$$\text{DPNH} \diagdown \text{Reductase}_{ox} \diagdown \text{Rubredoxin}_{red} \diagdown \begin{array}{l} \text{R—CH}_3 \\ \text{O}_2 \\ \text{H}_2\text{O} \end{array}$$
$$\text{DPN}^+ \diagup \text{Reductase}_{red} \diagup \text{Rubredoxin}_{ox} \diagup \text{R—CH}_2\text{OH}$$

2 Fe–2 S*
a. Adrenodoxin (Steroid 11-β-hydroxylation)

$$\text{TPNH} \diagdown \text{Adrenodoxin}_{ox} \diagdown \text{cyt P-450}_{red} \diagdown \begin{array}{l} \text{Steroid} \\ \text{O}_2 \\ \text{H}_2\text{O} \end{array}$$
$$\text{TPN}^+ \diagup \text{Adrenodoxin}_{red} \diagup \text{cyt P-450}_{ox} \diagup \text{Hydroxysteroid}$$

b. Putidaredoxin (Camphor hydroxylation)

$$\text{TPNH} \diagdown \text{Putidaredoxin}_{ox} \diagdown \text{cyt P-450}_{red} \diagdown \begin{array}{l} \text{Camphor} \\ \text{O}_2 \\ \text{H}_2\text{O} \end{array}$$
$$\text{TPN}^+ \diagup \text{Putidaredoxin}_{red} \diagup \text{cyt P-450}_{ox} \diagup \text{Hydroxycamphor}$$

c. Chloroplast Ferredoxin

$$4\,\text{Fd}_{ox} + 2\,\text{H}_2\text{O} \xrightarrow{\text{Chloroplasts},\, h\nu} 4\,\text{Fd}_{red} + \text{O}_2 + 4\,\text{H}^+$$

$$2\,\text{TPN}^+ + 4\,\text{Fd}_{red} \longrightarrow 2\,\text{TPNH} + 2\,\text{H}^+ + 4\,\text{Fd}_{ox}$$

$$2\,\text{TPN}^+ + 2\,\text{H}_2\text{O} \xrightarrow[\text{Ferredoxin}]{\text{chloroplast},\, h\nu} 2\,\text{TPNH} + \text{O}_2 + 6\,\text{H}^+$$

8 Fe–8 S*
   Clostridium Ferredoxin
$$\text{CH}_3\text{COCOOH} + \text{CoA} + \text{Fd}_{ox} \rightleftarrows \text{CH}_3\text{CO-SCoA} + \text{CO}_2 + \text{Fd}_{red}$$
$$\text{Fd}_{ox} + \text{H}_2 \rightarrow \text{Fd}_{red} + 2\,\text{H}^+$$
$$\text{Fd}_{red} + \text{NO}_2^- \rightarrow \text{Fd}_{ox} + \text{NH}_4^+$$

2-sulfur protein, although these differ chemically and physically in very significant ways. One is a one-electron acceptor and the other is a two-electron acceptor; the amount of iron and sulfur vary. The polypeptide portions of the proteins differ a great deal as well.

The reaction catalyzed by the clostridial pyruvate oxidase is a reversible reaction. In the presence of reduced ferredoxin, CO_2 may be fixed to form pyruvate, which then can be used in various metabolic reactions and is believed to be a very important reaction in certain of the photosynthetic bacteria where the reaction was first discovered (46). In the clostridia, however, the enzyme appears to function in the direction of pyruvate oxidation (47).

The isolation of the clostridial ferredoxin is remarkably simple (Table III). This is a very elegant isolation devised by Mortenson (48) and based on the fact that the protein is very small and soluble in acetone, whereas most soluble proteins will be precipitated. It is an extremely

Table III. Purification of Ferredoxin from *Clostridium pasteurianum*[a]

Fraction	Activity, Units per Mg.
Acetone extract	54
DEAE-cellulose	307
Ammonium sulfate (75–90%)	468

[a] Constructed from data in Ref. 48.

acidic protein, very readily absorbed on the anion exchanger, DEAE-cellulose, and can be eluted with high salt and crystallized by the addition of ammonium sulfate. The amount of ferredoxin in a clostridial cell varies, but one can obtain roughly about a milligram of crystalline protein per gram of cells.

The crystals of several clostridial ferredoxins are shown in Figure 2. It indicates some of the problems involved in the determination of the structure of the protein. The crystalline forms from different clostridial species differ, but all are very small crystals. The largest ones I have seen are these from *Clostridium acidi-urici,* but even these are very small as far as a crystallographer is concerned, and unfortunately, they are multiple crystals. There has been very little work done on the structure by x-ray crystallography. One problem is the small size of the crystals; the second difficulty is the fact that it has not been possible to obtain any heavy metal replacement for these proteins. However, Jensen has gotten some information on clostridial type ferredoxin that I will discuss a little later. The plant type ferredoxin has not been obtained in crystalline form that is suitable for any x-ray work.

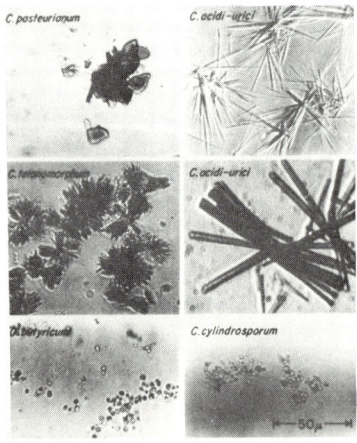

Figure 2. Crystals of several clostridial ferredoxins

The complete structure of clostridial rubredoxin has been solved by Jensen and his associates using x-ray analysis (49), and the structure of the iron–sulfur center of the "high potential iron protein," but not of the polypeptide, has been determined by x-ray analysis by Kraut and associates (50). Both of these are relatively atypical proteins in this class, either having no inorganic sulfide or having a very high redox potential. So far, we still have no solution from x-ray work for the 2-iron–2-sulfur or the 8-iron–8-sulfur type protein.

The iron and the sulfur can be removed from the chloroplast and clostridial type proteins fairly readily by a variety of procedures. These are based either on treating the protein with an iron chelating agent such as α,α-dipyridyl (51), a sulfhydryl reagent such as the mercurial mersalyl (52), or simply by acidifying the proteins at zero degrees (53). In all cases, both iron and sulfide are released.

It is possible to reconstitute ferredoxins from the isolated apoprotein by the addition of iron and sulfide (52). There are a variety of methods that can be used for reconstitution, but essentially it involves the conversion of the apoprotein to its sulfhydryl form and the addition of either ferrous or ferric salts, in the presence of sodium sulfide (53) or elemental sulfur (51). Normally, because of the difficulties involved in mixing the sulfide and the iron, it is done in the presence of the reducing agent, such as mercaptoethanol, which seems to complex with the iron and sulfide to yield some soluble form of these elements that react with the apoprotein to form a ferredoxin that is indistinguishable from the native material.

Attempts have been made to replace the iron with other metals. As far as I'm aware, these have all been unsuccessful, at least in terms of isolating a ferredoxin with a metal ion other than iron. However, it has been possible to replace the labile sulfur atom with a selenium atom (16, 54), and this derivative has been very valuable for determining the ligands of the iron of putidaredoxin. However, the stability of the selenium derivative is very low compared with the sulfur analog, and at room temperature this selenium derivative shows almost no stability.

I would like next to consider the known amino acid sequences of iron–sulfur proteins because of some interesting aspects of this subject. I realize that this is an inorganic chemistry symposium, but I feel that the organic part of the molecules is very pertinent to understanding the inorganic chemistry, and I would like to demonstrate that here.

I've shown the sequence of *C. acidi-urici* ferredoxin (32) in Figure 3 to illustrate some of the aspects of the molecule. The clostridial ferredoxins contain only 55 amino acids. This diagram has nothing to do with the structure; it is just a way of presenting the material in a legible manner. However, there seems to be a symmetry in the sequence. Starting from the amino terminal end of the molecule, the first half of the molecule

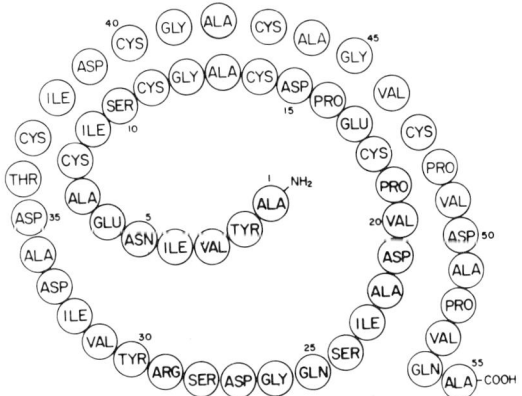

Figure 3.   Sequence of C. acidi-urici *ferredoxin*

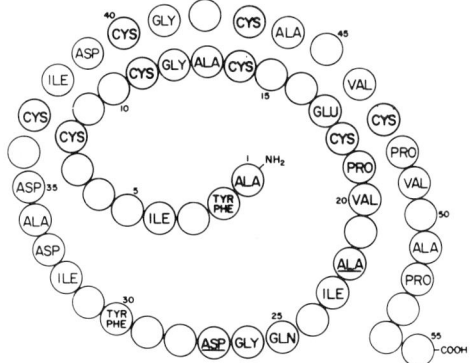

Figure 4.   Amino acids common to sequences of four clostridial ferredoxins

includes residues 1 through 29. The sequence residues tyrosine-2, valine-3, isoleucine-4, asparagine-5 is repeated as tyrosine-30, valine-31, isoleucine-32, and aspartate-33. The eight cysteine residues line up in the two halves in exactly analogous positions. In addition, the sequence cysteine-11, glycine-12, alanine-13 is identical in both halves, and another long sequence of cysteine-18, proline-19, valine-20, aspartic-21, alanine-22 is identical to residues 47–51 in the other half.

Figure 4 illustrates those amino acids common to four clostridial ferredoxins that have been sequenced and those residues that are common to all four sequences. The positions of the eight cysteine residues are common to all the proteins. There is no doubt the cysteine acts as a ligand for the iron as well as the sulfide on the basis of much experimental evidence.

The sequences of all the iron–sulfur proteins that have been determined are shown schematically in Figure 5. The number after the name indicates the number of proteins represented. The solid lines are the cysteine residues, the number under it is the position in the polypeptide chain going from the amino to the carboxyl end. The number at the end is the number of residues in each of these type ferredoxins. Three rubredoxins have been sequenced. It might interest crystallographers particularly that one of these, *Clostridium pasteurianum* rubredoxin, was sequenced by Jensen from the x-ray data (55) and still has not been done chemically. It is an expensive way to do a sequence. Five of the chloroplast ferredoxins have been sequenced. For a long time I thought about the various iron–sulfur proteins as very different proteins. Yet the people interested in protein evolution suggested that there is a relation among them (56, 57, 58). There is an over-all similarity among the proteins despite the fact that these represent 1-Fe–0-S, 2-Fe–2-S, and 8-Fe–8-S proteins. There are no cysteines other than those indicated by the white bars. They seem to occur in two areas with relatively large sections of other amino acids between them indicated by the number in the bar. The sequence for one chromatium ferredoxin has been done (38), and this is the one that had a spectrum very similar to the clostridial one. It has been reported in the literature that this ferredoxin contains only four or at most six iron (39, 59). Its sequence of cysteines is similar to that of the clostridial ferredoxin. If one makes a loop of the peptide region between residues 41 to 50 and includes one cysteine residue in the loop, the cysteine pattern is exactly analogous to the sequence in clostridial ferredoxin. Indeed, I have recently been informed (60) that chromatium ferredoxin contains eight iron and eight sulfide as well.

The structure for the clostridial rubredoxin that has been determined by x-ray analysis (49) has clearly shown that the iron is ligated to the four cysteine residues at positions 6, 9, 38, and 41. This involves, of

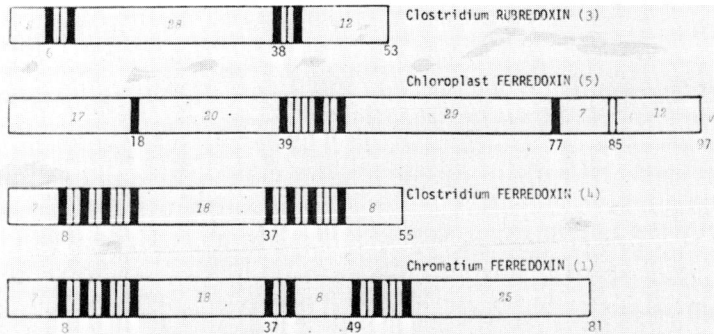

Figure 5. Locations of cysteine residues in iron–sulfur proteins

course, a folding of this polypeptide chain as shown in Figure 6 (*49*). The iron atom is surrounded by the cysteine atoms in a tetrahedral form with two above and two below, and there are no other ligands for the iron. There is nothing unusual about the bonding distances or bonding angles for the sulfurs to the iron. Thus this example clearly argues against the "entatic site" hypothesis discussed by Ulmer (*61*). Jensen has determined by x-ray analysis the structure of the reduced rubredoxin (*55*). Lovenberg has been able to show the reduction of the clostridial rubredoxin in a solid state (*62*), and Jensen has now shown that absolutely no change can be detected with respect to the position of the cysteine sulfurs around the iron in the reduced state.

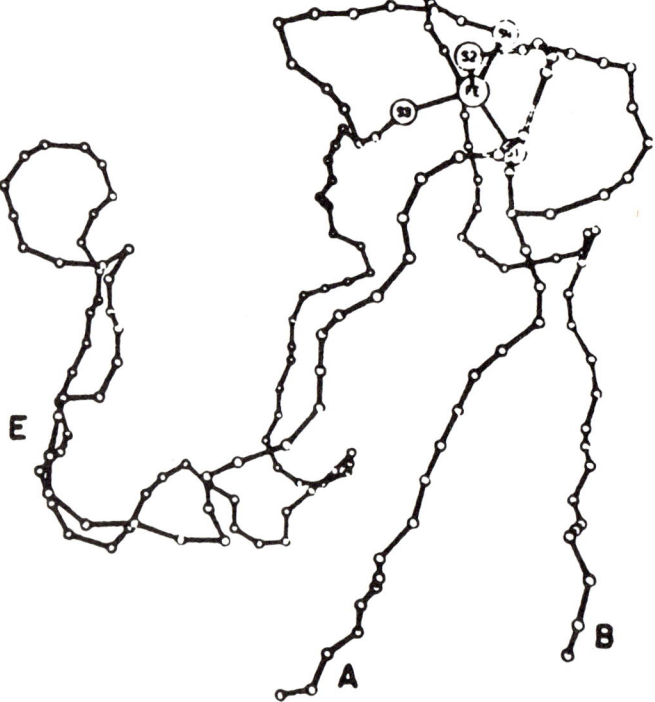

Figure 6. Folding of polypeptide chain

The structures of the other compounds are not known. In addition to the cysteine atoms, we know that there are sulfide ions involved as iron ligands in the other proteins. We don't really know whether the iron is ligated in tetrahedral or octahedral form. Personally, I feel that it will take a crystallographic solution to find these structures. The following figures are some of the structures that have been proposed.

A structure proposed by Blomstrom, Knight, Phillips, and Weiher (63) is shown in Figure 7. They suggest that the iron is ligated to four sulfur atoms, two of them from cysteine residues and two from inorganic sulfide, and that the iron atoms are in a linear array. Sieker and Jensen (64) have done some relatively low-resolution work on a clostridial ferredoxin and were unable to confirm the presence of a linear array of iron atoms in the clostridial ferredoxin molecule. They feel that if such a linear array was present, it would have been detected. On the other hand, if they were not absolutely linearly arranged, one might not have seen them, so data available so far do not necessarily eliminate this type of structure.

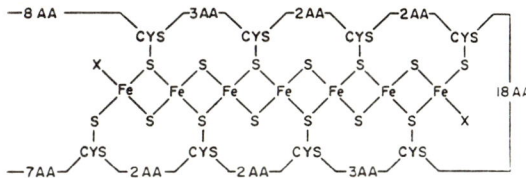

Figure 7. *Structure with iron ligated linearly*

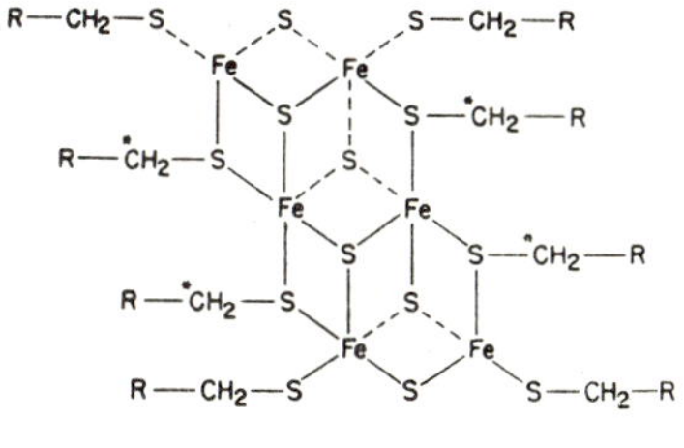

Figure 8. *Structure based on proton magnetic resonance studies*

In a more recent publication, Poe, Phillips, McDonald, and Lovenberg (65) suggested the structure shown in Figure 8. It is based on proton magnetic resonance studies. The data had suggested that the methylene groups of the cysteine residues occurred in two types of groups, four in each. They have indicated by the asterisks on methylene groups of cysteine residues those that are ligated to two iron atoms each and a second group, unmarked, which is ligated to a single iron atom each.

Several years ago Hong and I thought of another structure shown in Figure 9. It suggests that the iron is ligated to a cysteine sulfur and then a so-called persulfide sulfur and other unknown ligands. They could be residues in the polypeptide chain, or water, or who knows what. The

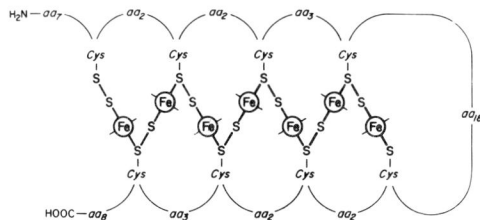

Figure 9. Structure with persulfide sulfur

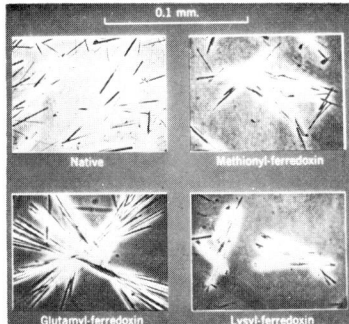

Figure 10. Crystals of Clostridium acidi-urici *ferredoxin and some derivatives*

only reason for doing this was that it seemed to me nobody was considering this type of linkage. The persulfide does occur in several biochemical systems (66), but there are only a few models for it (67); one could not do much about testing it, but we have tried.

While all of these models have emphasized the iron and sulfur part and have forgotten about the polypeptide part of it, I would like to describe some experiments that we have done that involve chemistry on the polypeptide portion because I think that the results are somewhat unexpected. Jen-Shiang Hong proposed that we try modifying the apoprotein to see what effect this would have on the reconstitution reaction, and I said to him, "Well, you know, that's sort of silly because what can we modify? We can modify the amino end." Luckily, the apoprotein in the *Clostridium acidi-urici* ferredoxin has no amino groups other than the N-terminal amino group, and it also has a particular sequence around the carboxyl group that enabled us to remove quantitatively the two terminal carboxy amino acids. My thoughts were that, if you are going to modify the end by adding an amino acid or by removing one or two, you are really not going to change the ferredoxin very much because the crucial part of the molecule is the arrangement of the cysteine residues and their bonding of the iron and sulfide.

He went ahead, nevertheless. He made derivatives by putting a methionine or other various amino acid residues on the amino terminal end and found that he could obtain crystalline derivatives, shown in Figure 10. Amino acid analyses showed that he had added only the one residue, and the crystals were very similar to the native form. So I said, "Well, see what I told you, Hong. I like crystals, too, but they are just like what we started with." It turned out that they were not.

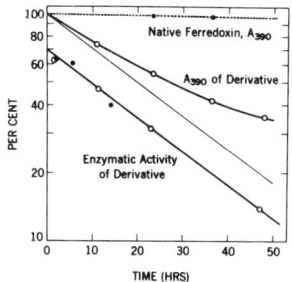

Figure 11. Stability of methionyl ferredoxin

Studies were carried out on the stability of these derivatives in solution, as shown in Figure 11. We studied either the rate of decay of the absorption maximum of 390 nm or the activity in the enzyme assay. When the iron and sulfur are removed from the protein, the absorption decreases. Native ferredoxin is very stable under the conditions we used, and no change in absorption is seen in 50 hours. The derivatives decay when examined by either test. The results of these experiments on stability of these derivatives are shown in Table IV.

We looked at activity and the half-life of the derivatives, most of which were much less stable than the native protein. The only derivative we could make that was not any less stable was the acetimido derivative which has a positive charge and is also very small. But if we put an acetyl group on the N-terminal end, which destroyed the potential positive charge, we obtained an extremely unstable derivative. Although you can incorporate the iron and sulfur into the apoprotein and can crystallize the derivative ferredoxin, when you put this derivative into solution, you find that it decays very readily. Similarly, if we take off the carboxy terminal alanine and glutamine, one obtains a ferredoxin with much lowered stability. I find these results surprising. I didn't expect to find any difference in the stability of these derivatives.

Another series of experiments was done by Hong to test the tritium exchange or the hydrogen exchange of the various forms of the ferredoxin.

Table IV. Enzymatic Activity and Stability of Modified Ferredoxins Prepared from *Clostridium acidi-urici* Apoferredoxin Derivatives

Apoferredoxin Derivative Used for Reconstitution	Enzymic Activity, %	Stability $(t_{1/2})$, Hours
Native	100	–
tert-BOC-	78	4.5
Regenerated from *tert*-BOC-apoferredoxin	100	–
Acetyl-	86	11
Acetimido-	78	132
Glycyl-	94	53
Methionyl-	85	31
Phenylalanyl-	68	20
Lysyl-	56	43
Glutamyl-	50	20
Des-(Ala^{55}-Gln^{54})-	77	31

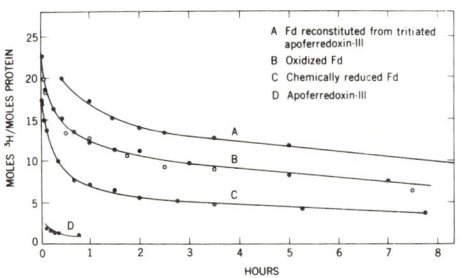

Figure 12. Rate of ^3H–H exchange

The procedure is to tritiate the various forms of the protein and to determine the rate of removal of these tritium ions. This is a function of the geometry and the shape of the protein molecule. Potentially, the apoprotein has 70 replaceable hydrogen atoms. We found that for the apoprotein (Figure 12) there are no detectable hydrogen atoms retained. In other words, it is a structureless polypeptide. The oxidized form of ferredoxin contains some slowly exchangeable hydrogen atoms. There are even fewer slowly exchangeable H atoms in reduced ferredoxin. Although the differences were small, they were very reproducible. This indicates that there is a change in the shape of the clostridial type ferredoxin on reduction and that the reduced form is a more compact form with fewer replaceable hydrogen atoms. We carried out the reconstitution with the fully tritiated ferredoxin and then measured the tritium exchange of that product. There are several more hydrogen atoms that are retained in the protein. It indicates that some of the residues become buried in the protein upon reconstitution and simply do not exchange. The oxidized and reduced forms of the protein appear to have different conformations by this criterion.

Hong also studied the immunological activity of the ferredoxins, and we were very surprised to find that these proteins were immunogenic in rabbits. These are relatively small proteins, and I think they are among the smallest that have been shown to elicit an antibody response. The antibody is very specific for the ferredoxin, and in this case we made one

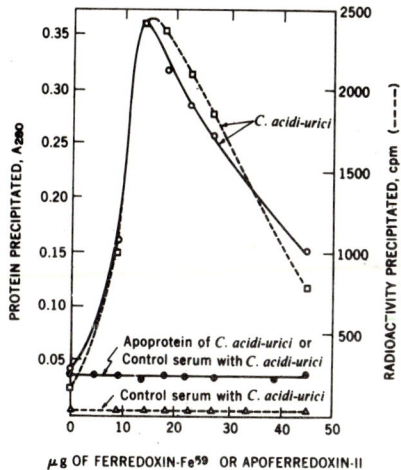

Figure 13. Precipitation reaction by antiserum to Clostridium acidi-urici ferredoxin

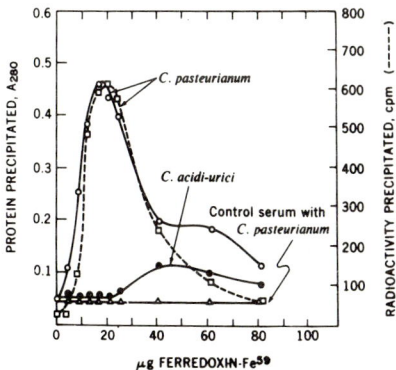

Figure 14. Precipitation reaction by antiserum to Clostridium pasteurianum ferredoxin

from a particular clostridial species and measured the reaction by using a ^{59}Fe–ferredoxin and measuring the precipitation of the antibody by measuring the ^{59}Fe or the protein in the precipitate as shown in Figure 13. The results are very similar. But interestingly enough, the apoprotein does not react with the ferredoxin antisera. One would expect the geometry of the apoprotein to be very different from that of the protein in which the iron and sulfur is contained, and this result confirms it. Figure 14 shows that the antibodies are also specific for the type of ferredoxin. That is, we prepared antibodies to the *Clostridium pasteurianum* ferredoxin and got very poor cross reactivity with the *Clostridium acidi-urici* ferredoxin.

The next series consists of some reconstitution experiments carried out by Hong. We can get yields of up to nearly 80% of ferredoxin, based on the amount of apoferredoxin we used, by adding back iron and sulfide in excess. We asked ourselves what would happen if we used less than an excess of iron or sulfide. Could we get forms of the protein with less than a full complement of iron or sulfide? Could we isolate ferredoxins with only four or two irons or sulfurs? I'll just go into a little experimental

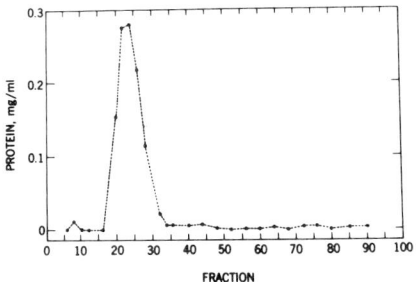

Figure 15. Apoferredoxin on DEAE-cellulose

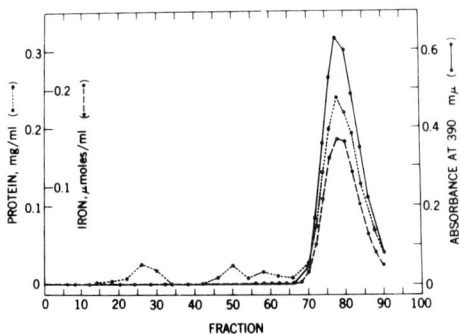

Figure 16. Ferredoxin on DEAE-cellulose

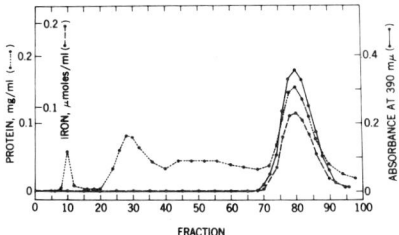

Figure 17. Absence of intermediate forms between apoferredoxin and ferredoxin

detail here. Figure 15 shows the chromatography of apoferredoxin on DEAE-cellulose, an anion exchanger. It comes off as a fairly sharp peak very early. Figure 16 shows the behavior of ferredoxin on such a chromatogram. Ferredoxin is eluted much later. There are 2-ml fractions, so there is a very good separation of the two proteins. The peaks were assayed for iron, protein, and absorption. We suspected that if we reconstituted with a quarter of the amount of the iron that is known to be present in the ferredoxin and intermediate forms of ferredoxin were formed, they would be eluted somewhere between the apoferredoxin and ferredoxin. As shown in Figure 17, we did not find any evidence for intermediate forms. There is some residue of apoferredoxin, but with this limiting amount of iron, the iron-containing product is found only in the position where ferredoxin with a full complement of eight iron atoms would be. However, the chromatogram differs from that expected because now the apoferredoxin is no longer obtained as a single sharp peak. There is significant tailing. The tail is protein, and it contains absolutely no iron. It was just an assumption that ferredoxins with intermediate contents of iron would appear in intermediate positions. It is also possible that they would all be in the ferredoxin peak, which would have a heterogeneous composition. In order to test for this, each of the tubes in this peak was analyzed. They were perfectly homogeneous. In the case where 25% of the iron was added, we found 6.5 iron atoms per mole of protein in each tube of the peak. If there were totally independent sites and noncooperativity in the iron and sulfide binding, one would not expect to find any ferredoxin molecules with eight irons in them. The experimental results gave us a derivative with less than eight atoms of iron, but this chromatogram clearly indicated that the peak might be contaminated with some apoprotein. These fractions were therefore combined and rechromatographed. The resulting peak was analyzed before crystallization and contained about a 50% yield of ferredoxin, with eight iron and eight sulfur atoms. From this we conclude that the addition of the iron and

Table V. Cyanolysis of Disulfides and Persulfides

$$R—S—S—CH_2R' + CN^- \rightarrow R—SCN + HS—CH_2R'$$
$$HS—S—CH_2R' + CN^- \rightarrow HSCN + HS—CH_2R'$$
$$HSH + NCS—CH_2R'$$

Table VI. Thiocyanate Formation by Cyanolysis of Clostridial Ferredoxin[a]

pH	Buffer, 0.61M	Fd Sulfide, mM	SCN⁻ Formed, mM
9	Tris-chloride	0.60	0.48
10	Glycine	0.60	0.49
10.6	Bicarbonate	0.60	0.41

[a]The reaction mixture contained 0.075 mM C. $acidi$-$urici$ ferredoxin and was treated with 49 mM KCN at 37° for 20 hours.

Table VII. Specificity of Thiocyanate Formation[a]

Reaction Component	Conc., mM	Thiocyanate Formed, mM
Ferredoxin	0.080	0.18
Apoferredoxin	0.095	0.0
Sodium sulfide	1.3	0.0

[a]The reaction was incubated at 37° for 7 hours in the presence of 52 mM KCN at pH 9.0.

Table VIII. Thiocyanate Formation from ³⁵S–Ferredoxin[a]

³⁵S–Ferredoxin Sulfide, mM	Na₂S, mM	Thiocyanate Formed	
		Amount, mM	Specific Act., cpm/μmole
0.48	0.0	0.169	418,000
0.48	1.13	0.393	208,000

[a]The reaction was incubated at 37° for 7.5 hours in the presence of 45 mM KCN at pH 9.0. The specific activity of the ³⁵S–ferredoxin sulfide was 335,000 cpm/μmole of sulfide.

sulfur in the reconstitution reaction is an all-or-none phenomenon and that iron-poor or sulfide-poor species do not exist. The experiments have been done with limiting sulfide with the same results. They have also been done by treating the ferredoxin with limiting amounts of an iron chelator like α,α-dipyridyl or a sulfhydryl reagent like mersalyl, and in each case one finds the only product that can be recovered is unreacted material, with a full complement of iron and sulfur.

We did think about the possibility of the occurrence of a persulfide linkage in the protein, and Ellen Wallace, a post-doctoral fellow in my laboratory, and I considered various ways of detecting persulfide. Being more chemically than physically oriented, we tried to find some color test for its detection. As shown in Table V, there is a reaction of cyanide with disulfides to give thiocyanate. Cyanide would react with the persulfide to give thiocyanate, and this can be determined colorimetrically. Based on this, we went ahead and treated ferredoxin with cyanide and attempted to detect thiocyanate. Table VI shows our first experiment. Clostridial ferredoxin, containing the amount of sulfide shown, was allowed to react with cyanide, and a very good yield of thiocyanate was detected. Table VII shows that no thiocyanate was formed under these conditions with apoferredoxin or sodium sulfide. This looked like a good control but was not good enough. We did the experiment with ^{35}S–ferredoxin in which the sulfide, not the cysteine sulfur, is labeled. As shown in Table VIII, the thiocyanate that was recovered had a specific activity that was very close to the starting sulfide. In fact, it was a little higher. We then did a control experiment in which some sulfide was added to the medium and found that the specific activity of the thiocyanate was diluted but, somewhat disturbingly, we got a much higher yield of thiocyanate. If sulfide is added to the apoprotein (Table IX), thiocyanate was also formed. There is not a direct reaction of the cyanide with the ferredoxin to give thiocyanate as indicated in Table X, but the ferredoxin

Table IX. Thiocyanate Formation from Apoferredoxin and Sulfide[a]

$Apo-Fd$, mM	Na_2S, mM	$Time$, $Days$	SCN^- $Formed$, mM	$\dfrac{[SCN^-]}{8 \times [Apo-Fd]}$
0.085	28.9	1	2.11	2.46
0.0	28.9	1	0.57	
0.085	28.9	5	6.68	8.52
0.0	28.9	5	1.34	

[a] The components were incubated at 37° in the presence of 58 mM KCN at pH 9.0.

Table X. Hypothetical Schemes for Thiocyanate Formation from Ferredoxin

1. Reaction of CN^- directly with persulfide moieties present in native ferredoxin:
Ferredoxin + 8 CN^- → 8 SCN^- + apo-Fd + [8 Fe^{2+-3+}]

2. Reaction of CN^- with persulfide formed in a secondary reaction from apo-Fd and S^{2-}:
Ferredoxin → apo-Fd + [8 S^{2-}] + [8 Fe^{2+-3+}]
Apo-Fd + S^{2-} → apo-Fd-S-S$^-$
Apo-Fd-S-S$^-$ + CN^- → apo-Fd + SCN^-

in the presence of cyanide is degraded to apoferredoxin and sulfide, and the sulfide is re-reacting with the apoferredoxin to form the persulfide. This subsequently reacts with cyanide to give the thiocyanate. We have tried various other experiments to measure the rate of the formation of thiocyanate and the initial specific activities, but these have all failed to provide any evidence suggestive of the presence of the persulfide structure in the native ferredoxin.

The big problem, I think, still is the determination of the molecular structure of these iron–sulfur proteins, and my hope is that some crystallographers will finally succeed in solving the problem.

Literature Cited

(1) Lovenberg, W., Sobel, B. E., "Rubredoxin: A New Electron Transfer Protein from *Clostridium pasteurianum*," *Proc. Natl. Acad. Sci. U.S.* (1965) **54**, 193.
(2) Bachmayer, H., Yasunobu, K. T., Peel, J. L., Mayhew, S., "Non-Heme Iron Proteins. V. The Amino Acid Sequence of Rubredoxin from *Peptostreptococcus elsdenii*," *J. Biol. Chem.* (1968) **243**, 1022.
(3) Bachmayer, H., Benson, A. M., Yasunobu, K. T., Garrard, W. T., Whiteley, H. R., "Non-Heme Iron Proteins. IV. Structural Studies of *Micrococcus aerogenes* Rubredoxin," *Biochemistry* (1968) **7**, 986.
(4) Palmer, G., Brintzinger, H., "Nature of the Non-haem Iron in Ferredoxin and Rubredoxin," *Nature* (1966) **211**, 189.
(5) Coon, M. J., Lode, E. T., personal communication.
(6) Peterson, J. A., Coon, M. J., "Enzymatic ω-Oxidation. III. Purification and Properties of Rubredoxin, a Component of the ω-Hydroxylation System of *Pseudomonas oleovorans*," *J. Biol. Chem.* (1968) **243**, 329.
(7) Matsubara, H., Sasaki, R. M., "Spinach Ferredoxin. II. Tryptic, Chymotryptic, and Thermolytic Peptides, and Complete Amino Acid Sequence," *J. Biol. Chem.* (1968) **243**, 1732.
(8) Benson, A. M., Yasunobu, K. T., "Non-Heme Iron Proteins. X. The Amino Acid Sequences of Ferredoxins from *Leucaena glauca*," *J. Biol. Chem.* (1969) **244**, 955.
(9) Rao, K. K., Matsubara, H., "The Amino Acid Sequence of Taro Ferredoxin," *Biochem. Biophys. Res. Commun.* (1970) **38**, 500.
(10) Sugeno, K., Matsubara, H., "The Amino Acid Sequence of *Scenedesmus* Ferredoxin," *J. Biol. Chem.* (1969) **244**, 2979.
(11) Keresztes-Nagy, S., Perini, F., Margoliash, E., "Primary Structure of Alfalfa Ferredoxin," *J. Biol. Chem.* (1969) **244**, 981.
(12) Tagawa, K., Arnon, D. I., "Ferredoxins as Electron Carriers in Photosynthesis and in the Biological Production and Consumption of Hydrogen Gas," *Nature* (1962) **195**, 537.
(13) Tagawa, K., Arnon, D. I., "Oxidation-Reduction Potentials and Stoichiometry of Electron Transfer in Ferredoxins," *Biochim. Biophys. Acta.* (1968) **153**, 602.
(14) Cushman, D. W., Tsai, R. L., Gunsalus, I. C., "The Ferroprotein Component of a Methylene Hydroxylase," *Biochem. Biophys. Res. Commun.* (1967) **26**, 577.
(15) Tsibris, J. C. M., Woody, R. W., "Structural Studies of Iron–Sulfur Proteins," *Coordination Chem. Rev.*, in press.

(16) Tsibris, J. C. M., Namtvedt, M. J., Gunsalus, I. C., "Selenium as an Acid-Labile Sulfur Replacement in Putidaredoxin," *Biochem. Biophys. Res. Commun.* (1968) **30**, 323.
(17) Tsibris, J. C. M., Tsai, R. L., Gunsalus, I. C., Orme-Johnson, W. H., Hansen, R. E., Beinert, H., "The Number of Iron Atoms in the Paramagnetic Center ($g = 1.94$) of Reduced Putidaredoxin, a Non-Heme Iron Protein," *Proc. Natl. Acad. Sci. U.S.* (1968) **59**, 959.
(18) Tanaka, M., Haniu, M., Yasunobu, K. T., "The Primary Structure of Bovine Adrenodoxin," *Biochem. Biophys. Res. Commun.* (1970) **39**, 1182.
(19) Kimura, T., Suzuki, K., "Components of the Electron Transport System in Adrenal Steroid Hydroxylase. Isolation and Properties of Non-Heme Iron Protein (Adrenodoxin)," *J. Biol. Chem.* (1967) **242**, 485.
(20) Estabrook, R. W., private communication.
(21) Orme-Johnson, W. H., Beinert, H., "Reductive Titrations of Iron–Sulfur Proteins Containing Two to Four Iron Atoms," *J. Biol. Chem.* (1969) **244**, 6143.
(22) Watari, H., Kimura, T., "Study of the Adrenal Non-Heme Iron Protein (Adrenodoxin) by Electron Spin Resonance," *Biochem. Biophys. Res. Commun.* (1966) **24**, 106.
(23) Dervartanian, D. V., Shethna, Y. I., Beinert, H., "Purification and Properties of 2 Iron–Sulfur Proteins from *Azotobacter vinelandii*," *Biochim. Biophys. Acta* (1969) **194**, 548.
(24) Dus, K., De Klerk, H., Sletten, K., Bartsch, R. G., "Chemical Characterization of High Potential Iron Proteins from *Chromatium* and *Rhodopseudomonas gelatinosa*," *Biochim. Biophys. Acta* (1967) **140**, 291.
(25) Flatmark, T., Dus, K., "Studies on the Chelate Structure of the High-Potential Iron Protein of *Chromatium*," *Biochim. Biophys. Acta* (1969) **180**, 377.
(26) Yoch, D. C., Benemann, J. R., Valentine, R. C., Arnon, D. I., "The Electron Transport System in Nitrogen Fixation by *Azotobacter*. II. Isolation and Function of a New Type of Ferredoxin," *Proc. Natl. Acad. Sci. U.S.* (1969) **64**, 1404.
(27) Shethna, Y. I., "Non-Heme Iron (Iron–Sulfur) Proteins of *Azotobacter vinelandii*," *Biochim. Biophys. Acta* (1970) **205**, 58.
(28) Lovenberg, W., Buchanan, B. B., Rabinowitz, J. C., "Studies on the Chemical Nature of Clostridial Ferredoxin," *J. Biol. Chem.* (1963) **238**, 3899.
(29) Tanaka, M., Nakashima, T., Benson, A., Mower, H., Yasunobu, K. T., "The Amino Acid Sequence of *Clostridium pasteurianum* Ferredoxin," *Biochemistry* (1966) **5**, 1666.
(30) Benson, A. M., Mower, H. F., Yasunobu, K. T., "The Amino Acid Sequence of *Clostridium butyricum* Ferredoxin," *Arch. Biochem. Biophys.* (1967) **121**, 563.
(31) Tsunoda, J. N., Yasunobu, K. T., Whiteley, H. R., "Non-Heme Iron Proteins. IX. Amino Acid Sequence of Ferredoxin from *Micrococcus aerogenes*," *J. Biol. Chem.* (1968) **243**, 6262.
(32) Rall, S. C., Bolinger, R. E., Cole, R. D., "The Amino Acid Sequence of Ferredoxin from *Clostridium acidi-urici*," *Biochemistry* (1969) **8**, 2486.
(33) Buchanan, B. B., Lovenberg, W., Rabinowitz, J. C., "A Comparison of Clostridial Ferredoxins," *Proc. Natl. Acad. Sci. U.S.* (1963) **49**, 345.
(34) Sobel, B. E., Lovenberg, W., "Characteristics of *Clostridium pasteurianum* Ferredoxin in Oxidation–Reduction Reactions," *Biochemistry* (1966) **5**, 6.
(35) Evans, M. C. W., Hall, D. O., Bothe, H., Whatley, F. R., "The Stoichiometry of Electron Transfer by Bacterial and Plant Ferredoxins," *Biochem. J.* (1968) **110**, 485.

(36) Uyeda, K., Rabinowitz, J. C., "Pyruvate:Ferredoxin Oxidoreductase. IV. Studies on the Reaction Mechanism," in preparation.
(37) Palmer, G., Sands, R. H., Mortenson, L. E., "Electron Paramagnetic Resonance Studies on the Ferredoxin from *Clostridium pasteurianum*," *Biochem. Biophys. Res. Commun.* (1966) **23**, 357.
(38) Matsubara, H., Sasaki, R. M., Tsuchiya, D. K., Evans, M. C. W., "The Amino Acid Sequence of *Chromatium* Ferredoxin," *J. Biol. Chem.* (1970) **245**, 2121.
(39) Bachofen, R., Arnon, D. I., "Crystalline Ferredoxin from the Photosynthetic Bacterium *Chromatium*," *Biochim. Biophys. Acta* (1966) **120**, 259.
(40) Fogo, J. K., Popowsky, M., "Spectrophotometric Determination of Hydrogen Sulfide. Methylene Blue Method," *Anal. Chem.* (1949) **21**, 732.
(41) Peterson, J. A., Basu, D., Coon, N. J., "Enzymatic ω-Oxidation. I. Electron Carriers in Fatty Acid and Hydrocarbon Hydroxylation," *J. Biol. Chem.* (1966) **241**, 5162.
(42) Arnon, D. I., "Role of Ferredoxin in Photosynthesis," *Naturwissenschaften* (1969) **56**, 295.
(43) Yocum, C. F., San Pietro, A., "Ferredoxin Reducing Substance (FRS) from Spinach," *Biochem. Biophys. Res. Commun.* (1969) **36**, 614.
(44) Mortenson, L. E., Valentine, R. C., Carnahan, J. E., "An Electron Transport Factor from *Clostridium pasteurianum*," *Biochem. Biophys. Res. Commun.* (1962) **7**, 448.
(45) Hardy, R. W. F., "Biochemistry of N_2 Fixation," ADVAN. CHEM. SER. (1971) **100**, 219.
(46) Bachofen, R., Buchanan, B. B., Arnon, D. I., "Ferredoxin as a Reductant in Pyruvate Synthesis by a Bacterial Extract," *Proc. Natl. Acad. Sci. U.S.* (1964) **51**, 690.
(47) Raeburn, S., Rabinowitz, J. C., "Pyruvate Synthesis by a Partially Purified Enzyme from *Clostridium acidi-urici*," *Biochem. Biophys. Res. Commun.* (1965) **18**, 303.
(48) Mortenson, L. E., "Purification and Analysis of Ferredoxin from *Clostridium pasteurianum*," *Biochim. Biophys. Acta* (1964) **81**, 71.
(49) Herriott, J. R., Sieker, L. C., Jensen, L. H., "Structure of Rubredoxin: An X-Ray Study to 2.5 Å Resolution," *J. Mol. Biol.* (1970) **50**, 391.
(50) Strahs, G., Kraut, J., "Low-Resolution Electron-Density and Anomalous Scattering-Density Maps of *Chromatium* High-Potential Iron Protein," *J. Mol. Biol.* (1968) **35**, 503.
(51) Bayer, E., Eckstein, H., Hagenmair, H., Josef, D., Koch, J., Krauss, P., Roder, A., Schretzmann, P., "Untersuchungen zur Struktur der Ferredoxine," *European J. Biochem.* (1969) **8**, 33.
(52) Malkin, R., Rabinowitz, J. C., "The Reconstitution of Clostridial Ferredoxin," *Biochem. Biophys. Res. Commun.* (1966) **23**, 822.
(53) Hong, J.-S., Rabinowitz, J. C., "Preparation and Properties of Clostridial Apoferredoxins," *Biochem. Biophys. Res. Commun.* (1967) **29**, 246.
(54) Orme-Johnson, W. H., Hansen, R. E., Beinert, H., Tsibris, J. C. M., Bartholomaus, R. C., Gunsalus, I. C., "On the Sulfur Components of Iron–Sulfur Proteins. I. The Number of Acid-Labile Sulfur Groups Sharing an Unpaired Electron with Iron," *Proc. Natl. Acad. Sci. U.S.* (1968) **60**, 368.
(55) Jensen, L. H., private communication.
(56) Matsubara, H., Jukes, T. H., Cantor, C. R., "Structural and Evolutionary Relationships of Ferredoxins," *Brookhaven Symp. Biol.* (1968) **21**, 201.
(57) Fitch, W. M., "Further Improvements in the Method of Testing for Evolutionary Homology among Proteins," *J. Mol. Biol.* (1970) **49**, 1.

(58) Fitch, W. M., "A Method for Estimating the Probability that a Specific Frameshift Mutation was Selected in the Course of Evolution," *J. Mol. Biol.* (1970) **49**, 15.
(59) Buchanan, B. B., Matsubara, H., Evans, M. C. W., "Ferredoxin from the Photosynthetic Bacterium, *Chlorobium thiosulfatophilum*—A Link to Ferredoxins from Nonphotosynthetic Bacteria," *Biochim. Biophys. Acta* (1969) **189**, 46.
(60) Buchanan, B. B., personal communication.
(61) Ulmer, D. D., Vallee, B. L., "Structure and Function of Metallo-Enzymes," ADVAN. CHEM. SER. (1971) **100**, 187.
(62) Lovenberg, W., Williams, W. M., "Further Observations on the Chemical Nature of Rubredoxin from *Clostridium pasteurianum*," *Biochemistry* (1969) **8**, 141.
(63) Blomstrom, D. C., Knight, E., Jr., Phillips, W. D., Weiher, J. F., "The Nature of Iron in Ferredoxin," *Proc. Natl. Acad. Sci. U.S.* (1964) **51**, 1085.
(64) Sieker, L. C., Jensen, L. H., "An X-Ray Investigation of the Structure of a Bacterial Ferredoxin," *Biochem. Biophys. Res. Commun.* (1965) **20**, 33.
(65) Poe, M., Phillips, W. D., McDonald, C. C., Lovenberg, W., "Proton Magnetic Resonance Study of Ferredoxin from *Clostridium pasteurianum*," *Proc. Natl. Acad. Sci. U.S.* (1970) **65**, 797.
(66) Flavin, M., "Microbial Transsulfuration: The Mechanism of an Enzymatic Disulfide Elimination Reaction," *J. Biol. Chem.* (1962) **237**, 768.
(67) Coucouvanis, D., Lippard, S. J., "Preparation and Structural Characterization of Six-Coordinate Iron (III) Complexes Containing the Fe–S–S Linkage," *J. Am. Chem. Soc.* (1968) **90**, 3281.

RECEIVED June 26, 1970.

16

Studies on Mechanism of Action of B_{12} Coenzymes

R. H. ABELES

Brandeis University, Waltham, Mass. 02154

> Dioldehydrase, in the presence of vitamin B_{12} coenzyme, catalyzes the conversion of D or L-1,2-propanediol and ethyleneglycol to propionaldehyde and acetaldehyde. Substrate C-1 tritium replaces the two C-5' hydrogens of the coenzyme and occupies the α position of the product aldehyde. When B_{12} coenzyme containing C-5' tritium is used with nonisotopic substrates, tritium is transferred to the α position of the product. The rate of these reactions is consistent with a mechanism in which the coenzyme functions as intermediate hydrogen carrier.
>
> $$SH + XH_2 \rightarrow I \cdot XH_3 \rightarrow PH + XH_2$$
>
> SH is the substrate, PH the product, XH_2 the vitamin B_{12} coenzyme, and $I \cdot XH_3$ an intermediate complex derived from the substrate and coenzyme. Formation of the intermediate may involve the breaking of the carbon–cobalt bond and the formation of 5'-deoxyadenosine from the adenosyl moiety of the coenzyme. One of the methyl hydrogens of 5'-deoxyadenosine is one of the C-1 hydrogens of the substrate.

Vitamin B_{12} is particularly fascinating since neither its biological role nor its chemical mode of action is understood. It prevents the occurrence of pernicious anemia. No satisfactory explanation is available as to why the absence of this vitamin leads to pernicious anemia. The coenzyme form of the vitamin is involved in a number of biochemical rearrangements. The mechanism of these rearrangements is also subject to considerable discussion and disagreement. This paper describes experiments which we have carried out in order to clarify the mode of action of this coenzyme.

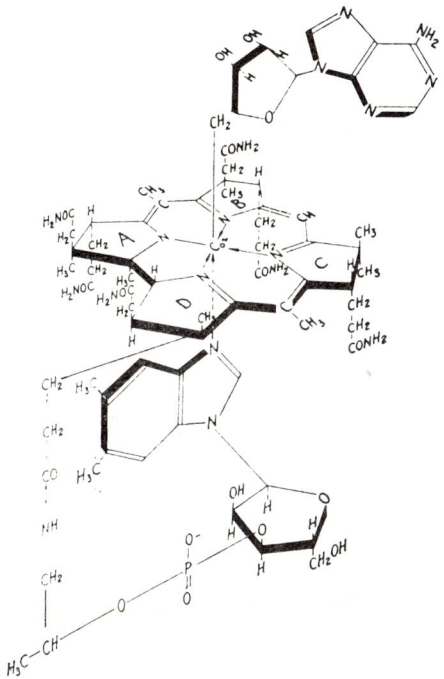

Figure 1. Structure of vitamin B_{12} coenzyme

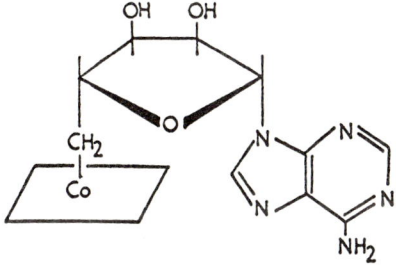

Figure 2. Partial structure of vitamin B_{12} coenzyme

Figure 1 shows the structure of vitamin B_{12} coenzymes. It is a modification of the vitamin, as all water-soluble coenzymes are. The unique feature of the structure is the covalent carbon–cobalt bond.

I have dissected out one portion of the coenzyme which I think is of particular interest (Figure 2). An important question is: Does that portion of the coenzyme just sit around and do nothing, or is it intimately involved in the reaction process? I think it plays an important role in the catalytic process. As a matter of nomenclature, the carbon next to the cobalt is the C-5' carbon, and I will frequently mention this carbon atom.

Table I is a summary of all of the reactions known to date which require vitamin B_{12} coenzyme (*1*). The reaction catalyzed by glutamate mutase, the first reaction in which vitamin B_{12} was known to be involved, was discovered by Barker and his coworkers. The second reaction is the only one of these which also occurs in mammals; all other reactions occur in bacteria. Much of the experimental data described here will be derived from the enzyme called dioldehydrase. An additional reaction which is not shown is the conversion of nucleotides to deoxynucleotides. The conversion of —CHOH— to —CH$_2$— is in some way very similar to the reaction catalyzed by dioldehydrase. Vitamin B_{12} coenzyme-depend-

Table I. Reactions Requiring Vitamin B_{12} Coenzyme

1) Glutamate mutase

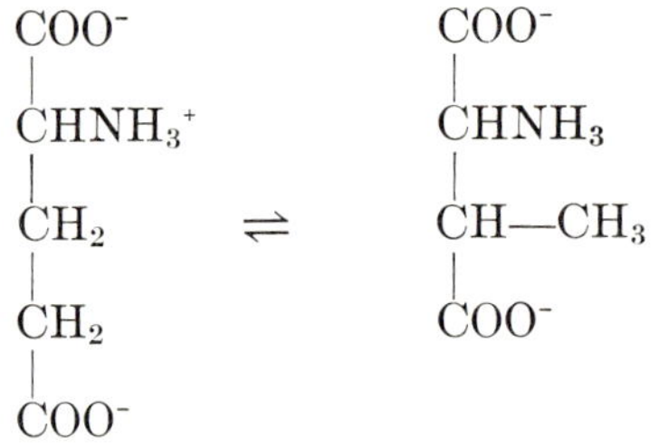

$$\begin{array}{c} COO^- \\ | \\ CHNH_3^+ \\ | \\ CH_2 \\ | \\ CH_2 \\ | \\ COO^- \end{array} \rightleftharpoons \begin{array}{c} COO^- \\ | \\ CHNH_3 \\ | \\ CH-CH_3 \\ | \\ COO^- \end{array}$$

2) Methylmalonyl-CoA isomerase

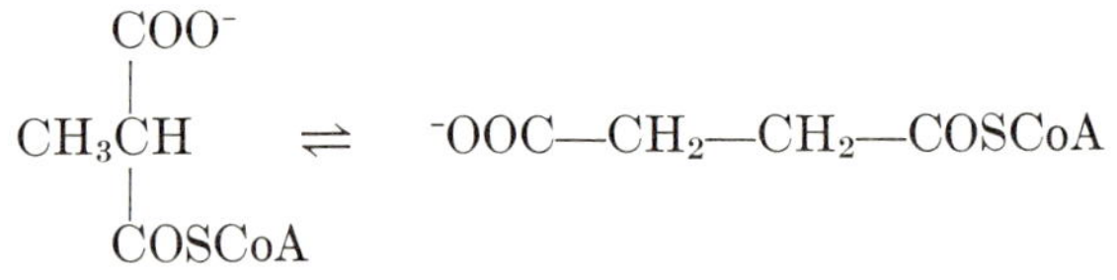

$$\begin{array}{c} COO^- \\ | \\ CH_3CH \\ | \\ COSCoA \end{array} \rightleftharpoons {}^-OOC-CH_2-CH_2-COSCoA$$

3) Glycerol-dehydrase

$$CH_2OH-CHOH-CH_2OH \rightarrow CH_2OH-CH_2-CHO$$

4) Dioldehydrase

$$CH_3-CHOH-CH_2OH \rightarrow CH_3-CH_2-CHO$$

$$CH_2OH-CH_2OH \rightarrow CH_3-CHO$$

5) Ethanol amine deaminase

$$CH_2NH_3^+CH_2OH \rightarrow CH_3-CHO + NH_4^+$$

6) β-Lysine mutase

$$CH_2NH_3^+(CH_2)_3CHNH_3^+ \rightarrow$$
$$CH_3-CHNH_3^+-(CH_2)_2-CHNH_3^+COO^-$$

ent ribonucleotide reductase has, so far, only been demonstrated in bacteria. Mammals seem to use a non-B_{12}-dependent reductase.

These seem like different and diverse reactions. The thing that concerns an enzymologist is what they have in common. As far as we know, whenever a coenzyme is functionally active in the reaction, it always does the same thing, basically by the same mechanism. When the fundamental mechanism of a coenzyme has been discovered, and this has always been done through model reactions so far, then the enzymatic reactions can be understood.

Table II. Hydrogen Transfer in Reactions Which Require Vitamin B_{12} Coenzymes

$$-\overset{|}{\underset{|}{C_1}}-\overset{|}{\underset{|}{C_2}}- \rightarrow -\overset{|}{\underset{|}{C_2}}-\overset{|}{\underset{|}{C_1}}-$$
$$\text{H} \quad \text{X} \qquad \text{X} \quad \text{H}$$

No exchange with H^+ of medium

Now what is the common feature in all these reactions? They look vastly different. Well, all of these reactions involve a rearrangement in which a hydrogen from one carbon migrates to the adjacent carbon, and some R group moves over (Table II). All of the B_{12} reactions that are known today involve a rearrangement in which an R group and the hydrogen exchange places. This is not immediately obvious for Reactions 3–5 in Table I, but it is the case here, too. The ribonucleotide reductase has to be modified a little, but it can fit into this pattern. At first glance, you might think these reactions are trivially simple; they involve a dehydration. However, dehydrations do not occur readily in biochemistry. By and large, they are very complex reactions. The only kinds of dehydration that occur relatively readily, with one or two exceptions, are those in which the water is removed alpha, beta- to a carbonyl group. Reactions in which an OH group is replaced by a hydrogen are generally complex reactions. There are no examples of direct displacement reactions of hydroxyl groups in biochemistry.

Since all of these reactions involve the exchange of the hydrogen with some group X, the nature of this hydrogen transfer became one of the first points which was investigated. The purpose of this investigation is to decide whether the hydrogen is transferred as a proton. If it is, then it should exchange with the hydrogens of the solvent. In none of the B_{12} coenzyme-dependent reactions (those reactions which have been examined so far) has any hydrogen exchange with the solvent been observed.

In enzymatic reactions there are some proton transfer reactions known in which no exchange occurs with the solvent (2). This is also the case in nonenzymatic reactions. However, the majority of all proton transfer reactions in biochemical systems proceeds with at least some exchange with the solvent. I know of only one reaction to date in which no exchange at all has been discovered. Now, this simply appears to be a matter of competition between interaction with the solvent and the rate of proton transfer, and in many of these reactions, the proton transfer is very fast. For instance, steroid isomerase is the fastest enzyme in the world, and there you see no proton exchange. Some B_{12} coenzyme reactions have been examined for any evidence of proton exchange with the solvent; there is none. Therefore, it is very likely that no protons are involved.

Table III. Reaction Catalyzed by Dioldehydrase

$$\begin{array}{ccc}
\overset{*}{\text{HO}}-\overset{\text{D}}{\underset{|}{\text{C}}}-\text{H} & & \overset{x}{\text{CHO}} \\
\text{H}-\overset{x}{\underset{|}{\text{C}}}-\text{OH} & \rightarrow & \text{D}-\underset{|}{\text{C}}-\text{H} \\
\text{CH}_3 & & \text{CH}_3 \\
1(R),2(R) & &
\end{array} \qquad
\begin{array}{ccc}
\overset{*}{\text{HO}}-\overset{\text{D}}{\underset{|}{\text{C}}}-\text{H} & & \overset{*}{\text{CDO}} \\
\overset{x}{\text{HO}}-\underset{|}{\text{C}}-\text{H} & \rightarrow & \text{H}-\underset{|}{\text{C}}-\text{H} \\
\text{CH}_3 & & \text{CH}_3 \\
1(R),2(S) & &
\end{array}$$

$$\begin{array}{ccc}
\text{HO}-\overset{\text{H}}{\underset{|}{\text{C}}}-\text{D} & & \text{CDO} \\
\text{H}-\underset{|}{\text{C}}-\text{OH} & \rightarrow & \text{H}-\underset{|}{\text{C}}-\text{H} \\
\text{CH}_3 & & \text{CH}_3 \\
1(S),2(R) & &
\end{array} \qquad
\begin{array}{ccc}
\text{HO}-\overset{\text{H}}{\underset{|}{\text{C}}}-\text{D} & & \text{CHO} \\
\text{HO}-\underset{|}{\text{C}}-\text{H} & \rightarrow & \text{HC}-\text{D} \\
\text{CH}_3 & & \text{CH}_3 \\
1(S),2(S) & &
\end{array}$$

Table III summarizes how the atoms move around in the conversion of diols to aldehydes. Very interesting stereospecificity becomes apparent here. If we start out with the R isomer, containing deuterium at C-1, the deuterium appears in the C-2 position of the product (3). If we start with the S isomer, the deuterium is not transferred and remains at C-1. A similar experiment was done by Arigoni (4). When he started out with the R isomer with ^{18}O in the C-2 position, all of the ^{18}O (within experimental error) ended up in C-1. This, of course, establishes that this is a group transfer reaction, just like the other reaction, because whatever else is going to happen now, the oxygen from C-2 ends up at C-1. If one starts out with the S-isomer and labels with ^{18}O, then the ^{18}O disappears and is released into the solvent. The stereochemistry of this reaction is controlled by the stereochemistry at C-2; one can rationalize this simply (both the deuterium results and the ^{18}O results) if one assumes a three-point attachment (3). If you make a model and say both hydroxyl groups have to be in the same spot and the methyl group of the substrate has to stick out in a certain direction, then you will find that the two hydrogens on C-1 are not in the same position (depending on whether you have R or S), and the ^{18}O results follow out of this same model.

Table IV shows how Arigoni interpreted his results (4). The diol is converted to the gem-diol through a series of rearrangements, in which it was postulated that the B_{12} coenzyme was involved as an intermediate hydrogen acceptor. The gem-diol is then stereospecifically dehydrated. If the enzyme has first seen the S isomer, it removes one of the —OH groups; if it has first seen the R-isomer, it removes the other —OH group. If we accept this intermediate, we have a similarity between all of the reactions—*i.e.*, a rearrangement. With the ethanolamine deaminase, there is some evidence which is in further agreement with this type of an intermediate.

Table IV. Conclusion From ^{18}O Data

$$CH_3-CH-CH_2 \to \to CH_3-CH_2-\overset{H}{\underset{OH}{C}}-OH \to CH_3-CH_2-C\overset{O}{\underset{H}{\diagup}}$$
$$OHOH$$

Regarding the hydrogen transfer, hydrogen from C-1 of the substrate ends up at C-2 of the product; this, of course, suggested a 1,2-hydride shift. However, the isotope effect in the conversion of this substrate to this product was between 10 and 12—*i.e.*, the deuterated substrate reacted at 1/10 to 1/12 the rate of the nondeuterated substrate. This struck us as a very high isotope effect for a 1,2-hydride shift. Therefore, we considered the possibility that these reactions may not be 1,2-hydride shifts. One could envision a two-step process whereby the hydrogen somehow goes on the enzyme (the coenzyme), and then gets transferred back to the product. This would, of course, look like a 1,2-hydride shift.

Table V illustrates an experiment to test this point. We started with tritiated propanediol and unlabeled ethylene glycol. The tritium from the C-1 position ought to, and does, end up on C-2 of propionaldehyde. However, tritium was found also in the alpha position of the acetaldehyde. Therefore, acetaldehyde has now acquired tritium from propanediol, or intermolecular hydrogen transfer has taken place. One can set up this experiment so that essentially all of the tritium of 1,2-propanediol

Table V. Intermolecular Hydrogen Transfer in the Conversion of Diols to Aldehydes by Dioldehydrase

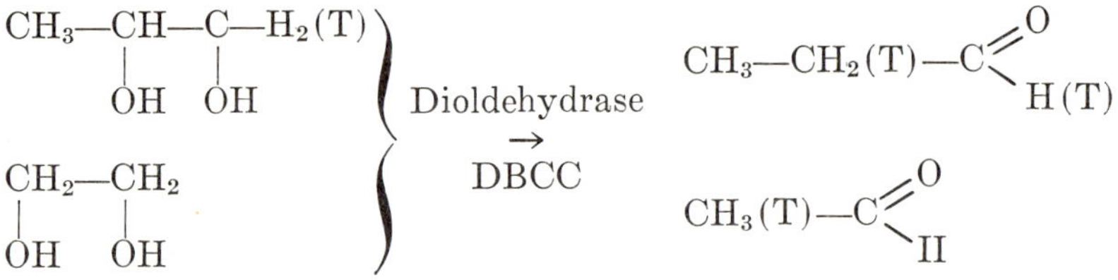

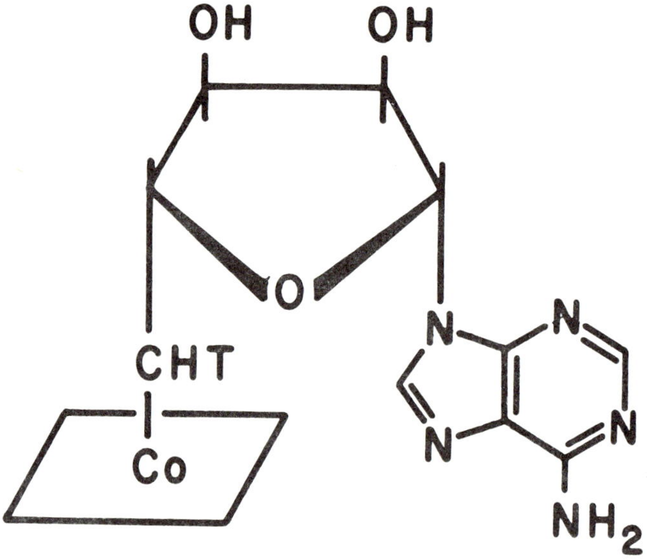

$$\left.\begin{array}{l}\underset{|}{CH_3}-\underset{|}{CH}-\underset{|}{C}-H_2(T)\\ OHOH\\ \\ \underset{|}{CH_2}-\underset{|}{CH_2}\\ OHOH\end{array}\right\} \begin{array}{c}\text{Dioldehydrase}\\ \xrightarrow{}\\ \text{DBCC}\end{array} \begin{array}{l} CH_3-CH_2(T)-C\underset{H(T)}{\overset{O}{\diagup}} \\ \\ CH_3(T)-C\underset{H}{\overset{O}{\diagup}} \end{array}$$

Figure 3. Coenzyme isolated after reaction of enzyme–coenzyme with tritiated substrate

ends up in acetaldehyde. This suggested to us a mechanism in which the coenzyme functions as a hydrogen carrier. If you grant this, there must be an enzyme-bound intermediate containing at least two equivalent hydrogens. A mechanism of this kind predicts that if you start with a tritiated substrate, you ought to find a tritiated enzyme–coenzyme complex. Well, we did this (5). We used a tritiated substrate, ran the reaction, re-isolated the enzyme–coenzyme, and there was tritium in the coenzyme in the C-5' position as shown in Figure 3. This was established by P. A. Frey in our laboratory by chemical degradation of the coenzyme and by synthesis. In either case, when we had the coenzyme labeled in this position, we could add it back to the apoenzyme and transfer all of the tritium to the reaction product. The fact that the synthetic coenzyme transferred all of the tritium to the reaction product surprised us, because, as you can see, these two hydrogens are not stereochemically equivalent, and therefore the enzyme ought to choose one or the other. So we began to worry that maybe when we synthesized the coenzyme we had fortuitously carried out a stereospecific synthesis. There are two possibilities: Either the coenzyme was originally nonstereospecifically synthesized, proving that these two hydrogens were equivalent somewhere during the course of the reaction, or we have synthesized a stereospecific coenzyme, and we have to prove that the enzyme uses both hydrogens.

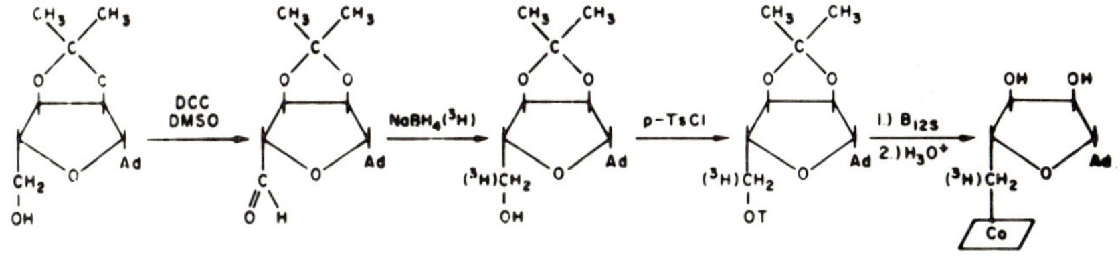

Figure 4. Synthesis of C-5' tritiated coenzyme with NaB^3H_4

Figure 4 shows how the C-5' tritiated coenzyme was synthesized with NaB^3H_4 (6). Assume this is stereospecifically labeled, but remember that this tritiated coenzyme transferred all of its tritium to its products. We photolyzed the coenzyme which gave us the aldehyde. We then used this tritiated aldehyde through the same synthesis with unlabeled $NaBH_4$. If we had made stereospecifically labeled coenzyme in the first place, we would now have the tritium in the opposite configuration. The coenzyme prepared in this way again transferred all of its tritium to the product. Therefore, these two C-5' hydrogens of the coenzyme must become equivalent at some point during the reaction.

Figure 5 portrays a three-dimensional model. The question frequently arises if those hydrogens are accessible. They stick out clearly on the model. These are the tritiated hydrogens, and it is apparent that they can be quite accessible to the substrate.

The introduction of tritium from substrate to coenzyme suggested the following reaction scheme to us. The reaction involves at least a two-step process in which tritium from the substrate is transferred to the coenzyme and then from the coenzyme to the reaction product. As with any mechanism, one ought to show that it is at least kinetically permissible, and we undertook to do this. If one starts with C-1 tritiated substrate, is sufficient tritium introduced into the coenzyme so that it can account for the tritium which is found at any time in the reaction product?

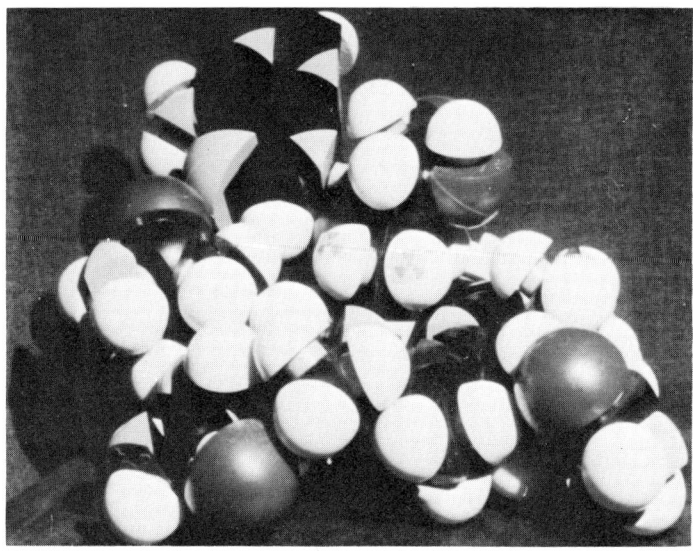

Figure 5. Three-dimensional model of a coenzyme

Table VI. The Relative Rates for the Conversion of Substrates to Products and for the Transfer of Tritium from DBC–^{3}H to the Products of the Dioldehydrase Reaction

Temp., °C.	Substrate	k_{cat} (sec^{-1})	k_2 (sec^{-1})	$\dfrac{k_{cat}}{k_2} \times 10^2$
10	dl-1,2-propanediol	48	0.20	2.4
10	ethylene glycol–^{14}C	11.8	0.038	3.1
10	ethylene glycol–d_4	3.6	0.26	0.14

We made several measurements. From the rate of the over-all reaction, we determined k_{cat}. Then we measured the rate of the tritium transfer from the enzyme–coenzyme (^{3}H) complex, containing coenzyme tritiated at C-5′, to the reaction product. This is a first-order process in the amount of tritium in the coenzyme, and we obtained a rate constant, k_2.

A very surprising thing happens here. Table VI shows that k_{cat} is 240 to 300 times as big as the rate of tritium transfer (k_2), which is a fantastic tritium discrimination. This means that in one out of 240 turnovers, a tritium is transferred from the coenzyme to the reaction product. There is a statistical factor involved here, if we assume that there are three equivalent hydrogens in the intermediate, but that still leaves a very large tritium isotope effect. The same thing was done with ethylene glycol—don't pay any attention to ^{14}C; it was put in for analytical convenience. Ethylene glycol is a poorer substrate; k_{cat} is smaller here, k_2 is smaller. The ratio, however, is in the same order of magnitude. When one did this experiment with deuterated ethylene glycol, the rate of tritium transfer from the coenzyme was actually faster than with the nondeuterated substrate; the ratio of k_{cat} to k_2 changed by a factor of 20. That means with a deuterated substrate the probability of tritium transfer from the coenzyme to the product is significantly enhanced.

We have a rate constant for transferring tritium from the enzyme–coenzyme complex to the product. The next thing to find out is how hot the enzyme–coenzyme complex gets during the course of the reaction. Then, with that rate constant, we should predict how hot the products can get.

Table VII. The Contribution of DBC–^{3}H to the Rate of Propionaldehyde-2-^{3}H Production During the Reaction of D-1,2-Propanediol-1-^{3}H

Time, Sec	DBC–^{3}H, cpm	Calculated Rates	Observed Rates
0	0.057×10^4	—	—
5	2.3×10^4	3.5×10^5	—
10	2.8×10^4	4.2×10^5	4.1×10^5
30	4.1×10^4	6.2×10^5	5.4×10^5
60	5.1×10^4	7.7×10^5	7.6×10^5
120	57.6×10^4	(86×10^5)	—

Table VII shows one way of treating the data. We measured the specific activity of the coenzyme during the course of the reaction. From this, we can calculate an instantaneous rate of tritium appearance in the product and compare it with the observed rate. This is in very good agreement. Therefore, a mechanism involving the coenzyme as an intermediate hydrogen acceptor is kinetically permissible. Whether it is correct or not is another question. Out of these experiments came another series of results.

The specific activity of the coenzyme, the residual substrate, and the product were examined at various time periods (Table VIII). We were concerned about the possibility that tritium might be introduced into the coenzyme by an equilibrium process. If an equilibration process occurred either with a substrate or a product, then the specific activity of the coenzyme could only exceed that of the substrate or the product by a small equilibrium isotope effect.

As you see, at all stages during the reaction, the specific activity of the coenzyme is approximately 20 times as high as that of the residual substrate, and it is from 200 to 700 times as high as that of the reaction product. This is a very important result because, I believe, it makes very

Table VIII. Relative Specific Activities of Vitamin B_{12} Coenzyme, Residual Substrate, and Product During the Conversion of D-1,2-Propanediol-1-^{3}H to Propionaldehyde-2-^{3}H

	Specific Activity (cpm/μmole) × 10^{-5}		
Time, Sec	B_{12} Coenzyme	Residual Substrate	Product
0	1.8	4.0	—
5	75	4.3	0.27
10	88	4.4	0.22
30	130	5.9	0.26
60	160	10.3	0.22

Table IX. Proposed Reaction Pathway

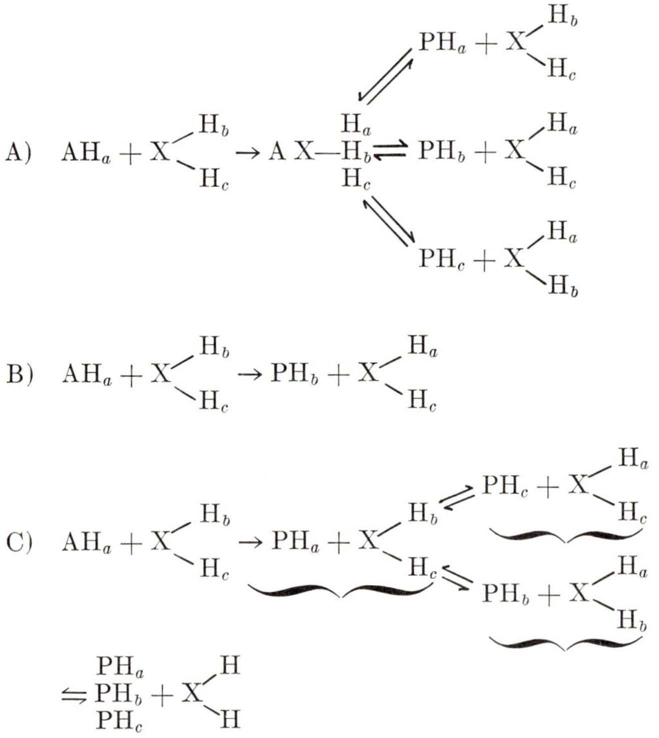

A) $AH_a + X{<}^{H_b}_{H_c} \rightarrow AX{-}H_b{<}^{H_a}_{H_c} \rightleftarrows PH_b + X{<}^{H_a}_{H_c}$

with branches to:
$PH_a + X{<}^{H_b}_{H_c}$
$PH_c + X{<}^{H_a}_{H_b}$

B) $AH_a + X{<}^{H_b}_{H_c} \rightarrow PH_b + X{<}^{H_a}_{H_c}$

C) $AH_a + X{<}^{H_b}_{H_c} \rightarrow PH_a + X{<}^{H_b}_{H_c} \rightleftarrows PH_c + X{<}^{H_a}_{H_c}$
$\rightleftarrows PH_b + X{<}^{H_a}_{H_b}$

$\rightleftarrows \begin{array}{c} PH_a \\ PH_b \\ PH_c \end{array} + X{<}^{H}_{H}$

unlikely any process of tritium introduction into the coenzyme by an equilibration process, or a process approaching equilibration.

Table IX illustrates three reaction schemes which have been proposed. The first one is the reaction scheme which we favor at the moment. A substrate comes along and reacts with the coenzyme. The two C-5′ hydrogens of the coenzyme are H_b, H_c. Hydrogen is transferred from the substrate to the coenzyme which leads to the formation of an intermediate containing three hydrogens (H_a, H_b, H_c), all of them more or less equivalent. Two of them are originally derived from the coenzyme, and one is contributed by the product. I think there is a considerable amount of evidence which supports this kind of intermediate.

A second mechanism, which Jack Richards (7) at Cal Tech has called the two-hydrogen merry-go-round, involves a hydrogen transfer without an intermediate. In this mechanism, you say, "Well, you don't want an intermediate like this. You simply have some kind of a scheme whereby you push a hydrogen from a substrate into the coenzyme and simultaneously another one comes out of the coenzyme." This is called the two-hydrogen merry-go-round because one would assume this mechanism would be highly stereospecific. To use both hydrogens, as the experimental data require, each displacement must involve an inversion. It takes two turnovers to get all of the hydrogen out of the coenzyme.

The third mechanism (8), proposed by Schrauzer, involves an entirely different principle. He has proposed that the conversion of substrate to product proceeds by way of a 1,2-hydride shift. The cobalt of B_{12} coenzyme participates in this reaction, but the adenosyl side chain participates in no direct way. It is further proposed, in order to account for the observed isotope exchange data, that the enzyme-bound product exchanges tritium with the enzyme-bound adenosyl portion of the coenzyme, and this is the only process whereby tritium is introduced into the coenzyme. The fundamental difference between that mechanism and the two previous mechanisms is that the adenosyl side chain of the coenzyme plays no fundamental role in the rearrangement. I object to this mechanism on several grounds. It is proposed that the transfer between the product and the coenzyme happens because the aldehyde has an activated hydrogen. This is obviously true. However, it is very likely that at this stage the aldehyde is not an aldehyde but is hydrated and so it does not have an activated hydrogen. Furthermore, there are a number of other reactions which require B_{12} coenzyme in which there is no possible way of activating this hydrogen; nevertheless, this tritium transfer occurs. Tritium transfer between coenzyme and substrate and product has been observed with all B_{12}-requiring reactions. I believe that all of these reactions probably go by the same mechanism. If the coenzyme became labeled by interaction with an enzyme-bound product, the maximum amount of label that could be introduced into the coenzyme is that which would be obtained at equilibrium, and we have shown that the specific activity of the coenzyme far exceeds that of the reaction product. We believe these results eliminate any process in which tritium is introduced into the coenzyme by exchange with the reaction product. No experimental support exists for any of the intermediates required by the third mechanism. No one has ever found the unsaturated adenosine product, and there is no evidence for the involvement of vitamin $B_{12(s)}$.

If vitamin $B_{12(s)}$ were involved, one ought to be able to do something about it. Schrauzer (9) has shown that vitamin $B_{12(s)}$ is the most nucleophilic thing in the world. I think it is several orders of magnitude more nucleophilic than I^-, and therefore one should be able to use a trapping agent to trap it. We have, and other people have, used a large variety of trapping agents that should very readily react with vitamin $B_{12(s)}$, such as acetylene and methyl iodide. No evidence was obtained that $B_{12(s)}$ was involved. This was done with radioactive materials, and very small amounts of reaction product could have been detected. We were not satisfied with this kind of approach, because we can always hide behind the folds of the enzyme and say that these reagents simply don't get in. We tried a lot of things and did some very clever experiments; none of them worked. One of the less clever ones worked. This experiment was based on an observation we had previously made. When the enzyme–coenzyme (^3H) complex is allowed to react with acetaldehyde, tritium is transferred from the coenzyme to acetaldehyde. We therefore decided to use a reactive aldehyde, chloroacetaldehyde, and to use whether it would acquire tritium from a tritiated coenzyme as a criterion as to whether or not it had gotten into the active site of the enzyme. Chloroacetaldehyde, when added to enzyme–coenzyme complex, became tritiated. The substrate, therefore, goes into the active site when the enzyme is in the reactive state, but no change takes place in the chloroacetaldehyde. This appears to us to be a very strong argument against the existence of $B_{12(s)}$. In view of the very high nucleophilicity of $B_{12(s)}$, one would expect that it should react with chloroacetaldehyde to displace chlorine. Since this did not happen, the presence of a highly nucleophilic species at the active center seems unlikely. In summary, although $B_{12(s)}$ appears as an attractive intermediate, all experiments carried out to date to obtain evidence for its existence have been negative.

The mechanism proposed by Schrauzer involves an activation process of the coenzyme in which the carbon–cobalt bond is broken. This is initiated by abstraction of a proton from C-4' of the coenzyme. We carried out an experiment to obtain evidence for such a proton abstraction. One would assume that if a proton were abstracted, it should become exchangeable with the solvent. We, therefore, carried out the conversion of propanediol to propionaldehyde in tritiated water (10). The experiment was carried out under conditions such that if 1% of the expected exchange had taken place, we should have detected it. However, we saw no tritium incorporation from the solvent into the coenzyme. Therefore, one must conclude that this experiment certainly does not support the proposed activation mechanism and is probably inconsistent with it.

Table X. Experimental Test for Intramolecular H-Transfer

Allow enzyme to react with
$$CH_3\text{—}\underset{OH}{CH}\text{—}\underset{OH}{\overset{T}{C}H} + \underset{OH}{CH_2}\text{—}\underset{OH}{CH_2}$$

At ∞ conc. $CH_2OH\text{—}CH_2OH$: intramolecular transfer $CH_3\text{—}CHT\text{—}CHO$
no intramolecular transfer $CH_5CH_2\text{—}CHO$

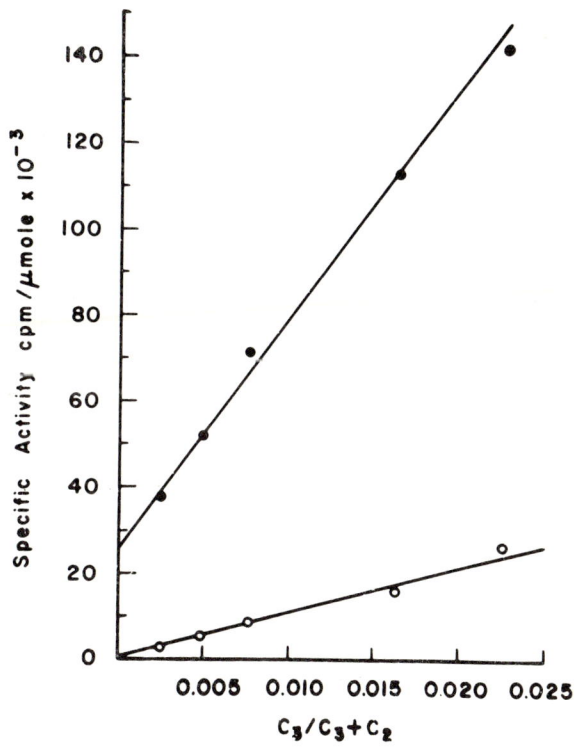

Figure 6. Evidence for intramolecular H-transfer

Available experimental results are consistent with Mechanism A. I will only briefly summarize some of these results rather than go through all of the details. One of the differences between Mechanisms A and B is that in a single turnover, with a tritiated substrate, no tritium could end up in the product by Mechanism B. That is implicit, because Mechanism B says that the first hydrogen to come out is that of the coenzyme. We tested this and showed that in a single turnover with a tritiated substrate, tritium does occur in the product. This eliminates Mechanism B. The way this was done is shown in Table X. We allowed tritiated propanediol to react in the presence of increasing quantities of unlabeled ethylene glycol and measured the specific activity of the resulting propionaldehyde. Figure 6 shows that at infinite concentration of ethylene glycol, the propionaldehyde still contains tritium. Therefore, this tritium could not have been acquired by intermolecular transfer. It must have been acquired by intramolecular turnover, which means that in a single turnover the tritium of the substrate can appear in the product. This is consistent with Mechanism A, but inconsistent with Mechanism B. Jack Richards (7) has reached the same conclusion with methylmalonyl CoA isomerase by a different argument.

Another kind of approach which we used was the following. What decides whether H_a, H_b, or H_c (Table IX) will be transferred to the reaction product? There may be some steric problems and isotope effect. If one of the hydrogens, say H_a, is replaced by tritium, then in any one turnover H_b and H_c, being hydrogen, will have a much higher probability of transfer to the product because of the difference in zero point energies. That explains our results with a deuterated substrate. When you have a deuterated substrate, very soon all of the hydrogens of the coenzyme exchange for deuterium. Then a tritium atom competes with two deuteriums, and that makes the probability of a tritium coming out much higher. This explains the increased rate of tritium release from the enzyme–coenzyme to the product when deuterated substrate is used (see Table VI).

Now, we extended this kind of experiment a little bit by asking the following question. Is there an intermediate in which the deuterium of the substrate really competes with the hydrogens of the coenzyme? If we allow a nondeuterated and a deuterated substrate to react with enzyme–coenzyme (^3H), does the deuterated substrate have a higher probability of getting tritium? According to Mechanism A, the intermediate derived from a deuterated substrate when coenzyme (^3H) is used will be $X_{TD}{}^H$ and that derived from a nondeuterated substrate will be $X_{TH}{}^H$. Therefore, the product derived from a deuterated substrate should have a higher probability of acquiring tritium from the coenzyme than that from a nondeuterated substrate. This prediction was tested experimentally by simultaneously incubating a deuterated and a nondeuterated substrate with enzyme in the presence of a tritiated coenzyme. In all cases, the specific activity of the product derived from the deuterated substrate was 4–7 times as high as that derived from the nondeuterated substrate. These results, therefore, are in accordance with the predictions made from Mechanism A, and it would be very difficult to obtain this type of discrimination from Mechanism B.

These are complex experiments, and they could well bear some more discussion. The results available at this time require an intermediate, derived from the coenzyme, containing three equivalent hydrogens. What is the structure of this intermediate? We carried out some experiments to attempt to trap such an intermediate with a "bad" substrate. If you postulate an intermediate in some reaction, and if you have a bad substrate, you could hope to form the intermediate. Then the enzyme sees a structure it really doesn't want and doesn't know what to do with it. Hopefully, the intermediate will accumulate, and you might have a chance to see it. This kind of approach has worked in many enzymatic studies.

Table XI. Reaction of E · DBCC With Glycolaldehyde

$$E \cdot DBCC + CH_2OH-CHO \rightarrow Complex$$

$$Complex \xrightarrow{TCA} CHO-CHO + B_{12(b)} + \text{5-deoxyadenosine}$$

We tried glycolaldehyde (*11*) (Table XI). Its reaction with the enzyme–coenzyme complex is competitive with the substrate, which indicates that it probably reacts at the active site. The reaction requires potassium, as does the substrate reaction. This compound behaves like the substrate in many respects. However, it forms a stoichiometric complex with enzyme–coenzyme. One mole reacts with 1 mole of enzyme–coenzyme and the complex is no longer catalytically reactive. If one looks at the spectrum of this complex, one sees exactly the same spectrum as you see in the presence of the substrate. This complex contains all the carbons of the substrate (glycolaldehyde), all of the coenzyme, and, of course, the protein. It is very difficult to see what is going on without taking this complex apart. This can be done by denaturing the protein under the mildest conditions. If one takes that complex apart, one sees glyoxal, which is an oxidation product derived from glycolaldehyde, and one finds 5′-deoxyadenosine. Furthermore, one of the hydrogens of glycolaldehyde is now in the methyl group of 5′-deoxyadenosine. B_{12} is also isolated, indicating that the carbon–cobalt bond was broken. This experiment was done under anaerobic conditions in order to eliminate the possibility of having oxidized glycolaldehyde by air. Under anaerobic conditions, one sees a reduced form of B_{12}; otherwise, one isolates the same products. In fact, 5′-deoxyadenosine was the only product derived from the coenzyme that we have ever seen under these conditions.

Other people have also found 5′-deoxyadenosine in several other reactions of this type (*12*). This compound occurs quite frequently and conceivably could fulfill the function of an intermediate.

We made a coenzyme analog (Figure 7), in which the ribose oxygen of the adenoxyl moiety of the coenzyme was replaced by carbon (*13*). This was done to test the mechanism which had been proposed, in which

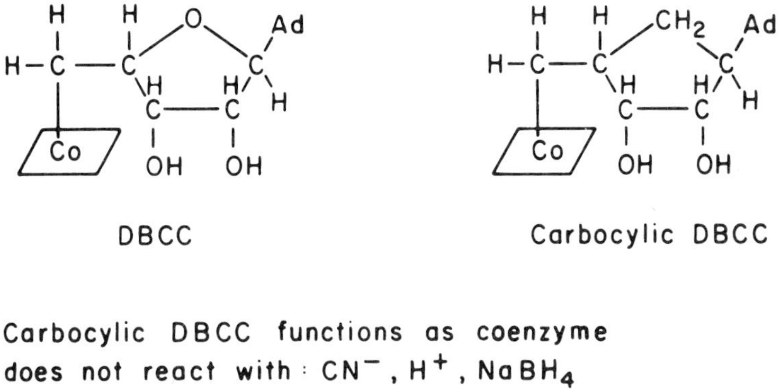

DBCC

H–C(H)–C(H)(O-ring)–C(H,H)–C(H)–C(H)
ring: C—C, OH, OH, Ad
Co—

Carbocylic DBCC

H–C(H)–C(H)(CH₂-ring)–C(H,H)–C(H)–C(H)
ring: C—C, OH, OH, Ad
Co—

Carbocylic DBCC functions as coenzyme does not react with: CN^-, H^+, $NaBH_4$

Figure 7. Coenzyme analog with carbon replacing the ribose oxygen of the adenoxyl moiety of the coenzyme

Table XII.   Reaction Sequence for Dioldehydrase Reaction

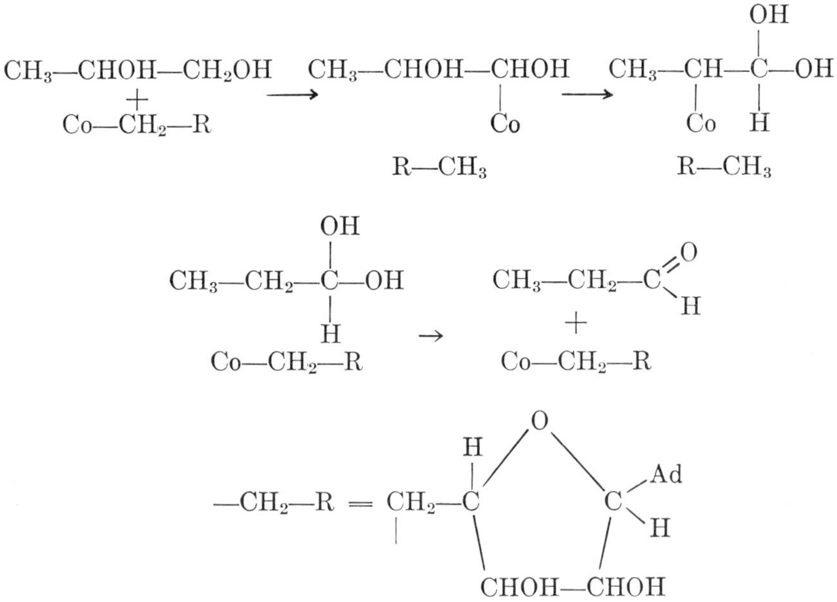

$$CH_3-CHOH-CH_2OH \atop + \quad Co-CH_2-R \longrightarrow {CH_3-CHOH-CHOH \atop \underset{R-CH_3}{|} \atop Co} \longrightarrow {CH_3-CH-\underset{H}{\overset{OH}{|}}C-OH \atop \underset{R-CH_3}{|} \atop Co}$$

$$CH_3-CH_2-\underset{H}{\overset{OH}{\underset{|}{C}}}-OH \atop \underset{}{|} \atop Co-CH_2-R \quad \rightarrow \quad CH_3-CH_2-C{\overset{O}{\underset{H}{\diagup\diagdown}}} \atop + \atop Co-CH_2-R$$

$$-CH_2-R = CH_2-\underset{|}{\overset{H}{C}}\diagup^{O}\diagdown\underset{CHOH-CHOH}{\diagup} C\diagup^{Ad}_{H}$$

the carbon–cobalt bond was envisioned to break by shifting the electrons towards the oxygen. Since the compound works, this and similar mechanisms can be eliminated.

Table XII shows a reaction scheme which is consistent with all of the experimental data. It suffers from the disadvantage that there are as yet no analogous chemical reactions known for the reactions proposed here.

This may simply reflect the fact that the proper chemistry in this area has not been examined so far.

This reaction mechanism involves the following. First, the substrate reacts with the coenzyme and a ligand exchange reaction takes place in which a new carbon–cobalt bond is formed between the substrate and cobalt of the coenzyme concomitantly. The carbon–cobalt bond of the coenzyme is broken and a hydrogen from the substrate is transferred to the adenosyl group forming a 5'-deoxyadenosine. This compound has actually been isolated in a number of reactions. It would also be consistent with the isotope data, since it is a compound which contains three equivalent hydrogens. Subsequently, we propose a rearrangement as indicated, in which a new carbon–cobalt bond is formed. This rearrangement might be analogous to the reactions of beta–cyano–ethyl–cobaloximes discussed by Schrauzer (*14*). The final step of the reaction again involves a ligand exchange reaction in which the coenzyme is reformed and the product is released. In this process, a hydrogen from the methyl group of 5'-deoxyadenosine is now transferred to the reaction product.

The reaction scheme will explain all of the known results to date. Perhaps as it stands now it should not be taken too seriously, but may well serve as a guide for future experiments. I fully agree that no enzyme mechanism can be considered elucidated until appropriate nonenzymic models are available.

Acknowledgment

The work described here was carried out by O. Wagner, P. A. Frey, M. Essenberg, and T. Finlay, supported by a grant from the National Institute of Health, 12633.

Literature Cited

(1) Hogencamp, H. P. C., *Ann. Rev. Biochem.* (1968) **37**, 225.
(2) Rose, I. A., *Ann. Rev. Biochem.* (1966) **35**, 23.
(3) Zagalak, B., Frey, P. A., Karabatsos, G. L., Abeles, R. H., *J. Biol. Chem.* (1966) **241**, 3028.
(4) Retey, J., Umani-Ronchi, A., Seibl, J., Arigoni, D., *Experientia* (1966) **22**, 502.
(5) Frey, P. A., Essenberg, M. K., Abeles, R. H., *J. Biol. Chem.* (1967) **242**, 5369.
(6) Frey, P. A., Kerwar, S. S., Abeles, R. H., *Biochem. Biophys. Res. Commun.* (1967) **29**, 873.
(7) Miller, W. W., Richards, J., *J. Am. Chem. Soc.* (1969) **91**, 1498.
(8) Schrauzer, G. N., Sibert, J. W., *J. Am. Chem. Soc.* (1970) **92**, 1022.
(9) Schrauzer, G. N., Deutsch, E., Windgassen, R. J., *J. Am. Chem. Soc.* (1968) **90**, 2441.

(10) Frey, P. A., Essenberg, M. K., Kerwar, S. S., Abeles, R. H., *J. Am. Chem. Soc.*, in press.
(11) Wagner, O. W., Lee, H. A., Jr., Frey, P. A., Abeles, R. H., *J. Biol. Chem.* (1966) **241,** 1751.
(12) Babior, B. M., *J. Biol. Chem.* (1970) **245,** 1755.
(13) Kerwar, S. S., Smith, T. A., Abeles, R. H., *J. Biol. Chem.* (1970) **245,** 1169.
(14) Schrauzer, G. N., ADVAN. CHEM. SER. (1971) **100,** 1.

RECEIVED June 26, 1970.

17

Structural Models for Iron and Copper Proteins Based on Spectroscopic and Magnetic Properties

HARRY B. GRAY

California Institute of Technology, Pasadena, Calif. 91109

The spectroscopic and magnetic properties of several iron(III) model systems have been investigated and compared with those of certain iron-containing proteins. Spectral data show that [Fe(III)O_6] octahedral coordination is present in the Saltman–Spiro iron(III) polymer and probably also in the ferritin core. A tetrahedral polymeric [Fe(III)O_4] coordination is compatible with the spectral and magnetic data found for iron(III) phosvitin. Extensive data have been collected on oxobridged Fe(III) dimers. An electronic structural model involving a moderate antiferromagnetic coupling of the two high-spin Fe(III) centers satisfactorily explains the ligand field spectra. Evidence for simultaneous pair electronic excitations is also presented. The spectral and magnetic data for hemerythrins are interpreted in terms of a dimeric Fe(III) model, and a general two-electron oxidative addition binding model is proposed for oxyhemerythrin, oxyhemoglobin, and oxyhemocyanin.

For several years my coworkers and I have been studying the electronic absorption spectra of models for metal proteins and, in certain cases, of the proteins themselves. In favorable cases, additional information has been gathered from infrared spectral and magnetic susceptibility measurements.

Our objective in this work is the development of a line of attack which will allow the construction of a reasonable model for the structure in the immediate vicinity of the metal, or the coordination structure, of the metal protein.

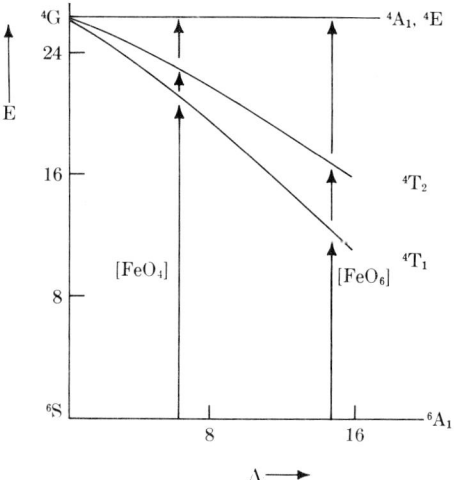

Figure 1. Relationship of the lowest LF energy levels for d^5 Fe(III) complexes in octahedral and tetrahedral coordination

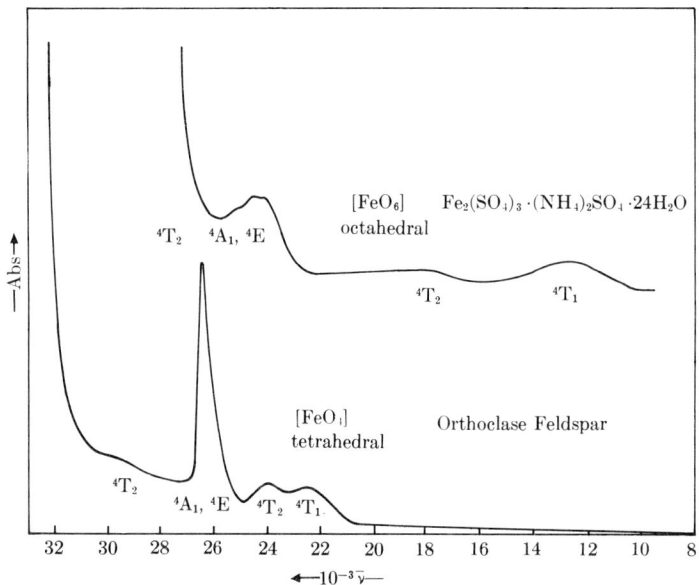

Figure 2. Electronic absorption spectra of $Fe(H_2O)_6^{3+}$ in ferric ammonium sulfate and $[Fe(III)O_4]_{tet}$ in orthoclase feldspar

High-Spin Fe(III) Complexes

A large part of our effort has been directed toward an understanding of the electronic spectra of d^5 high-spin complexes of the structural types $[Fe(III)O_6]_{oct}$ and $[Fe(III)O_4]_{tet}$. The interesting situation with respect to the ligand field (LF) transitions for these two types of complexes is shown in Figure 1. There are no spin-allowed LF transitions from a $^6A_1$ ground state, and the two lowest spin-forbidden transitions decrease in energy as the cubic LF splitting increases. Therefore, the first two spin-forbidden LF bands for $[Fe(III)O_6]_{oct}$ should occur at lower energies than the analogous bands for $[Fe(III)O_4]_{tet}$.

Experimental spectra of two structurally well-defined model systems illustrate this difference (1). The upper spectrum of Figure 2 is that of a single crystal of ferric ammonium sulfate which contains $Fe(H_2O)_6^{3+}$; the lower spectrum is that of an Fe^{3+}-doped sample of orthoclase feldspar. In this sample, Fe^{3+} ions replace Al^{3+} in a certain percentage of tetrahedral oxide donor sites. The two spectra show very clearly that the two lowest bands, $^6A_1 \rightarrow {}^4T_1(^4G)$ and $^6A_1 \rightarrow {}^4T_2(^4G)$, are substantially lower in energy in the $[Fe(III)O_6]_{oct}$ case. These spectra were selected because four LF bands are resolved in each case. The third band is fairly sharp in each spectrum and is assigned to the transition which does not require orbital promotion, $^6A_1 \rightarrow {}^4A_1, {}^4E(^4G)$. The fourth band in each case is assigned $^6A_1 \rightarrow {}^4T_2(^4D)$.

Summaries of the spectral data for $Fe(H_2O)_6^{3+}$ and $[Fe(III)O_4]_{tet}$ are given in Tables I and II, respectively. The molar extinction coefficients of the spin-forbidden LF bands are relatively small, as expected. The bands for $[Fe(III)O_4]_{tet}$ are approximately a factor of ten more intense than those for $Fe(H_2O)_6^{3+}$ because the LF transitions in the latter complex are both spin- and orbitally-forbidden. The computed energy

Table I. Electronic Spectral Data for $Fe(H_2O)_6^{3+}$ in $Fe_2(SO_4)_3 \cdot (NH_4)_2SO_4 \cdot 24H_2O$ [a]

$^6A_1 \rightarrow$	$\bar{\nu}$, cm^{-1} [b]	ε
$^4T_1$	12,600	0.05
$^4T_2$	18,200	0.01
$(^4A_1, {}^4E)$	24,200 24,600 25,400	1.3
$^4T_2$	27,700	1

[a] From Ref. 1.
[b] The spectrum may be fit satisfactorily using the following parameter values: $10Dq(oct) = 13{,}700\ cm^{-1}$; $C/B = 3.13$; $B = 945\ cm^{-1}$.

Table II. Electronic Spectral Data for $[Fe(III)O_4]$ in Orthoclase Feldspar[a]

$^6A_1 \rightarrow$	$\bar{\nu}, cm^{-1}$	ε	$Calcd$[b]
$^4T_1$	22,500	0.73	22,150
$^4T_2$	23,900	0.76	24,550
$(^4A_1, {}^4E)$	26,500	4.1	26,550
$^4T_2$	29,200	0.1	28,940

[a] From Ref. *1*.
[b] For Δ(tet) = 7350, B = 540, C = 4230 cm^{-1}.

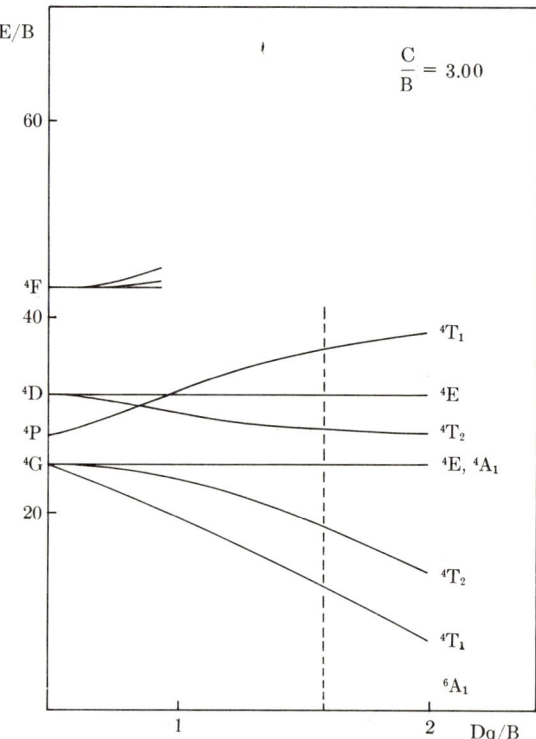

Figure 3. Calculated LF energy levels for $[Fe(III)O_6]_{oct}$

level diagrams shown in Figures 3 and 4 are based on the experimental data for the two model complexes. The value of Δ_{oct} is considerably larger than Δ_{tet}, in agreement with expectation. Of interest is the large difference in the Racah parameter ratio C/B—for the octahedral case, this ratio is approximately 3, whereas for the tetrahedral coordination it is about 8.

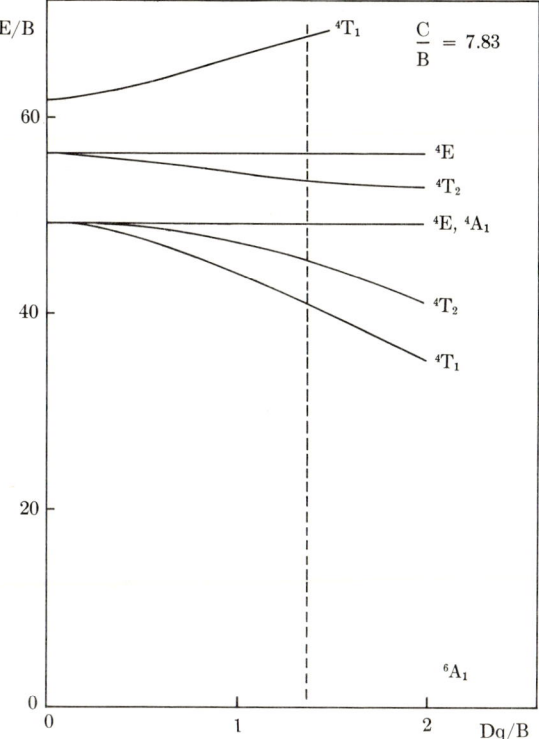

Figure 4. Calculated LF energy levels for $[Fe(III)O_4]_{tet}$

The Ferritin Core

An interesting biological system which contains Fe^{3+} in an $[Fe(III)\text{-}O_n]$ coordination structural environment is the iron-storage protein, ferritin. The core of ferritin consists of a phosphate-containing, iron(III) hydrated oxide or oxide–hydroxide (2). Models have been suggested for the ferritin core which feature Fe(III) in $[Fe(III)O_4]_{tet}$ (3), $[Fe(III)\text{-}O_6]_{oct}$ (4), and mixed $[Fe(III)O_4]_{tet}\text{-}[Fe(III)O_6]_{oct}$ (5) coordination environments.

A particularly interesting synthetic model for the ferritin core is the iron(III) polymer obtained by Spiro, Saltman, and coworkers from bicarbonate-hydrolyzed ferric nitrate solutions (6). The polymer has an approximate molecular weight of 150,000 and is about 70 Å in diameter. Analytical studies indicate a composition $Fe_4O_3(OH)_4(NO_3)_2 \cdot 1.5H_2O$ for the polymer. In physical size and shape, the Saltman–Spiro ball is remarkably similar to the ferritin core. The structure that has been suggested (3) for the Saltman–Spiro ball and implied for ferritin is shown in Figure 5. The Fe(III) ions are bound together in a network of bent oxo and hydroxo bridges with a tetrahedral coordination structure, $[Fe(III)O_4]_{tet}$. In this model, water molecules and NO_3^- ions coat the ball, leading to Saltman's graphic analogy with M and M's candy (7). The principal basis for the $[Fe(III)O_4]_{tet}$ coordination structural model is a low-angle x-ray scattering experiment and subsequent analysis of the radial distribution function (3). This is a difficult experiment to analyze, and in our opinion electronic spectroscopy offers a more foolproof route to the coordination structure.

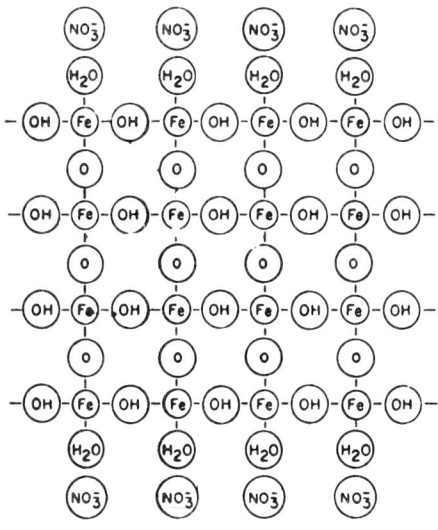

Biochemistry

Figure 5. Structural model proposed for the Saltman–Spiro ball, adapted from Ref. 3

We have examined (8), as did Brady and coworkers (3), the electronic absorption spectrum of the Saltman–Spiro ball. The spectrum (Figure 6) (8) shows a pattern of four weak bands with the lowest band at about 900 nm, in very good agreement with $[Fe(III)O_6]_{oct}$ coordination. The derived LF parameters are $\Delta_{oct} = 11{,}260$ cm^{-1}, $C/B = 3$, and $B = 815$ cm^{-1}. We can say with considerable confidence that most of the Fe(III) ions occupy octahedral coordination sites. There is no hint of bands attributable to $[Fe(III)O_4]_{tet}$, and as these bands are intrinsically more intense than are those assigned to $[Fe(III)O_6]_{oct}$, we can effectively rule out tetrahedral coordination in this synthetic model compound.

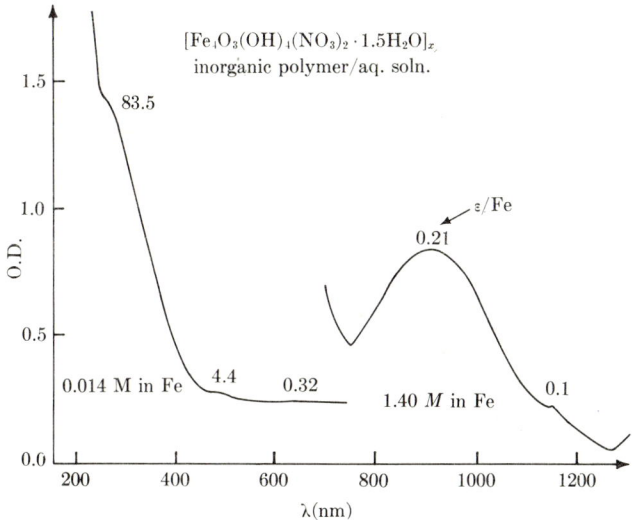

Figure 6. Electronic absorption spectrum of the Saltman–Spiro ball

Our electronic absorption spectral studies on the ferritin core itself are less definitive (8, 9). Because of strong tail absorption through the visible, we have been able to resolve only a 918-nm LF band. This band is diagnostic of $[Fe(III)O_6]_{oct}$ coordination, but without better spectra in the 400–700-nm region we cannot distinguish between exclusively $[Fe(III)O_6]_{oct}$ and mixed $[Fe(III)O_6]_{oct}$–$[Fe(III)O_4]_{tet}$ coordination structures. There is, however, no positive evidence for the presence of $[Fe(III)O_4]_{tet}$ units in the core and the analogy with the Saltman–Spiro ball leads us to prefer a model with $[Fe(III)O_6]_{oct}$ coordination.

Another important structural feature in both the Saltman–Spiro ball and the ferritin core that can be examined by spectroscopic means is the question of whether the polymeric Fe(III) network is held together through both oxo and hydroxo bridges, as suggested in Figure 5, or whether oxo (or hydroxo) bridges alone suffice. Infrared spectral studies of various hydrated iron oxides of known structure have shown that bridging hydroxide invariably absorbs in the 900–1200-cm^{-1} region (9). Neither the Saltman–Spiro ball nor the ferritin core shows an infrared band attributable to bridging hydroxide (9), although absorption owing respectively to NO_3^- and PO_4^{3-} in the region of interest casts some doubt on the conclusion. It seems highly probable, however, that both the Saltman–Spiro ball and the ferritin core contain disordered polymeric networks of $[Fe(III)O_6]_{oct}$ units primarily linked by bent oxo bridges.

Presumably, throughout these polymeric networks water molecules and NO_3^- (or PO_4^{3-}) are bound to some of the Fe(III) centers.

Iron(III) Phosvitin

Another iron(III) protein where [Fe(III)O_n] coordination is likely is iron(III) phosvitin. The fact that 46 Fe(III) ions are bound by one molecule of phosvitin has recently been established by Saltman and Multani (10). Electronic absorption spectra of the iron(III) protein are shown in Figure 7 (11). Three weak LF bands are resolvable in the 150°K spectrum, at 447, 426, and 400 nm. The relatively high energies of the first two LF bands suggest [Fe(III)O_4]$_{tet}$ coordination, and in fact a very satisfactory fit of the spectrum is obtained with the LF parameters $\Delta_{tet} = 5270$ cm^{-1}, B = 495 cm^{-1}, and C/B = 8. Notably, the spectrum is apparently devoid of peaks in the 800–1000-nm region where [Fe(III)O_6]$_{oct}$ invariably appears. Thus, we suggest that iron(III) phosvitin provides a biological example of the fairly rare [Fe(III)O_4]$_{tet}$ coordination.

Two more bits of information tempt us to build up the detail of the model. We have measured the magnetic susceptibility of iron(III) phosvitin over the temperature range 85°–300°K and find evidence for antiferromagnetic behavior (Figure 8) (11). The magnetic moment per iron drops from 4.45 to 3.39 B.M. over the temperature range investigated. Thus, it is highly probable that at least some of the [Fe(III)O_4]$_{tet}$ units are interacting through ligand bridges. As there are 100 serine phos-

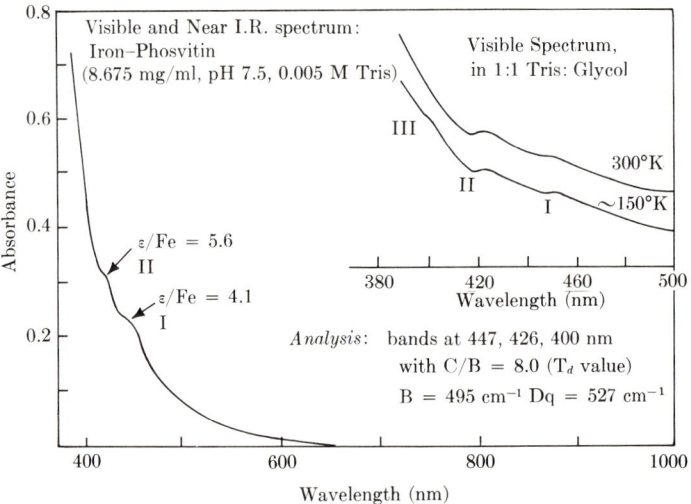

Figure 7. Electronic absorption spectra of iron(III) phosvitin

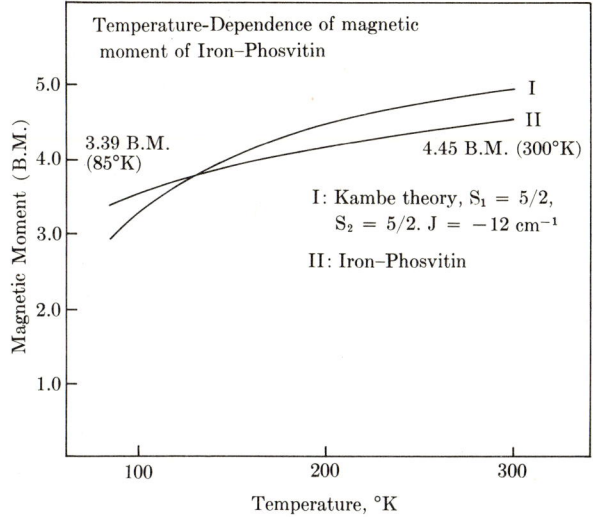

Figure 8. Magnetic moment data for iron(III) phosvitin

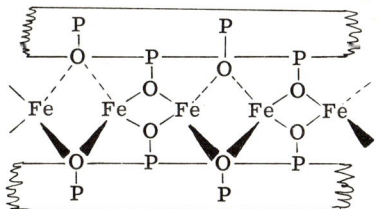

Figure 9. Model for iron(III) binding by phosvitin

phate residues in phosvitin, or roughly two for each Fe(III), an interesting possibility is a structure of the type shown in Figure 9.

A number of other models are possible, of course. For example, a small percentage of octahedrally coordinated Fe(III) ions could escape detection in our spectroscopic experiment. We should emphasize the major conclusions: Most of the Fe(III) ions are tetrahedrally coordinated and form some type of polynuclear structure.

Oxobridged Iron(III) Dimers

Our structural investigations of hydrolytically polymerized Fe(III) have not been limited to the high molecular weight systems such as the Saltman–Spiro ball. Much more of our research, in fact, has been concentrated in the area of relatively simple binuclear Fe(III) complexes. The polyfunctional ligands EDTA (ethylenediaminetetracetate) and

HEDTA (hydroxyethylethylenediaminetriacetate) chelate Fe(III) so effectively that upon base hydrolysis only one coordination position is available for bridging, and dimers of the type (EDTA Fe)$_2$O^{4-} and (HEDTA Fe)$_2$O^{2-}, but no higher polymers, form (*12, 13, 14*). As a result, EDTA and HEDTA are excellent solubilizing agents for Fe(III) over a wide pH range extending to very basic solutions.

Red crystalline salts can be obtained from the red solutions containing the dimeric complexes (*13*). In particular, we have investigated the enH$_2^{2+}$ (ethylenediammonium) salt of (HEDTA Fe)$_2$O^{2-}. The structure of this complex is known from x-ray crystallographic analysis (*15*). The most important features of the coordination structure are shown in Figure 10. The monomeric units [Fe(III)O$_4$N$_2$]$_{oct}$ share a nearly linear oxo bridge, the Fe(III)–O–Fe(III) angle being 165°. The Fe(III)-to-oxo distance is 1.8 Å, a number which might be considered an indication of $p\pi \rightarrow d\pi$ bonding were it not for the fact that a similarly short distance (1.68 Å) has been observed in an Al(III)–O–Al(III) structure (*16*).

The (HEDTA Fe)$_2$O^{2-} complex possesses an infrared absorption band at about 840 cm^{-1} which is characteristic of an almost linear Fe(III)–O–Fe(III) unit (*13, 14, 17, 18*). This spectral feature of the complex is shown in Figure 11 (*18*). The band is assigned to the antisymmetric stretching motion in the three-atom system. The fact that the 840-cm^{-1} band is present both in a crystal and in a D$_2$O solution of (enH$_2$)-[(HEDTA Fe)$_2$O] · 6H$_2$O establishes that the oxobridged structure persists in aqueous media. Magnetic measurements and electronic spectral

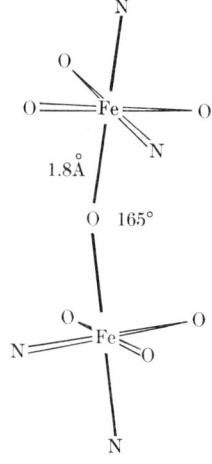

Figure 10. Coordination structure of (HEDTA Fe)$_2$O^{2-}

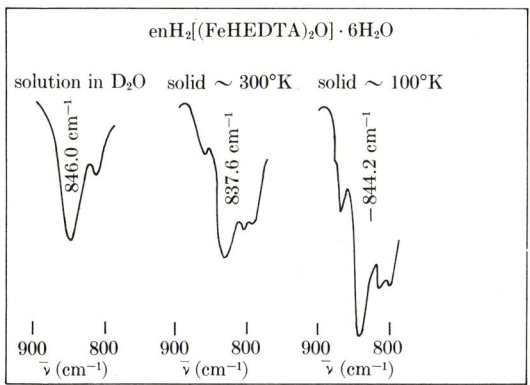

Figure 11. Infrared spectra of $(HEDTA\ Fe)_2O^{2-}$ in the 840-cm^{-1} region

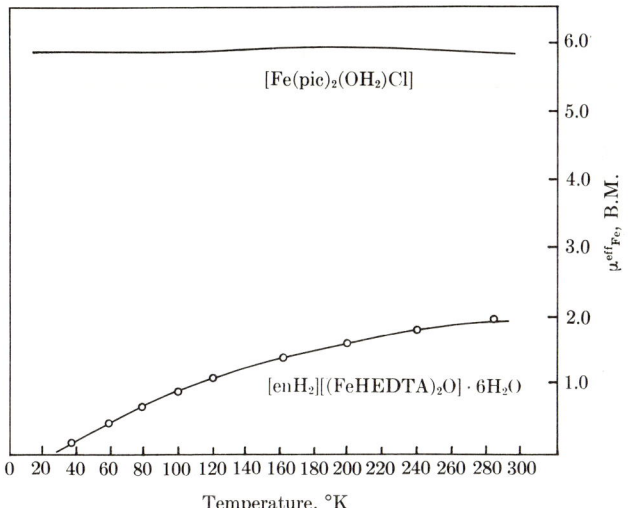

Figure 12. Magnetic moment data for $(enH_2)[(HEDTA\ Fe)_2O] \cdot 6H_2O$ and $[Fe(pic_2)(H_2O)Cl]$ (pic = picolinate)

data also confirm the essential identity of the structure in solids and aqueous solutions (18).

Turning to the magnetic data, Figure 12 shows μ_{eff} vs. T plots for a solid sample of $(enH_2)[(HEDTA\ Fe)_2O] \cdot 6H_2O$, and a reference high-spin ($S = 5/2$), monomeric Fe(III) complex. The binuclear complex exhibits antiferromagnetic behavior, the μ_{eff} per Fe(III) dropping from 1.9 B.M. at 300°K to effectively zero at about 30°K. An excellent fit of these data is produced by assuming that two $S = 5/2$ ions are spin coupled, with $J = -95$ cm^{-1}. Very strong supporting evidence for this model is provided by the electronic spectral data to be discussed later.

Examination of the literature in this field reveals (18) that oxobridging is well established in several binuclear Fe(III) complexes. The oxobridged Fe(III) complexes which have been carefully studied all exhibit J values in the range -85 to -105 cm^{-1}. In sharp contrast, the one dihydroxobridged Fe(III) dimer that has been isolated and fully characterized, [Fe(pic)$_2$OH]$_2$, displays a much weaker spin-spin coupling [$J = -8$ cm^{-1}, μ_{eff} (300°K) = 5.2 B.M.] (17).

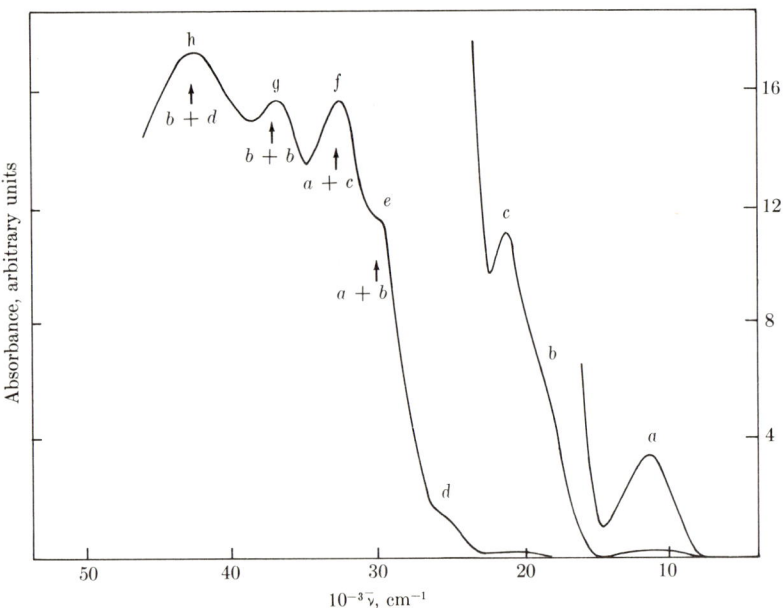

Figure 13. Electronic absorption spectrum of (HEDTA Fe)$_2$O^{2-}

Virtual proof that (HEDTA Fe)$_2$O^{2-} contains antiferromagnetically coupled octahedral $S = 5/2$ units may be extracted from its electronic absorption spectrum, which is shown in Figure 13 (18, 19). HEDTA is a good ligand for this study because it does not absorb appreciably in the near IR, visible, and near UV. No fewer than eight absorption peaks appear before intraligand and charge transfer absorption take over. The spectrum divides into two parts, a set of four low-energy peaks *a–d* with low to moderate extinction coefficients, and a group of more intense, higher-energy bands (*e–h*).

Bands *a–d* are positioned properly to be the first four LF transitions in each $S = 5/2$ [Fe(III)O$_4$N$_2$]$_{\text{oct}}$ unit (see Figure 3). Reasonable agreement between theory and experiment is obtained for the LF parameter values Δ_{oct} = 10,900 cm^{-1}, C/B = 2.4, and B = 950 cm^{-1}. The main difference between the LF spectrum in the oxobridged dimer and an octahedral monomeric Fe(III) complex is in the intensities of the bands. The molar extinction coefficients of the four LF bands in (HEDTA Fe)$_2$O^{2-}, for example, are two orders of magnitude larger than those for a monomeric complex such as Fe(H$_2$O)$_6^{3+}$. This intensity enhancement of the spin-forbidden LF bands is an expected consequence of the antiferromagnetic coupling in the oxobridged system; the coupling partially relaxes the spin selection rule and the transitions acquire some allowed character. The relationship of the spin–spin coupling to the intensity enhancement of spin-forbidden LF bands has been worked out in some depth in the case of MnF$_2$ and other Mn(II)-containing antiferromagnetic materials (20, 21). Our observation in the Fe(III)–O–Fe(III) system logically has the same explanation.

The four relatively intense UV bands (e–h) puzzled us for a long time. Based on our studies of related monomeric complexes, energetically they are too low to be attributable to ligand → Fe(III) charge transfer transitions; what is more, there are too many bands for a charge transfer explanation. The bands are much too intense to be the higher-energy LF bands, even after allowing for the intensity enhancement in an antiferromagnetically coupled dimer. Elimination of the usual types of transitions has led us to suggest that the UV spectrum consists of simultaneous pair electronic (SPE) excitations (19). That is, simultaneous LF transitions on the two Fe(III) centers can be coupled so that the pair excitation is spin-allowed. This explanation is not without precedent—Varsanyi and Dieke first proposed (22) a similar explanation for a portion of the fluorescence excitation spectrum of PrCl$_3$, and for this system Dexter has discussed (23) SPE excitations from a theoretical point of view. Such transitions in mixed Mn(II)–Ni(II) fluorides (24) and dimeric Cu(II) acetate (25) have also been suggested in interpretations of absorption spectra.

Assuming simple additivity, the positions of four SPE excitations in (HEDTA Fe)$_2$O^{2-} agree extremely well with the positions of bands e–h and provide considerable support for the interpretation. This agreement is shown in Figure 13 and is included in a summary of the interpretation of the electronic absorption spectrum of (HEDTA Fe)$_2$O^{2-} set out in Table III.

In summary, we are well on our way to an understanding of the spectral and magnetic properties of certain types of binuclear Fe(III) complexes. Oxobridged dimers of high-spin Fe(III) monomers show at least four distinctive characteristics: An infrared band around 840 cm^{-1}; antiferromagnetic behavior with J in the range −85 to −105 cm^{-1}; LF

Table III. Electronic Spectral Data for
enH$_2$[(HEDTA Fe)$_2$O] · 6H$_2$O[a]

Band	$(10^{-3})\bar{\nu}_{max}$, $cm^{-1}(\varepsilon)$	Assignment
a	11.2 (2.6)	$^6A_1 \rightarrow {}^4T_1$
b	18.2 (40)	$^6A_1 \rightarrow {}^4T_2$
c	21.0 (25)	$^6A_1 \rightarrow ({}^4A_1, {}^4E)$
d	24.4 (120)	$^6A_1 \rightarrow {}^4T_2$
e	29.2	$a + b = 29.4$
f	32.5	$a + c = 32.2$
g	36.8	$b + b = 36.4$
h	42.6	$b + d = 42.6$

[a] From Ref. *19*.

bands with enhanced intensity; and a rich series of UV bands, which we have interpreted as SPE excitations.

Hemerythrin

The spectroscopic and magnetic properties observed for the model oxobridged dimer, (HEDTA Fe)$_2$O^{2-}, have proved of value in attempts to understand the coordination structure of the protein hemerythrin, which binds oxygen reversibly. The protein from the sipunculid *Golfingia gouldii* has been studied extensively by Klotz and coworkers (*26*). Eight subunits, each containing two iron atoms capable of binding one oxygen molecule, associate in the full protein to give a molecular weight of about 108,000. Magnetic measurements show that the deoxy form of the protein contains high-spin Fe(II) (*26*). The electronic absorption spectrum of this form is relatively clear down to 300 nm; in the UV, the peak at about 280 nm characteristic of tyrosine appears.

Upon oxidation to the met [Fe(III)] protein, absorption peaks appear in the visible region. Each subunit of the met protein binds ligands such as Cl⁻, Br⁻, and N$_3^-$ in a 1:2 Fe(III) complex, and the visible peak positions of these met derivatives vary slightly (*27*). The electronic absorption spectrum of metchlorohemerythrin is shown in Figure 14 (*28*). In addition to the bands shown, there is some evidence for a peak owing to an electronic transition in the 900-nm region (*28*). This peak is heavily overlapped by bands attributable to vibrational overtones and is thus not well characterized. The bands at 668 and 500 nm, however, may be assigned as LF bands in an Fe(III) complex. Their moderate intensities (*27, 28*) are consistent with a model in which two Fe(III) ions are spin-coupled in an oxobridged dimer of the type [Cl–Fe(III)–O–Fe(III)]. Magnetic susceptibility data to be discussed momentarily support this proposal.

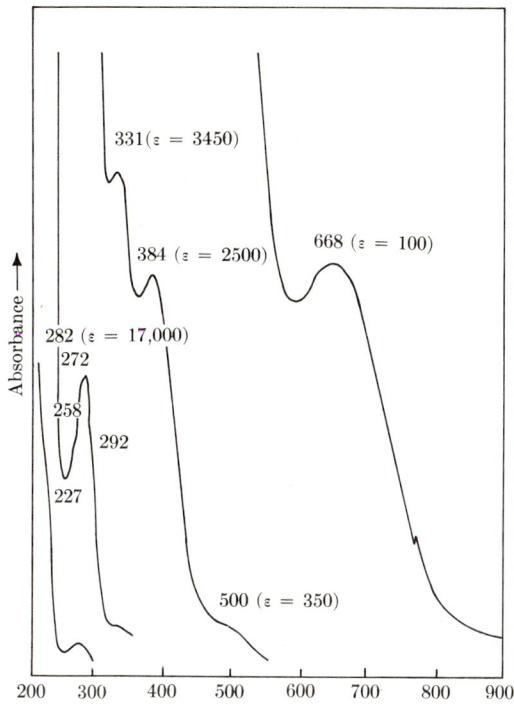

Figure 14. Electronic absorption spectrum of metchlorohemerythrin

The bands at 384 and 331 nm in the metchloro derivative must also be attributable to some kind of electronic transition involving one or both of the Fe(III) ions. Their positions and intensities can be rationalized if they are assigned to SPE excitations in an oxobridged dimer. Alternatively, ligand → Fe(III) charge transfer transitions are also possible explanations. At wavelengths lower than 331 nm, the tyrosine absorption takes over and additional peaks owing to Fe(III) are presumably buried.

The electronic absorption spectrum of oxyhemerythrin is shown in Figure 15 (28). This spectrum closely resembles that of metchlorohemerythrin, except for the much greater intensity in the band at about 500 nm. The spectrum can be considered as reasonable proof of a suggestion first made by Klotz (29) that in the oxy form both irons have been oxidized to Fe(III) and the O_2 reduced to O_2^{2-}. Large intensity at 501 nm can be easily understood in this model because a similar absorption band has been observed in the complex formed between [Fe(III)-EDTA]$^-$ and H_2O_2 in basic solution (30). Presumably, this band is attributable to an O_2^{2-} → Fe(III) or HOO$^-$ → Fe(III) charge transfer transition.

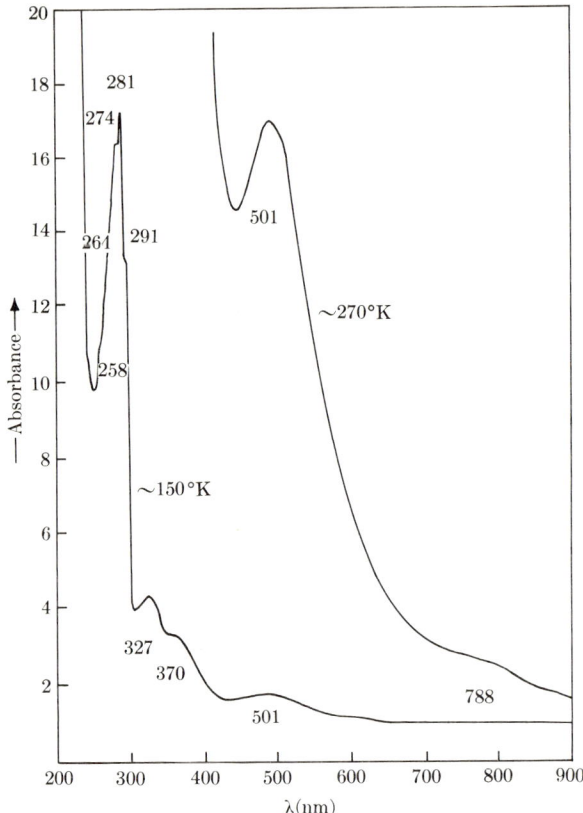

Figure 15. Electronic absorption spectrum of oxyhemerythrin

The bands at 370 and 317 nm in oxyhemerythrin are similar to the 384 and 331 nm bands in the metchloro derivative and in all probability analogous assignments are appropriate.

The spectral data for the metchloro and oxy proteins suggest coordination structures involving dimeric Fe(III) units. Magnetic susceptibility experiments over a limited temperature range for both proteins add considerable weight to this suggestion. These data are shown in Figure 16 (28). Although the error limits are large, it is apparent that in each case the μ_{eff} per Fe(III) is greatly reduced from the high-spin value of ~6 B.M. Furthermore, in the range 300° to 100°K there is some evidence that both proteins exhibit antiferromagnetic behavior.

The spectral and magnetic data point to an oxobridged Fe(III) dimer structure for the metchloro protein. An oxobridged model is also attractive for the oxyhemerythrin, but the alternative possibility of a

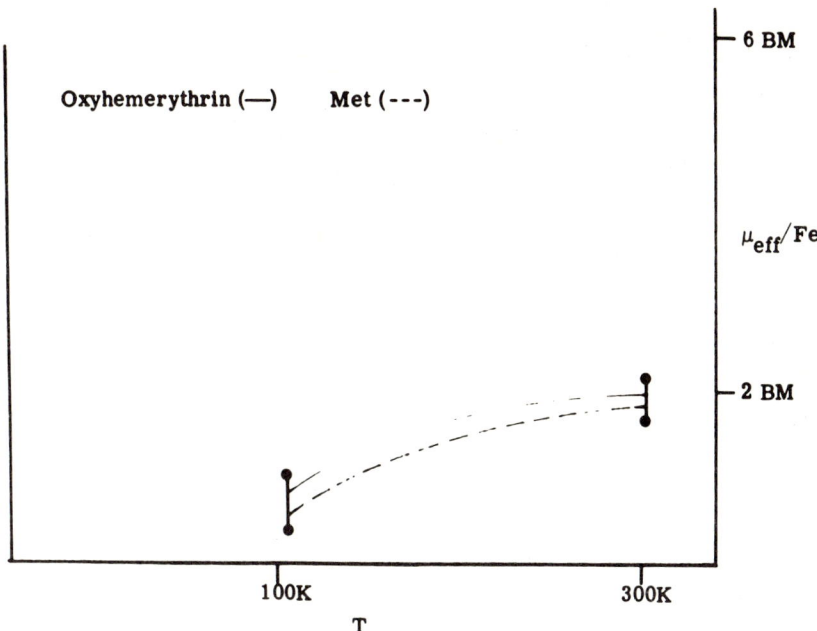

Figure 16. Magnetic moment data for metchlorohemerythrin and oxyhemerythrin

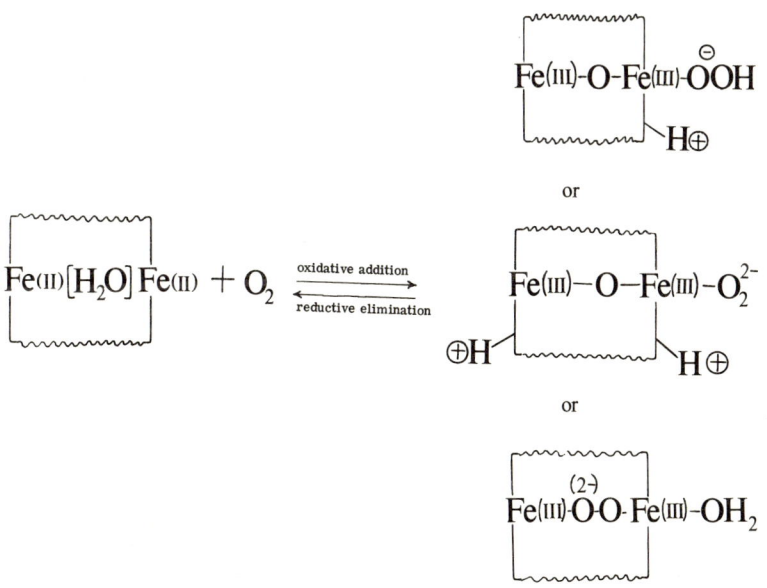

Figure 17. Oxidative addition model for O_2-binding by hemerythrin

peroxide-bridged Fe(III) cannot be ruled out (Figure 17). We slightly prefer the oxobridged model for the oxy form, because it falls naturally in the met series as a metperoxy or methydroperoxy derivative. We have obtained preliminary evidence that both oxy and met proteins show the 840-cm^{-1} absorption characteristic of an Fe(III)–O–Fe(III) structure (31). At the ambient temperature of our infrared cell there are serious decomposition problems with the oxy protein, and this result is highly tentative at this time.

Oxidative Addition Model of Oxygen Binding to Iron and Copper Proteins

The spectroscopic and magnetic data leave little doubt that a type of oxidative addition mechanism is operative in O_2 binding by hemerythrin. The two electrons necessary to reduce O_2 are furnished by the two Fe(II) ions, which from the magnetic and spectral data are antiferromagnetically coupled Fe(III) ions in the [O_2^{2-}–Fe(III)–Fe(III)] oxidative addition product. Two-electron reductive elimination of O_2 by this unit completes the reversible process. The oxidative addition process is shown in Figure 17.

The model is so simple in the case of hemerythrin that it is interesting to examine the problem of O_2 binding to hemoglobin in light of it. Oxyhemoglobin has μ_{eff} per Fe approximately equal to zero, indicating no unpaired electrons (32). Models for the electronic structure of oxyhemoglobin have been formulated primarily to explain this magnetic result. Because of the strong absorption of the porphyrin ligand in the visible region, little can be said about the oxidation state of the iron from electronic spectral measurements.

Four models for O_2 binding by hemoglobin are shown in Figure 18. The classic Pauling model (1936) involves delocalized bonding of O_2 to low-spin Fe(II), with a bent Fe–O–O unit (32). In 1956, Griffith pointed out that the bonding in oxyhemoglobin could be formulated along the same lines as in metal–ethylene π-complexes and suggested a model in which O_2 bonds symmetrically to a low-spin Fe(II) (33).

A third model, involving Fe(III) and O_2^-, aroused some controversy after it was proposed by Weiss in 1964 (34, 35, 36, 37, 38), but it has recently been supported by Peisach and coworkers (38). This binding model for oxyhemoglobin has become particularly attractive in view of the structural data now available on O_2 binding to Co(II) complexes. In a number of well established cases (39, 40, 41, 42), including the very interesting cobalt hemoglobin (or coboglobin) (43), O_2 is reduced to O_2^- and cobalt is oxidized to Co(III), so the mode of O_2 binding is

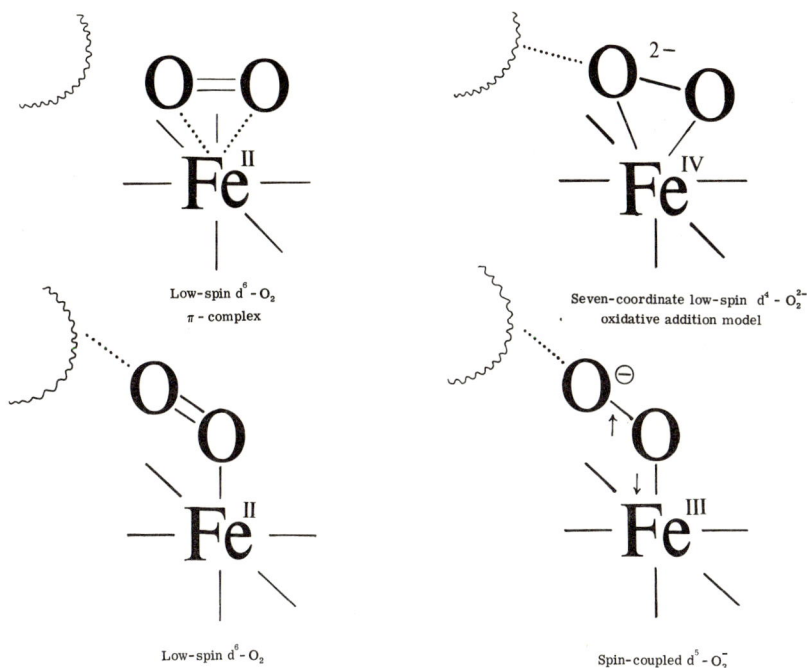

Figure 18. Four models for O_2-binding in oxyhemoglobin

$Co(II) + O_2 \rightleftharpoons [Co(III)(O_2^-)]$. Furthermore, the $Co(III)–O\overset{\ominus}{=}O$ unit is bent.

In the $[Fe(III)–(O_2^-)]$ model, the diamagnetism of oxyhemoglobin must be explained by postulating (38) an appropriately large antiferromagnetic coupling between low-spin Fe(III) and O_2^-. Although Peisach and coworkers (38) cite some spectral evidence in support of the presence of Fe(III), the strong analogy between met and oxy derivatives established from magnetic and spectral data for hemerythrin is lacking. Furthermore, there is one serious objection to the "metsuperoxide" model. Our spectral studies of several model Co(III) complexes containing O_2^- show that this ligand falls very close in the spectrochemical series to —NCS⁻ (44). Met thiocyanate hemoglobin has considerable high-spin ($S = 5/2$) character at room temperature (45), whereas at this temperature oxyhemoglobin has no unpaired electrons. On the basis of this correlation, it is difficult to understand why only low-spin Fe(III) should be present in an O_2^--bound product.

We have tried without success to locate the O—O stretching frequency which would be expected in the infrared spectrum for an unsymmetrically bound O_2 or O_2^- in oxyhemoglobin. We have measured (31) infrared spectra for oxyhemoglobin and carbonylhemoglobin in the region

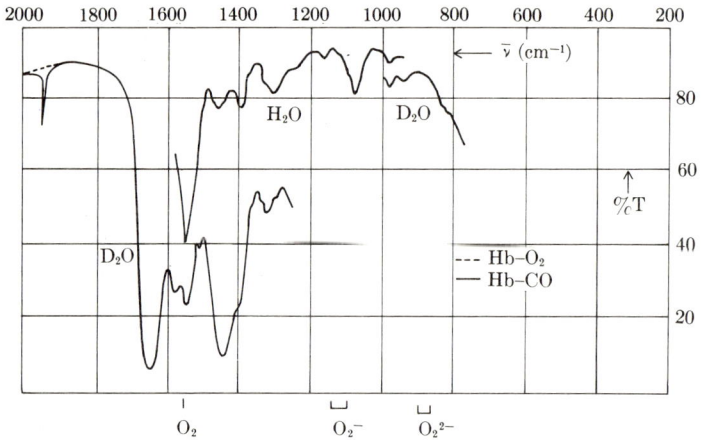

Figure 19. Infrared spectra of oxyhemoglobin and carbonyl-hemoglobin in the region 2000–700 cm^{-1}

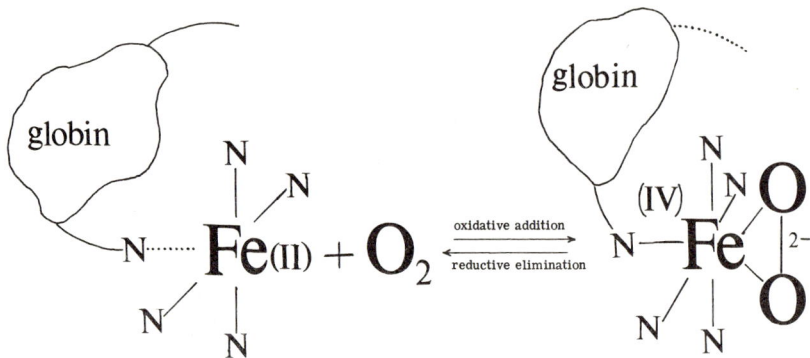

Figure 20. Oxidative addition model of O_2-binding to hemoglobin

2000–700 cm^{-1}. By use of both H_2O- and D_2O-hemoglobins, this range can be covered. As shown in Figure 19 (*31*), there is no observable difference in this region in the spectra of the two proteins, except for the sharp peak at 1952 cm^{-1} owing to the C≡O stretching mode. This bit of negative evidence suggests only that the absorption due to O—O stretching is not as strong as it is, for example, in [Co(acacen)O_2], which contains O_2^- bonded unsymmetrically to Co(III) (*41, 42*). Until the peak is located in oxyhemoglobin either in an infrared or a Raman experiment, nothing further can be said on this point.

We propose that the two-electron oxidative addition O_2-binding model developed for hemerythrin can be generalized to hemoglobin, as shown in Figure 20. The deoxy form is presumed to have weak axial bonding. Upon oxygenation, a seven-coordinate Fe(IV) complex is formed, which in a d^4 low-spin structure would be expected to bind strongly to the axial N-donor atom. Although in this model both ends of the O_2^{2-} are expected to bind to the Fe(IV), the two Fe–O bond lengths need not necessarily be the same. The heme group is required to donate the two electrons that two Fe(II) ions furnish in the hemerythrin case. The delocalized porphyrin system is capable of a two-electron oxidation (46), and in combination with the central Fe(II) such a two-electron oxidation should be facilitated. Important additional stabilization of the seven-coordinate Fe(IV) oxidative addition product could be provided by electron donation from the proximal imidazole.

The $d^6 \to d^4$ oxidative addition may seem unfamiliar because there are many more examples (47) of $d^8 \to d^6$ and $d^{10} \to d^8$ processes. However, ruthenocene, which is a d^6 ruthenium(II) complex with a delocalized electronic structure, undergoes two-electron oxidative addition by I_2 and Br_2 to give the Ru(IV) complexes $Ru(cp)_2I^+$ and $Ru(cp)_2Br^+$ (48). X-ray studies of $Ru(cp)_2I^+$ show that it is effectively a seven-coordinate complex (48).

$$d^6 \; Ru(cp)_2 \underset{-I^+}{\overset{+I^+}{\rightleftarrows}} d^4 \; Ru(cp)_2I^+$$

(6 coord) \hspace{2em} (7 coord)

The oxidative addition model is attractive because it suggests a natural trigger for the cooperative binding found in hemoglobin (49, 50). Steric considerations (51) lead us to suggest a seven-coordinate structure in which the Fe(IV) is pulled slightly out of the porphyrin plane toward the coordinated O_2^{2-} group. In such a low-spin Fe(IV) complex, the electron-donating proximal imidazole would be drawn by the [Fe(IV)-(O_2^{2-})] unit much closer to the heme group than in the high-spin deoxy protein. This substantial "trans" structural rearrangement generated by the oxidative addition could account for the alteration in the tertiary and quaternary structure of the protein in the "modified" allosteric transition model (52).

The oxidative addition model for reversible O_2 binding by metal proteins is also reasonable for hemocyanin. Hemocyanin is a copper protein which binds one O_2 molecule for every two copper atoms. The deoxy Cu(I) form has no appreciable absorption in the visible region. When oxygenated, the protein is blue and exhibits a rich visible spectrum, with bands at 700 (ϵ 75), 570 (ϵ 500), 440 (ϵ 65), and 347 nm (ϵ 8900) (53). The pattern of bands around 570 nm leaves little doubt that oxyhemocyanin contains Cu(II) (53). The enhanced LF band intensities further suggest a dimeric Cu(II) complex. For comparison,

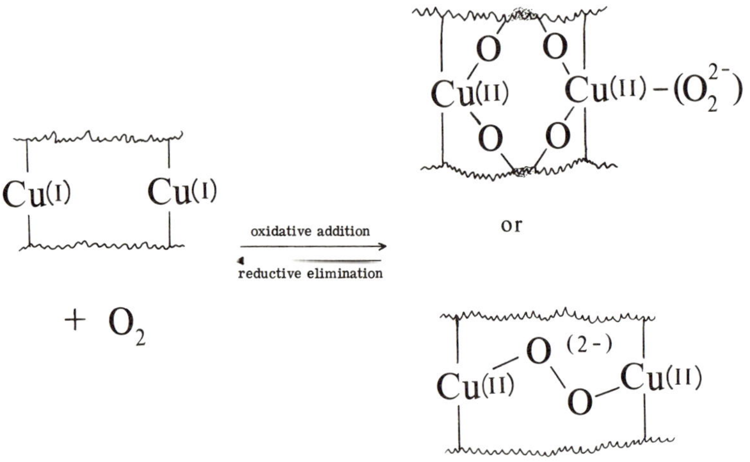

Figure 21. Oxidative addition model for O_2-binding by hemocyanin

dimeric Cu(II) acetate (54) has enhanced LF band intensities and also has an extra band at about 370 nm. The origin of this band has been much debated (54) but one possibility is an SPE excitation (25). The intense band at 347 nm in oxyhemocyanin could have a similar origin; more likely, however, with O_2^{2-} present at the binding site, we would expect a very large contribution in this region to come from $O_2^{2-} \to Cu(II)$ charge transfer.

The spectral data for oxyhemocyanin are thus entirely consistent with the oxidative addition O_2-binding model shown in Figure 21. As in the case of oxyhemerythrin, it is not possible from the spectral data to tell whether the O_2^{2-} is bridging or terminally bound; if it is the latter, presumably other bridging groups such as carboxylate must serve to bind the dimeric unit. A moderate antiferromagnetic coupling in the dimer is suggested by the lack of an ESR signal attributable to Cu(II).

Blue Copper Proteins

We turn finally to the interesting problem of the coordination structures of the so-called "blue" copper proteins (55). These proteins, which contain Cu(II), exhibit absorption bands in the visible region which are much more intense than the LF bands of model complexes such as $Cu(NH_3)_4^{2+}$. ESR measurements show that the blue Cu(II) proteins also have anomalously small copper hyperfine constants.

Formation of dimeric structures is ruled out as an explanation of the enhanced LF band intensities because some of the blue proteins contain only one Cu(II). In fact, the spectral problem is nicely illustrated by the protein stellacyanin, which contains a single Cu(II) ion (*56, 57*). The electronic absorption spectrum of stellacyanin consists of three visible bands, at 850 (ϵ 790), 604 (ϵ 4080), and 450 nm (ϵ 880) (*57*). It is difficult if not impossible to understand the intensity of the central band (604 nm) if either square planar or tetragonally distorted octahedral coordination is assumed. Furthermore, the LF band system is too high in energy to be explained by a distorted tetrahedral model.

We propose that the blue proteins have a five-coordinate copper(II) site (*58*). This proposal is based on the fact that the three-band system of stellacyanin, both in band positions and intensities, is very similar to the LF spectra of several low-spin Ni(II) model five-coordinate complexes (*59, 60*). In going from low-spin four-coordinate to low-spin five-coordinate sites in Ni(II), large enhancements ($\times$ 10) in LF band intensities are commonly observed (*59*).

Based on our experience in the Ni(II) systems, we hope to be able to synthesize five-coordinate Cu(II) complexes which will serve as useful models in further investigations of the nature of the coordination site of the blue proteins.

Acknowledgment

The research from our laboratory reported in this paper was supported by the National Science Foundation. I have been fortunate in having had Harvey J. Schugar of Rutgers University as a collaborator in all aspects of the iron chemistry. This manuscript draws heavily from Schugar's work as well as that of several of our coworkers whose names appear in the references. I am particularly grateful to George Rossman, Dana Powers, John Webb, Jack Thibeault, and Vinnie Miskowski for helpful discussions and for assistance in preparing the final manuscript. I should also like to thank Brian Hoffman, J. L. Hoard, and R. G. Shulman for interesting discussions and for sending preprints and prepublication comments which proved to be valuable in the formulation of certain parts of this paper. This is contribution No. 4165 from the Arthur Amos Noyes Laboratory of Chemical Physics, California Institute of Technology, Pasadena, Calif. 91109.

Literature Cited

(1) Rossman, G. R., Thibeault, J., Schugar, H. J., Gray, H. B., to be published.
(2) Michaelis, L., Coryell, C. D., Granick, S., *J. Biol. Chem.* (1943) **148**, 463.
(3) Brady, G. W., Kurkjian, C. R., Lyden, E. F. X., Robin, M. B., Saltman, P., Spiro, T., Terzis, A., *Biochemistry* (1968) **7**, 2185.
(4) Towe, K. M., Bradley, W. F., *J. Colloid Interface Sci.* (1967) **24**, 384.

(5) Harrison, P. M., Fischbach, F. A., Hoy, T. G., Haggis, G. H., *Nature* (1967) **216**, 1188.
(6) Spiro, T. G., Allerton, S. E., Renner, J., Terzis, A., Bils, R., Saltman, P., *J. Am. Chem. Soc.* (1966) **88**, 2721.
(7) Saltman, P., Lecture delivered at the California Institute of Technology, May 1970.
(8) Webb, J., unpublished data.
(9) Powers, D., Rossman, G. R., Schugar, H. J., unpublished data.
(10) Saltman, P., Multani, J., to be published.
(11) Webb, J., unpublished data.
(12) Gustafson, R. L., Martell, A. E., *J. Phys. Chem.* (1963) **67**, 576.
(13) Schugar, H. J., Walling, C., Jones, R. B., Gray, H. B., *J. Am. Chem. Soc.* (1967) **89**, 3712.
(14) Schugar, H. J., Hubbard, A. T., Anson, F. C., Gray, H. B., *J. Am. Chem. Soc.* (1969) **91**, 71.
(15) Lippard, S. J., Schugar, H. J., Walling, C., *Inorg. Chem.* (1967) **6**, 1825.
(16) Kushi, Y., Fernando, Q., *J. Am. Chem. Soc.* (1970) **92**, 91.
(17) Schugar, H. J., Rossman, G. R., Gray, H. B., *J. Am. Chem. Soc.* (1969) **91**, 4564.
(18) Schugar, H. J., Rossman, G. R., Barraclough, C. G., Gray, H. B., *J. Am. Chem. Soc.*, to be submitted.
(19) Schugar, H. J., Rossman, G. R., Thibeault, J., Gray, H. B., *Chem. Phys. Letters* (1970) **6**, 26.
(20) Lohr, L. L., Jr., McClure, D. S., *J. Chem. Phys.* (1968) **49**, 3516.
(21) Ferguson, J., Guggenheim, H. J., Tanabe, Y., *J. Chem. Phys.* (1966) **45**, 1134.
(22) Varsanyi, F. L., Dieke, G. H., *Phys. Rev. Letters* (1961) **7**, 442.
(23) Dexter, D. L., *Phys. Rev.* (1962) **126**, 1962.
(24) Ferguson, J., Guggenheim, H. J., Tanabe, Y., *Phys. Rev.* (1967) **161**, 207.
(25) Hansen, A. E., Ballhausen, C. J., *Trans. Faraday Soc.* (1965) **61**, 631.
(26) Okamura, M. Y., Klotz, I. M., Johnson, C. E., Winter, M. R. C., Williams, R. J. P., *Biochemistry* (1969) **8**, 1951, and references therein.
(27) Garbett, K., Darnall, D. W., Klotz, I. M., Williams, R. J. P., *Arch. Biochem. Biophys.* (1969) **103**, 419.
(28) Simon, S., Grube, B., Rossman, G. R., Gray, H. B., unpublished data.
(29) Klotz, I. M., Klotz, T. A., Fiess, H. A., *Arch. Biochem. Biophys.* (1957) **68**, 284.
(30) Walling, C., Kurz, M., Schugar, H. J., *Inorg. Chem.* (1970) **9**, 931.
(31) Rossman, G. R., Beach, N. A., Gray, H. B., unpublished data.
(32) Pauling, L., Coryell, C. D., *Proc. Natl. Acad. Sci.* (1936) **22**, 210.
(33) Griffith, J. S., *Proc. Roy. Soc.* (1956) **A235**, 23.
(34) Weiss, J. J., *Nature* (1964) **202**, 83.
(35) Pauling, L., *Nature* (1964) **203**, 182.
(36) Weiss, J. J., *Nature* (1964) **203**, 183.
(37) Viale, R. O., Maggiora, G. M., Ingraham, L. L., *Nature* (1964) **203**, 183.
(38) Peisach, J., Blumberg, W. E., Wittenberg, B. A., Wittenberg, J. B., *J. Biol. Chem.* (1968) **243**, 1871.
(39) Cristoph, G. G., Marsh, R. E., Schaefer, W. P., *Inorg. Chem.* (1969) **8**, 291.
(40) Walker, F. A., *J. Am. Chem. Soc.* (1970) **92**, 4235.
(41) Crumbliss, A. L., Basolo, F., *J. Am. Chem. Soc.* (1970) **92**, 55.
(42) Hoffman, B. M., Diemente, D. L., Basolo, F., *J. Am. Chem. Soc.* (1970) **92**, 61.
(43) Hoffman, B. M., Petering, D., *Proc. Natl. Acad. Sci.* (1970) **67**, 637.
(44) Miskowski, V., Treitel, I. M., Gray, H. B., unpublished data.
(45) Schoffa, G., *Advan. Chem. Phys.* (1964) **7**, 182.

(46) Dolphin, D., Felton, R. H., Borg, D. C., Fajer, J., *J. Am. Chem. Soc.* (1970) **92**, 743.
(47) Collman, J. P., *Accts. Chem. Res.* (1968) **1**, 136.
(48) Sohn, Y. S., Schlueter, A. W., Hendrickson, D. N., Gray, H. B., *Inorg. Chem.*, to be submitted.
(49) Antonini, E., Brunori, M., *Ann. Rev. Biochem.* (1970) **39**, 977.
(50) Perutz, M. F., *Proc. Roy. Soc. (London)* (1969) **B 173**, 113.
(51) Hoard, J. L., private communication.
(52) Ogawa, S., Shulman, R. G., *Science*, to be published.
(53) Van Holde, K. E., *Biochemistry* (1967) **6**, 93.
(54) Kokoszka, G. F., Allen, H. C., Jr., *J. Chem. Phys.* (1967) **46**, 3013.
(55) Vallee, B. L., Wacker, W. E. C., "The Proteins," H. Neurath, Ed., Vol. V, *Metalloproteins*, 2nd ed., Academic, New York, 1970.
(56) Peisach, J., Levine, W. G., Blumberg, W. E., *J. Biol. Chem.* (1967) **242**, 2847.
(57) Malmstrom, B. G., Reinhammer, B., Vanngard, T., *Biochim. Biophys. Acta* (1970) **205**, 48.
(58) Dawson, J. W., Miskowski, V. M., Gray, H. B., *Nature*, to be submitted.
(59) Preer, J. R., Gray, H. B., *J. Am. Chem. Soc.*, in press.
(60) Dawson, J. W., Venanzi, L. M., Preer, J. R., Hix, J. E., Gray, H. B., *J. Am. Chem. Soc.*, in press.

RECEIVED June 26, 1970.

18

Nuclear Relaxation Studies of the Role of Metals in Enzyme–Catalyzed Enolization and Elimination Reactions

A. S. MILDVAN

The Institute for Cancer Research, Fox Chase, Philadelphia, Pa. 19111 and
The University of Pennsylvania, Philadelphia, Pa. 19104

> *Nuclear relaxation studies of substrates and inhibitors have resulted in the detection of 10 enzyme–Mn–substrate and 4 enzyme–Mn–inhibitor bridge complexes possessing kinetic and thermodynamic properties consistent with their participation in enzyme catalysis. Three cases of α activation, by divalent cations, of enzyme-catalyzed enolization reactions (pyruvate carboxylase, yeast aldolase, D-xylose isomerase), and one case of δ activation of an enzyme-catalyzed elimination reaction (histidine deaminase) have thereby been established. Thus, in each proven case, the enzyme-bound Mn coordinates an electronegative atom (Z) of the substrate, which is attached to a carbon atom one or two bonds away from the carbon atom which is to be deprotonated:*

$$\begin{array}{c} H \\ | \\ C \stackrel{\curvearrowleft}{-} C \\ | \\ Z-Mn-\boxed{Enz} \end{array} \quad \text{or} \quad \begin{array}{c} H \\ | \\ C \stackrel{\curvearrowleft}{-} C = C \\ \quad\quad\searrow | \\ Z-Mn-\boxed{Enz} \end{array}$$

> *By σ and π electron withdrawal, the metal ion increases the acidity of this proton and facilitates its removal.*

Considering the role of metals in the mechanism of enzyme-catalyzed elimination and enolization reactions, so that we might understand the role of the metal, we will restrict ourselves to those cases in which both the substrate and the enzyme donate ligands to the metal, in the ternary complex; *i.e.*, in which an enzyme–metal–substrate bridge complex (E–M–S) has been established.

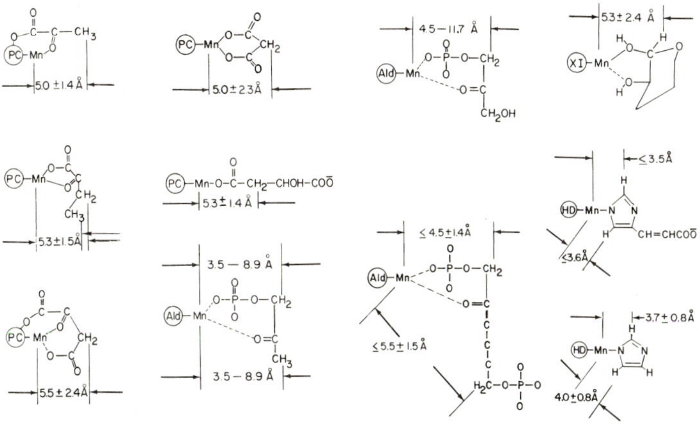

Figure 1. Distances in established enzyme–metal–substrate bridge complexes from longitudinal nuclear relaxation rates

In 1966, two methods were introduced for the detection and study of E–M–S complexes in solution: the EPR method (1) and the NMR method (2, 3). Both require the presence of a paramagnetic metal.

The EPR method, which was used to establish the structure of the metmyoglobin–Fe^{3+}–fluoride complex (1), consists of the detection of hyperfine splitting in the EPR spectrum of the metal by a magnetic nucleus of the ligand. Since hyperfine coupling operates only through chemical bonds (4), the EPR method rigorously demonstrates direct coordination, but is probably limited to complexes of Fe^{III} and Cu^{II} at liquid nitrogen temperatures.

The NMR method, first used to detect the pyruvate kinase–Mn^{2+}–O_3PF complex (2, 3), consists of the demonstration of an enhanced effect of the enzyme-bound paramagnetic metal on the nuclear relaxation rates of magnetic nuclei of the ligand. The NMR method, which is useful in the liquid state at ambient temperatures where the enzyme can function, has been used with Mn^{2+}, Co^{2+}, and Fe^{3+} (5, 6, 7) and in principle should be applicable to Cu^{2+} enzymes as well. A large number of enzyme–metal–substrate bridge complexes have been detected by the NMR method, which also permits the determination of their kinetic and thermodynamic properties (5). The metmyoglobin–Fe^{III}–fluoride complex, originally demonstrated by the EPR method, has been detected by the NMR method as well (6). Those metal bridge complexes detected in solution which possess kinetic and thermodynamic properties consistent with their participation in the catalytic process are summarized in Figure 1.

Parenthetically, the technique of x-ray diffraction, while providing a powerful method for determining the total structure of ternary E–M–S

Table I. Longitudinal

Solution	T_1, Sec	$\dfrac{1}{T_1}$, Sec^{-1}	$\dfrac{1}{T_{1p}}$, Sec^{-1}
H_2O, buffer	3.0	0.3	—
$10^{-4} M$ Mn $(H_2O)_6^{2+}$	0.9	1.1	0.8
$10^{-4} M$ PC–Mn $(H_2O)_3$	0.27	3.7	3.4
$10^{-4} M$ PC–Mn–Py	0.6	1.7	1.4
$10^{-4} M$ PC–Mn–Oxalate	>2.0	<0.5	<0.2

^a Conditions: 24.3 MHz, 25°.

[a] Conditions: 24.3 MHz, 25°.

complexes in the crystalline state, suffers from two limitations. First, there is no assurance that the complex observed in the Fourier synthesis is kinetically active. Second, in the formation of enzyme–metal–substrate bridge complexes, a water ligand has been replaced by a substrate ligand. Hence, the difference Fourier synthesis (E–M–S minus E–M–OH$_2$) reveals a minimum in the electron density (*i.e.*, in the signal/noise ratio) at the position of the coordinated atom of the substrate, often leading to equivocal results as to whether direct coordination has occurred.

Theory of the NMR Method

The theory of the nuclear relaxation method for detecting metal bridge complexes has been discussed in detail elsewhere (5). Here only a brief account will be given. Any atomic nucleus with an odd number of protons or neutrons has a net magnetic moment and its magnetic vector will tend to orient in a magnetic field. The longitudinal relaxation time (T_1) is the time constant for orientation of a population of magnetic vectors in a magnetic field.

Magnetic nuclei also have a property called spin. When a torque is applied to a spinning particle, it will tend to precess. By analogy, when a population of magnetic nuclei are placed in a magnetic field, their magnetic vectors experience a torque and precess about the lines of force of the field. Energy may be applied to this system to align the magnetic vectors of the nuclei to precess in phase with each other. Since each nucleus experiences a different magnetic microenvironment, each will precess at a different rate and phase coherence eventually will be lost with a characteristic time constant (T_2), the transverse relaxation time.

Magnetic nuclei undergo relaxation by interacting with and giving up magnetic energy to their magnetic environment. In an aqueous solution of diamagnetic salts and buffers, the predominant magnetic environment of a ligand (*e.g.*, water) are the water protons. Since protons are

Relaxation of Water Protons[a]

$\dfrac{1}{[Mn]T_{1p}}$	ε^*	$\dfrac{1}{\dfrac{[Mn]}{[L]}T_{1p}} = \dfrac{1}{pT_{1p}}$
$M^{-1}\ Sec^{-1}$		Sec^{-1}
0.8×10^4	1.0	0.4×10^6
3.4×10^4	4.2	1.9×10^6
1.4×10^4	1.7	0.8×10^6
$<0.2 \times 10^4$	<0.3	$<0.1 \times 10^6$

weak magnets, the relaxation times of water protons in aqueous solution are long, of the order of three seconds (Table I).

Unpaired electrons are much stronger magnets with magnetic moments which are three orders of magnitude greater than that of the proton. Hence, paramagnetic ions are exceedingly effective in shortening the relaxation times or increasing the relaxation rates of ligands. As indicated in Table I, Mn^{2+} (which has five unpaired electrons) at a level of $10^{-4}M$, is much more effective than protons at $111\ N$ in bringing about water proton relaxation, as seen by the paramagnetic contribution to the relaxation rate. The latter is calculated by subtracting the relaxation rate of the buffer alone from that observed in the Mn solution. The magnitude of the paramagnetic effect of Mn on the relaxation rate of the ligand protons is inversely related to the sixth power of the distance between Mn and the ligand protons and to the residence time of the ligand in the coordination sphere of the Mn. The effect of Mn is directly related to the coordination number for the ligand and to the correlation time for the Mn–proton interaction. The correlation time (τ_c) is the time constant of that process which modulates the interaction between the Mn and the ligand protons and is usually dominated by the time constant for the relative rotational motion of Mn and its coordinated water. For $Mn(H_2O)_6^{2+}$, τ_c is the tumbling time of the complex which is of the order of 3×10^{-11} sec. Upon binding of Mn to the surface of a macromolecule, the coordination number for water may halve, the water exchange rate may increase or decrease by an order of magnitude, but the tumbling will slow (i.e., the correlation time will increase) by as much as three orders of magnitude. Hence, the last mentioned effect usually outweighs the former, and an enhancement of the relaxation rate will be observed. The enhancement phenomenon discovered independently by Cohn and Leigh (8) and by Eisinger et al. (9) is illustrated in Table I by the data on pyruvate carboxylase, a manganese metalloenzyme (10, 11). The enzyme-bound manganese is four times as effective as inorganic man-

Table II. Enhancement of the Paramagnetic Effects of Mn^{2+} and Co^{2+}

Complex	Ligand Proton
Pyruvate carboxylase–Mn–pyruvate	C–3
Mn–pyruvate	C–3
Pyruvate carboxylase–Mn–oxalacetate	C–3
Mn–oxalacetate	C–3
Pyruvate carboxylase–Mn–α-ketobutyrate	C–3
	C–4
Mn–α-ketobutyrate	C–3
	C–4
Aldolase–Mn–acetol phosphate	C–1
	C–3
Mn–acetol phosphate	C–1
	C–3
Aldolase–Mn–DHAP	C–1
Mn–DHAP	C–1
Xylose isomerase–Mn–α-D-xylose	C–1
Mn–α-D-xylose	C–1
Xylose isomerase–Co–α-D-xylose	C–1
Co–α-D-Xylose	C–1
Histidine deaminase–Mn–imidazole	C–2
	C–4
Mn–imidazole	C–2
	C–4
Histidine deaminase–Mn–urocanate	C–2
	C–4
Mn–urocanate	C–2
	C–4

ganese in relaxing the protons of water ligands even though its coordination number for water is only 2 or 3.

Upon adding a substrate or an inhibitor ligand to an enzyme–Mn complex, water molecules would be replaced, and the effect of Mn on water would decrease (Table I); *i.e.*, the enhancement of the ternary complex (ϵ_T) is less than the enhancement of the binary complex (ϵ_b). As with any other physical technique, one may carry out titrations using the changes in the water relaxation rate to determine stoichiometries and stability constants of binary E–Mn (*12, 13*) and ternary E–Mn–substrate complexes (*14*).

A more conclusive experiment for detecting direct coordination of a substrate by the enzyme-bound manganese is to compare the effects of inorganic and enzyme-bound Mn on the relaxation rates of the protons of the substrate. Such an experiment on the pyruvate carboxylase–Mn–

on the Protons of Substrates in E–M–S Complexes

$Sec^{-1} \times 10^{-3}$		ε_1	ε_2	Ref.
$1/pT_{1p}$	$1/pT_{2p}$			
1.57	19.7	18.7	91.6	15
0.084	0.215	—	—	
0.920	2.18	20.4	1.7	19
0.045	1.30	—	—	
0.47	6.14	1.7	23.6	15
0.41	3.86	1.4	14.3	
0.27	0.26	—	—	
0.30	0.27	—	—	
0.90	4.3	0.36	0.7	22
0.90	5.8	4.5	5.3	
3.5	6.2	—	—	
0.20	1.10	—	—	
0.20	3.8	0.15	1.7	22
1.30	2.3	—	—	
1.06	43.0	3.7	23.0	7,30,31
0.29	1.87	—	—	
—	1.36	—	8.5	
—	0.16	—	—	
54.5	53.4	> 2.4	2.3	39
24.9	52.8	1.6	2.1	
<23.1	23.1	—	—	
15.4	24.9	—	—	
16.1	16.3	1.1	1.1	39
13.9	31.3	0.9	1.5	
14.5	14.5	—	—	
16.2	21.4	—	—	

pyruvate complex is exemplified in Table II (15) where the enzyme-bound manganese is seen to be 18.7 times more effective in causing longitudinal relaxation and 91.6 times more effective in causing transverse relaxation of the methyl protons of pyruvate than is inorganic manganese, providing proof of direct coordination of this ligand.

These qualitative interpretations of the relaxation rates may be extended by a quantitative treatment of the data as follows. For Mn complexes, Equations 1 and 2 relate the paramagnetic contributions to the

$$1/T_{1p} = pq/(T_{1M} + \tau_M) + 1/T_{o.s.} \quad (1)$$

$$1/T_{2p} = pq/(T_{2M} + \tau_M) + 1/T_{o.s.} \quad (2)$$

longitudinal $(1/T_{1p})$ and transverse $(1/T_{2p})$ relaxation rates to a term arising from the coordination sphere $pq/(T_{1M} + \tau_M)$ and an outer sphere term $(1/T_{o.s.})$. The latter contribution, which reflects the effect of Mn^{2+}

on layers of molecules beyond the inner coordination sphere, is usually small and will be neglected. The inner sphere term is a function of the coordination number for the ligand (q), the relative concentration of Mn and the ligand (p), the residence time of the ligand in the coordination sphere (τ_M), and the longitudinal (T_{1M}) and transverse (T_{2M}) relaxation times of coordinated ligands. The reciprocal of τ_M is the pseudo-first order rate constant for ligand exchange and thus provides kinetic information. The values of T_{1M} and T_{2M} provide structural information. From T_{1M}, interatomic distances between the paramagnetic ion and the magnetic nucleus of the coordinated ligand may be calculated using the Solomon-Bloembergen equation (see Ref. 5). Examples of such distance calculations for binary complexes are compared with crystallographically determined distances in Table III. From $1/T_{2M} - 1/T_{1M}$, the hyperfine coupling constant between the unpaired electron and the magnetic nucleus can, in certain cases, be calculated (15, 16). This parameter is related to the number and type of chemical bonds intervening between the Mn and the ligand nucleus under observation (15, 16). As discussed elsewhere (2, 3, 5, 15, 16), the contributions of τ_M, T_{1M}, and T_{2M} to the measured values of $1/T_{1p}$ and $1/T_{2p}$ may be determined by studies of the temperature and frequency dependences of the relaxation rates. When such studies have not been made, relaxation rates may often be used to set limiting values on τ_M, distances, and hyperfine coupling constants. The applications of such calculations to determine structural and kinetic properties of enzyme–metal–substrate bridge complexes in solution will be illustrated in the specific cases to follow.

Table III. Manganese-to-Ligand Distances in Solution from Relaxivity Measurements

Ligand		Distance From Relaxivity, Å	From Crystallography, Å
F[-]	Mn–F	2.1 ± 0.2[a]	$2.08 - 2.15$[b]
F–PO$_3^{2-}$	Mn- - -F	5.0 ± 0.8[a]	3.0, 4.2[c]
Imidazole	Mn- - -HC$_2$	≥ 3.1[d]	3.27[e]
	Mn- - -HC$_5$	3.4 ± 0.2[d]	3.24[e]
Urocanate	Mn- - -HC$_2$ (imidazole)	3.4 ± 0.2[d]	3.27[e]
	Mn- - -HC$_5$ (imidazole)	3.5 ± 0.2[d]	3.24[e]

[a] Ref. 2,3.
[b] Ref. 46.
[c] A. Perloff, personal communication (CaO$_3$PF).
[d] Ref. 39.
[e] Ref. 44 and 45 (Zn–histidine).

Enolization and Elimination Reactions

The mechanisms of enolization and elimination reactions, when generalized, are closely related.

$$\text{enolization} \qquad \text{elimination} \tag{3}$$

In both reactions, a proton is removed from a carbon atom (C_β) by a base, a double bond is formed, and an electron pair shifts from a carbon atom (C_α) to a basic (electronegative) atom. Elimination reactions have been more finely subdivided (17) depending on whether the atom removed initially is the proton (carbanion or E1CB mechanism), the base (carbonium ion or E-1 mechanism), or both (concerted or E-2 mechanism). The mode of activation of enzyme-catalyzed enolization and elimination reactions by metals will be classified according to the site on the substrate which is coordinated by the metal.

α Activation

In ternary enzyme–metal–substrate bridge complexes, metals might activate enolization and elimination reactions by coordination of the electronegative atom (X) attached to the α carbon atom ("α activation").

$$\tag{4}$$

Pyruvate Carboxylase

α Activation of pyruvate occurs with the enzyme pyruvate carboxylase which catalyzes the deprotonation (and carboxylation) of this substrate.

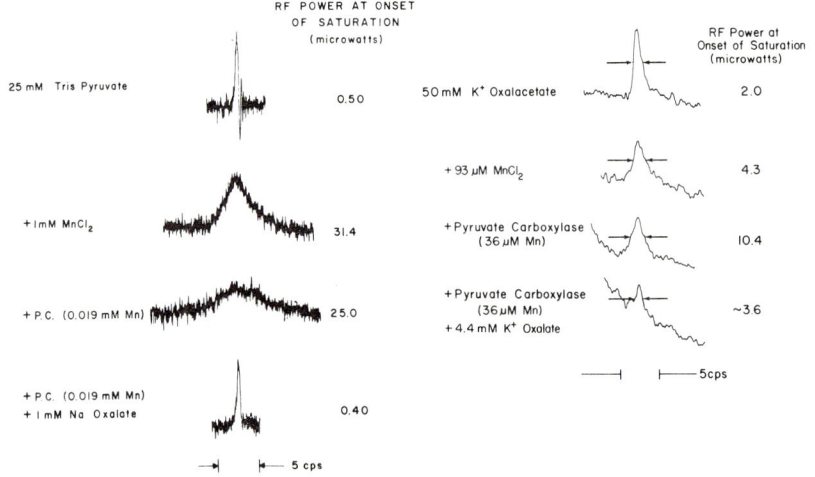

(5)

The over-all reaction catalyzed by this manganese metallo–biotin enzyme (10, 11) takes place in two steps (18):

$$\text{E-Biotin} + \text{HCO}_3 + \text{ATP} \underset{M^{2+}}{\overset{\text{Acetyl CoA}}{\rightleftarrows}} \text{E-Biotin-CO}_2 + \text{ADP} + P_i \quad (6)$$

$$\text{E-Biotin} + \text{Pyruvate} \rightleftarrows \text{E-Biotin} + \text{Oxalacetate} \quad (7)$$

The substrates of step 1 do not interact with the enzyme-bound manganese, as shown by the fact that they do not lower the enhanced effect of Mn on the relaxation rate of water (10, 11). The substrates of step 2 (pyruvate, oxalacetate, and α-ketobutyrate) are coordinated by the enzyme-bound Mn, as demonstrated indirectly by decreases in the water relaxation rates (Table I) (10, 11) and directly by enhancements of the relaxation rates of the carbon-bound protons of these substrates (Table

Biochemistry

Figure 2. Effect of inorganic Mn^{2+} and pyruvate carboxylase Mn^{2+} on the proton NMR spectra of (A) pyruvate (Ref. 15) and (B) oxalacetate (Ref. 19)

II, Figure 2) (15, 19). The temperature dependencies of the relaxation rates of the methyl protons of pyruvate have been fit by an exchange contribution $(1/\tau_M)$, a T_{1M} contribution, and an upper limit for T_{2M}, from which the kinetics of pyruvate binding and dissociation (see Table V) (Figure 3), the Mn-to-methyl distances (Figure 1), and a lower limit to the Mn-to-methyl coupling constant (Table IV) have been evaluated.

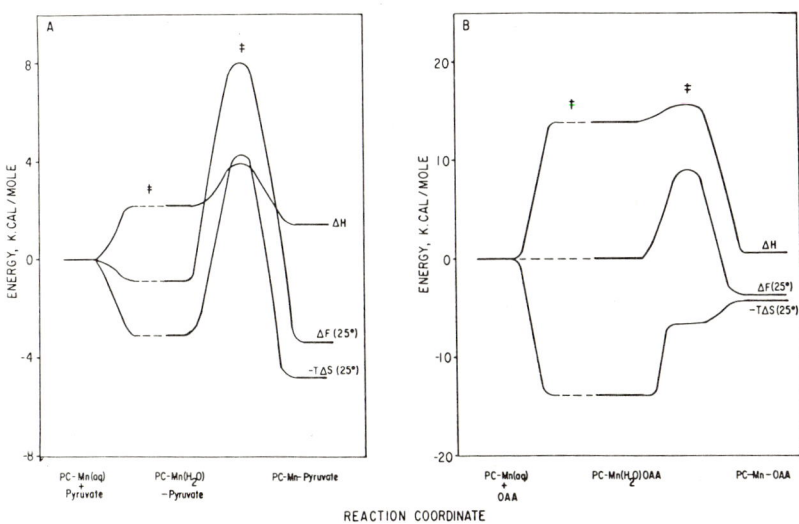

Figure 3. Energy diagrams for the formation of the pyruvate carboxylase–Mn–pyruvate and –oxalacetate bridge complexes based on the mechanism of Equation 8 (from Ref. 19)

Table IV. Hyperfine Coupling Constants (A/h) in E–M–S Complexes

Complex	Interaction	Log (A/h), Cps	Ref.
Pyruvate carboxylase–Mn–pyruvate	Mn·····HC-3	>5.1	15,16
Pyruvate carboxylase–Mn–oxalacetate	Mn·····HC-3	4.6–5.1	19
Pyruvate carboxylase–Mn–α–ketobutyrate	Mn·····HC-3 Mn·····HC-4	4.9–5.4 4.8–5.3	15,16
Aldolase–Mn–acetol phosphate	Mn·····HC-1 Mn·····HC-3	4.7–5.2 ≥4.8–5.3	22
Aldolase–Mn–DHAP	Mn·····HC-1	≥4.4–4.9	
Xylose isomerase–Mn–α–D–xylose	Mn·····HC-1	≥5.3	7,30,31
Histidine deaminase–Mn–imidazole	Mn·····HC-4	≥5.2	39
Histidine deaminase–Mn–urocanate	Mn·····HC-4	≥5.1	

Table V. Kinetic Parameters of

Complex	$Log\ (k_{off})$, Sec^{-1}
Pyruvate carboxylase–Mn–pyruvate	4.32
Pyruvate carboxylase–Mn–oxalacetate	4.11
Pyruvate carboxylase–Mn–α–ketobutyrate	4.0
Aldolase–Mn–acetol phosphate	≥3.76
Aldolase–Mn–DHAP	≥3.58
Aldolase–Mn–FDP	4.79
Xylose isomerase–Mn–α–D–xylose	≥4.63
Histidine deaminase–Mn–imidazole	≥4.72
Histidine deaminase–Mn–urocanate	≥4.49

[a] The rate constants are defined by the following equation:
$$E\text{–}Mn(H_2O) + L \underset{k_{off}}{\overset{k_{on}}{\rightleftarrows}} E\text{–}Mn\text{–}L + H_2O$$
and by Equation 8 in the text.
[b] Adjusted to the same conditions as used for the determination of $k_{3,4}$.
[c] 32° C.
[d] 27° C.

The coupling constant is inconsistent with carboxyl coordination but consistent with carbonyl coordination (15). Similar data for α-ketobutyrate (15) and oxalacetate (19) have been fit by exchange contributions ($1/\tau_M$) and inner sphere contributions (T_{1M} and T_{2M}). The rates of formation of these metal bridge complexes from an outer sphere complex ($k_{3,4}$) are limited predominantly by the rate of dissociation of a water molecule from the coordination sphere of the enzyme-bound manganese (Figure 3, Table V) (15, 19), as required by the S_N1-outer sphere mechanism of Eigen and Tamm (20),

$$E\text{–}Mn(H_2O) + S \overset{Fast}{\rightleftarrows} E\text{–}Mn(H_2O)S \underset{k_{off}}{\overset{k_{3,4}}{\rightleftarrows}} E\text{–}Mn\text{–}S + H_2O \quad (8)$$

The structural properties of these bridge complexes are summarized in Figure 1 and Table IV. A mechanism consistent with all of these observations is given in Figure 4. The role of the metal, in addition to α activation of the deprotonation of pyruvate, may be to promote the electrophilicity of the carboxyl group of carboxybiotin to render it more susceptible to nucleophilic attack. Consistent with a concerted mechanism, a kinetic isotope effect with tritiated pyruvate ($k_H/k_T = 4.2$) and "cis" carboxylation of pyruvate with retention of configuration have recently been found (21).

E–M–S Bridge Complexes at 25°[a]

$Log (k_{on})$, $M^{-1} Sec^{-1}$	$Log (k_{3,4})$ Sec^{-1}	$Log (k_{off}^{H_2O})$,[b] Sec^{-1}	Ref.
6.65	6.17	6.18	*15*
7.49	5.89	6.18	*19*
6.45	5.97	6.18	*15*
≥ 6.54[c]	≥ 4.94	5.23	*22*
≥ 5.83[c]	≥ 4.23	5.23	
7.79[d]	5.19	5.15	
≥ 7.29	≥ 7.29	≥ 5.77	*7,30,31*
≥ 6.18	≥ 6.18	≥ 6.33	*39*
≥ 7.19	≥ 6.71	≥ 6.33	

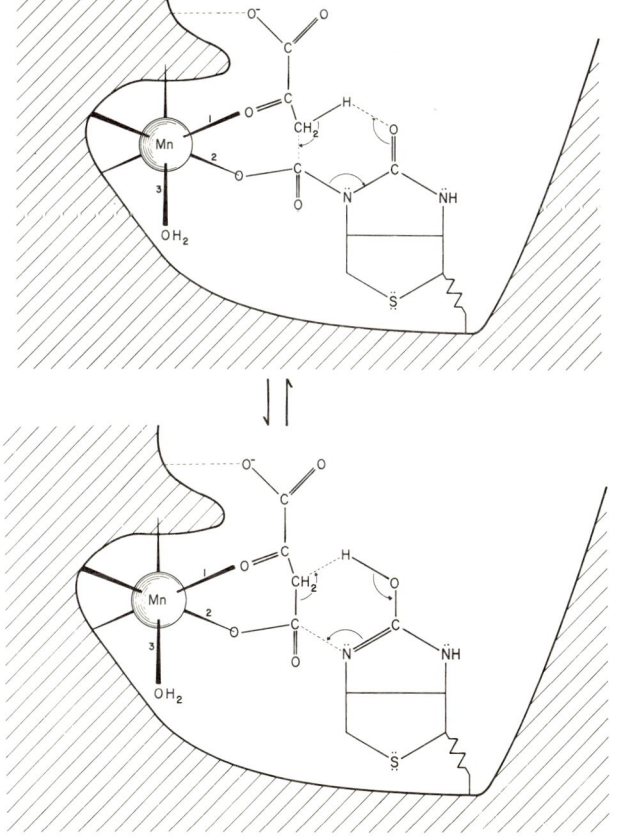

Journal of Biological Chemistry

Figure 4. Mechanism of the pyruvate carboxylase reaction (from Ref. 11 and 15)

Yeast Aldolase

Another example of α activation of enolization of the substrate, dihydroxyacetone phosphate (DHAP), occurs with yeast aldolase (22), a Zn metalloenzyme in which the Zn may be removed and replaced by Mn (23).

$$\text{(9)}$$

The enolate of the substrate then condenses with 3-phosphoglyceraldehyde (G–3–P) to form fructose-1,6-diphosphate (FDP). The over-all reaction is therefore:

$$\text{DHAP} + \text{G-3-P} \underset{}{\overset{\text{Aldolase}}{\rightleftharpoons}} \text{FDP} \quad (10)$$

The enzyme also catalyzes the exchange of a proton from the solvent into the C-3 position of DHAP and of acetol phosphate (24, 25).

$$\text{(11)}$$

$$\begin{array}{c}\text{CH}_2-\text{O}-\overset{\overset{\displaystyle O}{\|}}{\underset{\underset{\displaystyle O}{|}}{P}}-\text{O}\\|\\ \text{C}=\text{O}\\|\\ \text{HCH}_2\end{array} + \text{H}^*_2\text{O} \underset{}{\overset{\text{Aldolase}}{\rightleftharpoons}} \begin{array}{c}\text{CH}_2-\text{O}-\overset{\overset{\displaystyle O}{\|}}{\underset{\underset{\displaystyle O}{|}}{P}}-\text{O}\\|\\ \text{C}=\text{O}\\|\\ \text{H}^*\text{CH}_2\end{array} \qquad (12)$$

As seen from the data in Table II, the enzyme enhances the effect of bound Mn on the C-3 protons of acetol phosphate, indicating the formation of an enzyme–Mn–substrate bridge complex. However, the enzyme-bound Mn has a smaller effect on the C-1 protons, indicating that the presence of the enzyme alters the structure of the Mn coordination complexes. Thus, in the absence of enzyme, from distance calculations and from stability constants, inorganic Mn forms a monodentate phosphate complex with acetol phosphate.

$$\begin{array}{c}\text{CH}_2-\text{O}-\overset{\overset{\displaystyle O}{\|}}{\underset{\underset{\displaystyle O}{|}}{P}}-\text{O}-\text{Mn}\\|\\ \text{C}=\text{O}\\|\\ \text{CH}_3\end{array} \qquad (13)$$

Monodentate phosphate coordination provides the only structure which can accommodate the fact that Mn is 37% closer to the C-1 protons than to the C-3 protons in the binary complexes. In contrast to inorganic Mn, the enzyme-bound Mn is equidistant from the C-1 and C-3 protons in the ternary E–Mn–acetol phosphate complex (Figure 1), suggesting carbonyl coordination in addition to phosphoryl coordination. While the uncertainty in the absolute value of these distances in the ternary complexes is high (Figure 1), the error in the relative values of these distances is low ($\pm 4\%$). Hence, it may be concluded that the enzyme has forced the Mn^{2+} to coordinate the carbonyl group at C-2 as well as the phosphate at C-1 (Figure 1). The changes in the Mn-to-proton hyperfine coupling constants caused by the enzyme (Table IV) are consistent with this view. The exchange rates of DHAP and acetol phosphate into the coordination sphere of enzyme-bound Mn are at least two orders of magnitude greater than the rate of catalysis. The dissociation constants of these E–Mn–substrate complexes, as determined by displacement of the substrates in the NMR experiment by the competitive inhibitor arabinitol diphosphate, agree with their Michaelis constants. Hence, the kinetic and thermodynamic properties of these metal bridge complexes are consistent with their participation in catalysis. The results are consistent with an electrophilic role for the metal as proposed by Rutter (26). In addition, by coordinating the phosphate at C-1, the metal serves to orient the substrate (Figure 5). In animal aldolases, no metal is found. Rather, a protonated Schiff base functions as the electrophile to facilitate deprotonation and enolization of the substrate (27) (Figure 5C).

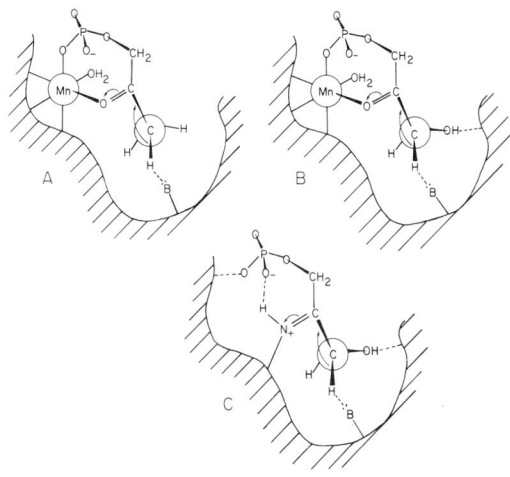

Figure 5. Structures and mechanisms in aldolase–Mn–substrate bridge complexes: A. acetol phosphate, B. dihydroxyacetone phosphate; Mechanism C is proposed for animal aldolases by analogy with A and B (from Ref. 22)

Xylose Isomerase

A third example of α activation occurs with D-xylose isomerase. The enzymes from lactobacillus (28) and streptomyces (29) require a divalent cation for activity. Both enzymes form metal bridge complexes to the C-1 hydroxyl group of α-D-xylose by the NMR method since these enzymes enhance the effect of Mn on the relaxation rates of the C-1 proton of α-D-xylose (Table II) (7, 30, 31). The calculated Mn^{2+}-to-proton distances (Figure 1) and coupling constants (Table IV) are consistent with the following structure.

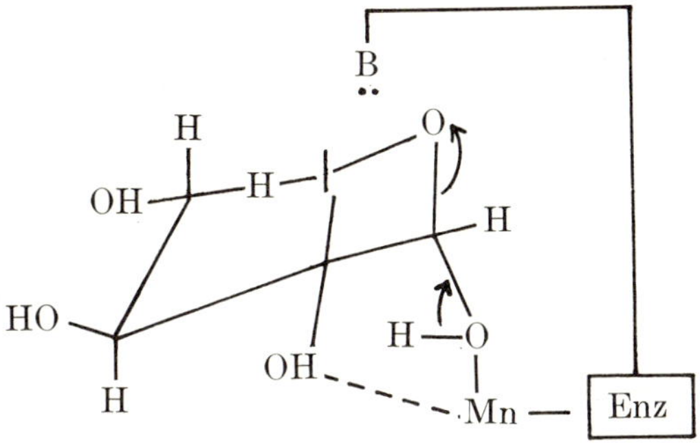

(14)

After ring opening, the resulting aldose would enolize to form a *cis*-enediol intermediate (32) and the metal would provide α activation of this process.

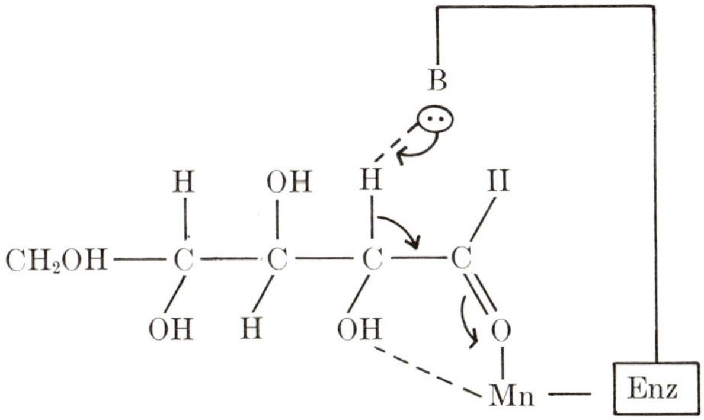

(15)

The proton which has been removed from C-2 to form the *cis*-enediol is conserved (7, 30, 31, 32) and transferred to C-1 from the same side of the double bond to form D-xylulose of the proper stereochemistry (32).

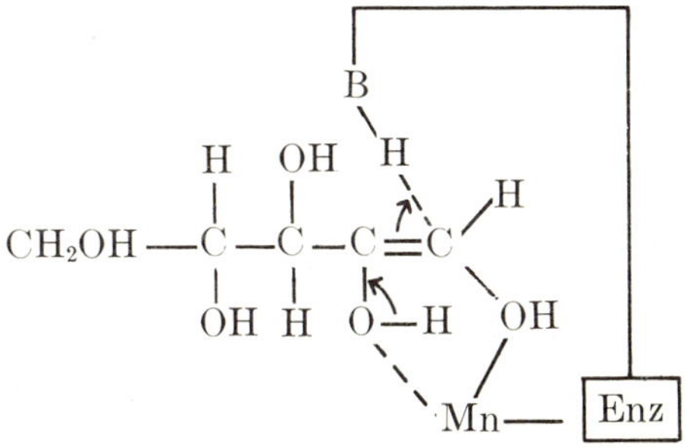

(16)

Closure of the furanose ring from above the plane of the carbonyl group to form α-D-xylulose would complete the reaction:

$$(17)$$

The thermodynamic and kinetic properties of the enzyme–Mn–α-D-xylose bridge complex (Table V) detected in the NMR experiment are consistent with its participation in the catalytic process. The inactive β anomer of D-xylose binds to the Mn-enzyme as detected by changes in the water relaxation rate but in a manner which differs in structure from that of the active (α) substrate since no effect is observed on the relaxation rates of the C-1 proton of the β form (*31*). Hence, the enzyme selects the α-anomer of the substrate from the mutarotated mixture.

β Activation

As judged by organic model reactions, β activation (*i.e.*, coordination of a group on the β carbon atom by a metal) should be an effective means of promoting an enolization or an elimination reaction.

$$(18)$$

Thus, the elimination of HBr from ethylene bromide is accelerated 200-fold by an electrophilic Br on the β carbon atom (*33*). Unfortunately, no clearcut cases of β activation have been established in enzyme-catalyzed enolization or elimination reactions. Indirect evidence exists for β activation (in addition to α activation) in the xylose isomerase reaction as depicted in the above figures. Thus, α-D-glucose, an alternative substrate for xylose isomerase, de-enhances the effect of the xylose isomerase–Mn complex on the relaxation rate of water protons more than D-2-deoxyglucose does, suggesting that glucose might donate an additional ligand to the Mn.

γ Activation

The coordination of a metal ion to a basic group (Y) on the γ carbon atom could activate an elimination reaction by an inductive effect (σ electron withdrawal)

$$\left[\begin{array}{c} B: \\ \\ H \\ | \\ -C \leftarrow C - C - \\ \downarrow | | \\ Y X \\ \searrow \\ M - \boxed{Enz} \end{array} \right] \quad (19)$$

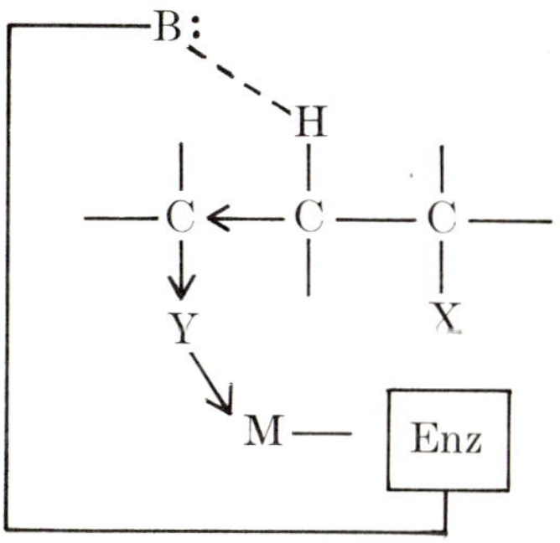

which would increase the acidity of the departing proton or by a resonance effect (π electron withdrawal)

$$\left[\begin{array}{c} B \odot \cdots H \\ \curvearrowleft | | \\ \diagdown C \equiv C - C \equiv \\ (\| ^\gamma |^\beta |^\alpha \\ Y - M - \boxed{Enz} X \end{array} \right] \longrightarrow \left[\begin{array}{c} BH \\ | \\ \diagdown C = C - C - \\ \diagup | \\ \odot Y X \\ \diagdown \\ M - \boxed{Enz} \end{array} \right] \quad (20)$$

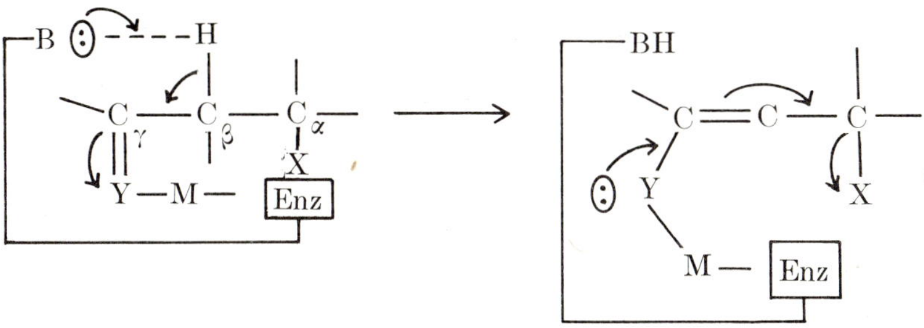

which would also stabilize the intermediate carbanion in an E1CB mechanism. The latter effect has been suggested, but not proven, for the metal-requiring elimination reaction catalyzed by β methyl aspartase (34).

[Scheme (21)]

A combination of α and γ activation of the elimination of water from citrate by the enzyme-bound Fe^{2+} in aconitase has been proposed (35).

[Scheme (22)]

Consistent with such a mechanism, an aconitase–Mn–citrate bridge complex has been detected by NMR (36), but its detailed structure is not yet known.

A similar combination of α and γ activation

[Scheme (23)]

has been proposed for the enolase reaction (37). Indirect evidence from water relaxation rate data suggests the presence of enzyme–Mn–substrate and enzyme–Mn–inhibitor bridge complexes (8, 38), but their detailed structures remain to be elucidated.

δ Activation

Activation of an elimination reaction by coordination at the δ carbon atom might be expected to be small, but might be amplified by intervening double bonds.

$$
\begin{array}{c}
\text{Enz}-\text{M}-\text{Y}-\overset{\delta}{\text{C}}=\overset{\gamma}{\text{C}}-\overset{\beta}{\text{C}}(\text{H})-\overset{\alpha}{\text{C}}(\text{X})- \quad \text{B:} \\
\end{array}
\tag{24}
$$

An example of δ activation has been established by the NMR method in the metal-activated histidine deaminase reaction (39).

$$
\text{Enz}-\text{Mn}-\text{N}\underset{\text{N}-\text{H}}{\overset{\text{CH}}{\diagdown}}-\text{CH}(\text{NH}_2)-\text{C}(=\text{O})\text{O}^- \quad \text{B:----H}
\tag{25}
$$

This enzyme utilizes a divalent cation (40) and also an electrophilic group, probably an activated dehydroalanine residue (41) to catalyze the elimination of ammonia from histidine to form urocanate.

$$
\text{Im}-\text{CH}(\text{NH}_3)-\text{CH}-\text{C}(=\text{O})\text{O}^- \longrightarrow \text{NH}_4^+ + \text{Im}-\text{C}(\text{H})=\text{C}(\text{H})-\text{C}(=\text{O})\text{O}^-
\tag{26}
$$

The enzyme binds 2 Mn^{2+} ions/mole with a dissociation constant which agrees with its kinetically determined activator constant. Chemical modification of the electrophilic group has no effect on the Mn–enzyme interaction. The enhanced effect of the E–Mn complex on the relaxation rate of water protons is greatly reduced by the substrate L-histidine, or by inhibitors (urocanate, imidazole, cysteine) which contain liganding groups at the γ or δ carbon atoms, suggesting coordination of imidazole by the enzyme-bound Mn. Such coordination is established by the enhanced effect of the enzyme-bound Mn on the relaxation rates of the protons of imidazole and of urocanate. Cysteine displaces imidazole and urocanate from their metal bridge complexes at a concentration consistent with the inhibitor constants of these ligands determined kinetically. From $1/T_2$, the rate of dissociation of the product urocanate from the bridge complex is more than 600 times V_{max} (Table V), consistent with its role in catalysis. The distances between the enzyme-bound Mn and the protons of urocanate and imidazole (Figure 1) are consistent with direct coordination of N-3 of urocanate (imidazole) by the Mn (δ activation) or, alternatively, with the formation of a π complex between the imidazole ring and the bound Mn (γ and δ activation). A mechanism of the histidine deaminase reaction consistent with the chemical (41) and NMR data (39) is given in Figure 6. The Mn coordinates the imi-

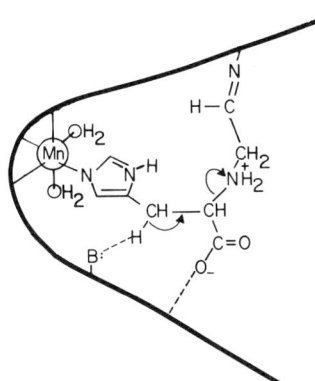

Figure 6. Mechanism of the histidine deaminase reaction (from Ref. 39 and 41)

dazole and the electrophilic center holds the leaving amino group of the substrate. The δ activation by the Mn would increase the acidity of the departing proton primarily by an inductive effect. A resonance effect is also possible,

(27)

as suggested by x-ray crystallographic studies of histidine and its Zn complex. All of the C–C and C–N bond lengths in the imidazole ring of histidine are shorter than single bonds, indicating a large amount of double bond character ($\geqslant 24\%$). Coordination of Zn to histidine produces small increases in the lengths of all of the bonds of the imidazole ligands (0.01–0.03 Å), suggesting small decreases in the double bond character (*42, 43, 44, 45*).

Another role of the Mn in histidine deaminase may be to mask the nucleophilic imidazole of the substrate to prevent its attacking the electrophilic center of the enzyme.

Acknowledgment

I am grateful to my collaborators, J. J. Villafranca, T. Nowak, M. C. Scrutton, R. D. Kobes, W. J. Rutter, K. Schray, I. A. Rose, I. Givot, and R. H. Abeles, whose work is described here. This project was supported in part by U.S. Public Health Service Grants AM-13351, AM-09760, GM-12246, CA-06927 and FR-05539, National Science Foundation Grant GB-8579, and an appropriation from the Commonwealth of Pennsylvania. This work was done during the tenure of an Established Investigatorship from the American Heart Association.

Literature Cited

(1) Kotani, M., Morimoto, H., *in* "Magnetic Resonance in Biological Systems," A. Ehrenberg, B. Malmstrom, and T. Vanngard, Eds., p. 135, Pergamon, New York, 1967.
(2) Mildvan, A. S., Cohn, M., Leigh, J. S., *in* "Magnetic Resonance in Biological Systems," A. Ehrenberg, B. Malmstrom, and T. Vanngard, Eds., p. 113, Pergamon, New York, 1967.
(3) Mildvan, A. S., Leigh, J. S., Cohn, M., *Biochemistry* (1967) **6**, 1805.
(4) Barfield, M., Karplus, M., *J. Am. Chem. Soc.* (1969) **91**, 1.
(5) Mildvan, A. S., Cohn, M., *Advan. Enzymol.* (1970) **33**, 1.
(6) Mildvan, A. S., Rumen, N. M., Chance, B., *Johnson Foundation Symp. Probes Macromolecular Struct. Function* (1970) in press.
(7) Mildvan, A. S., *Abstr. Middle Atlantic Regional Mtg., ACS, April* (1970) 22.
(8) Cohn, M., Leigh, J. S., *Nature* (1962) **193**, 1037.
(9) Eisinger, J., Shulman, R. G., Szymanski, B. M., *J. Chem. Phys.* (1962) **36**, 1721.
(10) Scrutton, M. C., Utter, M. F., Mildvan, A. S., *J. Biol. Chem.* (1966) **241**, 3480.
(11) Mildvan, A. S., Scrutton, M. C., Utter, M. F., *J. Biol. Chem.* (1966) **241**, 3488.
(12) Mildvan, A. S., Cohn, M., *Biochemistry* (1963) **2**, 910.
(13) Mildvan, A. S., Cohn, M., *J. Biol. Chem.* (1965) **240**, 238.
(14) Mildvan, A. S., Cohn, M., *J. Biol. Chem.* (1966) **241**, 1178.
(15) Mildvan, A. S., Scrutton, M. C., *Biochemistry* (1967) **6**, 2978.

(16) Mildvan, A. S., *Johnson Foundation Symp. Probes Macromolecular Struct. Function* (1970) in press.
(17) Hine, J., "Physical Organic Chemistry," 2nd ed., p. 186, McGraw-Hill, New York, 1962.
(18) Scrutton, M. C., Keech, D. B., Utter, M. F., *J. Biol. Chem.* (1965) **240**, 574.
(19) Scrutton, M. C., Mildvan, A. S., *Arch. Biochem. Biophys.* (1970) **140**, 131.
(20) Eigen, M., Tamm, K., *Z. Elektrochem.* (1962) **66**, 107.
(21) Rose, I. A., *J. Biol. Chem.* (1970) **245**, in press.
(22) Kobes, R. D., Mildvan, A. S., Rutter, W. J., *Abstr. 158th Meeting, ACS, New York*, September, 1969, Biol-58; *Biochemistry*, in press.
(23) Kobes, R. D., Simpson, R. T., Vallee, B. L., Rutter, W. J., *Biochemistry* (1969) **8**, 585.
(24) Richards, O. C., Rutter, W. J., *J. Biol. Chem.* (1961) **236**, 3185.
(25) Rose, I. A., O'Connell, E. L., *J. Biol. Chem.* (1969) **244**, 126.
(26) Rutter, W. J., *Federation Proc.* (1964) **23**, 1248.
(27) Grazi, E., Cheng, T., Horecker, B. L., *Biochem. Biophys. Res. Commun.* (1962) **7**, 250.
(28) Yamanaka, K., *Biochim. Biophys. Acta* (1968) **151**, 670.
(29) Takasaki, Y., Kosugi, Y., Kanbayashi, A., *Agr. Biol. Chem.* (1969) **33**, 1527.
(30) Mildvan, A. S., Rose, I. A., *Federation Proc.* (1969) **28**, 534.
(31) Schray, K., Rose, I. A., Mildvan, A. S., *J. Biol. Chem.*, to be published.
(32) Rose, I. A., O'Connell, E. L., Mortlock, R. P., *Biochim. Biophys. Acta* (1969) **178**, 376.
(33) Olivier, S. C. J., Weber, A. P., *Rec. Trav. Chim.* (1934) **53**, 1687.
(34) Bright, H. J., *Biochemistry* (1967) **6**, 1191.
(35) Glusker, J. P., *J. Mol. Biol.* (1968) **38**, 149.
(36) Villafranca, J. J., Mildvan, A. S., *J. Biol. Chem.*, to be published.
(37) Mildvan, A. S., in "The Enzymes," 3rd ed., P. D. Boyer, Ed., vol. 2, p. 445, Academic Press, New York, 1970.
(38) Nowak, T., Mildvan, A. S., *J. Biol. Chem.* (1970) **245**, 6057.
(39) Givot, I. L., Mildvan, A. S., Abeles, R. H., *Federation Proc.* (1970) **29**, 531.
(40) Peterkofsky, A., Mehler, L. N., *Biochim. Biophys. Acta* (1963) **73**, 159.
(41) Givot, I. L., Smith, T. A., Abeles, R. H., *J. Biol. Chem.* (1969) **244**, 6341.
(42) Donohue, J., Lavine, L. R., Rollett, J. S., *Acta Cryst.* (1956) **9**, 655.
(43) Donohue, J., Caron, A., *Acta Cryst.* (1964) **17**, 1178.
(44) Kretzinger, R. H., Cotton, F. A., Bryan, R. F., *Acta Cryst.* (1963) **16**, 651.
(45) Harding, M. M., Cole, S. J., *Acta Cryst.* (1963) **16**, 643.
(46) Griffel, M., Stout, J. W., *J. Am. Chem. Soc.* (1950) **72**, 4351.

RECEIVED June 26, 1970.

19

Rapid Reaction Kinetics Involving the Iron–Porphyrin Site of Horseradish Peroxidase

H. B. DUNFORD and B. B. HASINOFF

University of Alberta, Edmonton, Alberta, Canada

> *Kinetic studies of reactions of horseradish peroxidase (HRP) using stopped-flow and temperature-jump techniques are summarized. The reactions were studied intensively as a function of pH to establish the presence or absence of pH-dependences in the reaction rates. Minimum mechanisms are presented which cannot be proven to be correct. However, simpler mechanisms will not fit the data within experimental error. The reactions which have been studied are fluoride and cyanide binding by (dissociation from) HRP and the oxidation of ferrocyanide to ferricyanide by HRP Compounds I and II. From the pH profiles of the reaction rates, the pK values of acid groups which influence the rates are deduced. Trends in pK values can be explained qualitatively in terms of electrostatic effects.*

Horseradish peroxidase has played an historical role in the development of enzyme chemistry. In the 1920's, Willstatter believed that the enzymatic activity of horseradish peroxidase (HRP) was caused by traces of inorganic material, a conclusion which he thought was applicable to other enzymes. He thus remained skeptical of the results of Sumner, who obtained urease in crystalline form and reported no inorganic elements present. HRP is now known to have a prosthetic group, a porphyrin ring system with a ferric iron at its center known as ferriprotoporphyrin IX or hemin (Figure 1). This is the same group found in catalase, ferricytochrome C, and, except for the oxidation state of the iron, in myoglobin and hemoglobin.

The peroxidatic activity of HRP is the catalysis of reactions by hydrogen peroxide and certain other oxidizing agents (1) summarized in the following reaction scheme for H_2O_2 and the substrate ferrocyanide.

$$HRP + H_2O_2 \xrightarrow{k_{1app}} \text{Compound I}$$

$$\text{Compound I} + \text{ferrocyanide} \xrightarrow{k_{2app}} \text{Compound II} + \text{ferricyanide}$$

$$\text{Compound II} + \text{ferrocyanide} \xrightarrow{k_{3app}} HRP + \text{ferricyanide}$$

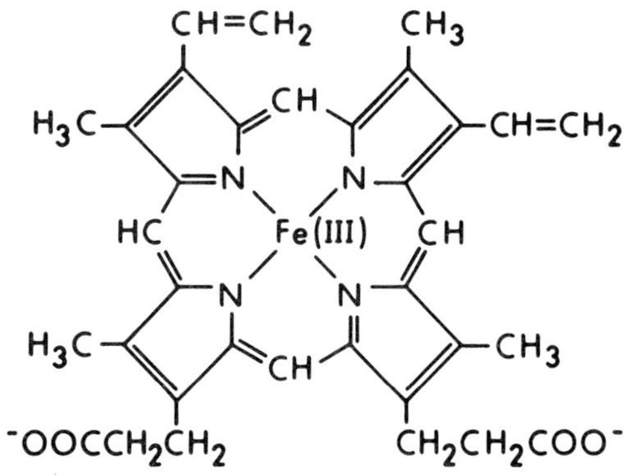

Figure 1. Ferriprotoporphyrin IX or hemin

The ferrocyanide may be replaced by hydrogen donors such as *p*-cresol, in which case free radicals are produced. Compounds I and II, which have unique spectral and chemical properties (2, 3, 4), were first observed by Theorell (5) and Keilin and Mann (6). Compound I was found by Chance to contain two oxidizing equivalents in excess of that present in HRP (7) and Compound II by George to have one oxidizing equivalent (8). The stoichiometries and reaction sequence have been verified (9) and early work has been well reviewed (1, 9, 10). The formation of Compound I occurs with a rate constant, k_{1app}, of $1 \times 10^7 M^{-1}$ sec^{-1} and is pH-independent over a considerable range (11). In addition, equilibrium studies of ligand binding by native HRP indicated a pH-independence of the binding constants in some cases if the acid form of the ligand were assumed to be the binding species (10). The arguments concerning the interpretation of equilibrium ligand binding data have been well summarized by Brill (10). These results, perhaps coupled with knowledge that the transition metal ion plays a role in the mechanism which could therefore be different from that for pure protein enzymes, have led to a widespread acceptance that HRP reactions are pH-independent, a reasonable conclusion on the basis of available evidence.

Fluoride and Cyanide Binding Kinetics by Native HRP

We undertook a kinetic study of fluoride binding by HRP in collaboration with Robert Alberty while one of us (H.B.D.) was spending a sabbatical leave at the University of Wisconsin. The problem appeared to be simple and was ideally suited for the temperature-jump technique. Details of the kinetic results are published elsewhere (12, 13) but the simplest mechanism which is in accord with the kinetic data and does not violate Onsager's principle of detailed balancing (14) is

$$\begin{array}{ccc}
& k_1 & \\
H_2P + F^- & \underset{k_{-1}}{\rightleftarrows} & H_2PF \\
K_1 \updownarrow & & \updownarrow K_{1c} \\
& k_2 & \\
HP + F^- & \underset{k_{-2}}{\rightleftarrows} & HPF \\
K_2 \updownarrow & & \updownarrow K_{2c} \\
& k_3 & \\
P + F^- & \underset{k_{-3}}{\rightleftarrows} & PF
\end{array} \qquad I$$

where H_2P and HP represent species of HRP with protons which influence the binding of fluoride and which might therefore be at the active site. The vertical arrows indicate proton dissociation reactions in which equilibrium is maintained. The constants K_1 and K_{1c} refer to an acid dissociation in the presence and absence of nearby complexed fluoride; similarly for K_2 and K_{2c}. The horizontal arrows represent three pathways between peroxidase and peroxidase complexed with fluoride (see Appendix A).

A similar study over a wider pH range was completed at our laboratory at the University of Alberta by William Ellis on the binding of cyanide by HRP (13). The minimum mechanisms are summarized in the following scheme

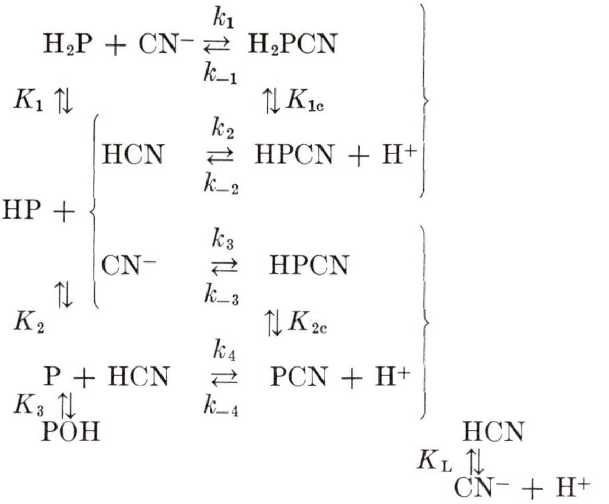

$$
\begin{array}{c}
\quad H_2P + CN^- \underset{k_{-1}}{\overset{k_1}{\rightleftarrows}} H_2PCN \\
K_1 \updownarrow \updownarrow K_{1c} \\
\left. \begin{array}{l} HCN \underset{k_{-2}}{\overset{k_2}{\rightleftarrows}} HPCN + H^+ \end{array} \right\} \\
HP + \left\{ \begin{array}{l} CN^- \underset{k_{-3}}{\overset{k_3}{\rightleftarrows}} HPCN \end{array} \right. \\
\updownarrow \updownarrow K_{2c} \\
K_2 P + HCN \underset{k_{-4}}{\overset{k_4}{\rightleftarrows}} PCN + H^+ \\
K_3 \updownarrow \\
POH HCN \\
 K_L \updownarrow \\
 CN^- + H^+
\end{array}
$$

in which two pairs of rate constants are sufficient to fit the kinetic data. These are indicated by the pairs of subscripts on rate constants labelled 1,3; 2,4; 2,3; and 1,4. The ambiguity concerning the mechanism arises because of the inability of the kinetic method to distinguish between two steps such as

$$
\left. \begin{array}{l} H_2P + CN^- \overset{k_1}{\rightarrow} \\ \\ HP + HCN \overset{k_2}{\rightarrow} \end{array} \right\}
$$

in which $k_2 = k_1 K_L / K_1$, unless one of the rate constants exceeds the diffusion-controlled limit. Although there is ambiguity as to whether the acid form of the ligand is the attacking species, there is no ambiguity beyond normal experimental error in the determination of the dissociation constants of acid groups which influence the kinetics of ligand binding and dissociation and which might therefore be at the active site of HRP. The values of K_1 and K_2 determined in the two studies agree within experimental error. All of the acid dissociation constants for both native and complexed HRP which could be determined in the above studies are listed in Table I.

The above two kinetic studies provide evidence that there is a pH dependence for ligand binding by native HRP which can be accounted for quantitatively by the presence of acid groups close to the hemin iron with pK_a values of 4.2, 6.2, and 10.8. The latter pK_a represents the equi-

Table I. pK Values for Ionizations in the Active Site of HRP Deduced from Kinetic Studies at 25° and $\mu = 0.11$

Species	pK_1	pK_2	pK_3
Compound I	–	5.3	–
Compound II	3.4	5.2	8.6
HRP	4.2[a]	6.2[a]	10.8
HRP–F	5.2	–[b]	
HRP–CN	6.7	10.7	

[a] Mean of two independent determinations (from fluoride and cyanide binding kinetics) which agree within experimental error.
[b] This pK could not be detected, perhaps because of the limited pH range of the study or the insensitivity of the dissociation rate data to pH or both.

librium between native HRP and HRP with a hydroxide group in the sixth coordination position of the ferric iron (15, 16).

These studies led us to begin kinetic investigations of the enzymatic reaction of HRP; the first of these is the oxidation of ferrocyanide.

Kinetics of the Oxidation of Ferrocyanide by Compounds I and II of HRP

The reactions of Compound I and Compound II with ferrocyanide were investigated independently on a stopped-flow apparatus, and the over-all reaction cycle was investigated using steady-state kinetics. The two approaches are self-consistent. Details of the investigation, now available in thesis form (17), will be published elsewhere (18).

Both Compounds I and II are labile, their lifetimes being affected by impurities in the enzyme preparation. Compound I was prepared in the stopped-flow apparatus by adding a large excess of buffered hydrogen peroxide to native HRP of P.N. 2.8 or greater. P.N. is an acronym for purity number, taken as the ratio of absorbances at 403 and 280 nm. Compound I formation was complete in less than 5 msec, the dead time of the apparatus. Ferrocyanide of appropriate concentration was added in the same syringe containing the HRP and its reaction with Compound I could be studied prior to the reaction of significant amounts of Compound II. Blank experiments indicated no interaction of HRP or H_2O_2 with ferrocyanide in the critical time intervals.

In order to isolate kinetically the Compound II reaction, a technique was used similar to that developed by George in his investigation of the number of oxidizing equivalents in Compound II (8, 19). The addition of 1.1 molar equivalents of H_2O_2 and 0.5 molar equivalents of p-cresol to 1.0 molar equivalents of native HRP produces a mixture of about 70% Compound II and 30% native HRP. The p-cresol, which reacts rapidly with Compound I, is completely consumed in catalyzing Compound II formation. Compound II gradually reverts to native HRP, but sufficient remains after 30 minutes to study its reaction with substrate. The maintenance of an isosbestic point at 411 nm during this time interval indicates the presence of only HRP and Compound II (Figure 2).

Both Compounds I and II react with ferrocyanide by second order processes, which indicates that complexes with ferrocyanide are not

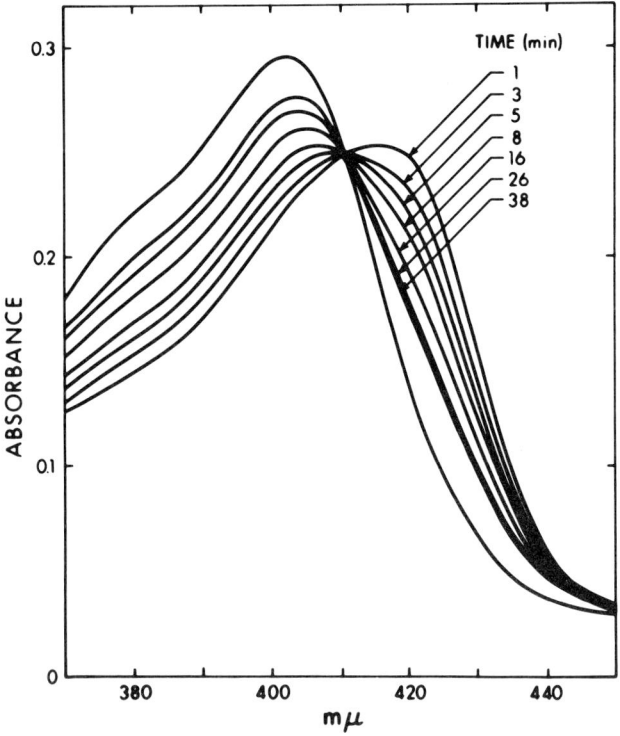

Figure 2. Absorbance of a solution of native HRP and Compound II as a function of time

formed in detectable amounts. The logs of the rate constants of both reactions as a function of pH are shown in Figure 3. The rate of reaction of Compound II with ferrocyanide varies by a factor of 50,000 over the pH range 2.8 to 10.3, compared with a factor of 100 for the Compound I–ferrocyanide rate over the pH range 3.7 to 11.3. The slope of the log plot for the Compound II–ferrocyanide rate becomes −1 above pH 9, whereas

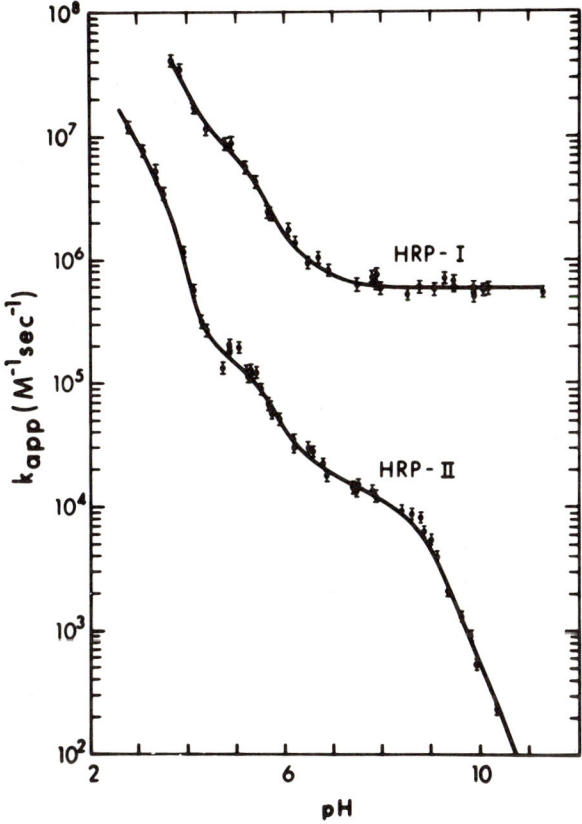

Figure 3. Log of the second order rate constants for the reactions of Compounds I and II of HRP with ferrocyanide as a function of pH

the plot for the reaction of Compound I becomes zero in the same region. This provides a clear indication of a group with a pK_a in the range of 8 to 9 in the active site of Compound II which does not influence the Compound I kinetics (20). Similarly, the shoulders in both curves in the region of pH 5 are accounted for most simply in terms of a group of pK_a near 5 present in both compounds.

Finally, at lower pH's for the Compound II reaction, a second shoulder can be seen. This shoulder is not apparent in the Compound I rate curve; the slope of the Compound I curve is significantly different from that of Compound II in the pH range 3.7 to 4.7. A detailed analysis of the results, taking into account a pK_L of 3.14 for the first protonation of ferrocyanide at ionic strength 0.11 (21) leads to the following minimum mechanisms for the two reactions.

$$\text{HP-I} + \begin{cases} \text{HFe (CN)}_6^{3-} \xrightarrow{k_1} \\ \text{Fe (CN)}_6^{4-} \xrightarrow{k_2} \end{cases}$$

$$K_2 = 4.9 \times 10^{-6} \updownarrow$$

$$\text{P-I} + \begin{cases} \text{HFe (CN)}_6^{3-} \xrightarrow{k_3} \\ \text{Fe (CN)}_6^{4-} \xrightarrow{k_4} \end{cases}$$

$$\text{H}_3\text{P-II} + \begin{cases} \text{HFe (CN)}_6^{3-} \xrightarrow{k_1} \\ \text{Fe (CN)}_6^{4-} \xrightarrow{k_2} \end{cases}$$

$$K_1 = 3.6 \times 10^{-4} \updownarrow$$

$$\text{H}_2\text{P-II} + \begin{cases} \text{HFe (CN)}_6^{3-} \xrightarrow{k_3} \\ \text{Fe (CN)}_6^{4-} \xrightarrow{k_4} \end{cases}$$

$$K_2 = 6.1 \times 10^{-6} \updownarrow$$

$$\text{HP-II} + \begin{cases} \text{HFe (CN)}_6^{3-} \xrightarrow{k_5} \\ \text{Fe (CN)}_6^{4-} \xrightarrow{k_6} \end{cases}$$

$$K_3 = 2.7 \times 10^{-9} \updownarrow$$

$$\text{P-II} + \begin{cases} \text{HFe (CN)}_6^{3-} \xrightarrow{k_7} \\ \text{Fe (CN)}_6^{4-} \xrightarrow{k_8} \end{cases}$$

The symbols P-I and P-II represent Compounds I and II; acid protons are indicated in the active site where required. In the kinetic analysis, indistinguishable steps such as

$$\left. \begin{array}{l} \text{HP-I} + \text{Fe (CN)}_6^{4-} \xrightarrow{k_2} \\ \text{P-I} + \text{HFe (CN)}_6^{3-} \xrightarrow{k_3} \end{array} \right\}$$

are combined. This reduces the Compound I mechanism to three rate terms and one equilibrium constant which were determined from 32 data points. In addition, the terms for $(k_2 + k_3K_L/K_1)$ and k_8 are insignificant in the Compound II–ferrocyanide reaction. Three rate terms and three equilibrium constants were determined for the latter reaction from 44 data points. The pK_a values are summarized in Table I. To the best of our knowledge, these two studies are the first in which Compound I and Compound II reactions with a substrate have been studied intensively as a function of pH (10).

Discussion

Our hope in carrying out the ligand binding and enzymatic oxidation studies outlined above is that the pK_a values so derived might help in the elucidation of the structure of the active site of HRP and the mechanism of its reactions. In order to do this, it is necessary to correlate the pK_a's found for the various derivatives of HRP. This is comparatively simple for the native enzyme and the fluoride and cyanide derivatives, since it may reasonably be assumed that in these species a molecule of water, fluoride, and cyanide occupies the same position, that of a ligand in the sixth coordination position of the iron. Our correlation is shown in the bottom three rows of Table I, in which it can be seen that the pK_a values follow a pattern which can be interpreted by a simple electrostatic picture. In each column, the substitution of a negatively charged fluoride or cyanide ligand progressively increases the pK_a value over that observed in native HRP.

The problem of detailed correlation is more difficult for Compounds I and II, since considerable uncertainty still surrounds the site of the oxidizing moiety. It might reasonably be expected that the presence either of a higher formal charge on the iron atom (3, 8, 22) or of some oxidizing group elsewhere in the enzyme (3, 23, 24, 25) might increase the acidity of the various groups whose pK_a's we have just discussed, and accordingly the values have been tentatively assigned in Table I as shown. In particular, this correlation implies that the pK_a of 8.6 observed for Compound II belongs to the protonation equilibrium

$$Fe^{IV}OH_2^{2+} \rightleftharpoons Fe^{IV}OH^+ + H^+$$

in which we have chosen to assign the extra oxidizing equivalent of Compound II to the iron. This corresponds to the equilibrium

$$Fe^{III}OH_2^+ \rightleftharpoons Fe^{III}OH + H^+$$

in the native enzyme to which the pK_a value of 10.8 is commonly assigned. In the foregoing equations, the contribution of two formal charges by the porphyrin ring has been shown. Under this scheme, the pK_1 and pK_2

values in Table I refer to ionizable protein groups, only one of which may be coordinated to the iron atom. Spectra of HRP as a function of pH do not provide clear evidence that either of the acid groups corresponding to pK_1 or pK_2 are bound directly to iron or are spectrally "heme-linked."

Another feature of our results worthy of discussion is the nonappearance in Compound I of one, or possibly two, of the pK_a's of Compound II, while the one pK_a which was observed in the former species lies very close to one of those found in the latter. The state of knowledge concerning these two compounds has been well reviewed (1, 10), but the picture is still incomplete and sometimes contradictory. Among the most recent contributions are those of Blumberg et al. and Pcisach et al. (26, 27), who postulate in particular on the basis of spectral comparisons that Compound I differs from Compound II in that the ligands in the fifth and sixth coordination positions are either completely absent or only weakly coordinated in the more highly oxidized compound. This proposal could account for the smaller number of pK_a's in Compound I, but it can only be reconciled with the results of a study showing that the two compounds have similar Mossbauer spectra (22) if such spectra are highly insensitive to ligand environment. An alternative explanation for our results, kindly suggested by R. J. P. Williams, is that in Compound I, one of the two oxidizing equivalents is stored on a group in the active site whose ionization is thus prevented:

HRP $\rightarrow$	Compound I $\rightarrow$	Compound II
Fe (III) X$^-$	Fe (IV) X·	Fe (IV) X$^-$
$\updownarrow$ H$^+$	no ionization	$\updownarrow$ H$^+$
XH	possible	XH

This scheme accounts for the absence of one ionization in Compound I. A quite different type of explanation is that if at least one of the oxidizing equivalents in Compound I is stored on the porphyrin ring, the reacting site of the ferrocyanide ion may be removed from the active site of Compound II and the reactivity accordingly influenced by different ionizing groups. On the other hand, the observation of a common pK_a close to 5.2 for both compounds may indicate that at least some features of the active site are similar, a point which accords with the Mossbauer data. Although the evidence so far available is insufficient to discriminate among alternative descriptions of the type presented above, our data provide an additional source of evidence which must be taken into account in future discussions of the structures of these two compounds.

In none of our studies of ligand binding by peroxidases or hemin have we obtained any evidence of inhibition of rate at high ligand concentration which would indicate either a dissociative or ion-pair mechanism (28, 29). This lack of evidence does not exclude such mechanisms, however. From a comparison of imidazole binding rate with solvent exchange rate by hemin in an ethanol–water solvent, our results favored an ion-pair mechanism (30, 31).

Appendix A: Analysis of Kinetic Results

At the suggestion of one reviewer, we offer a brief discussion of the techniques used in proposing a mechanism for a given reaction from kinetic data obtained by rapid reaction techniques. We use the case of the reaction of fluoride with HRP as an example. The proton reactions are too rapid to observe on the time scale of a conventional temperature-jump apparatus (32). These proton reactions would couple the three paths for fluoride binding and dissociation proposed in Mechanism I so that only one relaxation time, τ, is observed. At any given pH, the relation between τ and the constants in the above mechanism is

$$\frac{1}{\tau} = k_{1\mathrm{app}} [(\overline{\mathrm{P}})_{\mathrm{tot}} + (\overline{\mathrm{F}})_{\mathrm{tot}}] + k_{-1\mathrm{app}} \qquad (1)$$

where $(\overline{\mathrm{P}})_{\mathrm{tot}}$ and $(\overline{\mathrm{F}})_{\mathrm{tot}}$ are the equilibrium concentrations of all forms of free peroxidase (e.g., H_2P, HP, and P) and fluoride present in solution (32, 33). If a large excess of fluoride is used, the practical relation is

$$\frac{1}{\tau} = k_{1\mathrm{app}} (\mathrm{F})^\circ + k_{-1\mathrm{app}} \qquad (2)$$

where $(\mathrm{F})^\circ$ is the total concentration of fluoride added to the solution. Thus at each pH, the experimental rate constants for the binding reaction, $k_{1\mathrm{app}}$, and for the dissociation reaction, $k_{-1\mathrm{app}}$, are obtained by measuring τ as a function of $(\mathrm{F})^\circ$. The next step is to test the simplest possible mechanism for the reaction of fluoride with HRP. One attempts to choose values for the unknown rate and equilibrium constants in the proposed mechanism which would enable one to reproduce the experimental rate constants within their estimated errors (12). Provided an exhaustive search is made for the best-fit values of these parameters and provided the fit to the $k_{1\mathrm{app}}$ and $k_{-1\mathrm{app}}$ data remains unsatisfactory, then the proposed mechanism can be rejected. The next simplest mechanism is then tested. Of course, even though a given mechanism can be made to fit the experimental data within their estimated error, the mechanism may not be the correct one. The great power of the kinetic method lies in its use to disprove mechanisms. This places the onus on the investigator to vary the crucial parameters over the widest possible range. The most crucial and significant parameter in enzyme kinetics appears to be pH.

By use of the above technique, combined with the application of detailed balancing where it could be applied, mechanism I for the fluoride–HRP reaction was proposed (*13*). The relation between the apparent (experimental) rate constants and the individual rate and equilibrium constants of the mechanism is obtained as follows.

$$k_{1app}(P)_{tot}(F^-) = k_1(H_2P)(F^-) + k_2(HP)(F^-) + k_3(P)(F^-)$$

$$(P)_{tot} = (H_2P) + (HP) + (P)$$

$$k_{1app}(P)_{tot} = \frac{k_1(P)_{tot}}{1 + \frac{(HP)}{(H_2P)} + \frac{(P)}{(H_2P)}} + \frac{k_2(P)_{tot}}{\frac{(H_2P)}{(HP)} + 1 + \frac{(P)}{(HP)}}$$

$$+ \frac{k_3(P)_{tot}}{\frac{(H_2P)}{(P)} + \frac{(HP)}{(P)} + 1}$$

$$k_{1app} = \frac{k_1}{1 + \frac{K_1}{(H^+)} + \frac{K_1 K_2}{(H^+)^2}} + \frac{k_2}{\frac{(H^+)}{K_1} + 1 + \frac{K_2}{(H^+)}}$$

$$+ \frac{k_3}{\frac{(H^+)^2}{K_1 K_2} + \frac{(H^+)}{K_2} + 1}$$

$$k_{1app} = \frac{k_1 + k_2 K_1/(H^+) + k_3 K_1 K_2/(H^+)^2}{1 + \frac{K_1}{(H^+)} + \frac{K_1 K_2}{(H^+)^2}}$$

Similarly,

$$k_{-1app} = \frac{k_{-1} + k_{-2} K_{1c}/(H^+) + k_{-3} K_{1c} K_{2c}/(H^+)^2}{1 + \frac{K_{1c}}{(H^+)} + \frac{K_{1c} K_{2c}}{(H^+)^2}}$$

For systems with several relaxation times, we refer the reader to more advanced literature (*32, 34, 35, 36*).

If one is using a stopped-flow apparatus (37) to study either reversible (38) or irreversible reactions under pseudo-first-order conditions, then similar techniques can be applied (18). If the absorbance change of the solution is small during the course of the reaction, then the plot of voltage vs. time obtained from the stopped-flow apparatus is a simple exponential curve and a "relaxation time" is readily obtained. If the reaction is too fast to study as a pseudo-first-order process, it may be slowed down by changing reactant concentrations so that second order kinetics are obeyed. The analysis then becomes more complex; it is facilitated if the voltage at both zero and infinite time can be estimated accurately (18). The estimate of voltage at true zero time requires knowledge of the dead time of the stopped-flow apparatus.

Literature Cited

(1) Saunders, B. C., Holmes-Siedle, A. G., Stark, B. P., "Peroxidase," Butterworth, Washington, 1964.
(2) Chance, B., *Arch. Biochem. Biophys.* (1952) **41**, 404.
(3) Brill, A. S., Williams, R. J. P., *Biochem. J.* (1961) **78**, 253.
(4) Brill, A. S., Sandberg, H. E., *Biochemistry* (1968) **7**, 4254.
(5) Theorell, H., *Arkiv Kemi Mineral. Geol.* (1942) **49**, No. 24.
(6) Keilin, D., Mann, T., *Proc. Roy. Soc. (London)* (1937) **122B**, 119.
(7) Chance, B., *Arch. Biochem. Biophys.* (1952) **37**, 235.
(8) George, P., *Nature* (1952) **169**, 612.
(9) Yonetani, T., *J. Biol. Chem.* (1966) **241**, 2562.
(10) Brill, A. S., in "Comprehensive Biochemistry," M. Florkin and E. H. Stotz, Eds., Ch. 14, p. 447, Elsevier, Amsterdam, 1966.
(11) Chance, B., De Vault, D., Legallais, V., Mela, L., Yonetani, T., in "Fast Reactions and Primary Processes in Chemical Kinetics," p. 437, Interscience, Stockholm, 1967.
(12) Dunford, H. B., Alberty, R. A., *Biochemistry* (1967) **6**, 447.
(13) Ellis, W. D., Dunford, H. B., *Biochemistry* (1968) **7**, 2054.
(14) Onsager, L., *Phys. Rev.* (1931) **37**, 405.
(15) Keilin, D., Hartree, E. F., *Biochem. J.* (1951) **49**, 88.
(16) Ellis, W. D., Dunford, H. B., *Arch. Biochem. Biophys.* (1969) **133**, 313.
(17) Hasinoff, B. B., Ph.D. thesis, University of Alberta, 1970.
(18) Hasinoff, B. B., Dunford, H. B., *Biochemistry* (1970), in press.
(19) George, P., *Biochem. J.* (1952) **55**, 220.
(20) Dixon, M., Webb, E. C., "Enzymes," 2nd ed., pp. 116–145, Longmans, Green, London, 1964.
(21) Jordan, J., Ewing, G. J., *Inorg. Chem.* (1962) **1**, 587.
(22) Moss, T. H., Ehrenberg, A., Bearden, A. J., *Biochemistry* (1969) **8**, 4159.
(23) Chance, B., *Advan. Enzymol.* (1951) **12**, 153.
(24) Brown, S. B., Jones, P., Suggett, A., *Trans. Faraday Soc.* (1968) **64**, 986.
(25) Fajer, J., Borg, D. C., Forman, A., Dolphin, D., Felton, R. H., *J. Am. Chem. Soc.* (1970) **92**, 3451.
(26) Blumberg, W. E., Peisach, J., Wittenberg, B. A., Wittenberg, J. B., *J. Biol. Chem.* (1968) **243**, 1854.
(27) Peisach, J., Blumberg, W. E., Wittenberg, B. A., Wittenberg, J. B., *J. Biol. Chem.* (1968) **243**, 1871.
(28) Segal, R., Dunford, H. B., Morrison, M., *Can. J. Biochem.* (1968) **46**, 1471.

(29) Dolman, D., Dunford, H. B., Chowdhury, D. M., Morrison, M., *Biochemistry* (1968) **7**, 3991.
(30) Angerman, N. S., Hasinoff, B. B., Dunford, H. B., Jordan, R. B., *Can. J. Chem.* (1969) **47**, 3217.
(31) Hasinoff, B. B., Dunford, H. B., Horne, D. G., *Can. J. Chem.* (1969) **47**, 3225.
(32) Eigen, M., deMaeyer, L., *in* "Technique in Organic Chemistry," A. Weissberger, Ed., Vol. VIII, Part II, 2nd ed., pp. 895–1054, Interscience, New York, 1963.
(33) Caldin, E. F., "Fast Reactions in Solution," pp. 59–63, Wiley, New York, 1964.
(34) Amdur, I., Hammes, G. G., "Chemical Kinetics," pp. 138–143, McGraw-Hill, New York, 1966.
(35) Castellan, G. W., *Ber.* (1963) **67**, 898.
(36) Alberty, R. A., Yagil, G., Diven, W. F., Takahashi, M., *Acta Chem. Scand.* (1963) **17**, S34.
(37) Gibson, Q. H., Milnes, L., *Biochem. J.* (1964) **91**, 161.
(38) Frost, A. A., Pearson, R. G., "Kinetics and Mechanism," 2nd ed., p. 186, Wiley, New York, 1961.

RECEIVED June 26, 1970.

INDEX

A

Absorbance change	425
Absorption spectra	
of Co-LADH	207
of iron–sulfur proteins	323, 326
Acid-catalyzed decomposition	88
Activated energy state	198
Activation	
α	397
β	406
γ	407
δ	409
Activity of terminal double bond	264
Addition of nucleophiles	65
Adducts, oxygen	115
Adenine	137
Affinity of ligands for nitrogenase and inorganic complexes	241
Alcohol dehydrogenase from equine liver	199
Alcohols, method for coupling	95
Aldehydes, conversion of diols to, by dioldehydrase	352
Alkaline phosphatase	
cobalt content of	191
dimerization of	190
from $E.\ coli$	189
zinc content of	189
Alkylation, relative rates of	13
Alkylcobalamins, substituted	9
Alkyl–cobalt (I) complexes	76
Allosteric effect	164
Allyl alcohol	96
Alpha thalassemia	286
Aluminum	105
Amino acid sequences of iron–sulfur proteins	329
Ammonia	99
solutions	124
5′AMP	137
AMP isomers, broadening effects of	141
Analysis of kinetic results	423
Anemia	293
in rats, zinc-induced	315
pernicious	346
Aniline protons, relative shifts of	183
Anions, effect on rat ceruloplasmin	319
Anisole, chlorination of	42
Annulene ring systems, porphyrin and	65
Apoenzyme, metal-free	189
Apoferredoxin, thiocyanate formation from	341
Apophosphatase, dialysis of	193
Apoprotein, modification of	334
Apotransferrin, binding of iron(III) by	303
Artificial enzyme	32
Association–dissociation equilibrium	231
Atransferrinemia, congenital	304
Axial ligand	249
Azide, metal	88
Azotobacter molybdenum–iron protein, crystallization of	235
Azotobacter nitrogenase, characteristics of	232

B

Base binding	137
Benzoylhydrazine	86
Binary complex, enhancement of	394
Binding	
energies	176, 185
kinetics, fluoride and cyanide	415
of iron(III) by apotransferrin	303
Biochemical problems in nitrogen fixation	242
Biochemistry of nitrogen fixation	219
Biological	
function of rubredoxin	324
ligands of d^0 cations	158
role of copper	292
specificity	188
Biosynthesis, hemoglobin	293
Biphasic reactions	112
Bis(histidinato)cobalt(II), uptake of oxygen by	125
Bond cleavage, phosphodiester	150
Bridge complex, enzyme–metal–substrate	390
Broadening effects of AMP isomers	141

C

Calcium	
-binding compounds	159
enzymes	160
in biological systems	155
Campbellians	1
Carbon-17 chain, unsaturation in the	264
Carboxylase, pyruvate	397
Carboxymethylation	206
Carriers, oxygen	116
Catalysis	174

Catalyst, nucleophile–metal combination as a 26
Catalytic
  activity
    effect of cobalt on 193
    of ceruloplasmin 297
    efficiency of divalent cations ... 164
    electrolytic conversion 105
Catalytically active metalloenzyme,
  regeneration of 193
Cations, effect on ferroxidase
  activity 318
Ceruloplasmin 292
  catalytic activity of 297
  changes in developmental
    systems 315
  effect on iron mobilization 306
  ferroxidase activity of 297
  increase in plasma iron after
    injection of 308
  in the perfused liver, mobilization of iron by 310
  rat, effect of anions on 319
  substrate groups of 298
Charge density 157
Chelated ligands 122
Chelating agents 189, 223
  inhibition by 201
Chemical shift 263
  vs. concentration 251
Chlorination of anisole 42
Chloroplast ferredoxin 322
Cleavage specificity 150
Clostridial ferredoxin 221, 322
  crystals 328
  isolation of 327
Clostridial rubredoxin, structure for 331
Clostridia nitrogenase, characteristics of 232
Cobalamin chemistry 1
  models in 5
Cobaloximes 6
Cobalt 89
  absorption spectra 194
  (II)–ammonia–oxygen system .. 128
  (II)analog of hemoglobin 115
  –carbon bond 73
    cleavage reaction 15
  (II) complexes 111
  –oxygen interaction 115
  compounds 84
  content of alkaline phosphatase 191
  effect on catalytic activity 193
  in LADH, ligand environment of 209
  (I) nucleophilicity 12
  (II)–oxygen interaction 112
  physical–chemical probe 193
  (II)–trien system 130
  (II)trispyrazolylborate 178
Coenzyme B_{12} 5, 7
  dependent mutase reactions 18
  mechanism of action of 346

Coenzyme B_{12} *(Continued)*
  models, vitamin B_{12} 73
  rate of tritium transfer from ... 354
  structure of vitamin B_{12} 347
  three-dimensional model of a .. 354
  tritium introduction into 356
  vitamin B_{12}
    reactions requiring 348
    relative specific activities of .. 356
$Co_2(histidine)_4O_2$ 122
Co-LADH, absorption spectrum of 207
$Co(L-histidine)_2$ 126
Competitive inhibition 34
Congenital atransferrinemia 304
Controlled potential electrolysis .. 63
Conversion of
  diols to aldehydes by
    dioldehydrase 352
  nitrogen to hydrazine 109
  peroxidase compounds 68
Coordinated nitrogen, reactions of 89
Copolymerization 92
Copper 125, 139, 147
  and iron proteins 382
  oxidative addition model of
    oxygen binding to 382
  structural models for 365
  biological role of 292
  –bridged oxygen species 112
  deficiency 293
  effect on iron metabolism 305
  enzymes 292–4
  protein 293–4
  blue 386
Corrin ring 50
Cor system 50
Cotton effect 202
Coupling
  alcohols, method for 95
  constants, hyperfine 399
Crystal field parameters for ferric
  low-spin compounds of hemoglobin A 279
Crystallization
  of azotobacter molybdenum–iron
    protein 235
  of salts 168
Curtis' diene, iron complexes of .. 72
Curtis reaction 46
Cyanide and fluoride binding
  kinetics 415
Cyclic
  ligands 157
  voltammetry 64
Cyclodextrins 28
Cysteine residues in iron–sulfur
  proteins 331
Cytochrome c oxidase
  reactions of 257
  structure–function relationships
    in 248
Cytosine 137

INDEX

D

Decomposition	
acid-catalyzed	88
lability toward	124
Dehydration of propanediol, enzymatic	16
Denaturation and oxidation of hemoglobin	284
Deoxyhemoglobin	72
Deoxyribonucleic acid	136
Depolymerization	150
Destabilizing the double helix	146
Deuterohemins	249
Developmental systems, ceruloplasmin changes in	315
Dialysis of apophosphatase	193
Diamagnetism of oxyhemoglobin	383
Dibridged (O_2, OH) complex	123
Differential reactivity of the zinc atoms in LADH	205
Diffraction, x-ray	391
Dihistidine	
compound	277
hemichrome	284
Dimerization of alkaline phosphatase	190
Dimers	
oxobridged iron(III)	373
μ-oxo hemin	259
Dioldehydrase	348
conversion of diols to aldehydes by	352
reaction sequence for	362
Diol dehydratase, mechanism	15
Dipolar shift for the undissociated complex	185
Dissociation	
mechanism, fluoride binding and	423
of macrocyclic complexes	55
rate of	53
Distances in solution, manganese-to-ligand	396
Distorted geometry, relationship to enzymatic function	199
DNA	136
structure, effect of metal binding on	144
unwinding and rewinding of	146
Duhemists	1

E

E. coli, alkaline phosphatase from	189
Electrochemical oxidation	90
Electrolysis, controlled potential	63
Electrolytic conversion, catalytic	105
Electrolytic reduction of nitrogen	104
Electron	
activation	238
donors	221
Electronic absorption	
spectral studies on the ferritin core	371
Electronic absorption *(Continued)*	
spectrum of	
(HEDTA Fe)$_2$O^{2-}	376
iron(III) phosvitin	372
1-Electron reduction product	63
Electrophilic center	411
Elimination	
and enolization reactions, metals in enzyme-catalyzed	390
reactions, β	9
Energy	
binding	176, 185
required for nitrogenase activity	222
state, activated	198
Enhancement	393–4
Enolization	397
and elimination reactions, metals in enzyme-catalyzed	390
Entatic	
site hypothesis	198
state	56
Enzymatic	
activity	
and stability of modified ferredoxins	336
of horseradish peroxidase	413
dehydration of propanediol	16
function, relationship of distorted geometry to	199
Enzyme	2, 96, 151
artificial	32
binding of metal ions	57
calcium	160
catalysis, models for	22
-catalyzed	
enolization and elimination reactions, metals in	390
processes, selectivity of	38
copper	292–4
extracellular	160
-increased hydrolysis rate	34
–inhibitor complex	198
–metal–substrate bridge complex	390
models, studies on	21
multichain	188
potassium	161
reducing	285
single chain	188
stress	297
Enzymic sites, protein environment	188
Epoxide, potential terminal	96
EPR	
method	391
spectra of porphyrin samples	273
Equilibrium	
association–dissociation	231
constant	128
Equine liver, alcohol dehydrogenase from	199
Erythrocyte	283
ghosts	286
Ethylcobaloximes, β-substituted	9
Europium(II)	165

Evolution of ionic distribution	171
Exothermic interactions	112
Extracellular enzymes	160

F

Fermi contact interaction	177
Ferredoxin	
chloroplast	322
clostridial	322
isolation of	327
crystals, clostridium	328
^{3}H–H exchange	336
immunological activity of	337
modified, enzymatic activity and stability of	336
reconstitution of	329, 338
thiocyanate formation from	341
Ferric	
heme compounds	272
hemoglobin A	274
low-spin compounds of hemoglobin A, crystal field parameters for	279
Ferriprotoporphyrin IX	413
Ferritin core	
electronic absorption spectral studies on	371
synthetic model for	369
Ferrocyanide	414
ion, reacting site of	422
oxidation of	417
Ferroxidase	
activity	
effect of cations on	318
in plasma	301
kinetics of	316
method for determination of	299
of ceruloplasmin	297
hypothesis	303
inhibition by metal ions	317
role in iron metabolism	309
Fixation, nitrogen	1, 95
–reduction cycle	101
Fixedness, multiple juxtapositional	52
Flavodoxin	221
Flexibility	57
Fluorescence probes	166
Fluoride	
and cyanide binding kinetics	415
binding and dissociation mechanism	423
reaction with horseradish peroxidase	423
Folding of polypeptide chain	332

G

Genetic code	135
Geometric orientation of second sphere complex	183
Geometry, irregular	198
Ghosts, erythrocyte	286
Glycolaldehyde	361
Group IA cations, stability constants for	157
Group IA and IIA cations, biochemistry of	155
Group IIA cations, stability constants for	156
Guanine	137

H

HD formation	230
Health hazard, nitrite as	289
(HEDTA Fe)$_2$O^{2-}, electronic absorption spectrum of	376
Heinz bodies	286
Heme	
A	259
model of	265
structure of	263
compounds, ferric	272
proteins	248
low-spin compounds of	271
Hemerythrin	114, 378
Hemichrome	275
dihistidine	284
reversible	280
Hemoglobin	112, 274
A	
crystal field parameters	279
ferric	274
biosynthesis	293
cobalt(II) analog of	115
ferric, effect of oxidants on	288
formation	292
H	275
low-spin compounds of	282
model for oxygen binding by	382
oxidation and denaturation of	284
oxidative addition model of oxygen binding to	384
reaction of oxygen with	114
^{3}H–H exchange of ferredoxin	336
High-spin iron(III) complexes	367
Histidine deaminase	409
Horseradish peroxidase	
enzymatic activity of	413
iron–porphyrin site of	413
reaction of fluoride with	423
structure of the active site of	421
HRP	413
ΔH values	112
Hydrazine, conversion of nitrogen to	109
Hydroformylation reaction	174
Hydrogen	
exchange	
effect of metals on	211
in proteins, rates of	213
isotope exchange techniques	209
transfer	349
intermolecular	352, 359
Hydrogenation–dehydrogenation sequence	48
Hydrolysis	
of coordinated ligands, metal-catalyzed	23
rate, enzyme-increased	34

INDEX

Hydrophobic forces, substrate bound by 28
Hydrosulfite 222
Hydroxide compounds of myoglobin 282
Hyperfine coupling constants 399

I

Immunological activity of the ferredoxins 337
Infrared spectra of aqueous solutions 254
Inhibition
  by chelating agents 201
  competitive 34
  of ferroxidase by metal ions ... 317
  of rate 423
Inhibitors, reduction 231
Inner great rings 51
Inorganic
  complexes, affinity of ligands for 241
  models of nitrogen fixation 79
  sulfide 322
Insertion–reduction mechanism .. 227
Intermolecular hydrogen transfer .. 352
Intracellular ions 169
Intramolecular hydrogen transfer .. 359
Ion content of living systems 169
Ionic
  distribution 159
  evolution of 171
  model for chemical bonding ... 155
Ionization of the peptide hydrogens 125
Iron
  absorption 305
  and copper proteins
    oxidative addition model of oxygen binding to 382
    structural models for 365
  complexes
    of Curtis' diene 72
    of the synthetic macrocycles .. 69
  in plasma 303
  in protein 237
  oxidation state of 323
  metabolism 292
    effect of copper on 305
    in Wilson's disease 313
    role of ferroxidase in 309
  mobilization by ceruloplasmin in the perfused liver 310
    effect of ceruloplasmin on .. 306
  pool, mobilizable 312
  –porphyrin site of horseradish peroxidase 413
  -storage protein 369
  –sulfur protein 322
    absorption spectra of 323, 326
    amino acid sequences of ... 329
    cysteine residues in 331
    physiological role of 327

Iron(II)
  complexes, reaction of oxygen with 112
  nonenzymic oxidation of 301
Iron(III)
  binding by apotransferrin 303
  complexes, high-spin 367
  dimers, oxobridged 373
  formation, rate of 301
  phosvitin 372
Irregular geometry 198
Isolation of clostridial ferredoxin.. 327
Isomorphous replacement 161, 163
Isonitrile reduction, rate constants (M^{-1}) for 226
Isotope effect 351

J

Job's curve 205

K

Kinetic
  characteristics of the inhibition of ferroxidase 318
  parameters of E–M–S bridge complexes 400
  results, analysis of 423
Kinetics
  of ferroxidase activity 316
  rapid reaction 413

L

Labeling experiments, tritium ... 15
Lability toward decomposition .. 124
LADH 199
  differential reactivity of the zinc atoms in 205
  ligand environment of cobalt in 209
  zinc isotope exchange in 201
Lanthanides
  as probes 165
  chelates 166
  in biological systems 163
Ligand
  affinity of, for nitrogenase and inorganic complexes 241
  axial 249
  binding 414
  chelated 122
  environment of cobalt in LADH 209
  exchange reaction 175
  field transitions 367
  outer sphere 177
  phenoxo 251
  second coordination sphere, model for 181
Liver, perfused, mobilization of iron by ceruloplasmin in 310
Living systems, ion content of ... 169
Longitudinal relaxation time 392
Low-spin compounds
  of heme proteins 271
  of hemoglobin 282
  of the peroxidases 283
Lyotropic series 155

M

Macrocycles, synthetic iron complexes of 69
Macrocyclic
  complexes
    dissociation of 55
    natural 44
    tetradentate 60
  effect 56
Macromolecular conformation ... 215
  chemistry 158
  enzymes 164, 169
  in biological systems 155
Magnetic
  coupling 259
  susceptibility 380
  of iron(III) phosvitin 372
Manganese
  as a probe 163
  -to-ligand distances in solution .. 396
Mechanical constrictive effect 60
Mechanism of action of B_{12} coenzymes 346
Membrane binding 167
Metabolism, iron 292
  effect of copper on 305
Metal
  and protein, molar stoichiometry of 200
  azide 88
  -catalyzed hydrolysis of coordinated ligands 23
  complexes
    and metalloenzymes, spectral comparison of 197
    second coordination sphere of 174
  effect of, on hydrogen exchange 211
  enzyme-catalyzed enolization and elimination reactions 390
  -free apoenzyme 189
  -induced structural alterations .. 212
  ions
    and polynucleotide conformation 148
    effect of 135
    enzyme binding of 57
    inhibition of ferroxidase by .. 317
  nitrogen
    compounds, formation of 82
    distances 60
  stabilization of protein structure 209
Metalloenzyme
  properties of metals in 198
  regeneration of catalytically active 193
  structure and function of 187
  zinc 402
Metalloproteins 187
Metallothionein 209
Metastable adducts 115
Metchlorohemerythrin 378
Method for determination of ferroxidase activity 299

Mobilizable iron pool 312
Mobilization of iron by ceruloplasmin in the perfused liver 310
Models
  cobalamin chemistry 5
  enzyme
    catalysis 22
    studies on 21
  heme A 265
  ionic, for chemical bonding ... 155
  molecular, of the artificial enzyme 32
  motility 216
  nitrogenase 2
  nitrogen fixation, inorganic ... 79
  oxidative addition
    oxygen-binding to hemoglobin 384
    oxygen-binding to iron and copper proteins 382
  second coordination sphere ligand 181
  structural, for iron and copper proteins 365
  studies, philosophy of 1, 21
  systems, types of 22
  three-dimensional, of a coenzyme 354
  vitamin B_{12} coenzyme 73
Modification of apoprotein 334
Modified squalene oxides 96
Molar stoichiometry of protein and metal 200
Molecular model of the artificial enzyme 32
Molybdenum 87
  –iron protein 233
Monodentate phosphate coordination 403
Motility model 216
Multichain enzymes 188
Multiple juxtapositional fixedness .. 52
Muscle contraction 160
Mutase reactions, coenzyme B_{12} dependent 18
Myoglobin 254
  hydroxide compounds of 282
  reaction of oxygen with 112

N

Naphthalene pool 104
Natural macrocyclic complexes ... 44
Nerve membrane 161
New serum ferroxidase 300
Nichrome 105
Nickel(II) 53, 125
Nitrile hydrolysis, mechanisms for 24
Nitrite 288
  as health hazard 289
Nitrogen
  -bridged compounds, polymeric 92
  compounds, stable 81
  conversion to hydrazine 109
  fixation 1
    biochemical problems in 242
    biochemistry of 219

INDEX

Nitrogen *(Continued)*
  in air 101
  inorganic models of 79
  molecular 95
  research, potential utilities of 243
  ligands 53
Nitrogenase 219, 231
  activity, energy required for .. 222
  complexes, affinity of ligands for 241
  models 2
  reactions 220
  substrates reduced by 4, 223
p-Nitrophenyl acetate 31
NMR
  method 391
  shift 177
  spectra
    of aniline 179
    of pyridine 180
N–N stretching frequencies 81
Nonenzymic oxidation of iron(II) 301
Nuclear
  magnetic resonance line broadening studies 135
  relaxation 390
Nucleic acids, structure of 135
Nucleophile
  addition of 65
  –metal combination as a catalyst 26
  vitamin B_{12} as a 358
Nucleophilic displacement 83
Nucleophilicity, cobalt(I) 12
Nucleoside bases 135
Nucleotide
  complexes of 136
  sequence 150

O

Octopus hemocyanin 112
Olive mechanism 105
OP binding, stoichiometry of 203
Optical
  absorption spectra 14
  isomers 56
Osmium 89
Outer sphere ligands 177
Oxidants, effect on ferric hemoglobin 288
Oxidation
  and denaturation of hemoglobin 284
  electrochemical 90
  of ferrocyanide 417
  of iron(II), nonenzymic 301
  pyruvate 326
  –reduction properties of porphyrins and TAAB complexes .. 61
  selective enzymatic 38
  state of iron in proteins 323
Oxidative addition model of oxygen-binding
  to hemoglobin 384
  to iron and copper proteins ... 382
Oxides, modified squalene 96

Oxobridged iron(III) dimers 373
μ-Oxo hemin dimers 259
Oxygen
  adducts 115
  binding by hemoglobin, model for 382
  carriers 116
    successful 130
  uptake of 111
Oxygenation equilibrium 124
Oxyhemerythrin 379
Oxyhemocyanin 386
Oxyhemoglobin 285

P

Packing problem 157
Paramagnetic
  complexes 177
  effects, enhancement of 394
  shifts 251
Peptide hydrogens, ionization of .. 125
Pernicious anemia 346
Peroxidase
  compounds, conversion of 68
  derivatives 67
  horseradish
    enzymatic activity of 413
    iron–porphyrin site of 413
  low-spin compounds of 283
Peroxo species 124
Phenoxo ligand 251
Philosophy of model studies 1, 21
Phosphatase, stabilization of quaternary structure of 191
Phosphate 139
  binding 137
  coordination, monodentate 403
Phosphodiester bond cleavage 150
Phosphorylated protein 167
Phosvitin, iron(III), magnetic susceptibility of 372
Physical–chemical probe, cobalt as a 193
Physiological role of iron–sulfur proteins 327
Plasma
  ferroxidase activity in 301
  iron in 303
  iron, increase after injection of ceruloplasmin 308
Polymeric nitrogen-bridged compounds 92
Polynucleotide conformation, metal ions and 148
Polypeptide chain, folding of 332
Polyphosphate chelate 162
Porphyrin
  and annulene ring system 65
  complexes, oxidation–reduction properties of 61
  compounds 58
  ring 57, 421
  samples, EPR spectra of 273
  structure 248

Por system 50
Potassium
  enzyme 161
  in biological systems 155
Potential
  terminal epoxide 96
  utilities of nitrogen fixation
    research 243
Probe ions 160
Propanediol, enzymatic dehydration
  of 16
Properties of metals in metalloenzymes 198
Protein
  and metal, molar stoichiometry of 200
  azotobacter molybdenum–iron,
    crystallization of 235
  blue copper 386
  characteristics of molybdenum–
    iron 234
  copper 293–4
  environment at active enzymic
    sites 188
  heme, low-spin compounds of .. 271
  iron 237
  iron-storage 369
  iron–sulfur 322
    absorption spectra of 323, 326
    amino acid sequences of 329
    cysteine residues in 331
    physiological role of 327
  molybdenum–iron 233
  oxidation state of iron in 323
  phosphorylated 167
  rates of hydrogen exchange in .. 213
  respiratory 111
  structure, metal stabilization of 209
  synthesis 160
Pyruvate
  carboxylase 397
  oxidation 326

Q

Quadridentate imine condensates .. 126
Quaternary structure of phosphatase, stabilization of 191

R

Radius-ratio effect 156, 168
Rapid reaction kinetics 413
Rate
  constants (M^{-1}) for isonitrile
    reduction 226
  of dissociation 53
  of iron(III) formation 301
  of hydrogen exchange in proteins 213
  of tritium transfer from coenzyme 354
Rats, zinc-induced anemia in 315
Reacting site of the ferrocyanide ion 422
Reconstitution of ferredoxin ...329, 338
Reducing
  agents 85
  enzymes 285

Reduction
  inhibitors 231
  of unbranched substrates 230
  product, 1-electron 63
Regeneration of catalytically active
  metalloenzyme 193
Relationship of distorted geometry
  to enzymatic function 199
Relative
  shifts of aniline protons 183
  specific activities of vitamin B_{12}
    coenzyme 356
Relaxation
  rates, nuclear 391
  time, longitudinal and transverse 392
Respiratory proteins 111
Reticuloendothelial system 305
Reversible
  hemichromes 280
  oxidation 111
Rhenium 90
Rhombic field 281
Ribonucleic acid 136
Ring
  current field effects 251
  size 57
  systems, porphyrin and annulene 65
RNA 136
Rubredoxin
  biological function of 324
  clostridial, structure for 331
Ruthenium 83, 90

S

Saltman–Spiro ball 369
Salts, crystallization of 168
Schiff base condensation 46
Second coordination sphere
  complex, geometric orientation of 183
  ligand, model for 181
  of metal complexes 174
  structure 176
Selectivity of enzyme-catalyzed
  processes 38
Simultaneous pair electronic
  excitations 377
Single chain enzymes 188
Sodium in biological systems 155
Solomon–Bloembergen equation .. 396
1:1 Species 126
1:2 Species 121
Specificity
  biological 188
  cleavage 150
Spectra
  absorption
    cobalt 194
    of iron–sulfur proteins 326
  electronic absorption
    of iron(III) phosvitin 372
    of (HEDTA Fe)$_2$O^{2-} 376
  infrared, of aqueous solutions .. 254
  optical absorption 14

Spectral
　comparison of metal complexes
　　and metalloenzymes 197
　studies on the ferritin core,
　　electronic absorption 371
Sperm whale myoglobin 113
Squalene oxides, modified 96
Stability
　and enzymatic activity of modi-
　　fied ferredoxins 336
　constants
　　for group IA cations 157
　　for group IIA cations 156
　　thermodynamic 52
Stabilization
　of DNA 144
　of quaternary structure of
　　phosphatase 191
Stable nitrogen compounds 81
Stellacyanin 387
Stereospecificity 350
Steric
　hindrance 158
　limitations 228
　restrictions 167
Steroids 39
Sterol biosynthesis 96
Stoichiometry of OP binding 203
Stopped-flow technique 127
Strengths 157
Stress enzyme 297
Structural
　alterations, metal-induced 212
　models for iron and copper
　　proteins 365
Structure
　active site of horseradish
　　peroxidase 421
　and function of metalloenzymes 187
　clostridial rubredoxin 331
　–function relationships in cyto-
　　chrome c oxidase 248
　heme A 263
　nucleic acids 135
　second coordination sphere ... 176
　vitamin B_{12} coenzyme 347
Substituted alkylcobaloximes and
　　alkylcobalamins 9
β-substituted ethylcobaloximes ... 9
Substrate
　bound by hydrophobic forces .. 28
　groups of ceruloplasmin 298
　reduced by nitrogenase 223
　reduction 239
Sulfide
　inorganic 322
　thiocyanate formation from ... 341
Susceptibility, magnetic 380
Synthetic
　macrocycle chemistry 44
　macrocycles, iron complexes of .. 69
　model for the ferritin core 369

T

TAAB 47, 57
　complexes, oxidation–reduction
　　properties of 61
Terminal double bond, activity of 264
Ternary complex, enhancement of 394
Tetraanhydrotetramer 47
Tetrabutylammonium chloride 105
Tetradentate macrocycles 46
Tetradentate macrocyclic complexes 60
Thallium
　(I) complexes 162
　in biological systems 160
Thermodynamic stability 52
Thiocyanate formation
　from apoferredoxin and sulfide .. 341
　from ferredoxin 341
Three-dimensional model of a
　coenzyme 354
Thymidylic acid 137
Titanium 97
Titanocene 102
Titanocene fixation–reduction
　mechanism 107
5′TMP 137
Toluene 28
Transferrin 215
　role of 304
Transition metals 111
　relative activities of 4
Transverse relaxation time 392
Tritium
　introduction into the coenzyme .. 356
　labeling experiments 15
　transfer from coenzyme, rate of 354
Troponin 170
Trypsin 210
Tubercidin 139
Two-hydrogen merry-go-round ... 357

U

Unbranched substrates, reduction
　　of 230
Undissociated complex, dipolar shift
　for the 185
Unsaturation in the carbon-17 chain 264
Unwinding and rewinding of DNA 146
Uptake of oxygen 111
　by bis(histidinato)cobalt(II) .. 125
Uracil 137
Utilities of nitrogen fixation re-
　search, potential 243

V

Viologen dyes 222
Vitamin B_{12} 5, 14, 115, 127, 346
　as a nucleophile 358
　coenzyme
　　models 73
　　reactions requiring 348
　　relative specific activities of .. 356
　　structure of 347
Voltammetry, cyclic 64

W

Wilson's disease, iron metabolism in   313

X

X-ray diffraction   391
Xylose isomerase   404

Y

Yeast aldolase   402

Z

Zinc   147
   atoms in LADH, differential
      reactivity of   205
   content of alkaline phosphatase   189
   -induced anemia in rats   315
   isotope exchange in LADH   201
   metalloenzyme   402

QD
1
A355
#100

NOV 23 1971